Oscar Castillo, Patricia Melin, Janusz Kacprzyk, and Witold Pedrycz (Eds.)

Soft Computing for Hybrid Intelligent Systems

Studies in Computational Intelligence, Volume 154

Editor-in-Chief

Prof. Janusz Kacprzyk
Systems Research Institute
Polish Academy of Sciences
ul. Newelska 6
01-447 Warsaw
Poland
E-mail: kacprzyk@ibspan.waw.pl

Oscar Castillo
Patricia Melin
Janusz Kacprzyk
Witold Pedrycz
(Eds.)

Soft Computing for Hybrid Intelligent Systems

 Springer

Prof. Oscar Castillo
Tijuana Institute of Technology
Department of Computer Science
P.O. Box 4207
Chula Vista CA 91909
USA
Email: ocastillo@hafsamx.org

Prof. Patricia Melin
Tijuana Institute of Technology
Department of Computer Science
P.O. Box 4207
Chula Vista CA 91909
USA
Email: epmelin@hafsamx.org

Prof. Janusz Kacprzyk
Polish Academy of Sciences,
Systems Research Institute,
Newelska 6 01-447 Warszawa
Poland
Email: kacprzyk@ibspan.waw.pl

Prof. Witold Pedrycz
University of Alberta,
Dept. Electrical and Computer Engineering
Edmonton, Alberta T6J 2V4
Canada
Email: pedrycz@ece.ualberta.ca

ISBN 978-3-540-70811-7   e-ISBN 978-3-540-70812-4

DOI 10.1007/978-3-540-70812-4

Studies in Computational Intelligence   ISSN 1860949X

Library of Congress Control Number: 2008931160

Typeset & Cover Design: Scientific Publishing Services Pvt. Ltd., Chennai, India.

Printed in acid-free paper

9 8 7 6 5 4 3 2 1

springer.com

Preface

We describe in this book, new methods and applications of hybrid intelligent systems using soft computing techniques. Soft Computing (SC) consists of several intelligent computing paradigms, including fuzzy logic, neural networks, and evolutionary algorithms, which can be used to produce powerful hybrid intelligent systems. The book is organized in five main parts, which contain a group of papers around a similar subject. The first part consists of papers with the main theme of intelligent control, which are basically papers that use hybrid systems to solve particular problems of control. The second part contains papers with the main theme of pattern recognition, which are basically papers using soft computing techniques for achieving pattern recognition in different applications. The third part contains papers with the themes of intelligent agents and social systems, which are papers that apply the ideas of agents and social behavior to solve real-world problems. The fourth part contains papers that deal with the hardware implementation of intelligent systems for solving particular problems. The fifth part contains papers that deal with modeling, simulation and optimization for real-world applications.

In the part of Intelligent Control there are 5 papers that describe different contributions on achieving control of dynamical systems using soft computing techniques. The first paper, by Ricardo Martinez et al., deals with the Optimization of Interval Type-2 Fuzzy Logic Controllers for a Perturbed Autonomous Wheeled Mobile Robot Using Genetic Algorithms. The second paper, by Nohé Cázarez et al., deals with the Fuzzy Control for Output Regulation of a Servomechanism with Backlash. The third paper, by Jose Morales et al., studies the Stability on Type-1 and Type-2 Fuzzy Logic Systems. The fourth paper, by Alma Martinez et al., describes a Comparative Study of Type-1 and Type-2 Fuzzy Systems Optimized by Hierarchical Genetic Algorithms. The fifth paper, by Cristina Martinez et al., describes a Comparison between Ant Colony and Genetic Algorithms for fuzzy system optimization.

In the part of Pattern Recognition there are 5 papers that describe different contributions on achieving pattern recognition using hybrid intelligent systems. The first paper, by Denisse Hidalgo et al., describes Type-1 and Type-2 Fuzzy Inference Systems as Integration Methods in Modular Neural Networks for Multimodal Biometry and its Optimization with Genetic Algorithms. The second paper, by Olivia Mendoza et al., deals with Interval Type-2 Fuzzy Logic for Module Relevance Estimation in Sugeno Integration of Modular Neural Networks. The third paper, by Miguel Lopez et al., deals with the

Optimization of Response Integration with Fuzzy Logic in Ensemble Neural Networks using Genetic Algorithms. The fourth paper, by José M. Villegas et al., describes the Optimization of Modular Neural Network, Using Genetic Algorithms: The Case of Face and Voice Recognition. The fifth paper, by Pedro Salazar et al., describes a new biometric recognition technique based on hand geometry and voice using Neural Networks and Fuzzy Logic.

In the part of Intelligent Agents and Social Systems there are 5 papers that describe different contributions to solving real-world problems with the computing paradigms of agents and social behavior. The first paper by Cecilia Leal and Oscar Castillo, describes a hybrid model based on a cellular automata and fuzzy logic to simulate the population dynamics. The second paper, by Jose A. Ruz-Hernandez et al., deals with Soft Margin Training for Associative Memories: Application to Fault Diagnosis in Fossil Electric Power Plants. The third paper, by Manuel Castañón-Puga et al., describes Social Systems Simulation Person Modeling as Systemic Constructivist Approach. The fourth paper by Arnulfo Alanis et al., describes Modeling and Simulation by Petri Networks of a Fault Tolerant Agent Node. The fifth paper, by Eugenio Dante-Suarez et al., describes the new concept of Fuzzy Agents and its possible applications.

In the part of Hardware Implementations several contributions are described on the implementation of intelligent systems in hardware devices. The first paper, by Yazmin Maldonado et al., describes the Design and Simulation of the Fuzzification Stage Through the Xilinx System Generator. The second paper by Rogelio Serrano et al., describes High Performance Parallel Programming of a GA using Multi-Core Technology. The third paper by Martha Cardenas et al., describes the Scalability Potential of Multi-Core Architecture in a Neuro-Fuzzy System. The fourth paper by Jose A. Olivas et al., describes a Methodology to Test and Validate a VHDL Inference Engine Through the Xilinx System Generator. The fifth paper, by Gabriel Lizarraga et al., describes Modeling and Simulation of the Defuzzification Stage using Xilinx System Generator and Simulink.

In the part of Modeling, Simulation and Optimization several contributions are described on the application of soft computing techniques for achieving modeling, simulation and optimization of non-linear systems. The first paper, by Fevrier Valdez et al., describes a New Evolutionary Method Combining Particle Swarm Optimization and Genetic Algorithms using Fuzzy Logic. The second paper, by Juan R. Castro et al., describes a new Hybrid Learning Algorithm for Interval Type-2 Fuzzy Neural Networks and its application to the Case of Time Series Prediction. The third paper, by Salvador González-Mendivil et al., deals with the Optimization of Artificial Neural Network Architectures for Time Series Prediction using Parallel Genetic Algorithms. The fourth paper, by Qiang Wei et al., describes an Optimized Algorithm of Discovering Functional Dependencies with Degrees of Satisfaction based on the Attribute Pre-Scanning Operation. The fifth paper, by Tina Yu and Dave Wilkinson, deals with a Fuzzy Symbolic Representation for Intelligent Reservoir Well Logs Interpretation. The sixth paper, by Rostislav Horcik, describes how to Solve a System of Linear Equations with Fuzzy Numbers. The seventh paper by Ismael Millan et al., describes the design and implementation of a hybrid fuzzy controller using VHDL.

In conclusion, the edited book comprises papers on diverse aspects of soft computing and hybrid intelligent systems. There are theoretical aspects as well as application papers.

Tijuana Institute of Technology, Mexico   Oscar Castillo
Tijuana Institute of Technology, Mexico   Patricia Melin
Polish Academy of Sciences, Poland   Janusz Kacprzyk
University of Alberta, Canada   Witold Pedrycz

April 25, 2008

Contents

Part III: Intelligent Agents and Social Systems

Part IV: Hardware Implementations

Part V: Modeling, Simulation and Optimization

Part I

Intelligent Control

Optimization of Interval Type-2 Fuzzy Logic Controllers for a Perturbed Autonomous Wheeled Mobile Robot Using Genetic Algorithms

Ricardo Martínez[1], Oscar Castillo[1], and Luis T. Aguilar[2]

[1] Division of Graduate Studies and Research, Tijuana Institute of Technology, Tijuana México
   molerick@hotmail.com, ocastillo@hafsamx.org
[2] Instituto Politécnico Nacional, Centro de Investigación y Desarrollo de Tecnología Digital,
   2498 Roll Dr., #757 Otay Mesa, San Diego CA 92154,
   Fax: +52(664)6231388
   luis.aguilar@ieee.org

Abstract. We describe a tracking controller for the dynamic model of a unicycle mobile robot by integrating a kinematics and a torque controller based on Interval Type-2 Fuzzy Logic Theory and Genetic Algorithms. Computer simulations are presented confirming the performance of the tracking controller and its application to different navigation problems.

1 Introduction

Mobile robots have attracted considerable interest in the robotics and control research community, because they have nonholonomic properties caused by nonintegrable differential constrains. The motion of nonholonomic mechanical systems [3] is constrained by its own kinematics, so the control laws are not derivable in a straightforward manner (Brockett condition [4]).

Furthermore, most reported designs rely on intelligent control approaches such as Fuzzy Logic Control (FLC) [24][29][2][11][15][23][28] and Neural Networks[26][8]. However the majority of the publications mentioned above, have concentrated on kinematics models of mobile robots, which are controlled by the velocity input, while less attention has been paid to the control problems of nonholonomic dynamic systems, where forces and torques are the true inputs: Bloch and Drakunov [3] and Chwa [6], used a sliding mode control to the tracking control problem.

This paper is organized as follows: Section 2 presents the problem statement and the kinematics and dynamic model of the unicycle mobile robot. Section 3 introduces the posture and velocity control design where a genetic algorithm is used to select the parameters of the posture controller. Robustness properties of the closed-loop system are achieved with a type-2 fuzzy logic velocity control system using a Takagi-Sugeno model where the wheel input torques, linear velocity, and angular velocity will be considered as linguistic variables. Section 4 provides a simulation study of the unicycle mobile robot using the controller described in Section 3. Finally, Section 5 presents the conclusions.

O. Castillo et al. (Eds.): Soft Computing for Hybrid Intel. Systems, SCI 154, pp. 3–18, 2008.
springerlink.com   © Springer-Verlag Berlin Heidelberg 2008

2 Problem Statement

2.1 The Mobile Robot

The model considered in this paper is of a unicycle mobile robot (Figure 1), and it consists of two driving wheels mounted of the same axis and a front free wheel.

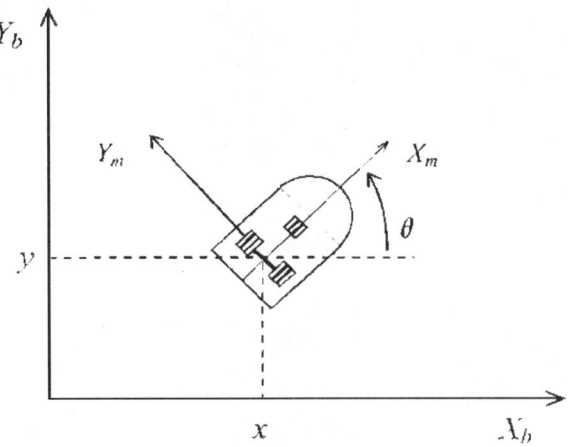

Fig. 1. Wheeled Mobile Robot

A unicycle mobile robot is an autonomous, wheeled vehicle capable of performing missions in fixed or uncertain environments. The robot body is symmetrical around the perpendicular axis and the center of mass is at the geometric center of the body. It has two driving wheels that are fixed to the axis that passes through C and one passive wheel prevents the robot from tipping over as it moves on a plane. In what follows, it is assumed that the motion of the passive wheel can be ignored in the dynamics of the mobile robot represented by the following set of equations [14]:

$$M(q)\dot{v} + C(q,\dot{q})v + Dv = \tau + P(t) \tag{1}$$

$$q = \underbrace{\begin{bmatrix} \cos\theta & 0 \\ \sin\theta & 0 \\ 0 & 1 \end{bmatrix}}_{J(q)} \begin{bmatrix} v \\ w \end{bmatrix}_{v} \tag{2}$$

where $q = (x, y, \theta)^T$ is the vector of the configuration coordinates; $v = (v, w)^T$ is the vector of velocities; $\tau = (\tau_1, \tau_2)$ is the vector of torques applied to the wheels of the robot where τ_1 and τ_2 denote the torques of the right and left wheel, respectively (Figure 1); $P \in R^2$ is the uniformly bounded disturbance vector; $M(q) \in R^{2 \times 2}$ is the positive-definite inertia matrix; $C(q,\dot{q})\vartheta$ is the vector of centripetal and coriolis forces; and $D \in R^{2x2}$ is a diagonal positive-definite damping matrix. Equation (2) represents the kinematics of the system, where (x, y) is the position in the X – Y (world) reference frame; θ is the angle between the heading direction and the x-axis; v and w are the linear and angular velocities, respectively. Furthermore, the system (1)-(2) has the following nonholonomic constraint:

$$\dot{y}\cos\theta - \dot{x}\sin\theta = 0 \qquad (3)$$

which corresponds to a no-slip wheel condition preventing the robot from moving sideways [16]. The system (2) fails to meet Brockett's necessary condition for feedback stabilization [4], which implies that no continuous static state-feedback controller exists that stabilizes the close-loop system around the equilibrium point.

The control objective is to design a fuzzy logic controller τ that ensures

$$\lim_{t\to\infty}\left\|q_d(t) - q(t)\right\| = 0, \qquad (4)$$

for any continuously, differentiable, bounded desired trajectory $q_d \in R^3$ while attenuating external disturbances.

2.2 Fuzzy Logic Control Design

This section illustrates the framework to achieve stabilization of a unicycle mobile robot around a desired path. The stabilizing control law for the system (1)-(2) can be designed using the backstepping approach [13] since the kinematics subsystem (2) is controlled indirectly through the velocity vector v. The procedure to design the overall controller consists of two steps:

1. Design a virtual velocity vector $\vartheta_r = \vartheta$ such that the kinematics model (2) be uniformly asymptotically stable.
2. Design a velocity controller τ by using FLC that ensures

$$\left\|\vartheta_r(t) - \vartheta(t)\right\| = 0, \quad \forall t \geq t_s \qquad (5)$$

where t_s is the reachability time.

In (5), it is considered that real mobile robots have actuated wheels, so the control input is τ that must be designed to stabilize the dynamics (1), without destabilizing the system (2), by forcing $\vartheta \in R^2$ to reach the virtual velocity vector $\vartheta_r \in R^2$ in finite-time. Roughly speaking, if (5) is satisfied asymptotically $(i,e;t_s = \infty)$ then ϑ along $t < \infty$, consequently the mobile robot will be neither positioned nor oriented at desired point. Figure 2 illustrates the feedback connection which involves the fuzzy controller.

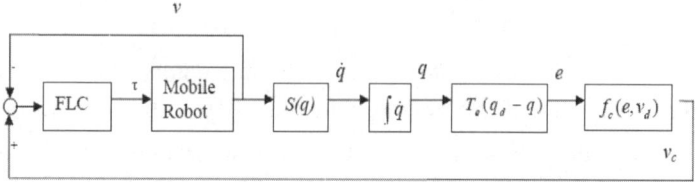

Fig. 2. Tracking control structure

Posture Control Design

First, we focus on the kinematics model by designing a virtual control (ϑ_r) such that the control objective (4) is achieved. To this end, let us consider the reference trajectory $q_d(t)$ as a solution of the following differential equation:

$$
\dot{q}_d = \begin{pmatrix} \cos\theta_d & 0 \\ \sin\theta_d & 0 \\ 0 & 1 \end{pmatrix} \begin{pmatrix} v_d \\ w_d \end{pmatrix}
\tag{6}
$$

where $\theta_d(t)$ is the desired orientation, and $v_d(t)$ and $w_d(t)$ denote the desired linear and angular velocities, respectively. In the robot's local frame, the error coordinates can be defined as

$$
\begin{pmatrix} \tilde{q}_1 \\ \tilde{q}_2 \\ \tilde{q}_3 \end{pmatrix} = \underbrace{\begin{pmatrix} \cos\theta & \sin\theta & 0 \\ -\sin\theta & \cos\theta & 0 \\ 0 & 0 & 1 \end{pmatrix}}_{T_e(\theta)} \begin{pmatrix} x_d - x \\ y_d - y \\ \theta_d - \theta \end{pmatrix}
\tag{7}
$$

where $(x_d(t), y_d(t))$ is the desired position in the world $X - Y$ coordinate system, $\tilde{q}_1$ and $\tilde{q}_2$ are the coordinates of the position error vector, and $\tilde{q}_3$ is the orientation error. The associated tracking error model is

$$
\begin{pmatrix} \dot{\tilde{q}}_1 \\ \dot{\tilde{q}}_2 \\ \dot{\tilde{q}}_3 \end{pmatrix} = \begin{pmatrix} w\tilde{q}_2 - v + v_d \cos\tilde{q}_3 \\ -w\tilde{q}_1 + v_d \sin\tilde{q}_3 \\ w_d - w \end{pmatrix}
\tag{8}
$$

which is in terms of the corresponding real and desired velocities, is then obtained by differentiating (7) with respect to time.

In order to present the main result of this subsection, we need first to recall the following theorems [4].

Theorem 1 [12](Uniform stability): Let $x = 0$ be an equilibrium point for $\dot{x} = f(x,t)$ and $D \subset R^n$ be a domain containing $x = 0$. Let $V : [0, \infty] \times D \to R$ be a continuously differentiable function such that

$$W_1(x) \le V(x,t) \le W2(x) \tag{9}$$

$$\frac{\partial V}{\partial t} + \frac{\partial V}{\partial x} f(x,t) \le 0 \tag{10}$$

for all $t \ge 0$ and for all $x \in D$, where $W_1(x)$ and $W_2(x)$ are continuous positive definite functions on D. Then, $x = 0$ is uniformly stable.

Theorem 2 [12](Uniform asymptotic stability): Suppose the assumptions of Theorem 1 are satisfied with inequality (10) strengthened to

$$\frac{\partial V}{\partial t} + \frac{\partial V}{\partial x} f(x,t) \le -W_3(x) \tag{11}$$

for all $t \ge 0$ and for all $x \in D$, where $W_3(x)$ is a continuous positive definite function on D. Then, $x = 0$ is uniformly asymptotically stable.

Theorem 3: Let the tracking error equations (8) be driven by the control law (virtual velocities)

$$\begin{aligned}
v_r &= v_d \cos \tilde{q}_3 + \gamma_1 \tilde{q}_1 \\
w_r &= w_d + \gamma_2 v_d \tilde{q}_2 + \gamma_3 \sin \tilde{q}_3
\end{aligned} \tag{12}$$

where γ_1, γ_2 and γ_3 are positive constants. If $v = v_r$ and $w = w_r$ for all $t \ge 0$ in (2), then the origin of the closed-loop system (8)-(12) is uniformly asymptotically stable.

Proof: Under the control (12), the closed-loop system takes the form:

$$\begin{pmatrix} \dot{\tilde{q}}_1 \\ \dot{\tilde{q}}_2 \\ \dot{\tilde{q}}_3 \end{pmatrix} = \begin{pmatrix} w_d \tilde{q}_2 + \gamma_2 v_d \tilde{q}_2^2 + \gamma_3 \tilde{q}_3 - \gamma_1 \tilde{q}_1 \\ -w_d \tilde{q}_1 - \gamma_2 v_d \tilde{q}_1 \tilde{q}_2 - \gamma_3 \tilde{q}_1 \sin \tilde{q}_3 + v_d \sin \tilde{q}_3 \\ -\gamma_2 v_d \tilde{q}_2 - \gamma_3 \sin \tilde{q}_3 \end{pmatrix} \tag{13}$$

Note that the origin $(\tilde{q}_1, \tilde{q}_2, \tilde{q}_3)^T = 0$ is an equilibrium point of the closed-loop system but not unique because $\tilde{q}_3$ can adopt several postures $(i.e., \tilde{q}_3 = 0, \pi, ..., n\pi)$. Genetic algorithms are applied for tuning the kinematics control gains $\gamma_i, i = 1,2,3$ to ensure that the error $\tilde{q} \in R^3$ converges to the origin. The asymptotic stability theorem is invoked as a guideline to obtain bounds in the values of γ_i, which shall guarantee convergence of the error $\tilde{q} \in R^3$ to zero. For this purpose, let us introduce the Lyapunov function candidate

$$V(\tilde{q}) = \frac{1}{2}\tilde{q}_1^2 + \frac{1}{2}\tilde{q}_2^2 + \frac{1}{\gamma_2}(1 - \cos\tilde{q}_3)$$ (14)

which is positive definite. Taking the time derivative of $V(\tilde{q})$ along the solution of the closed-loop system (13), we get

$$\dot{V}(\tilde{q}) = \tilde{q}_1\dot{\tilde{q}}_1 + \tilde{q}_2\dot{\tilde{q}}_2 + \frac{1}{\gamma_2}\dot{\tilde{q}}_3\sin\tilde{q}_3 = \gamma_1\tilde{q}_1^2 - \frac{\gamma_3}{\gamma_2}(\sin\tilde{q}_3)^2 \le 0$$ (15)

Thus concluding that for any positive constant γ_i, the closed-loop system is uniformly stable. To complete the proof it remains to note that $\tilde{q}_1, \tilde{q}_2, \tilde{q}_3 \in L_\infty^n$; $\dot{\tilde{q}}_1, \dot{\tilde{q}}_3 \in L_\infty^n$ and $\ddot{\tilde{q}}_1, \ddot{\tilde{q}}_2 \in L_\infty^n$ where

$$L_2^n = \left\{x(t): R_+ \mapsto R^n \left\|x(t)\right\|_\infty^2 = \int_0^\infty \left\|x(t)\right\|_2^2 dt < \infty\right\}$$

$$L_2^n = \left\{x(t): R_+ \mapsto R^n \left\|x(t)\right\|_\infty^2 = \sup\left\|x(t)\right\|_2^2 < \infty\right\}$$

hence we conclude, by applying Barbalat's lemma that $\tilde{q}_1$ and $\tilde{q}_3$ converge to the origin. Finally, by invoking the Matrosov's Theorem [21], convergence of $\tilde{q}_2$ to the origin can be concluded.

The genetic algorithm was codified with a chromosome of 24 bits in total, eight bits for each of the gains. Figure 3 shows the binary chromosome representation of the individuals in the population. Different experiments were performed, changing the parameters of the genetic algorithm and the best results were obtained by comparing the corresponding simulations. Changing the crossover rate and the number of crossover points used did not affect the results. Also, changing the mutation rate did not affect the optimal results. The advantage of using the genetic algorithm to find the gains is that time-consuming manual search of these parameters was avoided.

In the execution of the genetic algorithm for find the gains, we used the Lyapunov function for the stability, so that one is fulfilled, the result of this function must be a negative value for the optimal gains of the system.

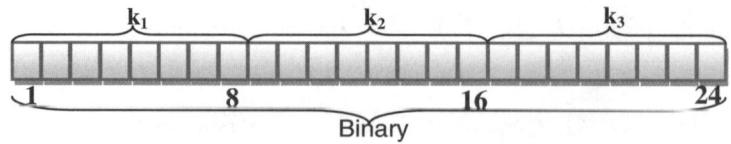

Fig. 3. Chromosome representation

where b_i, $i = 1, \ldots 24$ are binary values (0 or 1) representing the constants gains parameters.

Velocity Control Design

In this subsection a fuzzy logic controller is designed to force the real velocities of the mobile robot (1) and (2) to match those required in equations (12) of Theorem 3 to satisfy the control objective (4).

We design a Takagi-Sugeno fuzzy logic controller for the autonomous mobile robot, using linguistic variables in the input and mathematical functions in the output. The linear (v_d) and the angular (w_d) velocity errors were taken as input variables and the right (τ_1) and left (τ_2) torques as the outputs. The membership functions used in the input are trapezoidal for the Negative (N) and Positive (P), and triangular for the Zero (C) linguistics terms. The interval used for this fuzzy controller is [-50 50]. Figure 4 shows the input variables.

(a)   (b)

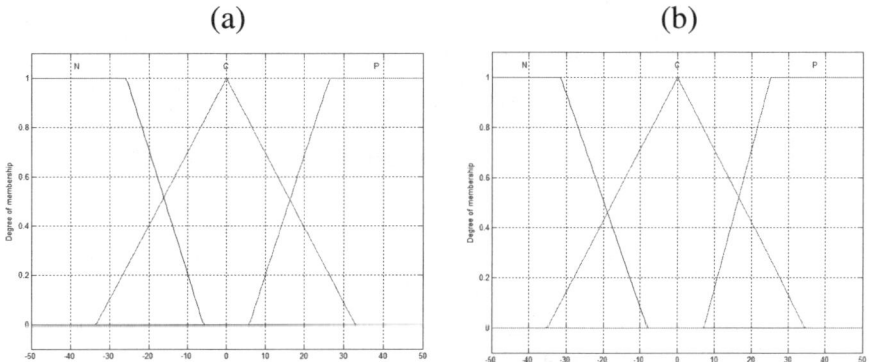

Fig. 4. (a) Linear velocity error (e_v). (b) Angular velocity error (e_w).

The rule set of the FLC contain 9 rules, which govern the input-output relationship of the FLC and this adopts the Takagi-Sugeno style inference engine [22], and we use a single point in the outputs (constant values), obtained using weighted average defuzzification procedure. In Table 1, we present the rule set whose format is established as follows:

Rule i: If e_v is G_1 and e_w is G_2 then F is G_3 and N is G_4

where $G_1..G_4$ are the fuzzy set associated to each variable and i= 1 ... 9.

To find the best fuzzy controller, we used the genetic algorithm to find the parameters of the membership functions. In figure 5 we show the chromosome with 28 bits (positions).

Table 1. Fuzzy rule set

e_v/e_w	N	C	P
N	N/N	N/C	N/P
C	C/N	C/C	C/P
P	P/N	P/C	P/P

Inputs
- Linear velocity error
  Negative, Zero, Positive
- Angular velocity error
  Negative, Zero, Positive

Outputs
- Torque 1
  Constant
  Negative, Zero, Positive
- Torque 2
  Constant
  Negative, Zero, Positive

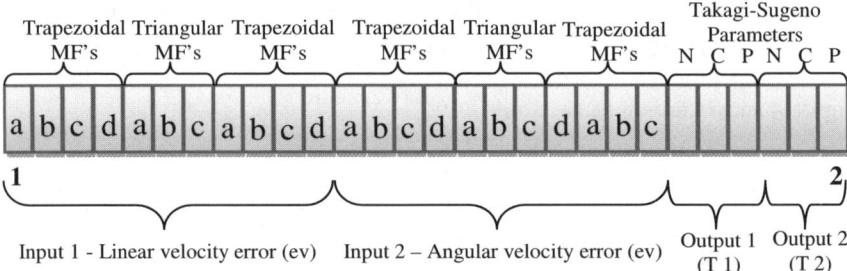

Fig. 5. Chromosome representation for the fuzzy logic controller

Table 2. Parameters of the membership functions

MF Type	Point	Minimum Value	Maximum Value
Trapezoidal	a	-50	-50
	b	-50	-50
	c	-15	-5.05
	d	-1.5	-0.5
Triangular	a	-5	-1.75
	b	0	0
	c	1.75	5
Trapezoidal	a	0.5	1.5
	b	5.05	15
	c	50	50
	d	50	50

Table 2 shows the parameters of the membership functions, the minimal and the maximum values in the search range for the genetic algorithm to find the best fuzzy controller system.

3 Simulation Results

In this section, we evaluate, through computer simulation performed in MATLAB® and SIMULINK®, the ability of the proposed controller to stabilize the unicycle mobile robot, defined by (1) and (2) where the matrix values

$$M(q) = \begin{bmatrix} 0.3749 & -0.0202 \\ -0.0202 & 0.3739 \end{bmatrix}, \quad C(q,\dot{q}) = \begin{bmatrix} 0 & 0.1350\,\theta \\ -0.1350\,\theta & 0 \end{bmatrix}, \quad D = \begin{bmatrix} 10 & 0 \\ 0 & 10 \end{bmatrix}$$

were taken from [3].

The desired trajectory is the following one:

$$\vartheta_d(t) = \begin{cases} v_d(t) = 0.2(1 - \exp(-t)) \\ w_d(t) = 0.4\sin(0.5t) \end{cases} \tag{16}$$

and was chosen in terms of its corresponding desired linear v_d and angular velocities w_d, subject to the initial conditions

$$q(0) = (0.1, 0.1, 0)^T \text{ and } \vartheta(0) = 0 \in R^2$$

The gains $\gamma_i, i = 1,2,3$ of the kinematics model (12) were tuned by using genetic algorithm approach resulting in $\gamma_1 = 5$, $\gamma_2 = 24$ and $\gamma_3 = 3$, the best gains that were found.

3.1 Genetic Algorithm Results for the Gains k1, k2 and k3

To find these gains we changed the number of generations, the mutation and the cross-over operators of the genetic algorithm represented in Table 3. Figure 6 shows the plot

Table 3. Results of the simulation to find the constants k1, k2 and k3

No.	Indiv.	Gen.	Cross.	Mut.	Average error	k1	k2	k3
1	100	70	0.8	0.3	14.1544	39	483	66
2	50	40	0.8	0.4	109.6417	451	416	80
3	20	15	0.7	0.4	41.4291	174	320	59
4	40	30	0.9	0.4	85.8849	354	311	51
5	60	80	0.9	0.4	116.6302	477	287	47
6	60	70	0.8	0.3	70.0033	293	382	59
7	70	100	0.7	0.2	31.4233	138	485	72
8	50	60	0.8	0.2	16.8501	74	402	59
9	40	20	0.5	0.2	101.0294	420	428	68
10	40	20	0.8	0.2	70.8432	299	481	78
11	55	25	0.8	0.3	15.0319	31	509	72
12	50	70	0.5	0.2	122.8667	507	406	65
13	30	50	0.8	0.3	103.7498	428	320	49
14	3	5	0.8	0.1	20.0183	27	507	98
15	2	10	0.8	0.1	31.4037	132	346	138
16	5	15	0.8	0.1	114.8469	477	219	27
17	30	50	0.8	0.1	4.1658	6	9	6
18	30	50	0.8	0.1	4.1844	6	9	5
19	30	50	0.8	0.1	4.0008	4	8	4
20	30	50	0.8	0.1	4.0293	4	5	4
21	50	40	0.8	0.4	3.8138	5	24	3
22	50	40	0.8	0.4	4.5141	10	16	6
23	50	40	0.8	0.4	6.9690	25	28	5
24	50	40	0.8	0.4	4.1202	6	15	5
25	70	80	0.8	0.3	3.9736	4	10	4
26	70	80	0.8	0.3	4.3726	7	10	4
27	30	20	0.8	0.1	4.0466	5	12	4
28	40	20	0.8	0.1	4.0621	5	11	4
29	70	80	0.8	0.3	4.7051	12	15	5
30	40	70	0.8	0.3	4.4199	8	11	5

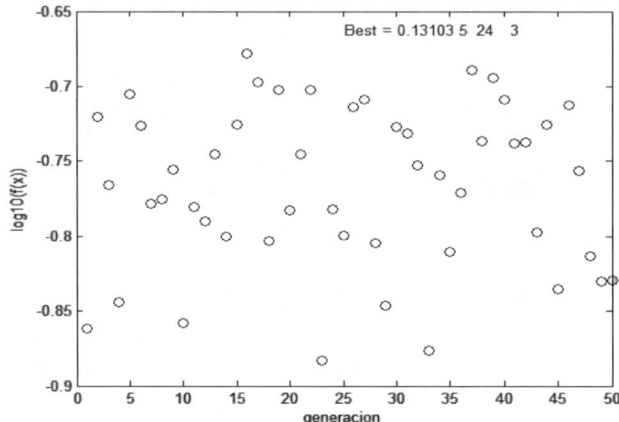

Fig. 6. Evolution of GA finding the optimal gains

Table 4. Best simulations results

Average error	k1	k2	k3
3.8138	5	24	3

of the evolution of the population on the genetic algorithm finding the best gains. Table 4 shows best error that was found and the values of the gains k1, k2 and k3.

Figure 7 shows the plot of the best simulation results.

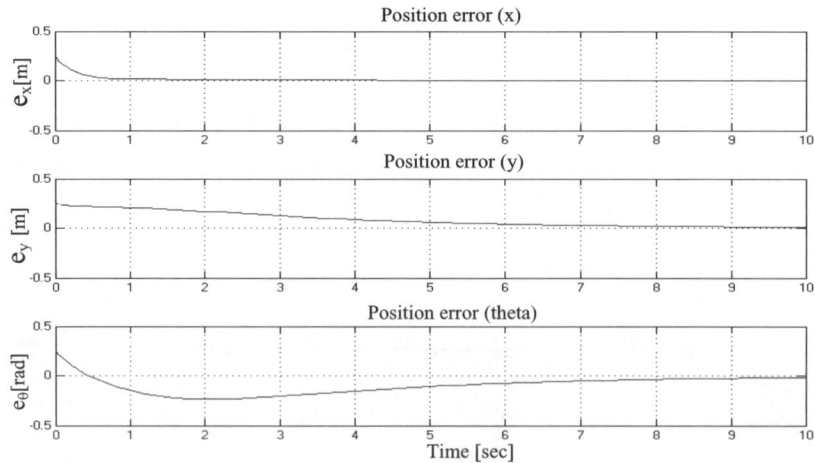

Fig. 7. Tracking errors

3.2 Genetic Algorithms Results for the Optimization of the Fuzzy Logic Controller (FLC)

Table 5 contains the results of the FLC, obtained by varying the values of generation number, percentage of replacement, mutation and crossover and Figure 8 shows the evolution of the GA.

Table 5. Genetic Algorithm results for FLC optimization

No.	Indiv.	Gen.	%Repl.	Cross.	Mut.	Selection Method	Average error	G.A. time
1	50	30	0.7	0.7	0.2	Roulette	0.4122618	7:18
2	100	25	0.7	0.8	0.3	Roulette	0.4212924	12:11
3	20	15	0.7	0.8	0.2	Roulette	0.5524043	1:26
4	10	20	0.7	0.8	0.2	Roulette	0.4899811	1:21
5	80	25	0.7	0.7	0.4	Roulette	0.4126189	9:57
6	150	50	0.7	0.6	0.3	Roulette	0.4094381	43:15
7	90	60	0.7	0.9	0.4	Roulette	0.4087614	44:43
8	10	25	0.7	0.8	0.2	Roulette	0.5703853	2:09
9	65	40	0.7	0.8	0.2	Roulette	0.4099531	22:52
10	30	25	0.7	0.9	0.5	Roulette	0.4086178	6:21
11	70	50	0.7	0.8	0.3	Roulette	0.4086729	29:17
12	80	50	0.7	0.9	0.3	Roulette	0.4099137	33:32
13	200	100	0.7	0.4	0.1	Roulette	0.4085207	2:43:28
14	15	10	0.7	0.8	0.5	Roulette	0.5669795	1:14
15	15	25	0.7	0.9	0.2	Roulette	0.4789307	3:03
16	30	40	0.7	0.7	0.2	Roulette	0.4108032	10:28
17	50	60	0.7	0.6	0.4	Roulette	0.4111103	1:05:14
18	20	50	0.7	0.6	0.2	Roulette	0.4339689	9:13
19	80	20	0.7	0.8	0.6	Roulette	0.4490967	13:13
20	100	80	0.7	0.8	0.3	Roulette	0.4083982	6:16
21	30	60	0.7	0.8	0.5	Roulette	0.4943807	14:43
22	25	40	0.7	0.8	0.6	Roulette	0.4247892	8:10
23	70	60	0.7	0.8	0.4	Roulette	0.4084446	34:44
24	35	40	0.7	0.7	0.3	Roulette	0.4099876	11:30
25	45	50	0.7	0.7	0.3	Roulette	0.4128472	18:19
26	26	30	0.9	0.6	0.3	Roulette	0.4082359	6:40
27	60	40	0.9	0.8	0.5	Roulette	0.4106830	20:38
28	80	50	0.9	0.4	0.1	Roulette	0.4095522	33:37
29	40	30	0.9	0.9	0.4	Roulette	0.4102437	10:07
30	100	35	0.9	0.7	0.4	Roulette	0.4094340	29:01
31	80	45	0.9	0.7	0.2	Roulette	0.4100034	29:50
32	5	20	0.9	0.9	0.2	Roulette	0.4377697	0:49
33	10	15	0.9	0.8	0.2	Roulette	0.4370570	1:16
34	15	15	0.9	0.7	0.4	Roulette	0.4769542	2:01
35	15	40	0.9	0.8	0.2	Roulette	0.4107797	5:12
36	5	30	0.9	0.9	0.3	Roulette	0.5089666	1:19
37	10	15	0.9	0.8	0.1	Roulette	0.5206810	1:55

Continuing with the results, the best simulation for control of tracking is shown in Figure 9, which shows the parameters optimized by the genetic algorithm for the input variables (e_v, e_w).

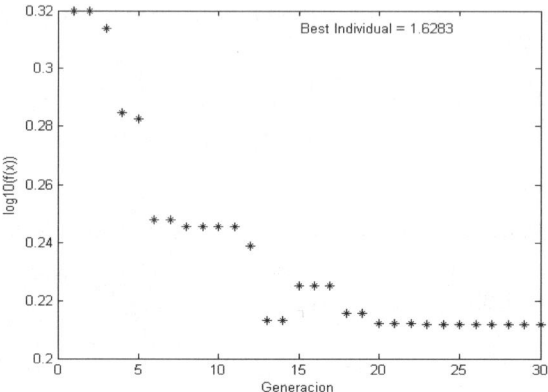

Fig. 8. Evolution of the GA for FLC optimization

(a)   (b)

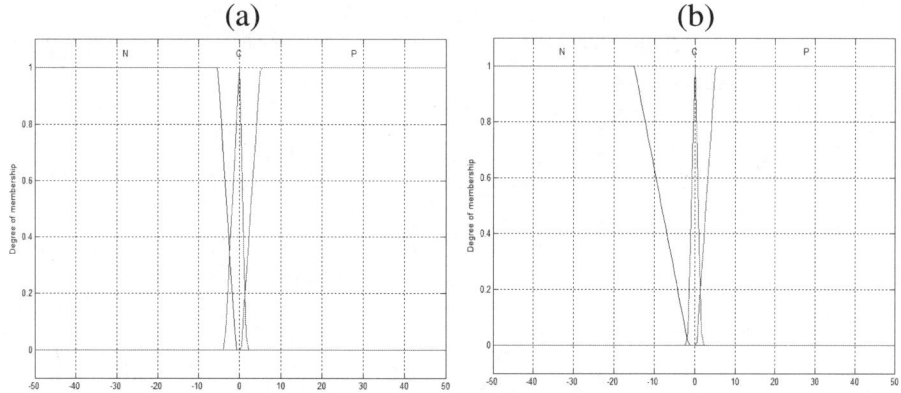

Fig. 9. (a) Linear velocity error, (b) Angular velocity error

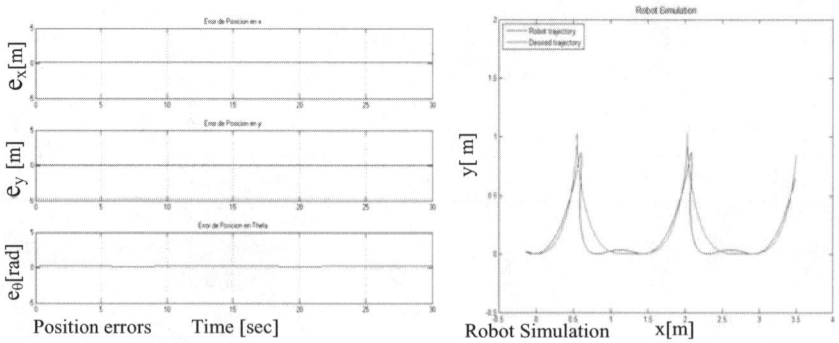

Fig. 10. Stabilization of the autonomous mobile robot with Type-1 FLC

Figure 10 shows the results of the linear and angular velocity errors, the input torques, the position errors and the obtained trajectory under desired trajectory of the robot autonomous mobile and we can observe the stability on the position error and how the robot mobile follows the trajectory.

3.3 Genetic Algorithms Results for the Optimization of the Type-2 Fuzzy Logic Controller (FLC)

Figure 11 shows the evolution of the GA and Table 6 contains the results of the Type-2 FLC, obtained by varying the values of generation number, percentage of replacement, mutation and crossover rate.

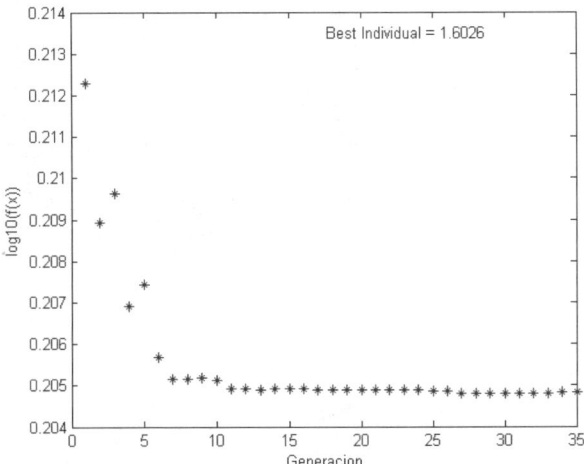

Fig. 11. Evolution of the GA for Type-2 FLC optimization

Table 6. Genetic Algorithm results for Type-2 FLC optimization

No.	Indiv.	Gener.	%Repl.	Cross.	Mut.	Selection Method	Average error	G.A. time
1	50	20	0.7	0.8	0.4	Roulette	0.3993130	4:52:08
2	20	15	0.7	0.8	0.5	Roulette	0.4008340	1:13:03
3	23	20	0.7	0.8	0.4	Roulette	0.3994720	02:56:23
4	40	25	0.7	0.8	0.5	Roulette	0.3993860	6:37:16
5	30	19	0.7	0.9	0.5	Roulette	0.3994950	3:02:35
6	35	10	0.7	0.8	0.5	Roulette	0.4111980	1:15:03
7	45	25	0.7	0.9	0.5	Roulette	0.4008810	7:22:52
8	38	18	0.7	0.7	0.3	Roulette	0.3991930	3:40:29
9	60	20	0.7	0.8	0.6	Roulette	0.3989860	6:40:59
10	45	20	0.7	0.8	0.6	Roulette	0.4007900	5:56:20
11	45	15	0.7	0.7	0.5	Roulette	0.4068480	3:22:18
12	58	25	0.9	0.6	0.4	Roulette	0.3995240	7:49:24
13	40	18	0.9	0.9	0.6	Roulette	0.3990670	3:29:21
14	58	45	0.9	0.8	0.6	Roulette	0.3989470	15:20:34
15	26	18	0.9	0.9	0.5	Roulette	0.4021550	3:48:35

Table 6. (*continued*)

16	10	15	0.9	0.8	0.5	Roulette	0.4028900	1:43:28
17	15	15	0.9	0.7	0.4	Roulette	0.4006630	1:45:03
18	25	22	0.9	0.8	0.5	Roulette	0.3995900	2:11:58
19	60	25	0.9	0.9	0.4	Roulette	0.4002830	9:38:31
20	30	15	0.9	0.8	0.4	Roulette	0.4110670	2:23:33
21	20	18	0.9	0.6	0.4	Roulette	0.3989810	1:28:53
22	46	28	0.9	0.6	0.4	Roulette	0.3991000	6:19:01
23	70	25	0.9	0.7	0.5	Roulette	0.3989980	10:36:57
24	54	20	0.9	0.8	0.6	Roulette	0.3992210	6:19:40
25	66	30	0.9	1	0.6	Roulette	0.3989810	12:54:21
26	**42**	**35**	**0.9**	**0.8**	**0.6**	**Roulette**	**0.3989410**	**7:11:52**
27	26	10	0.9	0.6	0.4	Roulette	0.4027990	2:29:51
28	40	20	0.9	0.6	0.4	Roulette	0.3990530	4:24:41
29	50	15	0.9	0.9	0.5	Roulette	0.4005340	3:33:39
30	80	12	0.9	0.9	0.6	Roulette	0.3997710	6:32:30
31	11	15	0.9	0.5	0.3	Roulette	0.4026380	1:55:02
32	28	18	0.9	0.8	0.3	Roulette	0.3997890	2:46:05
33	22	18	0.9	0.7	0.6	Roulette	0.4008280	2:11:57
34	15	12	0.9	0.8	0.9	Roulette	0.4109010	0:36:24
35	30	14	0.9	0.6	0.6	Roulette	0.4006320	2:32:30
36	60	18	0.9	0.7	0.5	Roulette	0.3990440	5:06:51
37	28	17	0.9	0.5	0.5	Roulette	0.3992410	3:43:54

(a)   (b)

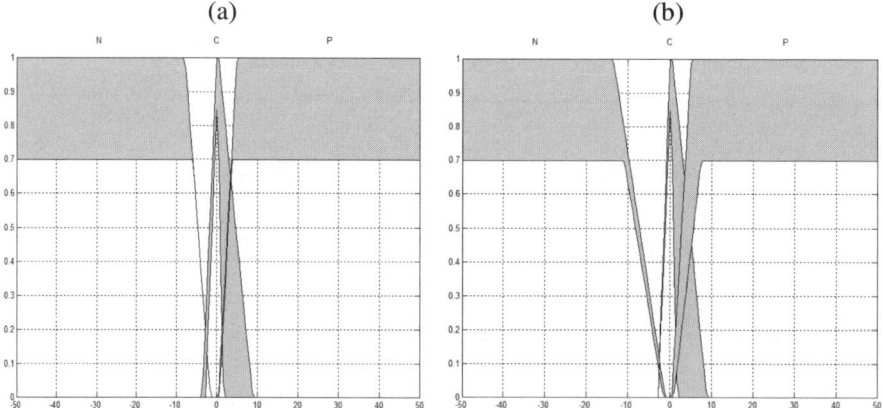

Fig. 12. (a) Linear velocity error, (b) Angular velocity error

Figure 12 shows the membership functions obtained by the genetic algorithm, and they are the inputs variables of the FLC Type-2 and Figure 13 shows the results of the position errors and the obtained trajectory under desired trajectory of the robot autonomous mobile and we can observe the stability on the position error and how the robot mobile follows the trajectory.

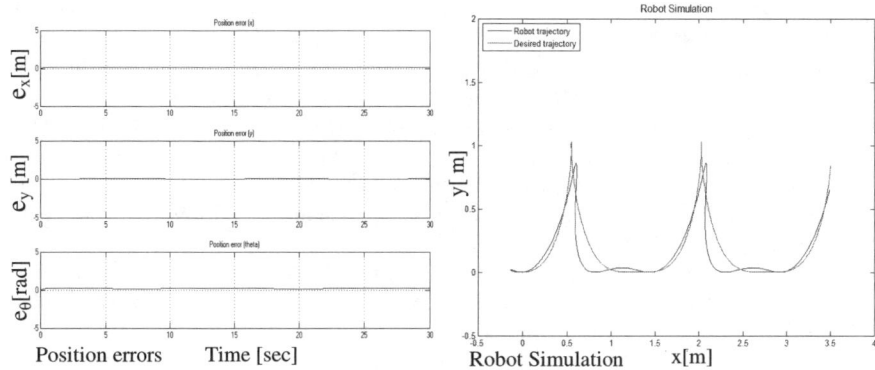

Fig. 13. Stabilization of the autonomous mobile robot with Type-2 FLC

4 Conclusions

We have designed a trajectory tracking controller taking into account the kinematics and the dynamics of the autonomous mobile robot using type-2 fuzzy logic and genetic algorithms.

Genetic algorithms are used for the optimization of the constants for the trajectory tracking and also for the optimization of the parameters of membership functions for fuzzy logic control.

Currently, the design of a type-2 fuzzy logic controller has been tested under a perturbed autonomous wheeled mobile robot, but more tests are in progress.

References

[1] Astudillo, L., Castillo, O., Aguilar, L.: Intelligent Control of an Autonomous Mobile Robot using Type-2 Fuzzy Logic. In: Proceedings of ICAI 2006, pp. 565–570 (2006)

[2] Bentalba, S., El Hajjaji, A., Rachid, A.: Fuzzy Control of a Mobile Robot: A New Approach. In: Proc. IEEE Int. Conf. On Control Applications, Hartford, CT, pp. 69–72 (October 1997)

[3] Bloch, A.M.: Nonholonomic mechanics and control. Springer, New York (2003)

[4] Brockett, R.W.: Asymptotic stability and feedback stabilization. In: Millman, R.S., Sussman, H.J. (eds.) Differential Geometric Control Theory, Birkhauser, Boston, pp. 181–191 (1983)

[5] Castillo, O., Melin, P.: Soft Computing for Control of Non-Linear Dynamical Systems. Springer, Heidelberg (2001)

[6] Chwa: Sliding-Mode Tracking Control of Nonholonomic Wheeled Mobile Robots in Polar coordinates. IEEE Trans. On Control Syst. Tech. 12(4), 633–644 (July 2004)

[7] Duc Do, K., Zhong-Ping, J., Pan, J.: A global output-feedback controller for simultaneous tracking and stabilizations of unicycle-type mobile robots. IEEE Trans. Automat. Contr. 30, 589–594 (2004)

[8] Fierro, R., Lewis, F.L.: Control of a Nonholonomic Mobile Robot Using Neural Networks. IEEE Trans. On Neural Networks 9(4), 589–600 (1998)

[9] Fukao, T., Nakagawa, H., Adachi, N.: Adaptive Tracking Control of a NonHolonomic Mobile Robot. IEEE Trans. On Robotics and Automation 16(5), 609–615 (2000)

[10] Hagras, H.: A Hierarchical type-2 Fuzzy Logic Control Architecture for Autonomous Mobile Robots. IEEE Transactions On Fuzzy Systems 12(4), 524–539 (2004)

[11] Ishikawa, S.: A Method of Indoor Mobile Robot Navigation by Fuzzy Control. In: Proc. Int. Conf. Intell. Robot. Syst., Osaka, Japan, pp. 1013–1018 (1991)

[12] Khalil, H.: Nonlinear systems, 3rd edn. Prentice Hall, New York (2002)

[13] Kristic, M., Kanellakopoulos, I., Kokotovic, P.: Nonlinear and adaptive control design. Wiley-Interscience, Chichester (1995)

[14] Lee, T.-C., Song, K.-T., Lee, C.-H., Teng, C.-C.: Tracking control of unicycle-modeled mobile robot using a saturation feedback controller. IEEE Trans. Contr. Syst. Technol. 9, 305–318 (2001)

[15] Lee, T.H., Leung, F.H.F., Tam, P.K.S.: Position Control for Wheeled Mobile Robot Using a Fuzzy Controller, pp. 525–528. IEEE, Los Alamitos (1999)

[16] Liberzon, D.: Switching in Systems and Control, Bikhauser (2003)

[17] Man, K.F., Tang, K.S., Kwong, S.: Genetic Algorithms, Concepts and Designs, pp. 5–10. Springer, Heidelberg (2000)

[18] Mendel, J.: Uncertain Rule-Based Fuzzy Logic Systems. Prentice Hall, Englewood Cliffs (2001)

[19] Mendel, J., Jhon, R.: Type-2 Fuzzy Sets Made Simple. IEEE Transactions on Fuzzy Systems 10, 117–127 (2002)

[20] Nelson, W., Cox, I.: Local Path Control for an Autonomous Vehicle. In: Proc. IEEE Conf. On Robotics and Automation, pp. 1504–1510 (1988)

[21] Paden, Panja, R.: Globally asymptotically stable PD+ controller for robot manipulator. International Journal of Control 47(6), 1697–1712 (1988)

[22] Passino, K.M., Yurkovich, S.: Fuzzy Control. Addison Wesley Longman, USA (1998)

[23] Pawlowski, S., Dutkiewicz, P., Kozlowski, K., Wroblewski, W.: Fuzzy Logic Implementation in Mobile Robot Control. In: 2nd Workshop On Robot Motion and Control, pp. 65–70 (October 2001)

[24] Sepulveda, R., Castillo, O., Melin, P., Montiel, O.: An Efficient Computational Method to Implement Type-2 Fuzzy Logic in Control Applications. Analysis and Design of Intelligent Systems Using Soft Computing Techniques, Advances in Soft Computing 41, 45–52 (2007)

[25] Sepulveda, R., Montiel, O., Castillo, O., Melin, P.: Fundamentos de Lógica Difusa. Ediciones ILCSA (August 2002)

[26] Song, K.T., Sheen, L.H.: Heuristic fuzzy-neural Network and its application to reactive navigation of a mobile robot. Fuzzy Sets Systems 110(3), 331–340 (2000)

[27] Takagi, T., Sugeno, M.: Fuzzy Identification of Systems and its application to modeling and control. IEEE Transactions on Systems, Man, and Cybernetics 15(1) (1985)

[28] Tsai, C.-C., Lin, H.-H., Lin, C.-C.: Trajectory Tracking Control of a Laser-Guided Wheeled Mobile Robot. In: Proc. IEEE Int. Conf. On Control Applications, Taipei, Taiwan, pp. 1055–1059 (September 2004)

[29] Ulyanov, S.V., Watanabe, S., Ulyanov, V.S., Yamafuji, K., Litvintseva, L.V., Rizzotto, G.G.: Soft Computing for the Intelligent Robust Control of a Robotic Unicycle with a New Physical Measure for Mechanical Controllability. In: Soft Computing, vol. 2, pp. 73–88. Springer, Heidelberg (1998)

[30] Zadeh, L.A.: Outline of a new approach to the analysis of complex systems and decision processes. IEEE Transactions on Systems, Man, and Cybernetics 3(1), 28–44 (1973)

Fuzzy Control for Output Regulation of a Servomechanism with Backlash

Nohé R. Cazarez-Castro, Luis T. Aguilar, Oscar Castillo, and Selene Cárdenas

Universidad Autonoma de Baja California, Facultad de Ciencias Quimicas e Ingenieria,
Tijuana, B.C. Mexico
nohe@ieee.org
Centro de Investigacion y Desarrollo de Tecnologia Digital, Tijuana, B.C., Mexico
luis.aguilar@ieee.org
Instituto Tecnologico de Tijuana, División de Estudios de Posgrado e Investigacion,
Tijuana, B.C., Mexico
ocastillo@tectijuana.mx

Abstract. A Fuzzy Logic Control System is designed to achieve the output regulation for a servomechanism with backlash. The problem is to design a fuzzy controller to obtain the closed-loop system in which the load of the driver is regulated to a desired position. The provided servomotor position as the only measurement available for feedback, the proposed method is far from trivial because of non-minimum phase properties of the system. Simulation results illustrate the effectiveness of the closed-loop system.

1 Introduction

A major problem in control engineering is a robust feedback design that asymptotically stabilizes a nominal plant while also attenuating the influence of parameters variation and external disturbances. In the last decade, this problem was heavily studied and considerable research efforts have resulted in the development of systematic design methodology for nonlinear feedback systems. A survey of the methods, fundamental in this respect, is given in [1].

This paper is in spirit of [2], where Aguilar et. al. extends a nonlinear H_∞ regulator for the output regulator of a nonminimum phase servomechanism with backlash, showing that using H_∞ as controller is sufficient to solve the problem in question. Our approach is to design and implement a fuzzy controller so as to obtain the closed-loop system in which all trajectories are bounded and the load of the driver is regulated to a desired position while also attenuating the influence of external disturbances.

Fuzzy controllers are used as compensators of other control strategies as PID [3-5] to solve the problem in question. Our approach is to use a fuzzy controller as the whole control strategy.

The paper is organized as follows. The backlash phenomenon and state equations of the drive system are introduced in Section 2. The output regulation problem is stated in Section 3. A fuzzy controller for output regulation is presented in Section 4.

O. Castillo et al. (Eds.): Soft Computing for Hybrid Intel. Systems, SCI 154, pp. 19–28, 2008.
springerlink.com © Springer-Verlag Berlin Heidelberg 2008

Performance issues of this regulator are illustrated in a simulation study in Section 5. Finally, Section 6 presents the conclusions.

2 Dynamic Model

The dynamic models of the angular position $q_i(t)$ of the DC motor and the $q_0(t)$ of the load is as follows

$$J_0 N^{-1} \ddot{q}_0 + f_0 N^{-1} \dot{q}_0 = T + w_0$$
$$J_i \ddot{q}_i + f_i \dot{q}_i + T = \tau_m + w_i \tag{1}$$

hereafter, J_0, f_0, $\ddot{q}_0$, and $\dot{q}_0$ are, respectively, the inertia of the load and the reducer, the viscous output friction, the output acceleration, and the output velocity. The inertia of the motor, the viscous motor friction, the motor acceleration, and the motor velocity are denoted by J_i, f_i, $\ddot{q}_i$, and $\dot{q}_i$, respectively. The input torque τ_m serves as a control action, and T stands for the transmitted torque. The external bances $w_i(t)$, $w_o(t)$ have been introduced into the driver equation (1) to account for destabilizing model discrepancies due to hard-to-model nonlinear phenomena, such as friction and backlash.

The transmitted torque T through a backlash with an amplitude j is typically modeled by a dead-zone characteristic ([6], p. 18)

$$T(\Delta q) = \begin{cases} 0 & |\Delta q| \leq j \\ k\Delta q - Kj\sin(\Delta q) & otherwise \end{cases} \tag{2}$$

where

$$\Delta q = q_i - N q_0 \tag{3}$$

K is the stiffness, and N is the reducer ratio. Such a model is depicted in Fig. 1. Provided the servomotor position $q_i(t)$ is the only available measurement on the system, the above model (1)-(3) appears to be nonminimum phase because along with the origin the unforced system possesses a multivalued set of equilibria (q_i, q_0) with $q_i = 0$ and $q_0 \in [-j, j]$.

To avoid dealing with a nonminimum phase system, we replace the backlash model (2) with its monotonic approximation (see Fig. 2):

$$T = K\Delta q - K\delta(\Delta q) \tag{4}$$

where

$$\delta = -2j \frac{1 - e^{-\left(\frac{\Delta q}{j}\right)}}{1 + e^{-\left(\frac{\Delta q}{j}\right)}} \tag{5}$$

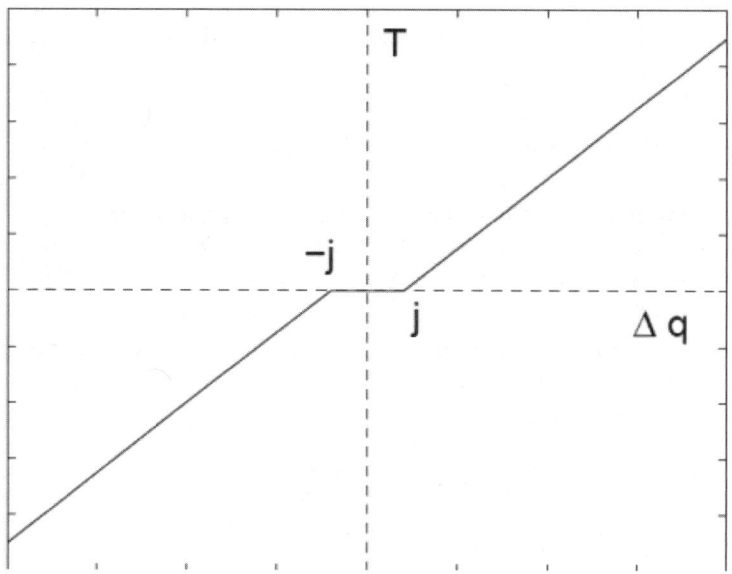

Fig. 1. The dead-zone model of backlash

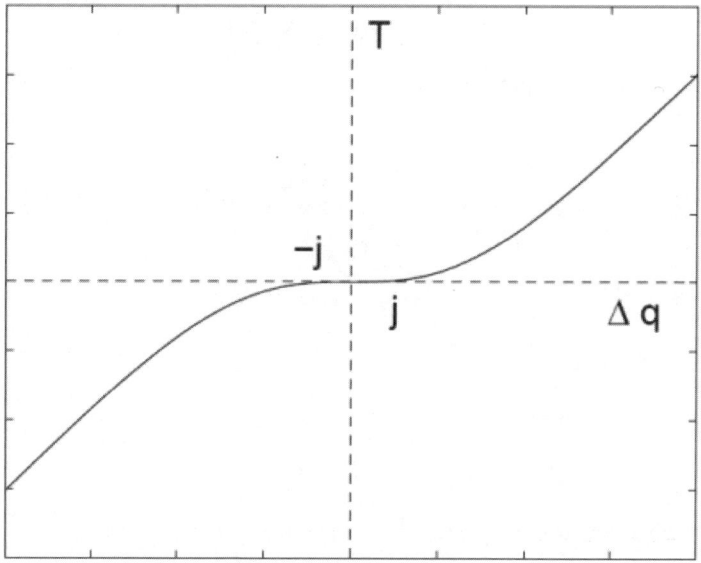

Fig. 2. The monotonic approximation of the dead-zone model

The present backlash approximation is inspired from [7]. Coupled to the drive system (1) subject to motor position measurements, it is subsequently shown to continue a minimum phase approximation of the underlying servomotor, operating under uncertainties $w_i(t)$, $w_0(t)$ to be attenuated. As a matter of fact, these uncertainties involve discrepancies between the physical backlash model (2) and its approximation (4) and (5).

3 Problem Statement

The objective of the Fuzzy Control output regulation of the nonlinear drive system (1) with backlash (4) and (5), is thus to design a Fuzzy Controller so as to obtain the closed-loop system in which all these trajectories are bounded and the output $q_0(t)$ asymptotically decays to a desired position q_d as $t \to \infty$ while also attenuating the influence of the external disturbances $w_i(t)$, $w_o(t)$. To formally state the problem, let us introduce the state deviation vector $x = [x_1, x_2, x_3, x_4,]^T$ with

$$x_1 = q_0 - q_d$$
$$x_2 = \dot{q}_0$$
$$x_3 = q_i - Nq_d$$
$$x_4 = \dot{q}_i$$

where x_1 is the load position error, x_2 is the load velocity, x_3 is the motor position deviation from its nominal value, and x_4 is the motor velocity. The nominal motor position Nq_d has been prespecified in such a way to guarantee that $\Delta q = \Delta x$, where

$$\Delta x = x_3 - Nx_1.$$

Then, system (1)-(5), represented in terms of the deviation vector x, takes the form

$$\dot{x}_1 = x_2$$
$$\dot{x}_2 = J_0^{-1}[KNx_3 - KN^2x_1 - f_0x_2 + KN\delta(\Delta q) + w_o]$$
$$\dot{x}_3 = x_4 \tag{6}$$
$$\dot{x}_4 = J_i^{-1}[\tau_m + KNx_1 - Kx_3 - f_ix_4 + K\delta(\Delta q) + w_i].$$

The zero dynamics

$$\dot{x}_1 = x_2$$
$$\dot{x}_2 = J_0^{-1}[-KN^2x_1 - f_0x_2 + KN\delta(-Nx_1)] \tag{7}$$

of the undisturbed version of system (6) with respect to the output

$$y = x_3 \tag{8}$$

is formally obtained (see [8] for details) by specifying the control law that maintains the output identically zero.

4 Fuzzy Controller

To solve the Fuzzy Control Output Regulation problem, two-inputs one-input rules will be used in the formulation of the knowledge base. The IF-THEN rules are of the following form:

$$\text{IF } x_1 \text{ is } A_1^l \text{ AND } x_2 \text{ is } A_2^l \text{ THEN } y \text{ is } B^l \tag{9}$$

where $[x_1\, x_2]^T = x \in U = U_1 \times U_2 \subset R^2$ and $y \in V \subset \mathbb{R}$. For each input fuzzy set A_j^l in $x_j \subset U_j$ and output fuzzy set B^l in $y \subset V$ exists an input membership function $\mu_{A_j^l}(x_j)$ and output membership function $\mu_{B^l}(y)$, respectively, with l being the number of membership functions associated to the input j. The number of rules M is defined by the number of membership functions of each input $M = N_1 N_2$.

The particular choice of each $\mu_{B^l}(y)$ will depend on the heuristic knowledge of the experts.

In our case, we select triangular membership functions for each input and output variables, membership functions can be seen in Figs. 3, 4 and 5 respectively, where you can see that we select to granulate each variable in three fuzzy sets n (negative), z (zero) and p (positive).

These input and output variables are combined in fuzzy rules in the form of (9), in our case, we select the seven fuzzy rules shown in Table 1.

For the inference process, we use the Mamdani [9-10] type of Fuzzy Inference, with minimum as disjunction operator, maximum as conjunction operator, minimum

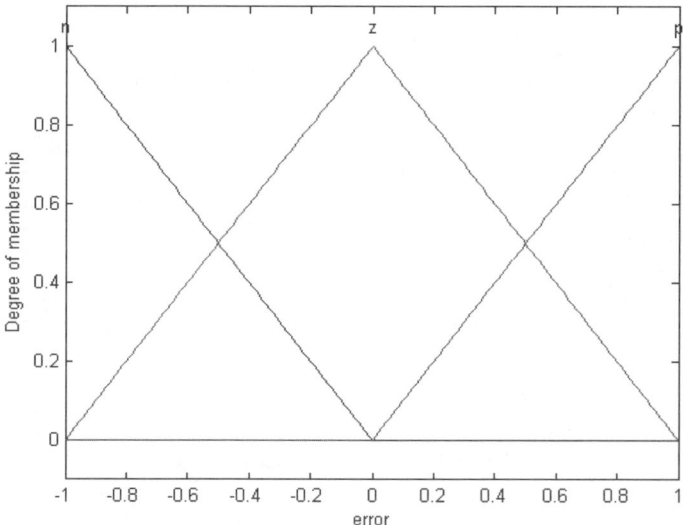

Fig. 3. Fuzzy input variable error

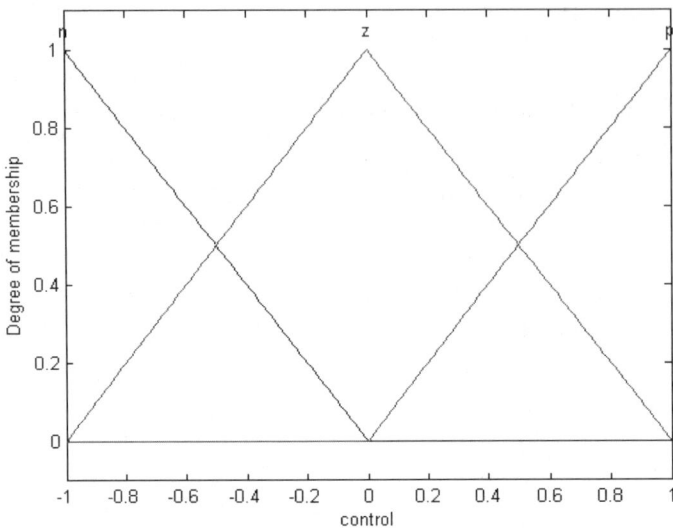

Fig. 4. Fuzzy output variable control

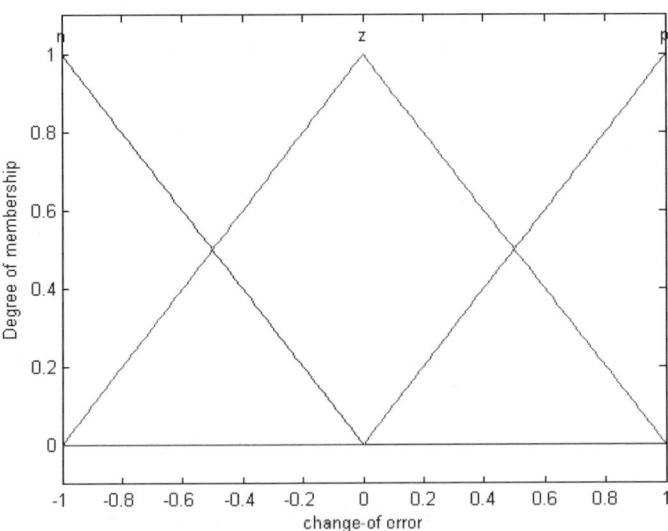

Fig. 5. Fuzzy input chance of error

as implication operator, maximum as aggregation operator and mean of maximums as our defuzzification method.

With the combination of the rules, variables, membership functions and fuzzy system parameters, we obtain the surface of fuzzy control shown in Fig. 6.

Table 1. Fuzzy Rules

No.	Error	change of error	control
1	N	n	p
2	N	p	z
3	N	z	p
4	P	p	n
5	P	n	z
6	P	z	n
7	Z	z	z

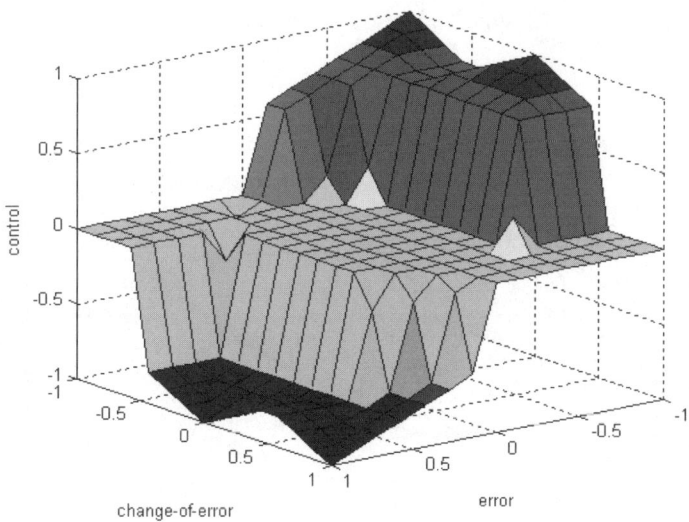

Fig. 6. Control surface

5   Simulation Study

5.1   Simulation Test Bench

For our simulations we use the dynamical model (1) of the experimental test bench installed in the Robotics & Control Laboratory of CITEDI-IPN, which involves a DC motor linked to a mechanical load through an imperfect contact gear train. Fig. 7 shows the physical test bench.

The input-output motion graph of Fig. 8 reveals the gear backlash effect.

In our simulations, the gear reduction ratio is of N=3, and it is the main source of friction, and the backlash level is of j=0.2 rad. The stiffness coefficient is of K=5 Nm/rad. Table 2 represents the parameters of the motor, taken from the manufacturer data specifications, and the nominal load parameters, and the load parameters, are taken from [2].

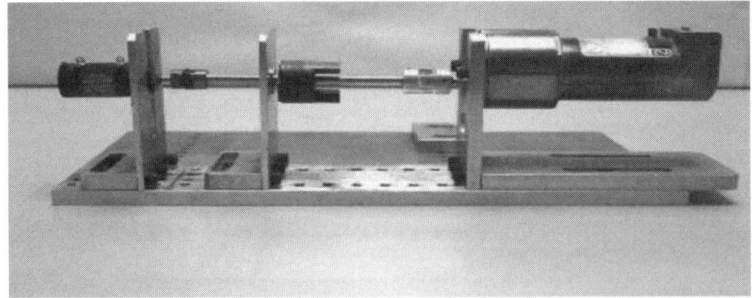

Fig. 7. Experimental test bench

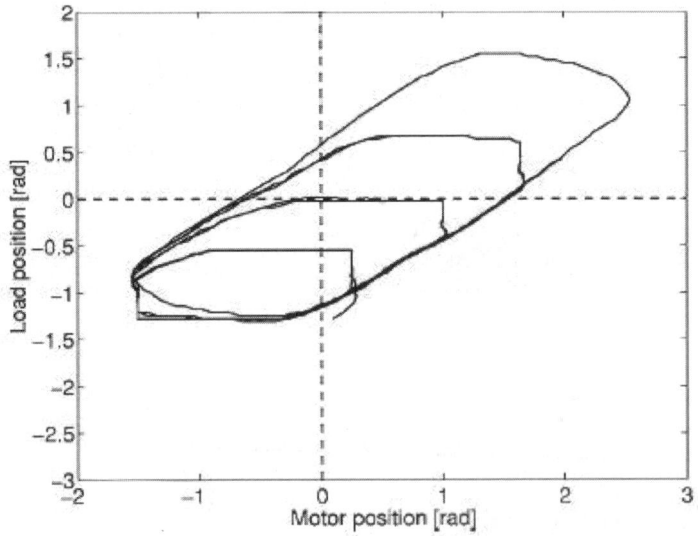

Fig. 8. Backlash hysteresis before compensation

Table 2. Nominal parameters

Description	Notation	Value
Motor inertia	J_i	2.8×10^{-6} Kgm²
Load inertia	J_0	1.07 Kgm²
Motor viscous friction	f_i	7.6×10^{-7} Nms/rad
Load viscous friction	f_0	1.73 Nms/rad

5.2 Simulation Results

The experiments were carried out for the closed-loop system, and we consider the angular motor position as the only information available for feedback. In the simulations, the load was required to move from the initial static position $q_0(0) = 0$ to the desired position $q_d = \pi/2$ rad. In order to illustrate the size of the attraction domain, the initial load position was chosen reasonably far from the desired position.

The resulting trajectories are depicted in Fig. 9. This figure demonstrates that the regulator stabilizes the disturbance-free load motion around the desired position and attenuates external load disturbances.

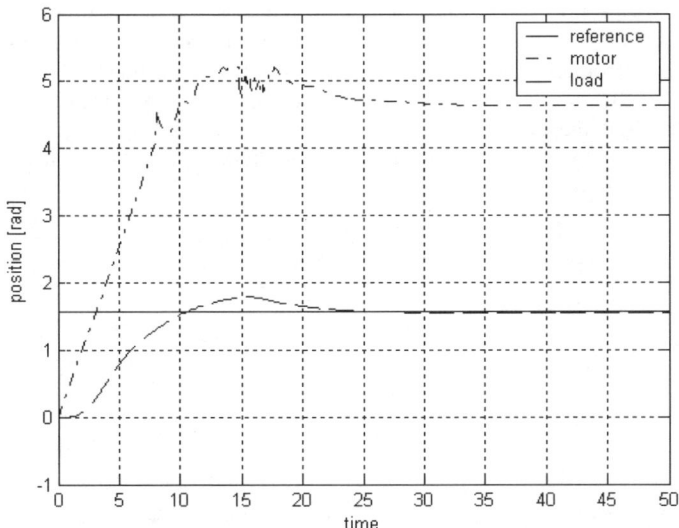

Fig. 9. Simulation results for the fuzzy regulator considering only motor position information available feedback

6 Conclusions

A partial state stabilization around a desired load position is developed for a servomotor with backlash. A simulation where the motor position is the only available measurement for feedback is under study. The fuzzy control output regulator proposed is shown to be eminently suited to locally solve the stabilization problem in question while also attenuating the backlash model discrepancies.

References

[1] Isidori, A.: A Tool for Semiglobal Stabilization of Uncertain Non-Minimum-Phase Nonlinear System via Output Feedback. IEEE Transactions on Automatic Control 4(10), 1817–1827 (2000)
[2] Aguilar, L.T., Orlov, Y., Cadiou, J.C., Merzouki, R.: Nonlinear -Output Regulation of a Nonminimum Phase Servomechanism With Backlash. Journal of Dynamic Systems, Measurement, and Control 129, 544–549 (2007)
[3] Woo, K.T., Wang, L.-X., Lewis, F.L., Li, Z.X.: A Fuzzy System Compensator for Backlash. In: Procc. Of the 1998 IEEE Int. Conf. on Robotics & Automation, Leuven, Belgium, pp. 181–186 (May 1998)
[4] Popovič, M.R., Gorinevsky, D.M., Goldenberg, A.A.: High-Precision Position of a Mechanism with Nonlinar Friction Using a Fuzzy Logic Pulse Controller. IEEE Trans. On Control System Technology 8(1), 151–158 (2000)
[5] Yoo, B.K., Ham, W.C.: Adaptive Control of Robot Manipulator Using Fuzzy Compensator. IEEE Trans. On Fuzzy Systems 8(2), 186–199 (2000)
[6] Nordim, M., Bodin, P., Gutman, P.O.: New Models and Identification Methods for Backlash and Gear Play. In: Gao, T., Lewis, F. (eds.) Adaptive Control of Nonsmooth Dynamic Systems, pp. 1–30. Springer, Berlin (2001)
[7] Merzouki, R., Cadiou, J.C., M'Sirdi, N.K.: Compensation of Friction and Backlash Effects in an Electrical Actuator. J. Systems and Control Engineering 218, 75–84 (2004)
[8] Isidori, A.: Nonlinear Control Systems, 3rd edn. Springer, Berlin (1995)
[9] Mamdani, E.H., Assilian, S.: An experiment in linguistic synthesis with fuzzy logic controller. Int. J. Man-Machine Studies 7, 1–13 (1975)
[10] Mamdani, E.H.: Advances in the Linguistic synthesis of fuzzy controllers. Int. J. Man-Machine Studies 8, 669–679 (1976)

Stability on Type-1 and Type-2 Fuzzy Logic Systems

Jose Morales, Oscar Castillo, and Jose Soria

Tijuana Institute of Technology, Tijuana, Mexico

Abstract. The fuzzy systems present some characteristics that the classical control systems (PI, PD and PID) don't have, like smoother control, noise immunity, important mathematical complexity reduction, little mathematical knowledge of the model work, and they can obtain results from imprecise data. Broadly stated, fuzzy logic control attempts to come to terms with the informal nature of the control design process. In its most basic form, the so-called Mamdani architecture is directly translating external performance specifications and observations of plant behavior into a rule-based linguistic control strategy. This architecture forms the backbone of the great majority of fuzzy logic control systems reported in the literature in the past years. This paper is based on the fuzzy Lyapunov synthesis, to determine the systems stability, which is based on the Lyapunov criterion; this concept was introduced by Margaliot to adjust the Lyapunov criteria by considering linguistic variables instead of numeric variables to determine the systems stability. The stability will be proving on Mamdani's architecture fuzzy logic systems type-1 and type-2 respectively.

Keywords: Fuzzy Systems, Control Systems, Imprecise Data, Fuzzy Lyapunov Synthesis, Systems Stability.

1 Introduction

Fuzzy logic has found applications in a incredibly wide range of areas in the relatively short period of time since its conception. Invented by Lofti Zadeh, a leading control expert, it is perhaps not surprising that system's theory is one of the areas in which fuzzy logic has made a profound impact.

This is because fuzzy logic, combined with the paradigm of *computing with words*, allows the use and manipulation of human knowledge and reasoning in the modeling and control of dynamical systems.

The ongoing research and applications in this field demonstrate the power and versatility of fuzzy logic. The fuzzy models considered include classical If-Then rule systems, cognitive maps, relational fuzzy equations, fuzzy-neuro models and much more, and have found applications in every field of engineering. Often, however, the design of these systems, be it fuzzy models or fuzzy controllers, is ad-hoc and based on heuristics, which also makes them difficult, if not impossible, to analyze mathematically [1].

Fuzzy logic has been widely studied ever since the very first introduction of this fundamental concept. Its main attraction undoubtedly lies in the unique characteristics that fuzzy logic systems possess. They are capable of handling complex, nonlinear solutions. Very often, fuzzy systems may provide a better performance than conventional non-fuzzy approaches with less development cost.

O. Castillo et al. (Eds.): Soft Computing for Hybrid Intel. Systems, SCI 154, pp. 29–51, 2008.
springerlink.com © Springer-Verlag Berlin Heidelberg 2008

2 Control Engineering

The term control engineering refers to a discipline whose main concern is with problems of regulating and generally controlling the behavior of physical systems. The term physical system refers to a physical objects or entities that together serve a specific purpose of function predictable in accordance with physical laws. The objective of control is to perform a particular task and this objective is realized, generally speaking, via feedback loop where some variable of the system is used to adjust the parameters needed to minimize or ideally eliminate the error that occurs [2]. A key component of a feedback control system is the controller, whose purpose is to accomplish the performance objectives one states at the outset of the formulation of the given control problem, in this paper, the problem consist to maintain the desired level of water in a water tank system. Broadly stated, fuzzy logic control attempts to come to terms with the informal nature of the control design process. In its most basic form, the so-called Mamdani architecture which will be discussed in the following paragraph, one may view fuzzy logic control as directly translating external performance specifications and observations of plant behavior into a rule-based linguistic control strategy. This architecture forms the backbone of the great majority of fuzzy logic control systems reported in the literature in the past 20 years.

2.1 Control Design Process

The design process in control engineering is generally a multistage process involving (i) selection of control design technique of methodology, (ii) determination of technical design objectives, and (iii) development of the plant model.

2.1.1 Selection of Design Methodology
Generally design methodologies in control engineering are categorized as either *time* or *frequency-domain*-based. Frequency-domain design methods range from the classical "loop Shaping" to the modern H_∞-based design techniques while time-domain design methods range from the simple PID design to linear quadratic optimal control.

2.1.2 Determination of Technical Design Objectives
This task requires interpreting, refining, and quantifying the given external performance objectives into a set of *technical design objectives* compatible with the given design methodology. In the context of the water tank system, for instance, one must translate such notions as "nearly desired level" or "quickly" into appropriate technical design objectives such as "inflow" denoted by $q(t)$, "nominal fixed flow" denoted by q_s, "outflow" denoted by $p(x)u$ or in the case of optimal control design as design to regulating the amount of water in a cylindrical water tank (figure 1).

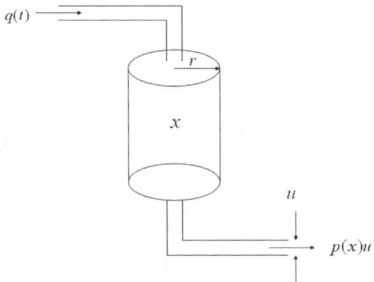

Fig. 1. Water tank system

This system's dynamical behavior is governed by the following differential equation:

$$\dot{x} = q(t) - p(x)u \qquad (1)$$

where x is the amount of water in the tank, $q(t)$ is the inflow to the tank (we assume that $q(t)$ varies around some nominal fixed flow q_s, $p(x)u$ is the outflow from de tank, where $p(x) = a\sqrt{2g\dfrac{x}{\pi r^2}}$ (a is a positive constant, r is the tank's radius, and $g = 9.8m/\sec^2$ is the acceleration due to gravity), and u the control variable, is the cross-section area of the drain opening. The control objective is: Design $u = (x; p_s, q_s)$ to regulate $x(t)$ to a desired nominal amount x_s. We assume that the functional relationship (1) is known but that $p(x)$ is *not* known explicitly and the only knowledge we have about $p(x)$ is:

- $p(x) \geq 0$ for all x
- The value $p_s = p(x_s)$ is known

Following, we assume that the fuzzy partition of the domain of x is already given, namely, the system has three operating modes: x is *low*, x is *normal*, and x is *high*. Qualitatively, as depicted in Fig. 2, x is *low* when x is much larger than x_s x is *high* when x is much larger than x_s, and x is *normal* when $|x - x_s|$ is small.

To determine the control rules in each of the three modes, we apply the fuzzy Lyapunov synthesis method using the Lyapunov function candidate $V = \dfrac{1}{2}(x - x_s)^2$. Differentiating V, we get:

$$\dot{V} = (x - x_s)\dot{x} = (x - x_s)(q(t) - p(x)u) \qquad (2)$$

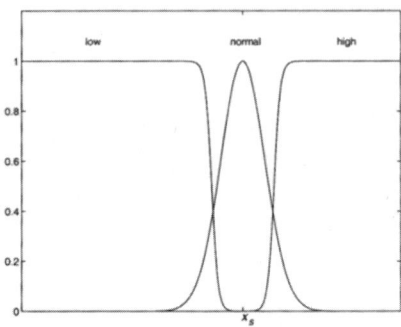

Fig. 2. Operating conditions

Equation (2) enables us to derive conditions for making $\dot{V}$ negative in each of the operating modes. If x is *low*, then $x - x_s < 0$. Hence, to make $\dot{V}$ negative, we require $q(t) > p(x)u$. But, we know that $q(t)$ and $p(x)$ are non-negative, therefore, we set $u = 0$.

Next, if x is *high*, then $x - x_s > 0$. Hence, to make $\dot{V}$ negative, we require $q(t) < p(x)u$ and set $u = u_{max}$ is the maximal opening of the drain.

Finally, the more difficult case is when x *normal*. Since we are designing a Mamdani fuzzy controller, we are based on a Takagi-Sugeno fuzzy controller, then Thenpart of each fuzzy rule -in TS fuzzy controller- is in the form $u = k_1 x + k_2$ for some constants k_1, k_2. Substituting this form in (2), we obtain:

$$
\begin{aligned}
\dot{V} &= (x - x_s)(q(t) - p(x)(k_1 x + k_2)) \\
&= (x - x_s)(q(t) - p(x)(k_1(x - x_s) + k_1 x_s + k_2)) \\
&= -k_1 p(x)(x - x_s)^2 + (x - x_s)(q(t) - p(x)(k_1 x_s + k_2))
\end{aligned}
\tag{3}
$$

The first term in (3) is non-positive for any $k_1 > 0$. Hence, to make $\dot{V}$ negative we would like the second term to vanish, that is, $q(t) - p(x)(k_1 x_s + k_2) = 0$ or $k_2 = \dfrac{q(t)}{p(x)} - k_1 x_s$. Since $q(t)$ and $p(x)$ are unknown, we approximate them using q_s and p_s, respectively (this is a logical approximation when x is *normal*). Thus, we obtain $k_2 = \dfrac{q_s}{p_s} - k_1 x_s$. So, when x is *normal,* $u = k_1 x + k_2 = k_1 x_s$ for some $k_1 > 0$. Note, however, that the drain opening is always non-negative, hence k_1 must satisfy: $k_1(0 - x_s) + \dfrac{q_s}{p_s} \geq 0$, or:

$$k_1 \leq \frac{q_s}{p_s x_s} \tag{4}$$

To summarize, using fuzzy Lyapunov synthesis we obtain the following TS-type control rules for the water tank system:

- If x is *low* Then $u = 0$
- If x is *normal* Then $u = k_1(x - x_s) + \dfrac{q_s}{p_s}$
- If x is *high* Then $u = u_{max}$

Now, using the fuzzy Lyapunov synthesis and Takagi-Sugeno fuzzy control rules type obtained for the Water Tank System, help us to get the rules where Mamdani type. In terms of input variables will be almost the same, the input variables are given as linguistic variables as x is the amount of water in the tank, which was seen as may be *low*, *normal* or *high*. Now in terms of the output variables we can notes there is a difference with the Takagi-Sugeno type controller where output variables are mathematical functions as:

If x is *normal* Then $u = k_1(x - x_s) + \dfrac{q_s}{p_s}$

Where now it will be linguistic variables on the output, as follows:

If x is *low* Then *valve_opening = close_fast*
If x is *normal* and *currenty_flow* is *positive* Then
valve_opening=close_fast
If x is *normal* Then *valve_opening=no_change*
If x is *normal* and *currenty_flow* is *negative* Then
valve_opening=open_slow
If x is *high* Then *valve_opening=open_fast*

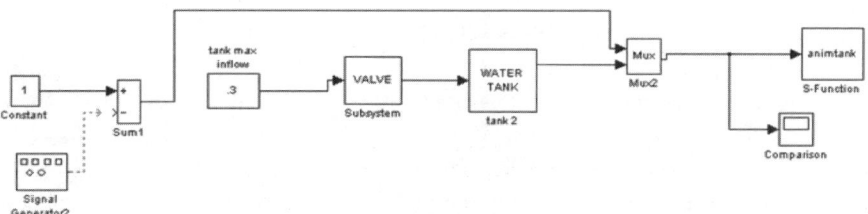

Fig. 3. Water tank system model plant

2.1.3 Development of the Plant Model

In order to properly design a control strategy, we must have a *predictive model* of the plant. In general such a model is a mathematical description of the behavior of the given physical system and is derived according to applicable physical laws. In the case of water tank system we use the plant depicted in figures 3 and 4.

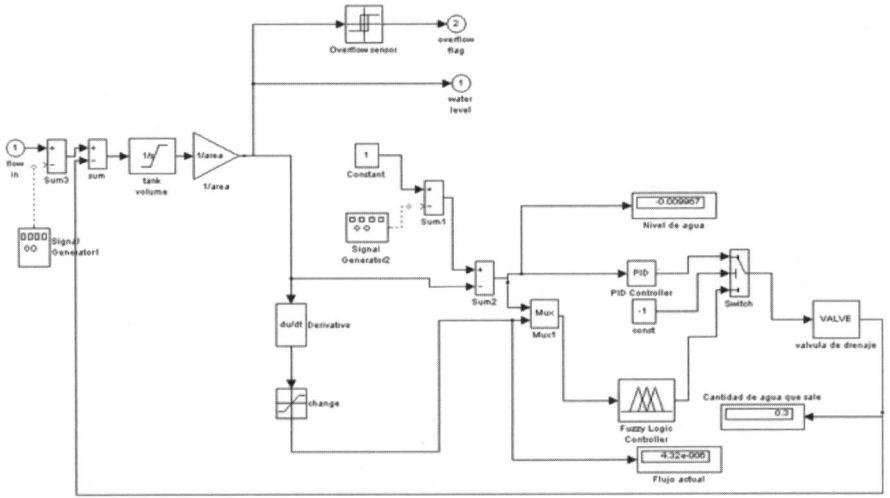

Fig. 4. Water tank system submask

3 Type-1 Fuzzy Inference Systems

The human brain interprets imprecise and incomplete sensory information provided by perceptive organs. Fuzzy set theory provides a systematic calculus to deal with such information linguistically, and it performs numerical computation by using linguistic labels stipulated by membership functions. Moreover, a selection of fuzzy If-Then rules forms the key component of a fuzzy inference system (FIS) that can effectively model human expertise in a specific application [3].

A classical set is a set with a crisp boundary. For example, a classical set A of real numbers greater than 6 can be expressed as

$$A = \{x \mid x > 6\} \tag{5}$$

Where there is a clear, unambiguous boundary 6 such that if x is greater than this number, then x belongs to the set A; otherwise x does not belong to the set. Although classical sets are suitable for various applications and have proven to be an important tool for mathematics and computer science, they do not reflect the nature of human concepts and thoughts, which tend to be abstract and imprecise.

As an illustration, we can express the set denoted by Equation (3.1), if we let A ="Tall person" and x="height". According to the theory of classical logic whole

"Tall men" is a set to belong to the man with a stature greater than a certain value, we can establish at 1.80 meters, for example, and all men with a height lower than this value would be out of joint. Well that would have a man measuring 1.81 meters in height belong to all tall men, and instead a man measuring 1.79 meters in height no longer belong to this group. But it does not seem very logical to say that a man is tall and the other is not when the height differs in two centimeters.

The approach of fuzzy logic believes that the set of "Tall men" is a set that does not have a clear boundary to belong or not to belong: with a function which defines the transition from "high" to "not high," he attached to each value of a high degree of membership in the joint between 0 and 1. For example, a man from 1.79 height could belong to a fuzzy collection "Tall men" with a grade of 0.8 of belonging, a person 1.81 tall with a degree 0.85, and a person 1.50 m tall with a grade 0.1.

Seen from this perspective can be seen that the classic logic is a limit case of fuzzy logic in which assigns a degree of belonging 1 to men with a height greater than or equal to 1.80 and a degree of belonging 0 to those with a lower altitude as shown in Figure 5.

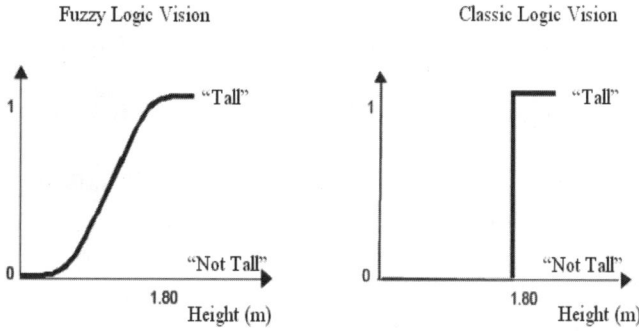

Fig. 5. Classic Logic and Fuzzy Logic

Well then, the fuzzy sets can be regarded as a classic joint generalization of the classical theory of sets only provides for the membership or non membership of an element to a set, but the fuzzy set theory provides for the membership of a partial element a whole, each item has a membership degree of a fuzzy collection can take any value between 0 and 1. This level of membership is defined by a function of membership associated with a fuzzy collection: for each value that it can take an element or variable input function x membership $\mu_A(x)$ provides the membership degree of the value of x to a fuzzy collection A.

Formally, a classic set A, in a universe of discourse U, can be defined in several ways: by listing the elements that belong to all, specifying the properties to be met by elements belonging to this group or, in terms of membership function $\mu_A(x)$:

$$\mu_A(x) = \begin{cases} 1 & if \quad x \in A \\ 0 & if \quad x \notin A \end{cases} \tag{6}$$

A fuzzy collection in the universe U is characterized by a membership function which takes values in the interval [0, 1], and can be represented as a set of ordered pairs of an element x and its value of membership to a set:

$$A = \{(x, \mu_A(x) | x \in U)\} \tag{7}$$

3.1 Mamdani Fuzzy Models

The basic assumption of underlying the approach to fuzzy logic control proposed by E.H. Mamdani in 1974[4] is that in the absence of an explicit plant model an/or clear statement of control design objectives, informal knowledge of the operation of the given plant can be codified in terms of *if-then*, or condition-action, rules and form the basis for a linguistic control strategy.

The basic paradigm for a fuzzy logic control that has emergent following Mamdani's original work is a linguistic or rule-based control strategy of the form

If OA_1 is --- and OA_2 is --- and --- Then CA_1 is --- and CA_2 is ---

If OA_1 is --- and OA_2 is --- and --- Then CA_1 is --- and CA_2 is ---

Which maps the observable attributes $(OA_1, OA_2, ...)$ of the given physical system into its controllable attributes $(CA_1, CA_2, ...)$. The controller structure in fig. 2.4, relates this architecture to than of a conventional feedback control system, where appropriately,

Output $\leftrightarrow$ *Observable Attribute*

Input $\leftrightarrow$ *Controllable Attribute*

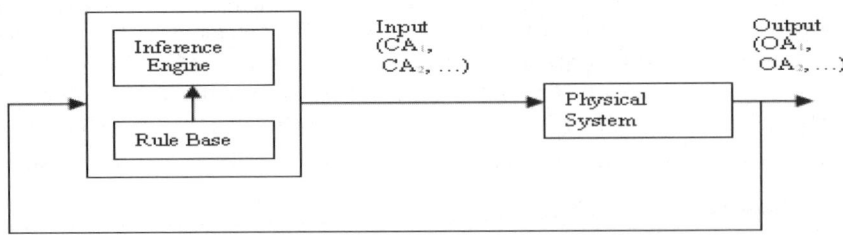

Fig. 6. Architecture for Fuzzy Control

In particular each OA_i, $i = 1, 2, ...$ is either a directly measurable variable and/or the difference between any such variable and its associated reference value.

4 Interval Type-2 Fuzzy Logic

The type-2 fuzzy sets are used to model the uncertainty and inaccuracy in real-world problems. This set was originally proposed by Zadeh in 1975 and are essentially "fuzzy-fuzzy" sets in which grades of membership are type-1 fuzzy sets.

A type-2 fuzzy set expresses the degree of non-determinist truth with vagueness and uncertainty with which an element belongs to the whole set.

The type-2 fuzzy set was originally proposed by Zadeh in 1975 and it is essentially a "fuzzy-fuzzy" set in which membership grades are a type-1 fuzzy set. A type-2 fuzzy [5, 6] expresses the non-deterministic truth degree with imprecision and uncertainty for an element that belongs to a set.

If $f_x(u) = 1, \forall u \in [\underline{J}_x^u, \overline{J}_x^u] \subseteq [0,1]$, the type-2 membership function $\mu_{\tilde{A}}(x,u)$ is expressed by a lower membership function $\underline{J}_x^u \equiv \underline{\mu}_A(x)$ and higher membership function $\overline{J}_x^u \equiv \overline{\mu}_A(x)$ of type-1 is labeled interval type-2 fuzzy set, denoted by:

$$\tilde{A} = \left\{ (x,u,1) \mid \forall x \in X, \forall u \in [\underline{\mu}_A(x), \overline{\mu}_A(x)] \subseteq [0,1] \right\} \tag{6}$$

or

$$\tilde{A} = \left\{ \int_{x \in X} \left[\int_{u \in [\underline{J}_x^u, \overline{J}_x^u] \subseteq [0,1]} 1/u \right] / x \right\} = \left\{ \int_{x \in X} \left[\int_{u \in [\underline{\mu}_A(x), \overline{\mu}_A(x)] \subseteq [0,1]} 1/u \right] / x \right\} \tag{7}$$

4.1 Type-2 Fuzzy Reasoning

Assuming a fuzzy system with M rules, p input variables and one output variable, we have that the antecedent and consequent are type-2 fuzzy sets.

$$R^l : IF \ x_1 \ is \ F_1^1 \ and \dots and \ x_p \ is \ F_p^1 \ THEN \ y \ is \ G^1$$
$$H : \quad x_1 \ is \ \Lambda_{x_1} \ and \dots and \ x_p \ is \ A_{x_p} \tag{8}$$
$$C : \quad\quad\quad\quad\quad\quad\quad\quad\quad\quad\quad\quad\quad y \ is \ \hat{y}$$

This reasoning evaluation is:

The k-th rule relation its

$$R^l = F_1^l x \dots x F_p^l \rightarrow G^l = F_1^l \rightarrow G^l = A^l \Pi G^l \tag{9}$$

The Fact relation is:

$$A_x = A_{x_1} \, x \ldots x \, A_{x_p} = A_{x_1} \Pi \ldots \Pi A_{x_p} \tag{10}$$

$B^l = A_x \circ R^l$, Generalized, fuzzy reasoning

$$\mu_{B^l}(y) = \mu_{A_x \circ R^l}(y) = \coprod_{\in X} \left[\mu_{A_x}(x) \Pi \mu_{A^l \to G^l}(x, y) \right]$$

$$\mu_{B^l}(y) = \mu_{G^l}(y) \Pi \left\{ \Pi_{i=1}^p \left[\mu_{A_{x_i}}(x_i) \Pi \mu_{F_i^l}(x_i) \right] \right\} = \left[\underline{\mu}_{B^l}(y), \overline{\mu}_{B^l}(y) \right]$$

Where

$$\underline{\mu}_{B^l}(y) = \left[\overset{p}{\underset{i=1}{\tilde{*}}} \left(\underline{\mu}_{A_{x_1}}(x_i) \tilde{*} \underline{\mu}_{F_i^l}(x_i) \right) \right] \tilde{*} \underline{\mu}_{G^l}(y)$$

$$\overline{\mu}_{B^l}(y) = \left[\overset{p}{\underset{i=1}{\tilde{*}}} \left(\overline{\mu}_{A_{x_1}}(x_i) \tilde{*} \overline{\mu}_{F_i^l}(x_i) \right) \right] \tilde{*} \overline{\mu}_{G^l}(y)$$

Aggregation

$$\mu_B(y) = \coprod_{l=1}^M \mu_{B^l}(y) = \coprod_{i=1}^M \left(\mu_{G^l}(y) \Pi \left\{ \Pi_{i=1}^p \left[\mu_{A_{x_i}}(x_i) \right] \right\} \right) = \left[\underline{\mu}_B(y), \overline{\mu}_B(y) \right]$$

Where

$$\underline{\mu}_B(y) = \overset{M}{\underset{l=1}{\vee}} \left(\underline{\mu}_{B^l}(y) \right) = \overset{M}{\underset{l=1}{\vee}} \left(\left[\overset{p}{\underset{i=1}{\tilde{*}}} \left(\underline{\mu}_{A_{x_i}}(x_i) \tilde{*} \underline{\mu}_{F_i^l}(x_i) \right) \right] \tilde{*} \underline{\mu}_{G^l}(y) \right)$$

$$\overline{\mu}_B(y) = \overset{M}{\underset{l=1}{\vee}} \left(\overline{\mu}_{B^l}(y) \right) = \overset{M}{\underset{l=1}{\vee}} \left(\left[\overset{p}{\underset{i=1}{\tilde{*}}} \left(\overline{\mu}_{A_{x_i}}(x_i) \tilde{*} \overline{\mu}_{F_i^l}(x_i) \right) \right] \tilde{*} \overline{\mu}_{G^l}(y) \right)$$

The interval type-2 fuzzy reasoning is depicted in figure 7.

4.2 Type-2 Rule Based Fuzzy Logic System

A rule based Fuzzy Logic System (FLS) contains four components: Rules, fuzzifier, inference engine, and output processor that are interconnected, as shown in Figure 8.

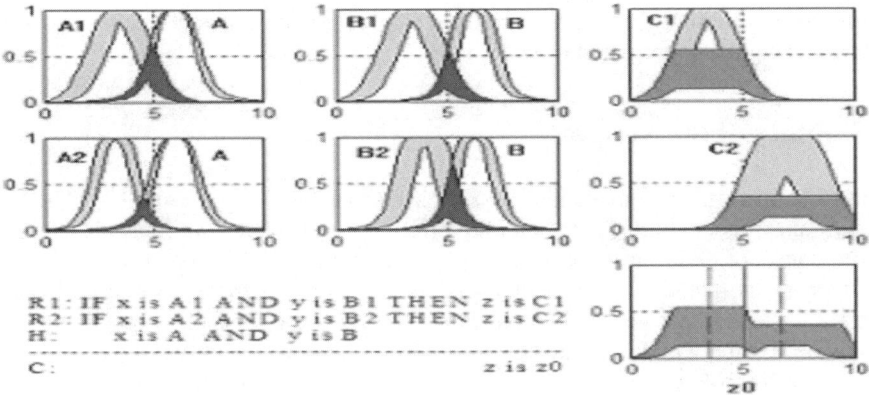

Fig. 7. Interval type-2 fuzzy reasoning

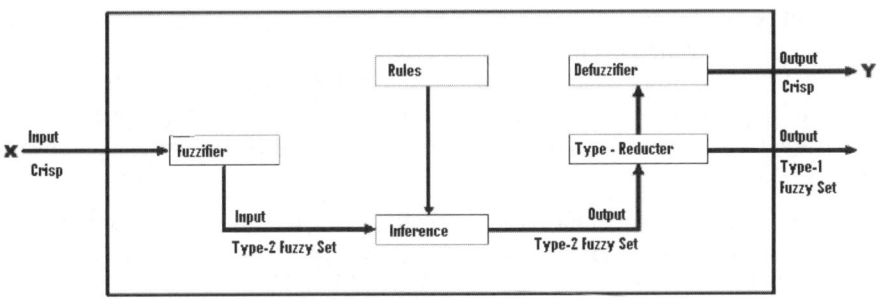

Fig. 8. Type-2 Fuzzy Logic System

A Fuzzy Inference System is a system based on fuzzy rules, instead Boolean logic, to data analysis [7, 8, 9]. Its basic structure includes four principal components, as shown in Figure 8.

1. Fuzzifier. Map inputs (crisp values) into fuzzy values
2. Inference System. Applies a fuzzy reasoning to obtain a type-2 fuzzy output.
3. Defuzzifier/Type Reducer. The defuzzifier maps an output to a crisp values; the type reducer transform a type-2 fuzzy set into a type-1 fuzzy set.
4. Knowledge Base. Contains a fuzzy rule set, known as the base of rules, and a membership function set known as a database.

The decision process is conducted by an inference system using the rules from the base of rules. These fuzzy rules define the connection between input and output fuzzy variables. A fuzzy rule has the form: IF <antecedent> THEN <consequent>, where antecedent is fuzzy-logic expressions consist of one of more simple fuzzy expressions connected by fuzzy operators, and consequent its an expression which assigns fuzzy values to a output variables. The inference system values all of the rules from the base of rules and combining weights of consequents of all the relevant rules in an only fuzzy set using the aggregation operation.

5 Stability of the System

Stability studies three fundamental problems that are very important [10]. The first problem, absolute stability, it is qualitative in nature, in which we look for a simple answer -"yes" or "not"- in relation at the system stability. The second problem, relative stability, is quantitative in nature and its associated with the problem of determines how stable a system is. The third one discusses the robustness qualities to determine how much we can perturb the plant maintaining the stability.

In fuzzy control systems, given the linguistic characteristics, the stability problem becomes somewhat controversial. For the people that based their work on intelligent control, it's not necessary to establish formal criterions to demonstrate that the control systems are stables, besides relying this on the pioneer approaches [15]:

- Zadeh: The fuzzy control is accepted because is a task control oriented, instead traditional control which is characterized because is oriented toward achieving a reference, therefore fuzzy control doesn't need a mathematical stability analysis .
- Sugeno: In general, in most of the industrial applications, the control stability is not always guaranteed and its most important the reliability of the mechanisms created than its stability

These justifications have not been accepted by all researches, for example Margaliot in [1] and [16] in which is based this paper, proposed a method to guarantee stability of fuzzy systems.

6 Fuzzy Lyapunov Synthesis

Margaliot in [11] introduces the concept, of *"Fuzzy Lyapunov Synthesis"* based on the Lyapunov Criterion to determine the stability of a system. The adjustment Margaliot does is to consider linguistic variables instead of numeric variables to find the system's stability, which means using the Lyapunov Criterion [12] and the Computing With Words paradigm introduced by Zadeh in [13].

To guarantee stability, a Lyapunov candidate function V must be proposed which represents the problem to solve following the Lyapunov Criterion, once we have V , we calculate $\dot{V}$ which also, and according the Lyapunov Criterion is known it must have the follow characteristics:

definite (11) or semi definite negative (12) to guarantee asymptotic stability or stability respectively.

$$\dot{V} < 0 \tag{11}$$

$$\dot{V} \leq 0 \tag{12}$$

At this moment, the equations are the same as followed by traditional methods. To guarantee stability on fuzzy control systems, we must find the necessary linguistic restrictions to ensure compliance of (11) or (12), for this we must properly granulate the linguistic variables [14], [3], [2], hence, find the adequate number and type of membership functions to ensure compliance of (11) or (12). Given the linguistic characteristics of the fuzzy systems, in general we look for satisfying (5.1) and make stable the system, although is very common the system tends to satisfy condition (12) instead (11).

7   Simulation Results

In this paper, the measure of the error is given by the equation (13).

$$IAE = \frac{\sum_{i=1}^{n}|r_i - u_i|}{n} \tag{13}$$

Where:

r_i = Reference value in each sampled point

u_i = Control value in each sampled point

n = Total sampled points

7.1   Type-1 Mamdani

7.1.1   Generic Type-1 Mamdani FIS
The Generic Type-1 Mamdani FIS, its inputs, output and the set of rules are shown in Figures 9 and 10. Then we show the results of the simulation and the error achieved.

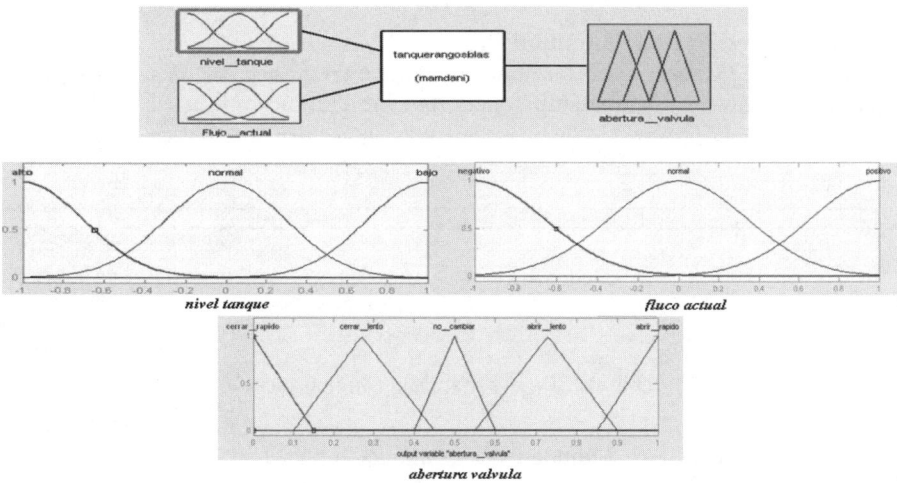

Fig. 9. Generic Type-1 Mamdani FIS

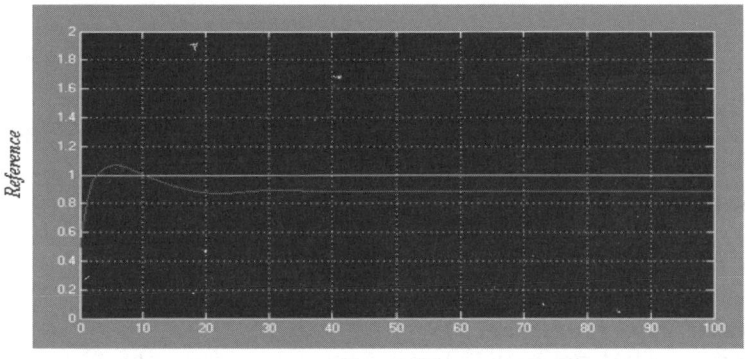

```
1. If (nivel__tanque is bajo) then (abertura__valvula is cerrar__rapido) (1)
2. If (nivel__tanque is normal) and (Flujo__actual is positivo) then (abertura__valvula is cerrar__lento) (1)
3. If (nivel__tanque is normal) then (abertura__valvula is no__cambiar) (1)
4. If (nivel__tanque is normal) and (Flujo__actual is negativo) then (abertura__valvula is abrir__lento) (1)
5. If (nivel__tanque is alto) then (abertura__valvula is abrir__rapido) (1)
```

If   and   Then
nivel__tanque is   Flujo__actual is   abertura__valvula

Fig. 10. Generic Type-1 Mamdani FIS, Rule Base

Figure 11 shows the result of the Generic Type-1 Mamdani FIS, this fuzzy system, achieved a **0.1099** error.

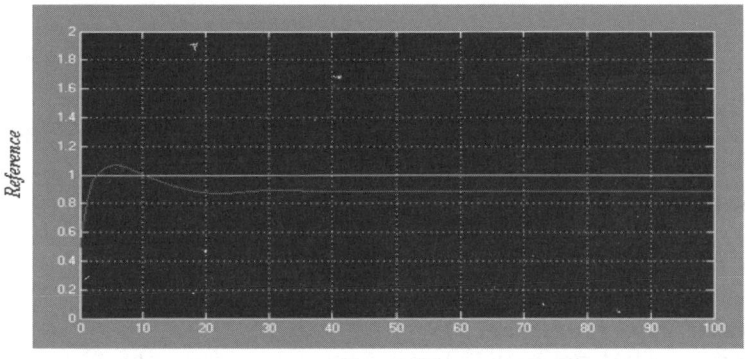

Fig. 11. Generic Type-1 Mamdani FIS, Simulation

7.1.2 Optimized Type-1 Mamdani FIS

To optimize the Generic Type-1 Mamdani FIS shown before, we used Genetic Algorithms for tuning the membership functions. The chromosome used in this case is shown in Figure 12.

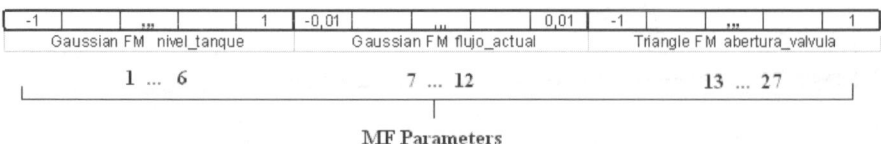

Fig. 12. Type-1 Mamdani Chromosome

The Optimized Type-1 Mamdani FIS, and its inputs and output, are shown in Table 1. Then we show the table of results, the simulation of the best result and the error achieved. Each row of the table is a different test.

Table 1. Results for Optimized Type-1 Mamdani FIS

INDIVIDUOS	GENERACIONES	GAP	CRUCE	MUTACION	MEJOR	ERROR	TIEMPO
60	20	0.8	0.1	0.01	Gen 21 Ind 31	0.119344	35 min
70	20	0.8	0.2	0.02	Gen 21 Ind 9	0.112635	42 min
70	20	0.5	0.2	0.02	Gen 21 In2 23	0.109199	19 min
100	20	0.5	0.1	0.01	Gen 21 Ind 2	0.169588	27 min
50	30	0.7	0.2	0.03	Gen 31 Ind 14	0.075536	39 min
50	30	0.8	0.2	0.03	Gen 31 Ind 4	0.115666	7 min
80	30	0.6	0.6	0.02	Gen 31 Ind 21	0.186388	15 min
100	30	0.6	0.2	0.01	Gen 31 Ind 17	0.127444	1:19 min
80	40	0.7	0.2	0.05	Gen 41 Ind 42	0.075484	2:12 hrs
100	40	0.8	0.3	0.02	Gen 41 Ind 7	0.037613	2:35 hrs
40	50	0.8	0.2	0.02	Gen 51 Ind 1	0.041473	48 min
80	50	0.6	0.6	0.02	Gen 51 Ind 41	0.057939	2:27 hrs
40	60	0.8	0.5	0.06	Gen 60 Ind 10	0.148233	8 min
50	60	0.8	0.2	0.03	Gen 61 Ind 31	0.282288	27 min
50	60	0.5	0.4	0.05	Gen 61 Ind 25	0.078345	1:23 hrs
60	60	0.6	0.1	0.01	Gen 61 Ind 17	0.063807	2:29 hrs
60	60	0.8	0.1	0.01	Gen 61 Ind 30	0.028155	2:18 hrs
90	60	0.7	0.2	0.02	Gen 61 Ind 15	0.134888	57 min
30	70	0.8	0.3	0.03	Gen 71 Ind 10	0.089791	1:08 hrs
150	70	0.7	0.1	0.01	Gen 71 Ind 29	0.065528	8:06 hrs
60	80	0.8	0.1	0.01	Gen 81 Ind 15	0.040161	1:49 hrs
100	80	0.5	0.1	0.02	Gen 81 Ind 12	0.164711	1:02 hrs
30	90	0.7	0.7	0.1	Gen 91 Ind 7	0.116799	42 min
60	100	0.6	0.1	0.01	Gen 101 Ind 11	0.100933	2:39 hrs
150	100	0.8	0.7	0.1	Gen 101 Ind 47	0.043268	6:26 min
30	120	0.7	0.1	0.7	Gen 121 Ind 15	0.126722	3:15 hrs
90	120	0.7	0.2	0.02	Gen 121 Ind 1	0.098665	37 min
60	140	0.6	0.1	0.01	Gen 139 Ind 28	0.120066	2:18 min

We show in Figure 13 an example of an optimized fuzzy system.

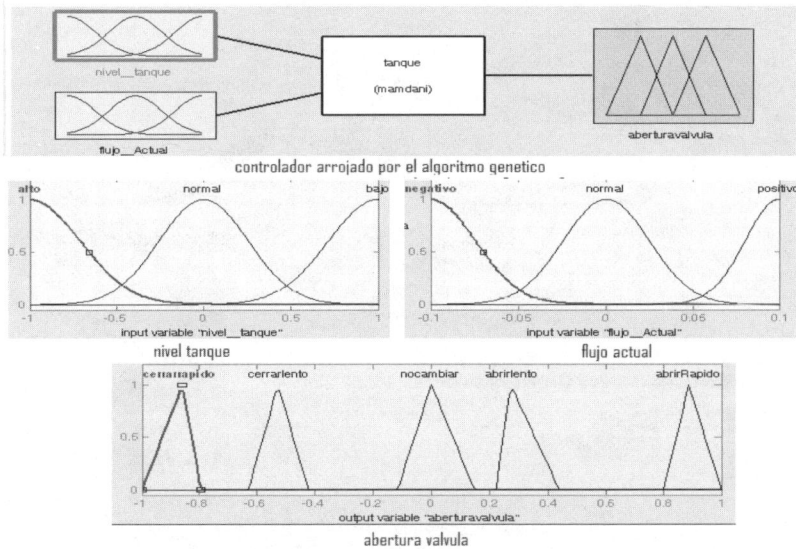

Fig. 13. Optimized Type-1 Mamdani FIS

Figure 14 shows the result of the Optimized Type-1 Mamdani FIS of Figure 13, this fuzzy system, achieved an error of **0.028155**.

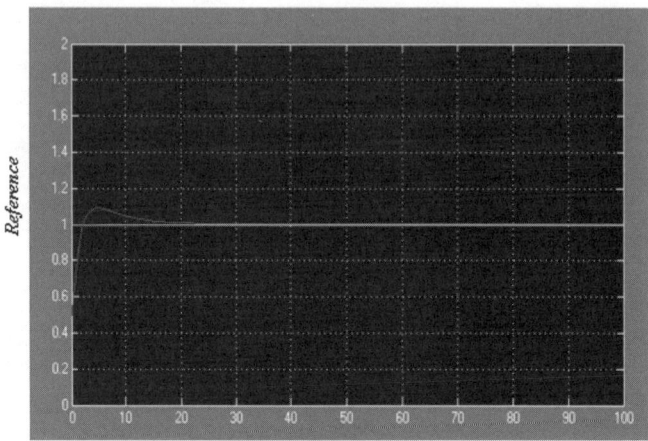

Time units

Fig. 14. Optimized Type-1 Mamdani Simulation

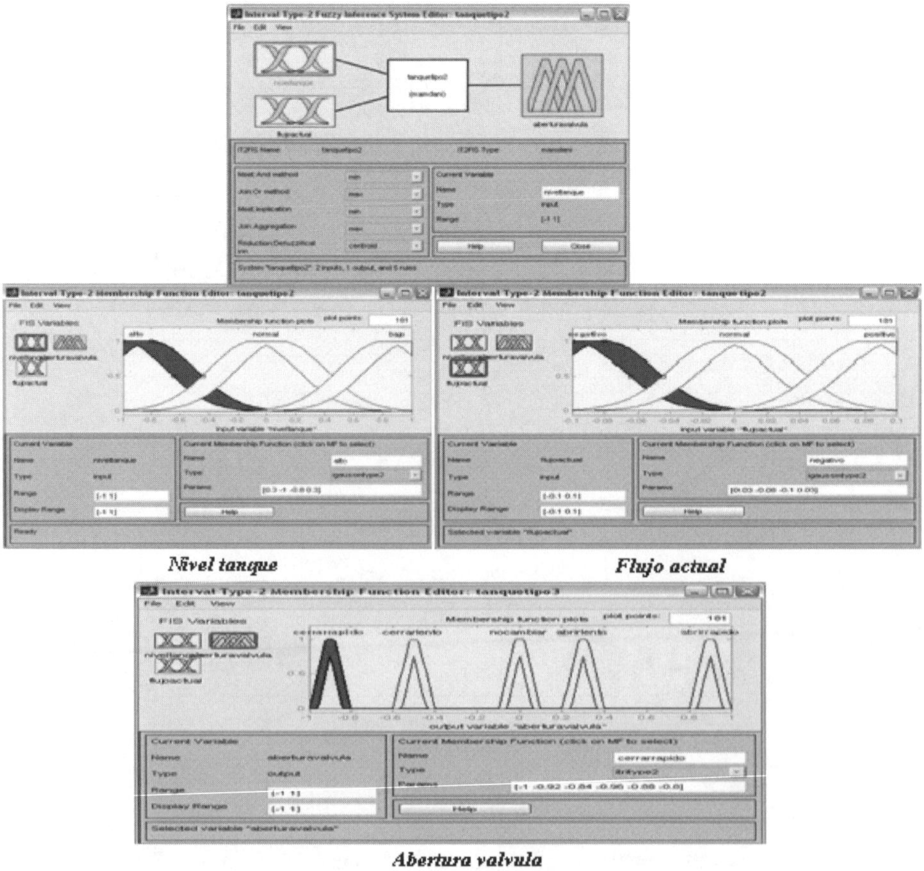

Nivel tanque

Flujo actual

Abertura valvula

Fig. 15. Generic Type-2 Mamdani FIS

7.2 Type-2 Mamdani

7.2.1 Generic Type-2 Mamdani FIS

The Generic Type-2 Mamdani FIS, its inputs, output and the set of rules, are shown in Figures 15 and 16.

1. If (niveltanque is alto) then (aberturavalvula is abrirrapido) (1)
2. If (niveltanque is normal) then (aberturavalvula is nocambiar) (1)
3. If (niveltanque is normal) and (flujoactual is negativo) then (aberturavalvula is cerrarlento) (1)
4. If (niveltanque is normal) and (flujoactual is positivo) then (aberturavalvula is abrirlento) (1)
5. If (niveltanque is bajo) then (aberturavalvula is cerrarrapido) (1)
If
niveltanque is
and
flujoactual is
Then
aberturavalvula is

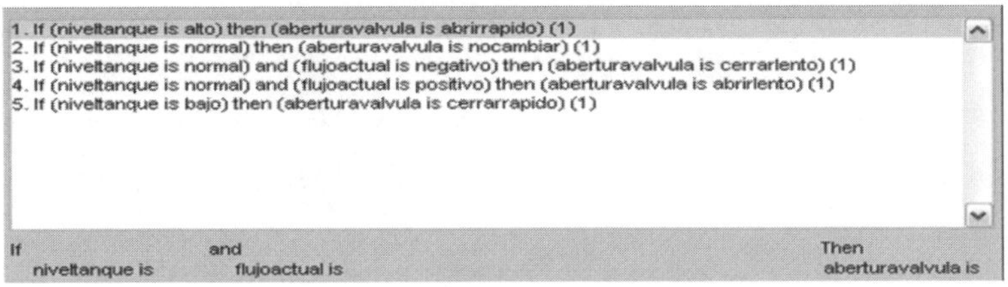

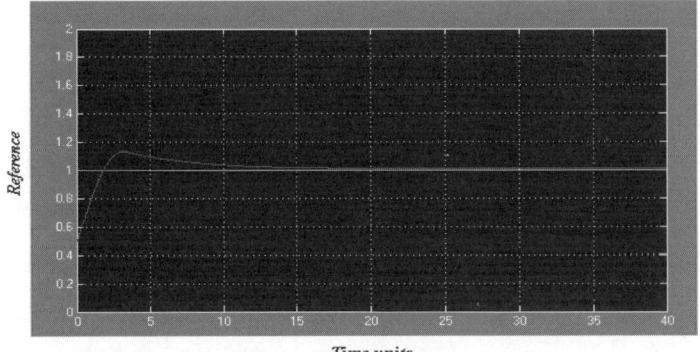

Fig. 16. Generic Type-2 Mamdani FIS, Rule Base

Figure 17 shows the result of the Generic Type-2 Mamdani FIS, this fuzzy system, achieved a **0.043581** error.

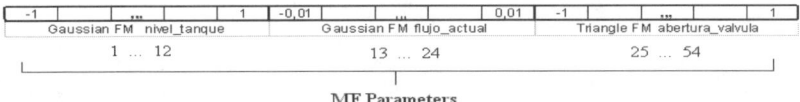

Reference
Time units
0
0.2
0.4
0.6
0.8
1
1.2
1.4
1.6
1.8
2
0
5
10
15
20
25
30
35
40

Fig. 17. Generic Type-2 Mamdani FIS, Simulation

7.2.2 Optimized Type-2 Mamdani FIS

To optimize the Generic Type-2 Mamdani FIS show before, we used Genetic Algorithms to tuning the membership functions. The chromosome used in this case is shown in Fig. 18.

-1
...
1
-0,01
...
0,01
-1
...
1
Gaussian FM nivel_tanque
Gaussian FM flujo_actual
Triangle FM abertura_valvula
1 ... 12
13 ... 24
25 ... 54
MF Parameters

Fig. 18. Type-2 Mamdani Chromosome

The Optimized Type-2 Mamdani FIS, and its inputs and output, are shown in Table 2. Then we show the table of results, the simulation of the best result and the error achieved. Each row of the table is a different test.

Table 2. Results for Optimized Type-2 Mamdani FIS

INDIVIDUOS	GENERACIONES	GAP	CRUCE	MUTACION	MEJOR	ERROR	TIEMPO
15	10	0.8	0.01	0.3	GEN 11 IND 1	0.028078	2:06 hrs
30	20	0.8	0.4	0.07	GEN 21 IND 6	0.017038	8:02 hrs
30	40	0.7	0.6	0.2	GEN 41 IND 8	0.015347	9:03 hrs
40	20	0.8	0.4	0.07	GEN 41 IND 36	0.027177	10:20 hrs
40	30	0.8	0.2	0.1	GEN 31 IND 16	0.018914	10:41 hrs
40	30	0.8	0.1	0.01	GEN 31 IND 25	0.017411	15:43 hrs
40	30	0.8	0.4	0.7	GEN 31 IND 8	0.018725	10:41 hrs
50	20	0.8	0.3	0.3	GEN 21 IND 33	0.015931	12:56 hrs
50	30	0.8	0.3	0.3	GEN 31 IND 29	0.015723	7:11 hrs
50	30	0.8	0.2	0.02	GEN 31 IND 5	0.024399	13:16 hrs
50	40	0.8	0.3	0.03	GEN 41 IND 33	0.015562	29:00 hrs
50	40	0.8	0.2	0.02	GEN 57 IND 33	0.015506	20:00 hrs

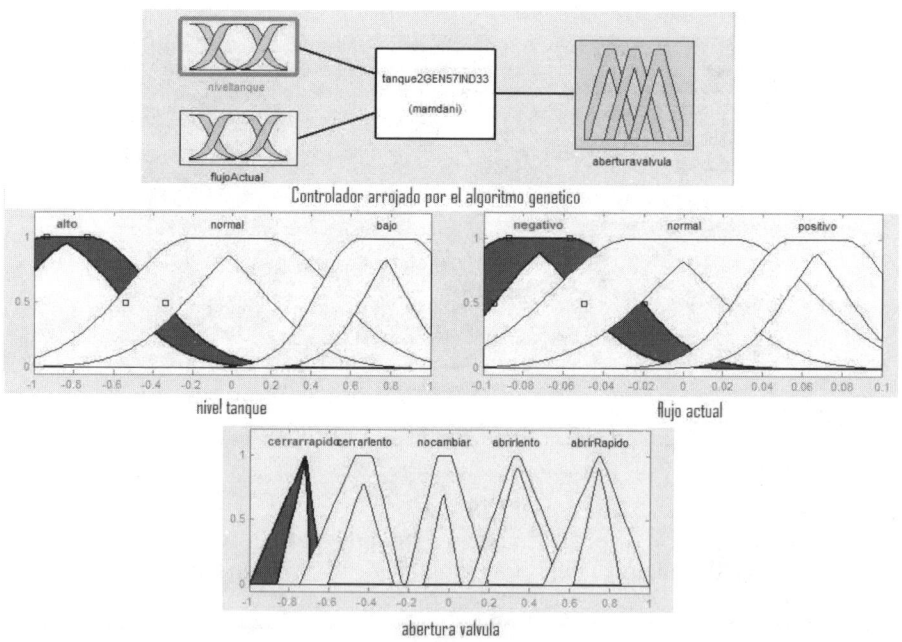

Fig. 19. Optimized Type-2 Mamdani FIS

Figure 20 shows the simulation results of the Optimized Type-2 Mamdani FIS of Figure 19, this fuzzy system, achieved an error of **0.015506**.

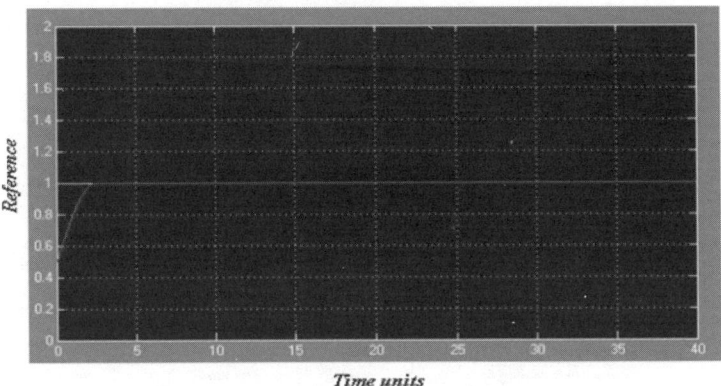

Time units

Fig. 20. Optimized Type-2 Mamdani FIS, Simulation

7.3 Type-1 Mamdani with Reference Change

The Optimized Type-1 Mamdani FIS, and its inputs and output, are shown in Table 3. Then we show the table of results, the simulation of the best result and the error achieved. Each row of the table is a different test. In this case we introduced one pulse on the input signal, which simulate unexpected increase of water in the nominal fixed flow.

Table 3. Results for Optimized Type-1 Mamdani

INDIVIDUOS	GENERACIONES	GAP	CRUCE	MUTACION	MEJOR	ERROR	TIEMPO
60	20	0.8	0.1	0.01	Gen 21 Ind 31	0.11934	35 min
70	20	0.8	0.2	0.02	Gen 21 Ind 9	0.112635	42 min
70	20	0.5	0.2	0.02	Gen 21 In2 23	0.10919	19 min
100	20	0.5	0.1	0.01	Gen 21 Ind 2	0.16958	27 min
50	30	0.7	0.2	0.03	Gen 31 Ind 14	0.13838	29 min
50	30	0.8	0.2	0.03	Gen 31 Ind 4	0.11566	7 min
80	30	0.6	0.6	0.02	Gen 31 Ind 21	0.18638	15 min
100	30	0.6	0.2	0.01	Gen 31 Ind 7	0.14971	59 min
80	40	0.7	0.2	0.05	Gen 41 Ind 52	0.14684	1:52 hrs
100	40	0.8	0.3	0.02	Gen 41 Ind 17	0.10123	2:06 hrs
40	50	0.8	0.2	0.02	Gen 51 Ind 1	0.10299	48 min
80	50	0.6	0.6	0.02	Gen 51 Ind 31	0.10077	1:57 hrs
40	60	0.8	0.5	0.06	Gen 60 Ind 1	0.11079	8 min
50	60	0.8	0.2	0.03	Gen 61 Ind 41	0.204/1	7 min
50	60	0.5	0.4	0.05	Gen 61 Ind 18	0.1204	43 min
60	60	0.6	0.1	0.01	Gen 61 Ind 7	0.11373	1:49 hrs
60	60	0.8	0.1	0.01	Gen 61 Ind 4	0.10853	58 min
90	60	0.7	0.2	0.02	Gen 61 Ind 5	0.15728	37 min
30	70	0.8	0.3	0.03	Gen 71 Ind 1	0.12591	48 min
150	70	0.7	0.1	0.01	Gen 71 Ind 89	0.11284	6:06 hrs
60	80	0.8	0.1	0.01	Gen 81 Ind 20	0.10941	1:29 hrs
100	80	0.5	0.1	0.02	Gen 81 Ind 12	0.17766	42 min
30	90	0.7	0.7	0.1	Gen 91 Ind 17	0.11901	22 min
60	100	0.6	0.1	0.01	Gen 101 Ind 1	0.13581	1:49 hrs
150	100	0.8	0.7	0.1	Gen 101 Ind 5;	0.10201	5:26 min
30	120	0.7	0.1	0.7	Gen 121 Ind 1	0.13372	2:15 min
90	120	0.7	0.2	0.02	Gen 121 Ind 1	0.11227	27 min
60	140	0.6	0.1	0.01	Gen 139 Ind 2	0.15126	2:18 min

Figure 21 shows the details of the optimal fuzzy system of Table 3.

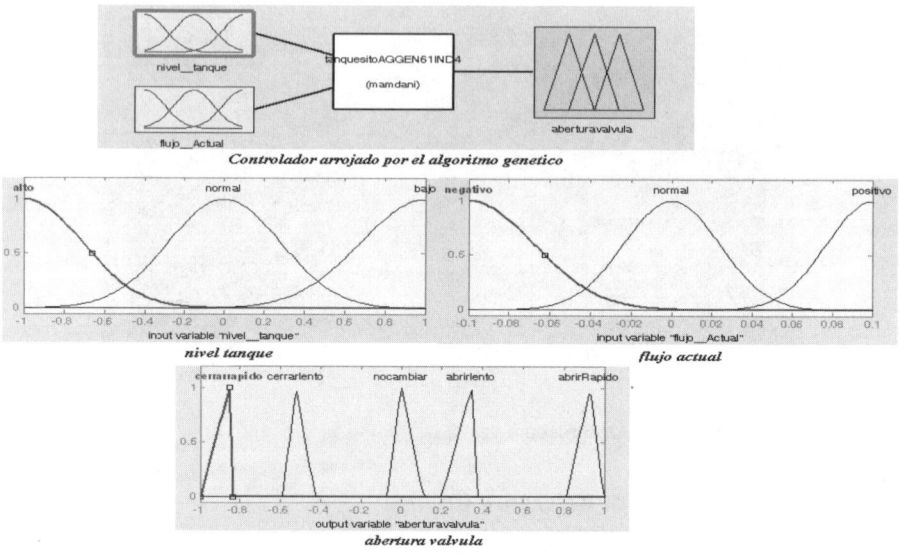

Fig. 21. Optimized Type-1 Mamdani FIS with reference change

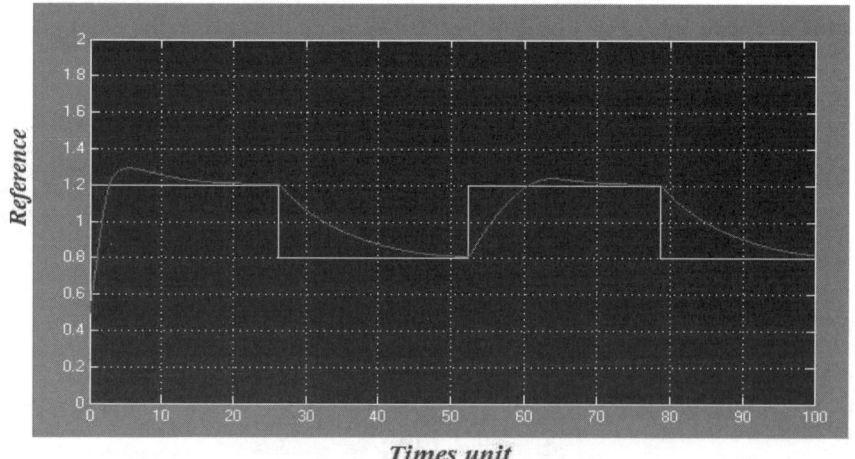

Fig. 22. Optimized Type-1 Mamdani FIS with reference change

Table 4. Results of optimized Type-2 Mamdani FIS (reference change)

INDIVIDUOS	GENERACIONES	GAP	CRUCE	MUTACION	MEJOR	ERROR	TIEMPO
40	20	0.7	0.1	0.03	gen 21 ind 25	0.043575	24 Horas
50	20	0.8	0.1	0.01	gen 21 ind 1	0.053759	14 Horas
60	20	0.8	0.2	0.02	gen 21 ind 47	0.061522	21 Horas
60	22	0.8	0.3	0.03	gen 22 ind 31	0.051644	19 Horas
30	30	0.9	0.3	0.03	gen 31 ind 13	0.05455	16 Horas
60	30	0.7	0.1	0.01	gen 31 ind 30	0.052733	21 Horas
40	35	0.8	0.5	0.03	gen 36 ind 31	0.052353	13 Horas
30	40	0.8	0.2	0.02	gen 31 ind 4	0.04483	28 Horas
30	40	0.8	0.6	0.2	gen 41 ind 24	0.056962	11 Horas
20	45	0.9	0.7	0.1	gen 46 ind 7	0.055411	12 Horas

Figure 22 shows the simulation results of the Optimized Type-1 Mamdani FIS with reference change, this fuzzy system, achieved a **0.10853** error.

7.4 Type-2 Mamdani with Reference Change

The Optimized Type-2 Mamdani FIS, and its inputs and output, are shown in Table 4. Then we show the table of results, the simulation of the best result and the error

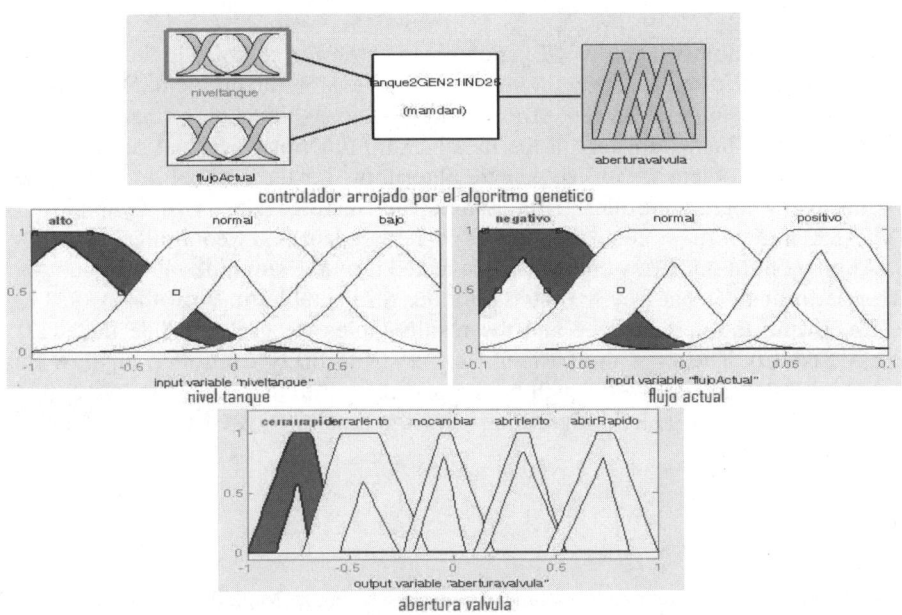

Fig. 23. Optimized Type-2 Mamdani FIS with reference change

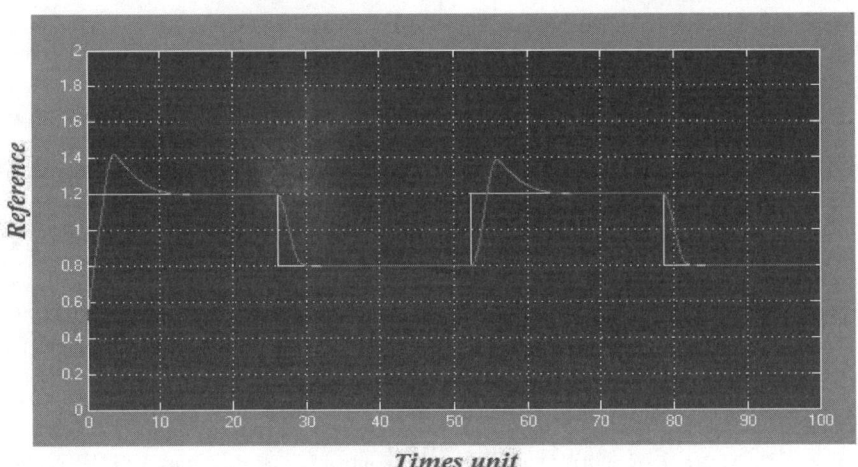

Fig. 24. Optimized Type-1 Mamdani FIS with reference change

achieved. Each row of the table is a different test. In this case we introduced one pulse on the input signal, which simulate unexpected increase of water in the nominal fixed flow.

Figure 24 shows the result of the Optimized Type-2 Mamdani FIS (shown in Figure 23) with reference change, this fuzzy system, achieved a **0.043575** error.

8 Conclusions

Based on the Stability Theory of Lyapunov, an extension to fuzzy logic was proposed by Margaliot. We built a fuzzy controller for a tank water problem. We notice that fuzzy controller for a tank water problem don't keep stable its performance. The main reason was that the parameters of the membership functions were not suitable chosen. To solve this problem we utilize genetic algorithms for the optimal set of parameters for the fuzzy logic controller. This assured the Stability Theory of Lyapunov for a Type-1 Mamdani fuzzy controller and Type-2 Mamdani fuzzy controller.

After we built our fuzzy controller, we tested it on the simulink model to obtain the error performance, that way we could built our results tables to show the errors.

Reviewing the performance and the results tables we can conclude that with the Fuzzy Stability Theory is the basis to built a stable fuzzy controller, anyhow this is not assured that this controller will be the optimal but to test the theory we optimized the controllers with genetic algorithms and the results showed that the theory was confirmed.

References

[1] Margaliot, M., Langholz, G.: New Approaches to Fuzzy Modeling and Control. In: Design and Analysis. World Scientific Co. Pte. Ltd, Singapore (2000)
[2] Yen, J., Langari, R.: Fuzzy Logic Intelligence, control, and Information. Prentice Hall, Englewood Cliffs (1999)
[3] Jang, J.S.R., Sun, C.T., Mizutani, E.: Neuro-Fuzzy and Soft Computing. In: A Computational Approach to Learning and Machine Intelligence. Prentice-Hall, Inc., Englewood Cliffs (1997)
[4] Mamdani, E.H.: Application of fuzzy algotithms for control of simple dynamic plant. IEEE proceedings 121(12) (1974)
[5] Karnik, N.N., Mendel, J.M.: An Introduction to Type-2 Fuzzy Logic Systems, Univ. of Southern Calif., Los Angeles, CA (June 1998)
[6] Zadeh, L.: Fuzzy logic = computing with words. IEEE Transactions on Fuzzy Systems 2, 103–111 (1996)
[7] Zadeh, L.A.: The concept of a linguistic variable and its application to approximate reasoning, Parts 1, 2, and 3, Information Sciences, 8:199–249, 8:301–357, 9:43–80 (1975)
[8] Mendel, J.: Uncertain Rule-Based Fuzzy Logic Systems: Introduction and New Directions. Prentice-Hall, NJ (2001)
[9] Yager, R.: On a general class of fuzzy connectives. Fuzzy Sets and Systems 4, 235–242 (1980)
[10] Rohrs, E., Melsa, J., Schultz, D.: Sistemas de Control Lineal. McGraw-Hill, USA (1994)

[11] Margaliot, M., Langhoiz, G.: Adaptive fuzzy controller design via fuzzy Lyapunov sinthesys. In: Proceedings of the International Conference on Fuzzy Systems (FUZZY-IEEE 1998), Alaska, USA (1998)
[12] Khalil, H.: Nonlinear Systems. Macmillan Publishing company, USA (1992)
[13] Zadeh, L.: Fuzzy Logic=Computing With Words. IEEE Transactions on Fuzzy Systems 4(2), 103–111 (1996)
[14] Chen, G., Phan, T.T.: Introduction to Fuzzy Sets, Fuzzy Logic and Fuzzy Control Systems. CRC Press, USA (2000)
[15] Dotoli, M., Jantzen, J.: Debate: Fuzzy Control vs. Conventional Control, http://fuzzy.iau.dtu.dk/debate.nsf
[16] Margaliot, M., Langholz, G.: A new approach to the design of fuzzy control rules. In: Proceedings of the International Conference on Fuzzy Logic and Applications (Fuzzy 1997), Israel, pp. 248–254 (1997)
[17] Cazarez, N., Castillo, O., Tupak, L.: Estabilidad en Sistemas de Control Difuso Tipo-2, Instituto Tecnologico de Tijuana, Octubre (2005)
[18] Cardenas, S., Castillo, O., Aguilar, L., Cazarez, N.: Intelligent Control of Dynamical Systems with Type-2 Fuzzy Logic and Stability Study. In: Proceedings of the International Conference on Artificial Intelligence (IC' AI' 2005), Las Vegas, USA, June 27-30 (2005)
[19] Cardenas, S., Castillo, O., Aguilar, L., Cazarez, N.: Tracking Control For Unicycle Mobile Robot Using A Fuzzy Logic Controller. In: Proceedings of the International Conference on Fuzzy Logic, Neural Networks and Genetic Algorithms 2005 (FNG 2005), Tijuana, Mexico (2005)
[20] Cazarez, N.: Estabilidad en Sistemas de Control con Logica Difusa Tipo-2. Boletin del IEEE-CIS-Mexico (Sociedad de Computacion Inteligente de la IEEE, section Mexico) 1(2), 2 (2005)
[21] Cazarez, N., Castillo, O., Aguilar, L., Cardenas, S.: From Type-1 to Type-2 Fuzzy Logic Control: A Stability and Robustness Study. In: Proceedings of the International Conference on Fuzzy Logic, Neural Networks and Genetic Algorithms 2005 (FNG 2005), Tijuana, Mexico (2005)
[22] Cazarez, N., Castillo, O., Aguilar, L., Cardenas, S.: Lyapunov Stability on Type-2 Fuzzy Logic Control. In: Proceedings of the IEEE International Seminar on Computation Intelligence 2005 (IEEE-ISCI 2005), Mexico Distrito Federal, Mexico, 17–18 de Octubre del (2005)

Comparative Study of Type-1 and Type-2 Fuzzy Systems Optimized by Hierarchical Genetic Algorithms

Alma I. Martinez, Oscar Castillo, and Mario Garcia

Division of Graduate Studies and Research, Tijuana Institute of Technology,
Tijuana, Mexico

Abstract. This paper describes a comparative study of type-1 and type-2 fuzzy controllers that are optimized using hierarchical genetic algorithms. Fuzzy controllers of Sugeno and Mamdani form are studied. The hierarchical genetic algorithms optimize the membership functions and the rules of the fuzzy controllers.

Keywords: Mamdani Fuzzy Inference System, Sugeno Fuzzy Inference System, Interval Type-2 Fuzzy Inference System, Hierarchical Genetic Algorithms.

1 Introduction

Fuzzy logic is an area of great interest now, not only in terms of its theory scope but also by the variety of applications that can be made with the techniques of this area of computing. In recent years, systems based on fuzzy rules have become one of the main applications of fuzzy models [1]. This paper presents an analysis of the application of genetic algorithms to optimize fuzzy control systems, to collect behavior data about the fuzzy system without optimization and compare them with data from the optimized version of every system built.

This work consist of building different fuzzy systems, to solve the same problem, Mamdani fuzzy systems and Sugeno fuzzy systems, using Type-1 and Type-2 fuzzy logic in all cases. Once the fuzzy systems are built they also optimized through hierarchical genetic algorithms, information is gathered and a comparison is made of fuzzy systems and their optimized versions of its with hierarchical genetic algorithms. Currently, there is an increasing interest to augment fuzzy systems with learning and adaptation capabilities. One of the most successful approaches to hybridize fuzzy systems with learning methods have been made in the realm of soft computing. Genetic fuzzy systems hybridize the approximate reasoning method of fuzzy systems with the learning capabilities of evolutionary computing. A genetic fuzzy system is basically a fuzzy systems augmented by a learning process based on a generic algorithm (GA). GAs are search algorithms, based on natural genetics, that provide a robust search capabilities in complex spaces, and thereby offer a valid approach to problems requiring efficient an effective search processes [2, 3, 4].

O. Castillo et al. (Eds.): Soft Computing for Hybrid Intel. Systems, SCI 154, pp. 53–70, 2008.
springerlink.com © Springer-Verlag Berlin Heidelberg 2008

2 Type-1 Fuzzy Set Theory

In contrast to a classical set, a fuzzy set, as the name implies, is a set without a crisp boundary. That is, the transition of "belong to a set" to "not belong to a set" is gradual, and this smooth transition is characterized by membership functions that give fuzzy sets flexibility in modeling commonly used linguistic expressions. Unlike the characteristic function of a classical set, a fuzzy set A, of the universe of discourse X, is characterized by a *membership function*

$$\mu_A : X \rightarrow [0,1] \tag{1}$$

This function associates every element x of X a number $\mu_A(x)$ on the interval $[0,1]$ represented by a membership grade of x in A.

$$\forall x \in X, \mu_A(x) = \begin{cases} y & if \quad x \in A \\ 0 & if \quad x \notin A \end{cases} \tag{2}$$

Where y can take any value on the interval $(0,1]$.

Therefore we can say that the definition of a classical set, according to equation (2), is a special case in the fuzzy set because the characteristic function can take only two values $\{0,1\}$, and the membership function of a fuzzy set covers any possible value in the interval $[0,1]$.

Analyzing equation (2), we have that if X is a collection of objects already generically denoted by x, then a fuzzy set A in X is defined like a ordered pair set:

$$A = \{x, \mu_A(x) x \in X\} \tag{3}$$

Where $\mu_A(x)$ is called membership function of the set A. The membership functions map every element of X in a membership grade between 0 and 1.

The membership functions can represent discrete universes, or continuous universes [11]. If a fuzzy set A has a finite number of elements $\{x_1, x_2, ..., x_n\}$ then the equation (3) can be represented by the summation:

$$A = \mu_1 + \mu_2 + ... + \mu_n / x_n \tag{4}$$

Or

$$A = \sum_{i=1}^{n} \mu_i / x_i \tag{5}$$

Where summation sign indicates the individual values union and it is considered a discrete discourse universe.

On continuous discourse universe, the fuzzy set A can be represented as:

$$A = \int_X \mu_A(x)/x \qquad (6)$$

Where the integral indicates the union of the fuzzy individual values. Both the equation 5 and 6, the symbol "/" is only a separator and does not involve a division.

3 Type-2 Fuzzy Set Theory

A type-2 fuzzy set [26, 27] expresses the non-deterministic truth degree with imprecision and uncertainty for an element that belongs to a set. A type-2 fuzzy set denoted by $\tilde{A}$, is characterized by a type-2 membership function $\mu_{\tilde{A}}(x,u)$, where $x \in X$, $u \in J_x^u \subseteq [0,1]$ and $0 \leq \mu_{\tilde{A}}(x,u) \leq 1$ is defined in equation 7.

$$\tilde{A} = \left\{ \left(x, \mu_{\tilde{A}}(x) \right) \big| x \in X \right\} = \left\{ \int_{x \in X} \left[\int_{u \in J_x^u \subseteq [0,1]} f_x(u)/u \right] / x \right\} \qquad (7)$$

An example of a type-2 membership function constructed in the IT2FLS Toolbox developed by our search group [28] is shown in Fig. 1 was composed by a lower and upper type-1 membership functions, and then is called Interval type-2 fuzzy set is denoted by equation 8.

$$\tilde{A} = \left\{ (x,u,1) \big| \forall x \in X, \forall u \in \left[\underline{\mu}_{\tilde{A}}(x), \overline{\mu}_{\tilde{A}}(x) \right] \subseteq [0,1] \right\} \qquad (8)$$

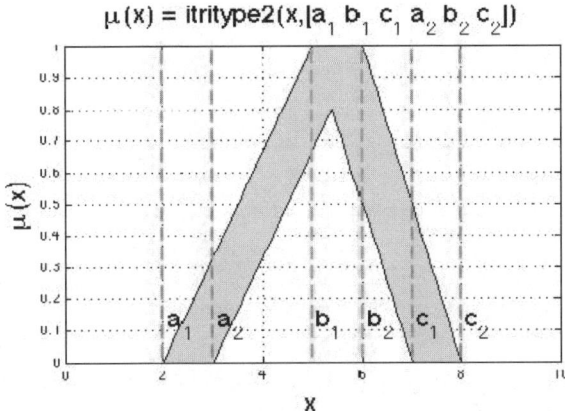

Fig. 1. Triangular Interval Type-2 Membership Function

4 Fuzzy Inference Systems

The fuzzy inference system is a popular computing framework based on the concepts of fuzzy set theory, fuzzy if-then rules, and fuzzy reasoning. We show in Figure 2 the structure of a fuzzy inference system.

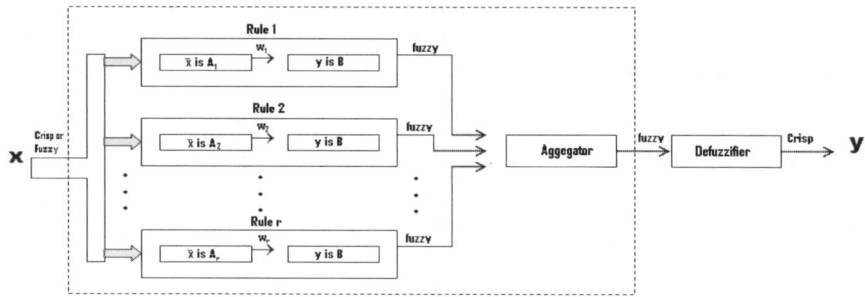

Fig. 2. Block diagram for a fuzzy inference system

4.1 Mamdani Fuzzy Inference System

The Mamdani fuzzy inference system [12] was proposed as the first attempt to control a steam and boiler combination by a set of linguistic control rules obtained from experienced human operators.

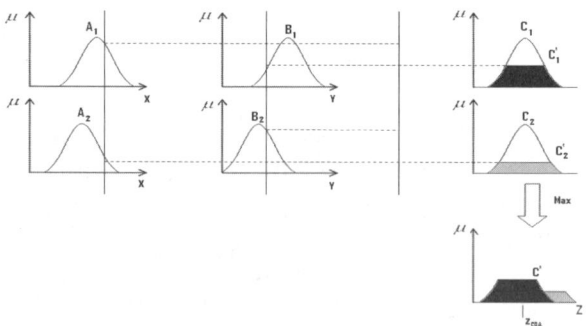

Fig. 3. The Mamdani fuzzy inference system using min and max for T-norm and T-conorm operators, respectively.

Figure 3 shows an illustration of how a two-rule Mamdani fuzzy inference system derives the overall output z when subjected to two crisp inputs x and y. Since the physical plant takes only crisp values as inputs, we have to use a defuzzifier to convert a fuzzy set to a crisp value.

The most widely adopted defuzzification strategy is:

Centroid of area z_{COA} :

$$^zCOA = \frac{\int_z \mu A(Z)Z \ dZ}{\int_z \mu A(Z)Z \ dZ},$$

(9)

Where $\mu A(Z)$ is aggregated output MF. This strategy is reminiscent of the calculation of expected values in probability distributions.

4.2 Sugeno Fuzzy Model

The Sugeno fuzzy model (also known as the TSK fuzzy model) was proposed by Takagi, Sugeno, and Kang [13, 14] in an effort to develop a systematic approach to generating fuzzy rules from a given input-output data set. A typical fuzzy rule in a Sugeno fuzzy model has the form

$$\text{If } x \text{ is } A \text{ and } y \text{ is } B \text{ then } z = f(x, y),$$

(10)

Where A and B are fuzzy sets in the antecedent, while $z = f(x, y)$ is a crisp function in the consequent.

Figure 4 shows the fuzzy reasoning procedure for a Sugeno fuzzy model.

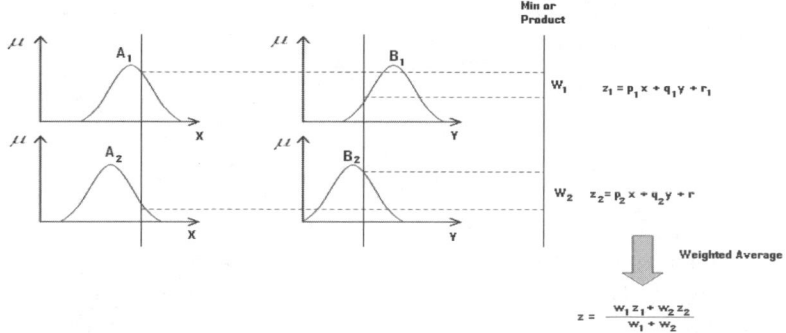

Fig. 4. The Sugeno Fuzzy Model

5 Hierarchical Genetic Algorithms

Genetic Algorithms are general purpose search algorithms which use principles inspired by natural evolution to generate solutions to problems [19, 20, 21].

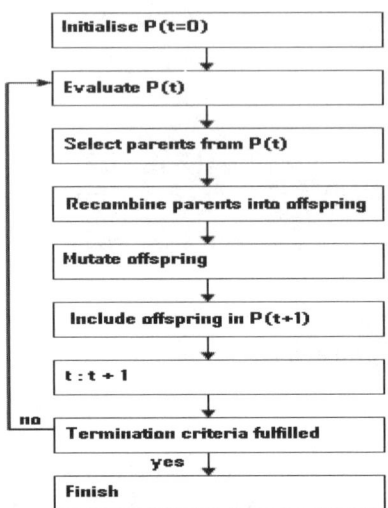

Fig. 5. Principal Structure of a genetic algorithm

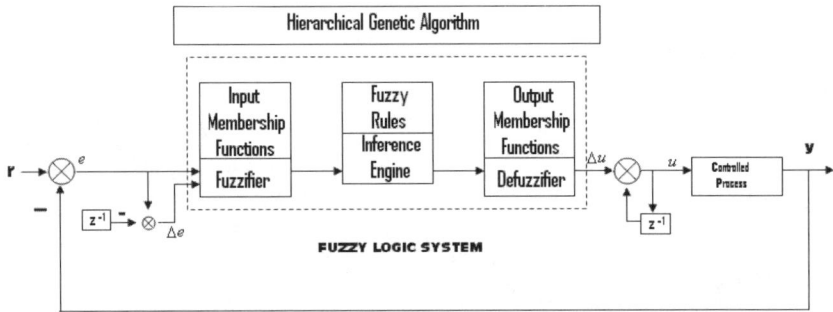

Fig. 6. HGA fuzzy logic control system

A GA starts with a population of randomly generated *chromosomes*, and advances towards better chromosomes by applying genetic operators modeled on the genetic processes occurring in nature. The population undergoes evolution in a form of natural selection. During successive iterations, called *generations*, chromosomes in the population are rated for their adaptation as solutions, and on the basis of these evaluations, a new population of chromosomes is formed using a selection mechanism and specific genetic operators such as *crossover* and *mutation*. An *evaluation* or *fitness function* must be devised for each problem to be solved. Given a particular chromosome, a possible solution, the fitness function returns a single numerical value, which is supposed to be proportional to the utility or adaptation of the solution represented by that chromosome. The operation mode if a GA is illustrated in Fig. 5.

To bring out the best use of the GA, we should explore further the study of genetic characteristics so that we can fully understand that the GA can be used with the approach of the hierarchical genetic structure for engineering purposes. The advantages that can be obtained from this method for solving typical fuzzy systems topology designs are given below.

The operational procedure of the fuzzy logic controller (FLC) examines the receiving input variables e and Δe in a fuzzzifying manner so that an appropriate actuating signal is derived to drive the system control input (u) in order to meet the ultimate goal of control [30].

Considering that the main attribute of the HGA is its ability to solve the topological structure of an unknown system, then the problem of determining the fuzzy membership functions and rules could also fall into this category.

The conceptual idea is to have an automatic and intelligent scheme to tune the fuzzy membership functions and rules, in which the closed loop fuzzy control strategy remains unchanged, as indicated in Fig. 6.

6 Simulation Results

We describe in this section the benchmark problem that will be considered to test the genetic algorithm optimization.

The problem to solve is the following: "We place a ball on a beam where it's allowed to roll with certain liberty along the beam, adding a lever arm and a servo-gear at one of the ends of the beam. As the servo-gear rotates an angle θ, the lever arm changes its angle α. When the angle trends to a vertical position the gravity makes the ball roll along the beam."

The simulink control scheme to test the different fuzzy inference systems is shown in Fig. 7.

We present results of a comparative analysis of the ball and beam problem, using Mamdani Type-1 fuzzy system, Optimized Mamdani Type-1 fuzzy system, Sugeno Type-2 fuzzy system and Optimized Sugeno Type-2 fuzzy system, respectively. The measure of the error is given by equation (11).

$$IAE = \frac{\sum_{i=1}^{n} |r_i - u_i|}{n} \tag{11}$$

Where: r_i = Reference value in each sampled point

u_i = Control value in each sampled point

n = Total sampled points

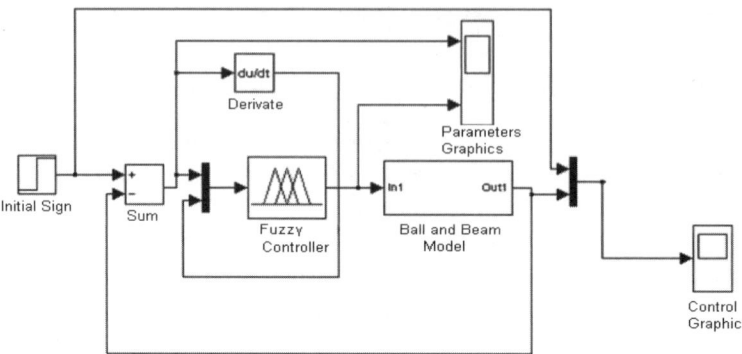

Fig. 7. Ball and beam simulink control scheme

6.1 Type-1 Mamdani Fuzzy Systems

In this section we present the simulation results of the type-1 Mamdani fuzzy systems, both with and without optimization.

6.1.1 Generic Type-1 Mamdani FIS

The Generic Type-1 Mamdani FIS, its inputs, output and the set of rules after the optimization, are shown as follows. Then we show the results of the simulation and the error achieved. In Figure 8 we show the structure of fuzzy system and the membership functions. In Figure 9 we show the fuzzy rules, and in Figure 10 a simulation of the fuzzy controller.

Figure 10 shows the result of the Generic Type-1 Mamdani FIS, this fuzzy system, achieved a **0.1406** error.

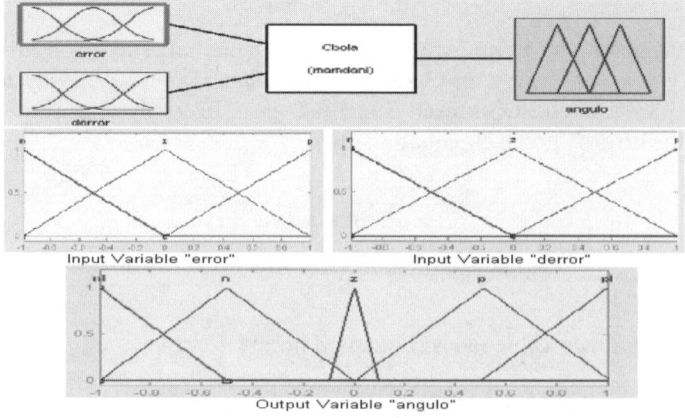

Fig. 8. Generic Type-1 Mamdani FIS

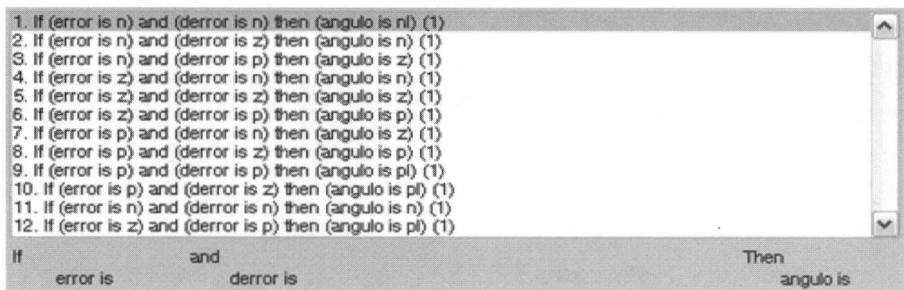

Fig. 9. Generic Type-1 Mamdani FIS, Rule Base

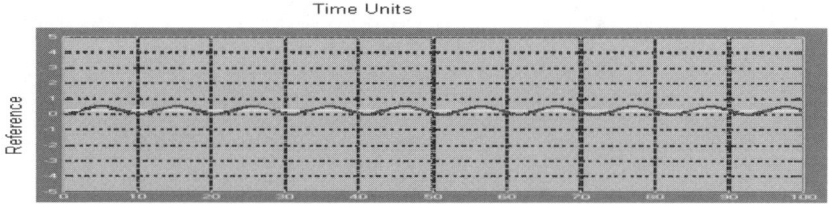

Fig. 10. Generic Type-1 Mamdani FIS, Simulation

6.1.2 Optimized Type-1 Mamdani FIS

To optimize the Generic Type-1 Mamdani FIS show after, we used Hierarchical Genetic Algorithms for tuning the membership functions and activate/deactivate the rules. The chromosome used in this case is shown in Figure 11

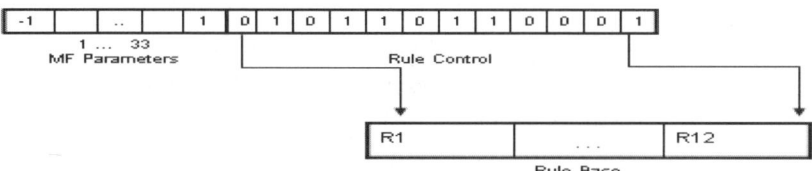

Fig. 11. Type-1 Mamdani Chromosome

The Optimized Type-1 Mamdani FIS, and its inputs, output and the set of rules before the optimization, are shown as follows. Then we show the table of results, the simulation of the best result and the error achieved (Table 1). Each row of the table is a different test, and each test was performed 5 times to obtain average results.

Figure 14 shows the result of the Optimized Type-1 Mamdani FIS, this fuzzy system, achieved a **0.061491** error.

Table 1. Average results for Optimized Type-1 Mamdani FIS

INDIVIDUALS	GENERATIONS	GGAP	CROSSOVER	MUTATION	BEST ERROR	AVERAGE ERROR	STANDARD DESVIATION
30	50	0.7	0.7	0.05	0.25	0.254	0.001472
15	50	0.3	0.4	0.1	0.071032	0.04111	0.06264779
90	50	0.5	0.2	0.01	0.02358	0.02548	0.01214
15	50	0.6	0.8	0.2	0.156	0.194	0.00147
20	50	0.3	0.8	0.2	0.1236	0.1354	0.129
35	50	0.4	0.8	0.2	0.235	0.2145	0.01478
40	50	0.6	0.8	0.2	0.142	0.14659	0.00201
50	70	0.5	0.8	0.2	0.1781	0.1658	0.10564
80	30	0.5	0.6	0.02	0.11176	0.10248	0.011423
80	70	0.1	0.02	0.03	0.082977	0.08265	0.3259
80	70	0.6	0.7	0.05	0.07551	0.07925	0.2589
80	70	0.8	0.6	0.01	0.085293	0.07959	0.00527
80	70	0.5	0.02	0.3	0.082977	0.08425	0.005489
90	50	0.3	0.2	0.01	0.042148	0.04628	0.325
100	30	0.4	0.6	0.05	0.094599	0.10002	0.0147
100	70	0.5	0.7	0.1	0.0842	0.0842	0.0025449
70	80	0.3	0.4	0.001	0.1058	0.1098	0.001254
60	25	0.7	0.6	0.05	0.098705	0.1056	0.00362853
60	80	0.7	0.6	0.05	0.1007	0.1056	0.00287792
80	89	0.5	0.7	0.1	0.081853	0.0826	0.002359
60	25	0.3	0.8	0.08	0.11216	0.1125	0.005416
15	**50**	**0.3**	**0.4**	**0.1**	**0.061491**	**0.06256**	**0.002183**
10	60	0.5	0.6	0.1	0.13024	0.1254	0.002659
20	70	0.3	0.8	0.05	0.11117	0.10256	0.001428
30	18	0.3	0.3	0.05	0.10866	0.10958	0.0032658
20	80	0.2	0.8	0.1	0.10131	0.09584	0.0016528
40	80	0.4	0.6	0.05	0.10557	0.10568	0.004069
30	50	0.4	0.3	0.01	0.25	0.235	0.002168

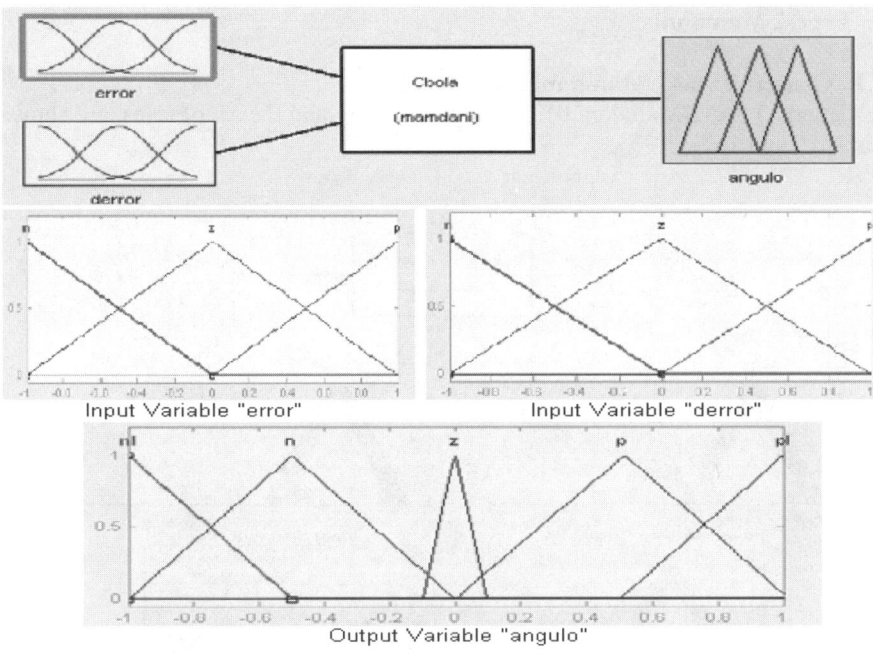

Fig. 12. Optimized Type-1 Mamdani FIS

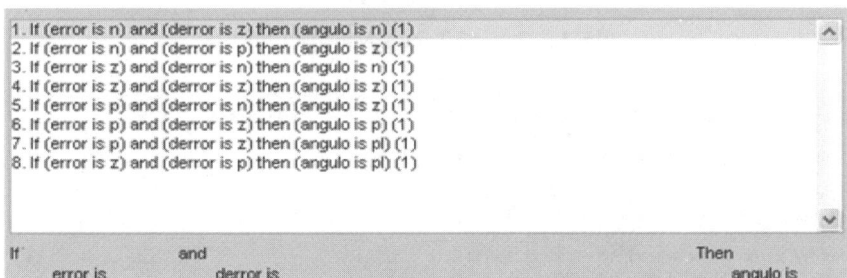

1. If (error is n) and (derror is z) then (angulo is n) (1)
2. If (error is n) and (derror is p) then (angulo is z) (1)
3. If (error is z) and (derror is n) then (angulo is n) (1)
4. If (error is z) and (derror is z) then (angulo is z) (1)
5. If (error is p) and (derror is n) then (angulo is z) (1)
6. If (error is p) and (derror is z) then (angulo is p) (1)
7. If (error is p) and (derror is z) then (angulo is pl) (1)
8. If (error is z) and (derror is p) then (angulo is pl) (1)

Fig. 13. Optimized Type-1 Mamdani, fuzzy rules

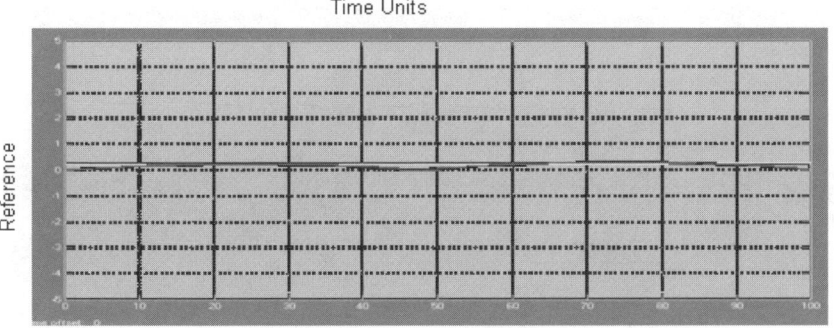

Fig. 14. Optimized Type-1 Mamdani, Simulation

6.2   Type-2 Mamdani

6.2.1   Generic Type-2 Mamdani FIS

The Generic Type-2 Mamdani FIS, its inputs, output and the set of rules, are shown in Figures 15 and 16.

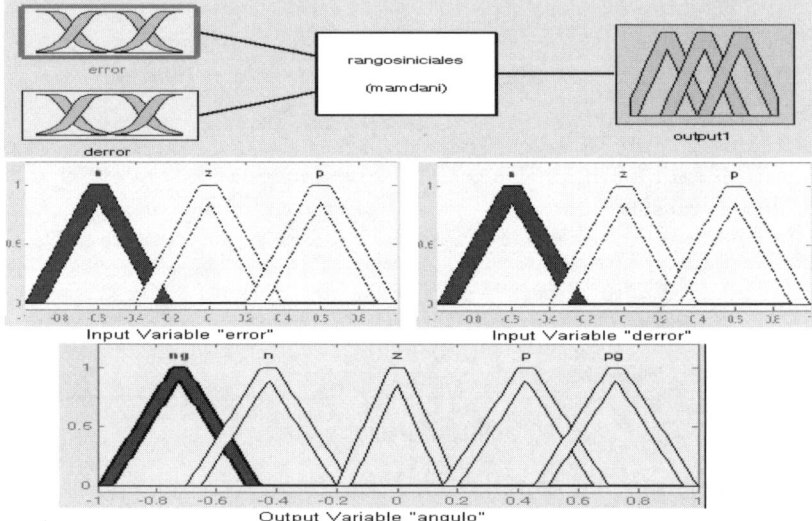

Fig. 15. Generic Type-2 Mamdani FIS

1. If (error is n) and (derror is n) then (output1 is ng) (1)
2. If (error is n) and (derror is z) then (output1 is n) (1)
3. If (error is n) and (derror is p) then (output1 is z) (1)
4. If (error is z) and (derror is n) then (output1 is n) (1)
5. If (error is z) and (derror is z) then (output1 is z) (1)
6. If (error is z) and (derror is p) then (output1 is p) (1)
7. If (error is p) and (derror is n) then (output1 is z) (1)
8. If (error is p) and (derror is z) then (output1 is p) (1)
9. If (error is p) and (derror is p) then (output1 is pg) (1)
10. If (error is p) and (derror is z) then (output1 is pg) (1)
11. If (error is n) and (derror is n) then (output1 is n) (1)
12. If (error is z) and (derror is p) then (output1 is pg) (1)

If	and	Then
error is	derror is	output1 is

Fig. 16. Generic Type-2 Mamdani FIS, Rule Base

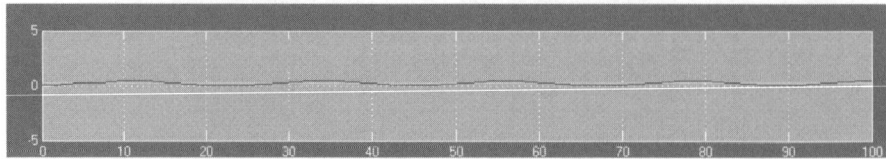

Fig. 17. Generic Type-2 Mamdani FIS, Simulation

Figure 17 shows the result of the Generic Type-2 Mamdani FIS, this fuzzy system, achieved a **0.1116** error.

6.2.2 Optimized Type-2 Mamdani FIS

To optimize the Generic Type-2 Mamdani FIS show after, we used Hierarchical Genetic Algorithms to tuning the membership functions and activate/deactivate the rules. The chromosome used in this case is shown in Figure 18

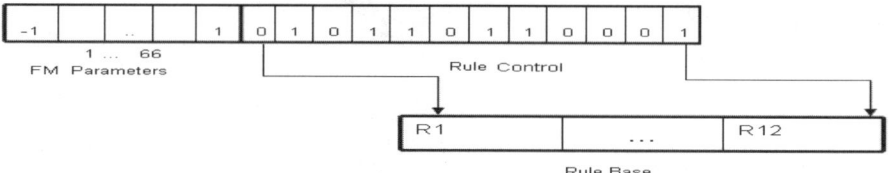

Fig. 18. Type-2 Mamdani Chromosome

The Optimized Type-2 Mamdani FIS, and its inputs, output and the set of rules before the optimization, are shown as follows. Then we show the table of results, the simulation of the best result and the error achieved (Table 2). Each row of the table is a different test, and each test was performed 5 times to obtain average results. Figures 19 and 20 show the optimal type-2 controller.

Table 2. Average Results for Optimized Type-2 Mamdani FIS

INDIVIDUALS	GENERATIONS	GGAP	CROSSOVER	MUTATION	BEST	BEST ERROR	AVERAGE ERROR	STANDARD DESVIATION
100	70	0.5	0.7	0.1	gen71 ind45	0.0842	0.0842	0.0025449
70	80	0.3	0.4	0.001	gen81 ind1	0.1058	0.1098	0.001254
60	25	0.7	0.6	0.05	gen26 ind34	0.098705	0.1056	0.00362853
60	80	0.7	0.6	0.05	gen81 ind38	0.1007	0.1056	0.00287792
80	89	0.5	0.7	0.1	gen89 ind26	0.081853	0.0826	0.002359
60	25	0.3	0.8	0.08	gen26 ind8	0.11216	0.1125	0.005416
15	**50**	**0.3**	**0.4**	**0.1**	**gen51 ind4**	**0.061491**	**0.06256**	**0.002183**
10	60	0.5	0.6	0.1	gen61 ind2	0.13024	0.1254	0.002659
20	70	0.3	0.8	0.05	gen71 ind5	0.11117	0.10256	0.001428
30	18	0.3	0.3	0.05	gen19 ind1	0.10866	0.10958	0.0032658
20	80	0.2	0.8	0.1	gen81 ind3	0.10131	0.09584	0.0016528
40	80	0.4	0.6	0.05	gen81 ind11	0.10557	0.10568	0.004069
30	50	0.4	0.3	0.01	gen51 ind6	0.25	0.235	0.002168
30	50	0.7	0.7	0.05	gen51 ind13	0.25	0.254	0.001472
15	50	0.3	0.4	0.1	gen51 ind13	0.071032	0.04111	0.06264779
90	50	0.5	0.2	0.01	gen51 ind13	0.02358	0.02548	0.01214
15	50	0.6	0.8	0.2	gen51 ind8	0.156	0.194	0.00147
20	50	0.3	0.8	0.2	gen51 ind10	0.1236	0.1354	0.129
35	50	0.4	0.8	0.2	gen51 ind1	0.235	0.2145	0.01478
40	50	0.6	0.8	0.2	gen51 ind3	0.142	0.14659	0.00201
50	70	0.5	0.8	0.2	gen71 ind2	0.1781	0.1658	0.10564
80	30	0.5	0.6	0.02	gen28 ind16	0.11176	0.10248	0.011423
80	70	0.1	0.02	0.03	gen70 ind3	0.082977	0.08265	0.3259
80	70	0.6	0.7	0.05	gen71 ind28	0.07551	0.07925	0.2589
80	70	0.8	0.6	0.01	gen71ind20	0.085293	0.07959	0.00527
80	70	0.5	0.02	0.3	gen71 ind10	0.082977	0.08425	0.005489
90	50	0.3	0.2	0.01	gen51 ind4	0.042148	0.04628	0.325
100	30	0.4	0.6	0.05	gen31 ind7	0.094599	0.10002	0.0147

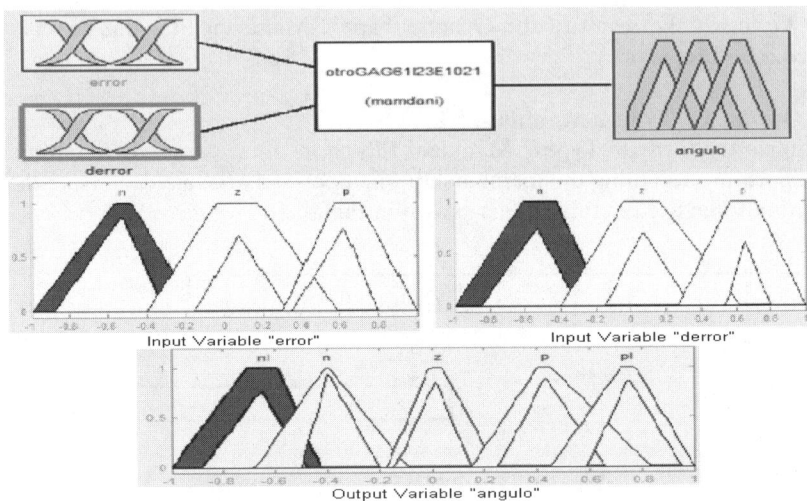

Fig. 19. Optimized Type-2 Mamdani FIS

1. If (error is n) and (derror is n) then (angulo is nl) (1)
2. If (error is n) and (derror is z) then (angulo is n) (1)
3. If (error is z) and (derror is z) then (angulo is z) (1)
4. If (error is z) and (derror is p) then (angulo is p) (1)
5. If (error is p) and (derror is n) then (angulo is z) (1)
6. If (error is p) and (derror is p) then (angulo is pl) (1)
7. If (error is p) and (derror is z) then (angulo is pl) (1)

If	and	Then
error is	derror is	angulo is

Fig. 20. Optimized Type-2 Mamdani FIS, Rule Base

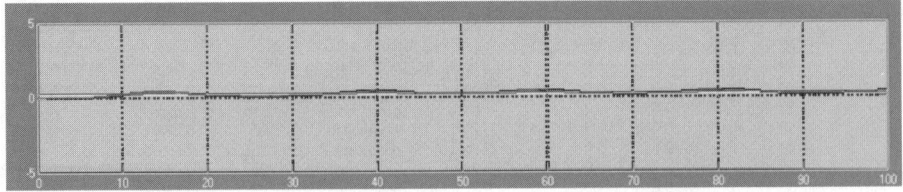

Fig. 21. Optimized Type-2 Mamdani FIS, Simulation

Figure 21 shows the result of the Optimized Type-2 Mamdani FIS, this fuzzy system, achieved a **0.099384** error.

6.3 Type-1 Sugeno

6.3.1 Generic Type-1 Sugeno

The Generic Type-1 Sugeno FIS, its inputs, output and the set of rules, are shown as follows in Figures 22 and 23.

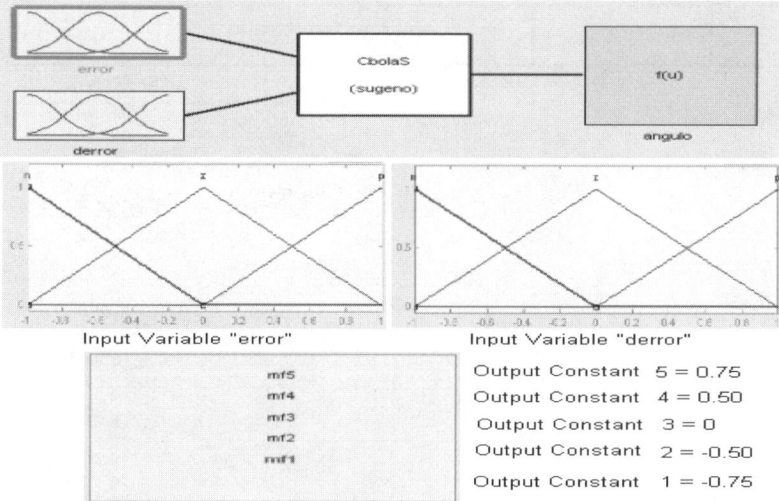

Fig. 22. Generic Type-1 Sugeno FIS

1. If (error is n) and (derror is n) then (angulo is mf1) (1)
2. If (error is n) and (derror is z) then (angulo is mf2) (1)
3. If (error is n) and (derror is p) then (angulo is mf3) (1)
4. If (error is z) and (derror is n) then (angulo is mf2) (1)
5. If (error is z) and (derror is z) then (angulo is mf3) (1)
6. If (error is z) and (derror is p) then (angulo is mf4) (1)
7. If (error is p) and (derror is n) then (angulo is mf3) (1)
8. If (error is p) and (derror is z) then (angulo is mf4) (1)
9. If (error is p) and (derror is p) then (angulo is mf5) (1)
10. If (error is p) and (derror is z) then (angulo is mf5) (1)
11. If (error is n) and (derror is n) then (angulo is mf2) (1)
12. If (error is z) and (derror is p) then (angulo is mf5) (1)

If	and	Then
error is	derror is	angulo is

Fig. 23. Generic Type-1 Sugeno FIS, Rule Base

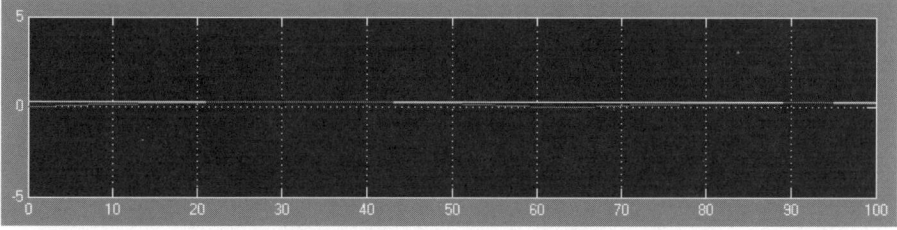

Fig. 24. Generic Type-1 Sugeno FIS, Simulation

Figure 24 shows the result of the Generic Type-1 Sugeno FIS, this fuzzy system, achieved a **0.092009** error.

6.3.2 Optimized Type-1 Sugeno FIS

Figures 25, 26 and 27 describe the type-1 Sugeno fuzzy system.

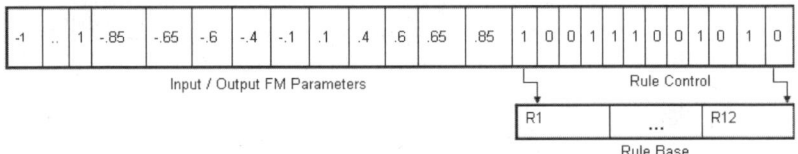

Fig. 25. Type-1 Sugeno Chromosome

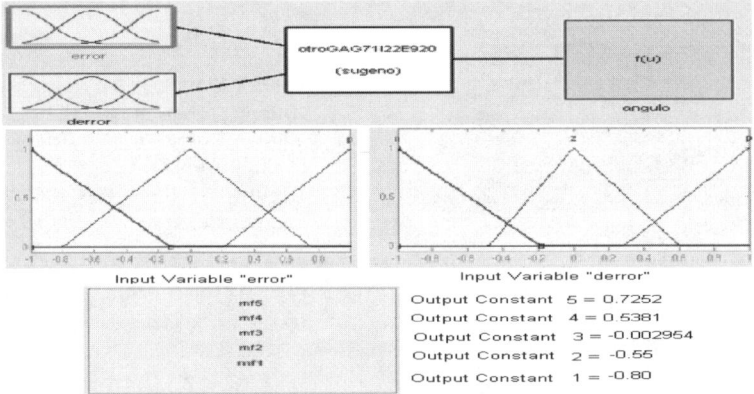

Fig. 26. Optimized Type-1 Sugeno FIS

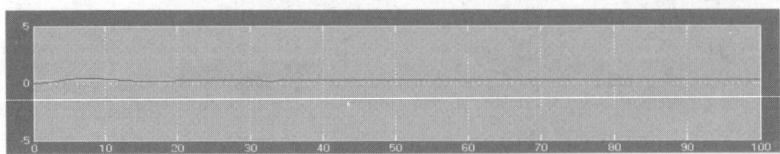

Fig. 27. Optimized Type-1 Sugeno FIS, Rule Base

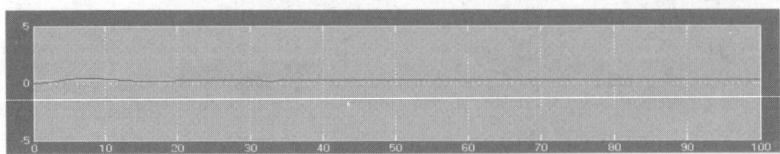

Fig. 28. Optimized Type-1 Sugeno FIS, Simulation

Figure 28 shows the result of the Generic Type-1 Sugeno FIS, this fuzzy system, achieved a **0.048685** error.

7 Conclusions

In this paper we applied tests on different fuzzy systems, and the results show that when using the Mamdani model, we will get an error higher than when using the Sugeno model; also that by using type-2 we will get an even lower error, in this case we show evidence for Type-1 and Type-2 Mamdani models and evidence only for Type-1 Sugeno Model. With the results, we can conclude that Sugeno model helps maintain better control. Genetic algorithms help to optimize the system to outperform those that might be find by the developers by trial and error, since the user experience serves to construct a general outline of control but the parameters of the membership functions and the number of rules that the system needs to function properly are very difficult to find.

References

[1] Sepulveda, R., Montiel, O., Castillo, O., Melin, P.: Fundamentos de Logica Difusa, Ediciones ILCSA 2002 (2002)
[2] Goldberg, D.E.: Genetic Algorithms in Search, Optimization, and Machine Learning. Addison-Wesley, Reading (1989)
[3] Goldberg, D.E.: The Design of Competent Genetic Algorithms: Steps Toward a Computational. In: Theory of Innovation. Kluwer Academic Publishers, Dordrecht (2002)
[4] Holland, J.H.: Adaptation in Natural and Artificial Systems. University of Michigan Press, Ann Arbor (1975)
[5] De Jong, K.: Learning with genetic algorithms: an overview. Mach. Learning 3(3), 121–138 (1988)
[6] Zadeh, L.A.: Fuzzy Sets. Information and Control 8, 338–353 (1965)
[7] Zadeh L.A,: A definition of Soft Computing-adapted for L. A. Zadeh (2002), http://www.soft-computing.de/def.html
[8] Kalman, R.E.: A new approach to linear filtering and prediction problems. Journal of Basic Engineering, 35–45 (march 1960)
[9] Sugeno, M., Tang, G.T.: Structure identification of fuzzy model. Fuzzy Sets and Systems 28, 15–33 (1988)
[10] Takagi, T., Sugeno, M.: Fuzzy identification of systems and its applications to modeling and control. IEEE Transactions on Systems, Man, and Cybernetics 15, 116–132 (1985)
[11] Fogel, L.J.: Artificial Intelligence through Simulated Evolution. John Wiley, New York (1966)
[12] Fogel, L.J.: Artificial Intelligence through Simulated Evolution. In: Forty Years of Evolutionary Programming. John Wiley & Sons, Inc., New York (1999)
[13] Bäck, T.: Evolutionary Algorithms in Theory and Practice. Oxford University Press, New York (1996)
[14] Cello Cello, A.C.: Introduction a la Computación Evolutiva, Notas de Curso. CINVESTAV-IPN -Departamento de Ingeniería Eléctrica - Sección de Computacion (April 2006)

[15] Goldberg, D.E.: Genetic Algorithms in Search, Optimization, and Machine Learning. Addison-Wesley, Reading (1989)
[16] Goldberg, D.E.: The design of Competent Genetic Algorithms: Steps Toward a Computational Theory of Innovation. Kluwer Academic Publishers, Dordrecht (2002)
[17] Holland, J.H.: Adaptation in Natural and Artificial Systems. University of Michigan Press, Ann Arbor (1975)
[18] Ishibuchi, H., Murata, T., Türksen, I.B.: Single-objective and two objective genetic algorithms for selecting linguistic rules for pattern classification problems. Fuzzy Sets and Systems 89, 135–150 (1997)
[19] Ishibuchi, H., Nakashima, T., Murata, T.: Performance evaluation of fuzzy classifier systems for multidimensional pattern classification problems. IEEE Trans. System Man Cybernet 29, 601–618 (1999)
[20] Jin, Y.: Fuzzy modeling of high-dimensional systems: complexity reduction and interpretability improvement. IEEE Trans. Fuzzy Systems 8(2), 212–220 (2000)
[21] Castillo, L., González, A., Pérez, R.: Including a simplicity criterion in the selection of the best rule in a fuzzy genetic learning algorithm. Fuzzy Sets and Systems 120(2), 309–321 (2001)
[22] Karnik, N.N., Mendel, J.M.: An Introduction to Type-2 Fuzzy Logic Systems, Univ. of Southern Calif., Los Angeles, CA (June 1998b)
[23] Zadeh, L.: Fuzzy Logic = computing with words. IEEE Transactions on Fuzzy Systems 2, 103–111 (1996)
[24] Castro, J.R.: Interval Type-2 Fuzzy Logic Toolbox In: Proceedings of International Seminar on Computational Intelligence, pp. 100-108. Tijuana Institute of Technology (October 2006)
[25] Fuzzy Sets and Systems. An international journal in information science and engineering 141(1) (January 2004)
[26] Man, K.F., Tang, K.S., Kwong, S.: Genetic Algorithms, concepts and designs. Springer, London (1999)

Comparison between Ant Colony and Genetic Algorithms for Fuzzy System Optimization

Cristina Martinez[1], Oscar Castillo[1], and Oscar Montiel[2]

[1] Tijuana Institute of Technology, [2] CITEDI-IPN
  Tijuana, Mexico

Abstract. In this paper we show some of the results that we obtain with different evolutionary methods on a Mamdani Fuzzy Inference System (FIS); we work with Hierarchical Genetic Algorithms (HGA) and the Ant Colony Optimization (ACO), the fuzzy inference system controls a benchmark problem which is "The Ball and Beam" system, optimizing the fuzzy rules of the system. Firs, we work to optimize the FIS that is structured by two inputs (the error and the derived error), an output (the angle of the beam so that we can get the ball position on it); and the 44 fuzzy rules that we used to be reduced with the evolutionary methods (HGA, ACO), so that we could make the comparisons between them via average and standard deviation, and concluding with the best evolutionary method for a fuzzy system optimization control problem.

1 Introduction

A large number of papers have been published regarding the combination of fuzzy logic (FL) and genetic algorithms (GA's) [24, 25 and 26] and also fuzzy logic and ant colony optimization (ACO) [23 and 26]. Fuzzy logic is a useful tool for modeling complex systems and deriving useful fuzzy relations or rules. However, it is often complicated for human experts to define the fuzzy sets and fuzzy rules used by these systems. GA's have proven to be a useful method for optimizing the membership functions of the fuzzy sets and the fuzzy rules used by these fuzzy systems; ACO has the ability to solve combinatorial optimization problems that has been inspired by the foraging behavior of ant colonies.

Most of the papers in optimization of fuzzy systems consider only the optimization of membership functions; for this reason, we decided to optimize the fuzzy rules of a fuzzy system using GA's and ACO. As we said before it is very difficult for human experts to define a fuzzy system even though we create a FIS based on two inputs one output using triangular membership function for all the fuzzy sets, 44 fuzzy rules (first we started at 25 then we choose to rebuild the initial fuzzy rules set and extend more the search space so the evolutionary algorithms would have more experimental space).

2 Genetic Algorithms

Our lives are essentially dominated by genes. They govern our physical features, our behavior, our personalities, our health, and indeed our longevity. The recent greater

O. Castillo et al. (Eds.): Soft Computing for Hybrid Intel. Systems, SCI 154, pp. 71–86, 2008.
springerlink.com   © Springer-Verlag Berlin Heidelberg 2008

understanding of genetic has proven to be a vital tool for genetic engineering applications in many disciplines, in addition to medicine and agriculture. It is well known that genes can be manipulated, controlled and even turned on and off in order to achieve desirable amino acid sequences of a polypeptide chain. This significant discovery has led to the use of genetic algorithms (GA) for computational engineering. GA has proven to be unique approach for solving various mathematical intangible problems which other gradient type of mathematical optimizers have failed to solve. The basic principles of the GA were first proposed by Holland [2]. Thereafter, a series of literature [3, 4, and 5] and reports [6, 7, 8, 9, 10, and 11] became available. GA is inspired by the mechanism of natural selection where stronger individuals are likely the winners in a competing environment. Here, the GA uses a direct analogy of such natural evolution. Through the genetic evolution method, an optimal solution can be found and represented by the final winner of the genetic game. The GA presumes that the potential solution of any problem is an individual and can be represented by a set of parameters. These parameters are regarded as the genes of a chromosome and can be structured by a string of values in binary form. A positive value, generally known as a fitness value, is used to reflect the degree of "goodness" of the chromosome for the problem which would be highly related with its objective value. Throughout a genetic evolution, the fitter chromosome has a tendency to yield good quality offspring, which means a better solution to any problem. In figure 1 we show the genetic GA cycle.

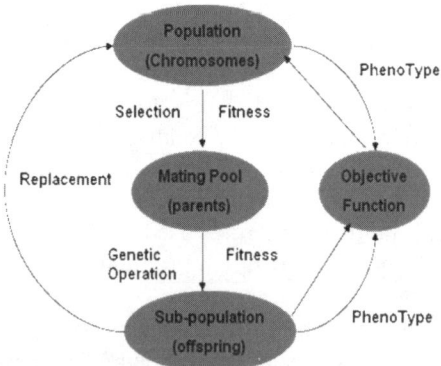

Fig. 1. A genetic algorithm cycle

The concept of applying a GA to solve engineering problems is feasible and sound. However, despite the distinct advantages of a GA for solving complicated, constrained and multi-objective functions where other techniques may have failed, the full power of the GA in engineering application is yet to be exploited and explored.

Hierarchical Genetic Algorithms (HGA) are known for the tree structure that generates, called "dendogram" in which every level is a set of possible solutions of the collection [7]. Each node of the tree (first level) is structured just by one set that contains all the elements. Each leaf of the last level of the tree is a set composed by one element (there are many leafs as objects on the collection). In the intermediate levels every node of the "n" level is divided for the offspring levels "n+1". This type of algorithms combine the notion of the fittest survival and the random and structured exchange of the characteristics between individuals of a population of possible solutions, conforming a search algorithm that applies for solving diverse optimization fields. We show in figure 2 an example of HGA with a chromosome of three levels.

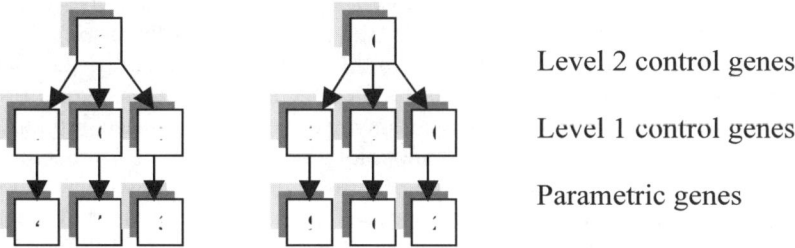

Level 2 control genes

Level 1 control genes

Parametric genes

Fig. 2. An example of 3-level chromosome

3   Ant Colony Optimization

Ant Colony Optimization (ACO) is part of a larger field of research termed based on social behavior of animals swarm intelligence that deals with algorithmic approaches. Swarm intelligence is a relatively new approach to problem solving that takes inspiration from the social behaviors of insects and of other animals. In particular, ants have inspired a number of methods and techniques among which the most studied and the most successful is the general purpose optimization technique known as ant colony optimization [12] – [14], [15], [16]. ACO takes inspiration from the foraging behavior of some ant species. These ants deposit pheromone on the ground in order to mark some favorable path that should be followed by other members of the colony. ACO exploits a similar mechanism for solving optimization problems.

   In ACO, the discrete optimization problem considered is mapped onto a graph called a *construction graph* in such a way that feasible solutions to the original problem correspond to paths on the construction graph. Then, artificial ants can generate feasible solutions by moving on the construction graph. In practice, colonies of artificial ants search for good solutions for several iterations. Every (artificial) ant of a given iteration builds a solution incrementally by taking several probabilistic decisions. The artificial ants that find a good solution mark their paths on the construction graph by putting some amount of pheromone on the edges of the path they followed. The ants in the next iteration are attracted by the pheromones, i.e., their decision probabilities are biased by the pheromones: in this way, they will have a higher probability of building paths that are similar to paths that correspond to good solutions.

   ACO has been applied successfully to a large number of difficult combinatorial optimization problems including traveling salesman problems, quadratic assignment problems, and scheduling problems, as well as to dynamic routing problems in telecommunication networks. Unfortunately, it is difficult to analyze ACO algorithms theoretically, the main reason being that they are based on sequences of random decisions (taken by a colony of artificial ants) that are usually not independent and whose probability distribution changes from iteration to iteration. Accordingly, most of the ongoing research in ACO is of an experimental nature, as this is also reflected by the content of most of papers published in the literature.

Deneubourg et al. [17] thoroughly investigated the pheromone laying and following behavior of ants. The model proposed by Deneubourg and co-workers for explaining the foraging behavior of ants was the main source of the inspiration for the development of ant colony optimization. In ACO, a number of artificial ants build solutions to the considered optimization problem at hand and exchange information on the quality of these solutions via a communication scheme that is reminiscent of the one adopted by real ants. Different ant colony optimization algorithms have been proposed. The original ant colony optimization algorithm is known as Ant-System [18]-[20] and was proposed in the early nineties. Since then, a number of other ACO algorithms have been introduced. All ant colony optimization algorithms share the same basic idea.

ACO has been formalized into a metaheuristic for combinatorial optimization problems by Dorigo and co-workers [21], [22]. A metaheuristic is a set of algorithmic concepts that can be used to define heuristic methods applicable to a wide set of different problems. In other words, a metaheuristic is a general-purpose algorithmic framework that can be applied to different optimization problems with relatively few modifications. In order to apply ACO to a given a combinatorial optimization problem, an adequate model is needed.

A model P=(S, Ω, f) of combinatorial optimization problem consists of:

- *A search space S defined over a finite set of discrete decision variables X_i, i=1,...,n;*

- *A set Ω of constraints among the variables; and*
- *An objective function f: S $\rightarrow$ R_0^+ to be minimized.*

The generic variable X_i takes values in $D_i = [v_i^1 ... v_i^{|D_i|}]$. A feasible solution s $\in$ S is a complete assignment of values to variables that satisfies all constraints in Ω. A solution s^* $\in$ S is called a global optimum if and only if: $f(s^*) \leq f(s)$ $\forall s \in S$.

The model of a combinatorial optimization problem is used to define the pheromone model of ACO. A pheromone value associated with each possible *solution component*; that is, with each possible assignment of a value to a variable. Formally, the pheromone value τ_{ij} is associated with the solution component c_{ij}, which consists of the assignment $X_i= v_i^j$ the set of all possible solution components is denoted by **C**.

In ACO, an artificial ant builds a solution by traversing the fully connected construction graph G_C(**V, E**), where **V** is a set of vertices and **E** is a set of edges. This graph can be obtained from a set of solution components **C** in two ways: components may be represented either by vertices and edges. Artificial ants move from a vertex along the edges of the graph, incrementally building a partial solution. Additionally, ants deposit a certain amount of pheromone on the components; that is, either on the vertices or on the edges that they traverse. The amount $\Delta\tau$ of pheromone deposited may depend on the quality of the solution found. Subsequent ants use the pheromone information as a guide toward promising regions of search space.

The Ant Colony Optimization Algorithm:
Set parameters, initialize pheromone trails
While termination condition not met **do**
    ContructAntSolutions
    ApplyLocalSearch (optional)
    UpdatePheromones
endwhile

4 Experiment Results

The problem we're about to describe is a benchmark in the control area called "The Ball and Beam System" and it's primarily based on the scheme shown in figure 3.

In this system we place a ball on a beam and it is allowed to roll with certain liberty along the beam, adding a lever arm and a servo-gear at one of the ends of the

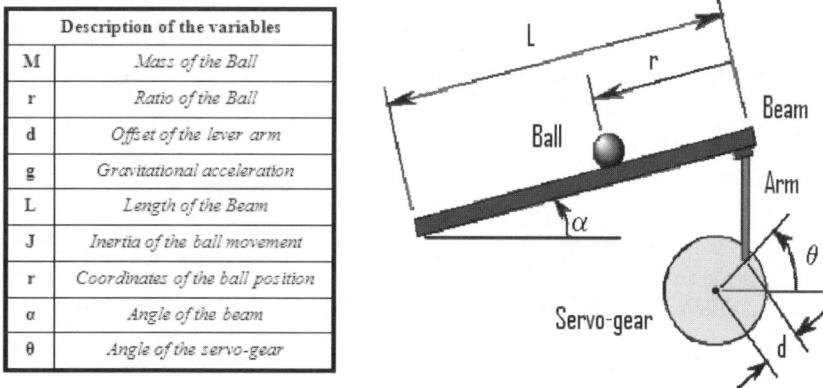

Description of the variables	
M	*Mass of the Ball*
r	*Ratio of the Ball*
d	*Offset of the lever arm*
g	*Gravitational acceleration*
L	*Length of the Beam*
J	*Inertia of the ball movement*
r	*Coordinates of the ball position*
α	*Angle of the beam*
θ	*Angle of the servo-gear*

Fig. 3. Scheme of the Ball and Beam system

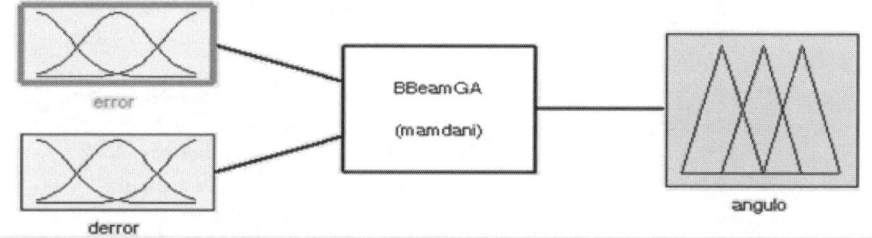

Fig. 4. Structure of the Fuzzy Controller

beam. As the servo-gear rotates an angle θ, the lever arm changes its angle α. When the angle trends to a vertical position the gravity makes the ball roll along the beam.

We have now to establish a knowledge base of fuzzy rules to be considered the initial fuzzy controller to be optimized by evolutionary methods. We have two inputs (error and change of error, Fig. 4) and one output (α angle) to know the position of the ball on the beam. We show in table 1 a set of initial fuzzy rules.

As we can see in figure 4 the error and the change of error as the inputs and the angle as the output, structure the fuzzy inference system of Mamdani type.

Table 1. Initial set of Fuzzy rules (Base of knowledge)

NO.	INDEXED	RULES
1	1 1 1 (1): 1	if error=NL and derror=NL then angulo=NL
2	1 2 1 (1): 1	if error=NL and derror=N then angulo=NL
3	1 3 1 (1): 1	if error=NL and derror=Z then angulo=NL
4	1 4 2 (1): 1	if error=NL and derror=P then angulo=N
5	1 5 3 (1): 1	if error=NL and derror=PL then angulo=Z
6	2 1 1 (1): 1	if error=N and derror=NL then angulo=NL.
7	2 2 1 (1): 1	if error=N and derror=N then angulo=NL
8	2 3 2 (1): 1	if error=N and derror=Z then angulo=N
9	2 4 3 (1): 1	if error=N and derror=P then angulo=Z
10	2 5 4 (1): 1	if error=N and derror=PL then angulo=P
11	3 1 1 (1): 1	if error=Z and derror=NL then angulo=NL
12	3 2 2 (1): 1	if error=Z and derror=N then angulo=N
13	3 3 3 (1): 1	if error=Z and derror=Z then angulo=Z
14	3 4 4 (1): 1	if error=Z and derror=P then angulo=P
15	3 5 5 (1): 1	if error=Z and derror=PL then angulo=PL
16	4 1 2 (1): 1	if error=P and derror=NL then angulo=N
17	4 2 3 (1): 1	if error=P and derror=N then angulo=Z
18	4 3 4 (1): 1	if error=P and derror=Z then angulo=P
19	4 4 5 (1): 1	if error=P and derror=P then angulo=PL
20	4 5 5 (1): 1	if error=P and derror=PL then angulo=PL
21	5 1 3 (1): 1	if error=PL and derror=NL then angulo=Z

22	5 2 4 (1): 1	if error=PL and derror=N then angulo=N
23	5 3 5 (1): 1	if error=PL and derror=Z then angulo=PL
24	5 4 5 (1): 1	if error=PL and derror=P then angulo=PL
25	5 5 5 (1): 1	if error=PL and derror=PL then angulo=PL
26	1 2 2 (1): 1	if error=NL and derror=N then angulo=NL
27	1 3 2 (1): 1	if error=NL and derror=Z then angulo=N
28	1 4 3 (1): 1	if error=NL and derror=P then angulo=Z
29	1 5 2 (1): 1	if error=NL and derror=PL then angulo=N
30	1 5 4 (1): 1	if error=NL and derror=PL then angulo=P
31	2 1 2 (1): 1	if error=N and derror=NL then angulo=N
32	2 2 2 (1): 1	if error=N and derror=N then angulo=N
33	2 5 3 (1): 1	if error=N and derror=PL then angulo=Z
34	3 1 2 (1): 1	if error=Z and derror=NL then angulo=N
35	3 2 3 (1): 1	if error=Z and derror=N then angulo=Z
36	3 5 4 (1): 1	if error=Z and derror=PL then angulo=P
37	4 1 3 (1): 1	if error=P and derror=NL then angulo=Z
38	4 4 4 (1): 1	if error=P and derror=Pthen angulo=P
39	4 5 4 (1): 1	if error=P and derror=PL then angulo=P
40	5 1 2 (1): 1	if error=PL and derror=NL then angulo=N
41	5 1 4 (1): 1	if error=PL and derror=NL then angulo=P
42	5 2 3 (1): 1	if error=PL and derror=N then angulo=Z
43	5 3 4 (1): 1	if error=PL and derror=Z then angulo=P
44	5 4 4 (1): 1	if error=PL and derror=P then angulo=P

4.1 Genetic Algorithm Experiments

Starting with the GA paradigm, we can apply a GA to generate different FIS' combining the rules of the knowledge base structured previously, with 44 fuzzy rules creating the search space for the evolutionary method. We based our implementation on the simple genetic algorithm, with some improvements to adequate it to our problem, and as a fitness function we have a function that evaluates as many Fuzzy Systems as individuals has the populations; they are created with the same inputs and outputs but every FIS has different active rules depending on the chromosome. We use the chromosome as an array of active rules; where 1 means "ON" and 0 means "OFF", as we can appreciate in figure 5 the chromosome representation in our GA.

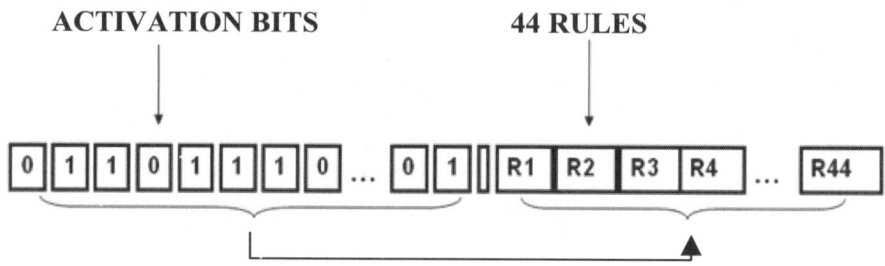

Fig. 5. Representation of the GA chromosome

Evaluation of the FIS is based on a simulation in order to get an average of the error obtained from the reference and the control given by the fuzzy controller. To have an average of the error we're using the Integral of the Absolute Error (IAE) equation:

$$\text{IAE} = \frac{\sum_{i=1}^{n}|\gamma_i - \mu_i|}{n} \quad \frac{\sum_{i=1}^{n}|\gamma_i - \mu_i|}{n} \tag{1}$$

Where γ_i marks the reference value, μ_i is the control value and n is the total sample points. We test the algorithm varying the mutation, crossover, with very low and very high levels and also the individuals, generations and the percentage of new individual per generations, all just to test the performance of the genetic algorithm. We obtain very good controllers for the Ball and Beam system; every individual was simulated with a plant that we previously described, where was simulated the reference as the input and we got a fuzzy controller where the inputs are taken by the error and the change of error, the α angle as the output and at the same time it's the input for the Ball and Beam Model, where the equations are computed and the output is the feedback for the plant shown in figure 6:

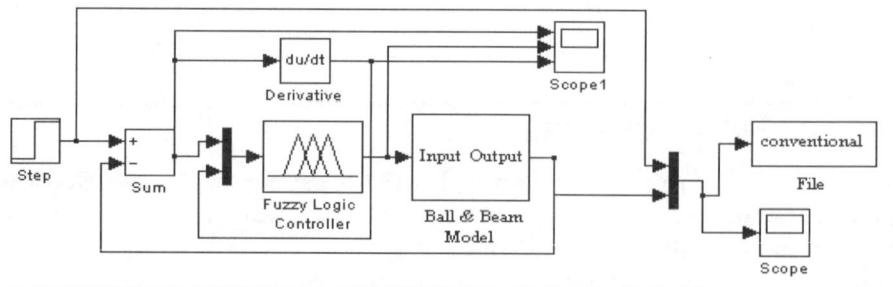

Fig. 6. Simulink Ball and Beam plant

As you can see the plant has a "Step" reference, we used 0.25 as the reference but, we also used some generators to change it, amplifying the frequency and the amplitude, and we'll see it in the graphics later. As was mentioned before we tested the GA with variations on the genetic operations; in the table 2 we can appreciate the results of the GA.

Table 2. Results obtaned with GA tests

NIND	MAXGEN	GGAP	MUT	XOVER	TYPE	SELECT	ERROR	STAND_DEV
25	10	0.5	0.1	0.2	XOVMP	RWS	0.11942	0.07856663
30	10	0.5	0.05	0.5	XOVMP	RWS	0.11942	0.06037278
35	15	0.5	0.3	0.3	XOVMP	RWS	0.03404	0.00632153
30	15	0.5	0.02	0.5	XOVMP	RWS	0.02510	0.01187232
50	20	0.5	0.001	0.01	XOVMP	RWS	0.00831	0.00217789
100	40	0.5	0.001	0.03	XOVMP	RWS	0.00523	0.00157685
50	30	0.5	0.001	0.02	XOVMP	RWS	0.00300	0.00375474
10	10	0.5	0.01	0.5	XOVSP	RWS	0.00831	0.00864084
20	15	0.5	0.2	0.5	XOVDP	RWS	0.02053	0.00864084
25	15	0.5	0.001	0.02	XOVDPRS	RWS	0.00831	0.07856663
30	15	0.5	0.001	0.02	XOVDPRS	RWS	0.00831	0.01187232
120	40	0.7	0.002	0.1	XOVDPRS	RWS	0.00831	0.03120741
150	50	0.4	0.05	0.01	XOVDPRS	RWS	0.00831	0.02893024
30	50	0.3	0.01	0.2	XOVMP	RWS	0.00012	0.00579120
30	90	0.3	0.01	0.2	XOVMP	RWS	0.00831	0.34414887
100	150	0.5	0.2	0.2	XOVMP	RWS	0.49501	0.18193150
100	50	0.3	0.05	0.3	XOVMP	RWS	0.75230	0.34632676
100	60	0.5	0.4	0.3	XOVMP	RWS	115.690	81.7960618
30	50	0.5	0.01	0.1	XOVMP	RWS	0.01290	0.67196357
100	70	0.5	0.01	0.2	XOVMP	RWS	0.96320	0.57818707
150	150	0.5	0.2	0.3	XOVMP	RWS	0.14552	0.02165161
50	100	0.5	0.1	0.2	XOVMP	RWS	0.11490	0.09482302
50	100	0.5	0.1	0.2	XOVMP	RWS	0.24900	0.06817924
100	100	0.5	0.01	0.01	XOVMP	RWS	0.34542	0.28770761
50	70	0.7	0.1	0.2	XOVMP	RWS	0.34542	0.16300226
100	150	0.5	0.1	0.2	XOVMP	RWS	0.34542	0.23837277
100	150	0.5	0.4	0.2	XOVMP	RWS	0.34542	0.23837277
25	50	0.5	0.2	0.2	XOVMP	RWS	38.1080	26.70217639

We see that the best result (the one marked) was obtained with a population of 30 individuals, 50 generations with a 30% of new individuals per generation, 1% of mutation and the 20% of crossover using a multipoint crossover; we obtained an error of 0.00012907 with a very low standard deviation. In the simulation the fuzzy controller has 27 fuzzy rules. Later in the comparison we could appreciate more clearly the results.

4.2 Ant Colony Optimization Experiment

The ant colony optimization (ACO) has to be started with the same base of knowledge. Describing the method, we can say that there's a colony which is going to explore a search space for the best trails that are saved in a matrix; then when we got the trails (real numbers) we make some adjustments to the numbers so that we can work with them. Also a Fitness Function is needed for the ACO algorithm and we can use the same function that we used in GA, but with some modifications that must to be made; as we can remember in a GA a chromosome was taken as the vector of active rules, now we only have a real number; 6 digits to be more accurate, and we're going to need 44 bits, so multiplying the real number by 100,000,000,000 and converting the result to binary code we should have a vector for the active rules. It's very important to mention that the ACO algorithm that is used in this paper was previously improved. Including some modifications that make the algorithm faster then the original one. Basically from the nest to the food is a line marked and the best trail is the one that makes a diagonal from the nest to the food.

If we describe the pseudo code of the algorithm we have that first of all we need to initialize some of the variables like the initial pheromone, the percentage of evaporation we want, we also have to construct a graph (with the dimensions for the search space). Once the variables are initialized every ant has to construct its trails and these trails are updated so the ant could not walk the same trail, the best trails of the ants are the ones we convert to binary for the active rules of the fuzzy controllers when its created the FIS we got the rules average then is the simulation that we have to do so we can obtain the error average and compare the best fuzzy controller of every epoch of the algorithm. Some of the results that were obtained with the ACO algorithm tests are shown in table 3.

Table 3. Results obtained with ACO algorithm tests

ANTZ	EPOCAS	PHINICIAL	EVAP	ERROR	DEV_STAN
10	10	0.500	0.200	0.017585	0.011071737
10	10	0.300	0.150	12904096.58	122527149.7
10	10	0.350	0.120	186183653.4	122527149.7
10	10	0.500	0.250	0.000010293	0.012427194
10	10	0.200	0.050	0.000010293	0.001355458
10	10	0.200	0.150	0.0019272	0.042361598
10	30	0.400	0.330	0.015318	0.098563147
15	30	0.350	0.120	0.12795	0.085241015
15	10	0.150	0.005	0.007401	0.003870561
20	40	0.400	0.300	0.0019272	0.085639810
25	25	0.500	0.200	855915630.3	605223745.9
25	100	0.500	0.200	0.50000	0.342435551
30	15	0.200	0.025	0.23361	0.074712903
30	30	0.500	0.200	0.000010293	0.012598710
30	30	0.350	0.120	0.000132510	0.010230548
30	50	0.800	0.675	0.00030885	0.078596314
30	50	0.700	0.436	0.0019272	0.001144346
40	80	0.100	0.075	13782.9115	9745.988823
50	25	0.500	0.250	0.015723	0.094681235

In the best result given by ACO, the algorithm was executed with very low level of ants (10), 10 epochs with a 20% of initial pheromone and a 5% of evaporation, the error reached is very low as the standard deviation is.

4.3 Comparison between the Evolutionary Methods GA vs. ACO

In this section will appreciate the differences between the methods by comparing the results obtained with the evolutionary algorithms. In the graphics we can appreciate how the fuzzy systems had an evolution during the different tests that have been realized. Is also notable how in some cases one method is better than the other and this is shown with the error averages in the result tables previously shown.

Most of the samples were realized with a linear reference, but we also tested with a change of reference so we could know the efficiency of the fuzzy controllers, at the same time another samples were realized from the beginning with a certain change of reference so the fuzzy controller could reach more easily a linear reference without any constraint.

1. If (error is NL) and (derror is NL) then (angulo is NL) (1)
2. If (error is N) and (derror is NL) then (angulo is NL) (1)
3. If (error is N) and (derror is Z) then (angulo is N) (1)
4. If (error is Z) and (derror is Z) then (angulo is Z) (1)
5. If (error is P) and (derror is Z) then (angulo is P) (1)
6. If (error is PL) and (derror is N) then (angulo is P) (1)
7. If (error is PL) and (derror is P) then (angulo is PL) (1)
8. If (error is NL) and (derror is P) then (angulo is Z) (1)
9. If (error is NL) and (derror is PL) then (angulo is N) (1)
10. If (error is N) and (derror is NL) then (angulo is N) (1)
11. If (error is N) and (derror is PL) then (angulo is Z) (1)
12. If (error is Z) and (derror is N) then (angulo is Z) (1)
13. If (error is P) and (derror is NL) then (angulo is Z) (1)
14. If (error is P) and (derror is P) then (angulo is P) (1)
15. If (error is PL) and (derror is NL) then (angulo is N) (1)
16. If (error is PL) and (derror is N) then (angulo is Z) (1)
17. If (error is PL) and (derror is Z) then (angulo is P) (1)
18. If (error is NL) and (derror is PL) then (angulo is P) (1)
19. If (error is N) and (derror is N) then (angulo is N) (1)
20. If (error is N) and (derror is PL) then (angulo is Z) (1)
21. If (error is Z) and (derror is PL) then (angulo is PL) (1)
22. If (error is P) and (derror is NL) then (angulo is Z) (1)
23. If (error is P) and (derror is P) then (angulo is P) (1)
24. If (error is PL) and (derror is P) then (angulo is P) (1)
25. If (error is P) and (derror is P) then (angulo is P) (1)
26. If (error is PL) and (derror is P) then (angulo is P) (1)

1. If (error is NL) and (derror is P) then (angulo is N) (1)
2. If (error is NL) and (derror is PL) then (angulo is Z) (1)
3. If (error is N) and (derror is NL) then (angulo is NL) (1)
4. If (error is N) and (derror is P) then (angulo is N) (1)
5. If (error is Z) and (derror is NL) then (angulo is NL) (1)
6. If (error is Z) and (derror is Z) then (angulo is Z) (1)
7. If (error is P) and (derror is N) then (angulo is Z) (1)
8. If (error is P) and (derror is Z) then (angulo is P) (1)
9. If (error is P) and (derror is PL) then (angulo is PL) (1)
10. If (error is PL) and (derror is N) then (angulo is P) (1)
11. If (error is PL) and (derror is P) then (angulo is PL) (1)
12. If (error is NL) and (derror is N) then (angulo is N) (1)
13. If (error is NL) and (derror is PL) then (angulo is N) (1)
14. If (error is NL) and (derror is PL) then (angulo is P) (1)
15. If (error is N) and (derror is N) then (angulo is N) (1)
16. If (error is P) and (derror is NL) then (angulo is N) (1)
17. If (error is P) and (derror is P) then (angulo is P) (1)
18. If (error is P) and (derror is PL) then (angulo is P) (1)
19. If (error is PL) and (derror is NL) then (angulo is P) (1)
20. If (error is PL) and (derror is Z) then (angulo is P) (1)
21. If (error is PL) and (derror is P) then (angulo is P) (1)

Fig. 7. Active rules GA sample   **Fig. 8.** Active rules ACO sample

We can see in figures 7 and 8 the active rules of the best simulations with each method; there's a difference of 6 rules where GA has more rules than ACO, at a first sight we could say ACO is better and it is in some cases, but having more active rules could bring a better control in different cases, and we'll appreciate it later. In figures 9-10 the surfaces of the both cases are shown, the differences are due to the active rules in the fuzzy controllers.

Once the fuzzy controllers were created the simulations have to be the next step for the experiments. The simulations in figures 11-12 represent the fuzzy controllers obtained with each method GA and ACO with a linear reference both samples were tested with a 0.25 of linear reference and as we can see the ACO reach perfectly the reference but GA is not.

Modifying the reference the following simulations have a saw tooth reference varying the amplitudes and frequencies the best results were obtained at a 50% of amplitude and a frequency of 10% (Figures 13 and 14).

In the saw tooth case we can see that the GA obtained a better fuzzy controller, and ACO is good but in some units of time the control is not as perfect as the GA is; this is result of the difference between the fuzzy rules that each evolutionary method has activated in each sample.

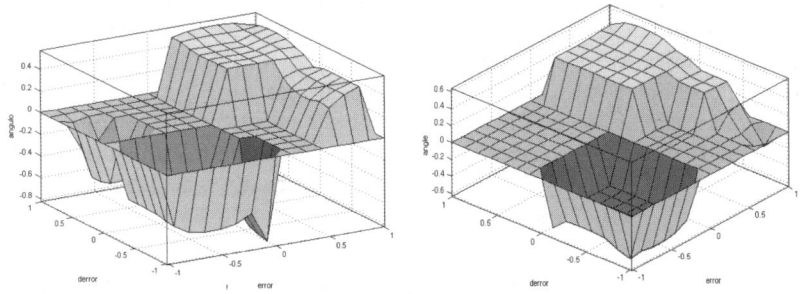

Fig. 9. FIS' Surface, GA sample   **Fig. 10.** FIS surface, ACO sample

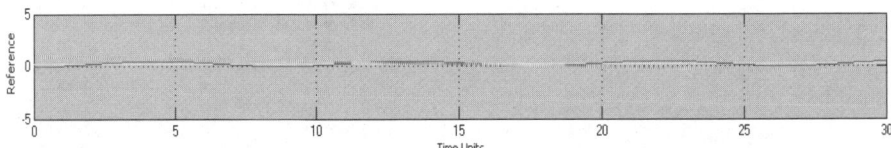

Fig. 11. Best GA simulation with 27 fuzzy rules, 30 individuals, 50 generations

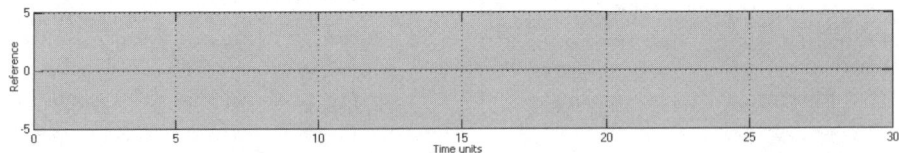

Fig. 12. Best ACO simulation with 21 fuzzy rules, 10 ants, 10 epochs

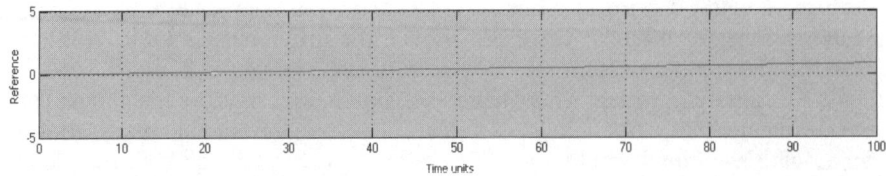

Fig. 13. Best GA simulation, saw tooth reference, 0.5 amplitude and a frequency of 0.1

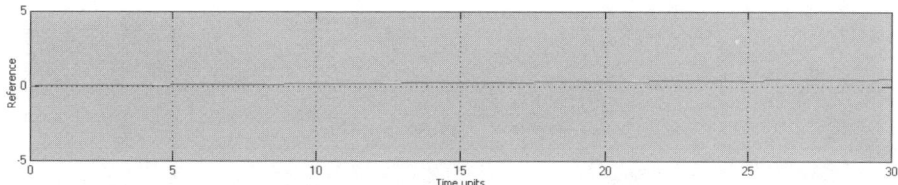

Fig. 14. Best ACO simulation, saw tooth reference, 0.5 amplitude and 0.1 frequency

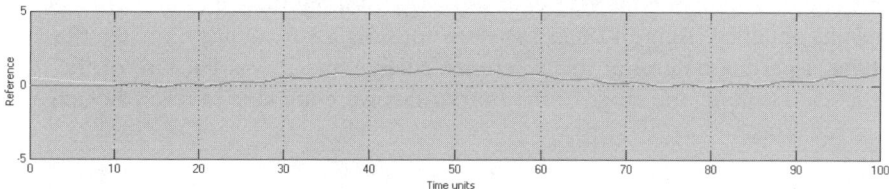

Fig. 15. GA with sinusoidal reference, 0.5 amplitude, 0.5 frequency

In figures 15-16 the simulations with a sinusoidal reference are shown, the amplitude is maintained in a 50% but frequency is increased at a 50% comparing with the saw tooth reference. The best result with sinusoidal reference was obtained in the GA test with 30 individuals, 50 generations as maximum, 1% of mutation and 20% for the crossover. The ACO best fuzzy controller is outrageously bad at this case of change of reference, as we can see. The controller tries to follow the reference and we could say it has sequence but control it's never reached.

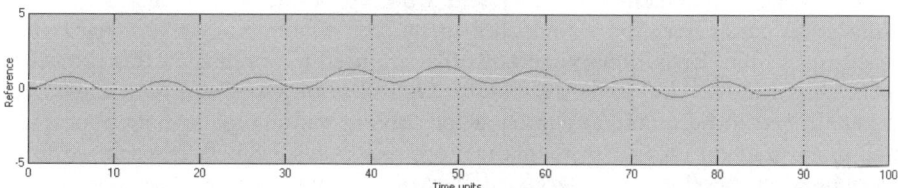

Fig. 16. ACO with sinusoidal reference, 0.5 amplitude, 0.5 frequency

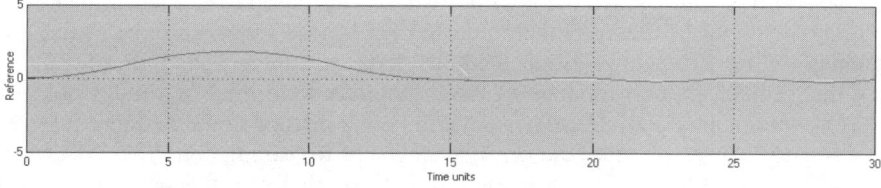

Fig. 17. GA with square reference, 0.5 amplitude, frequency 0.5

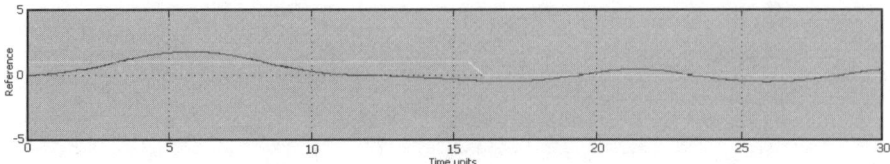

Fig. 18. ACO with square reference, 0.5 amplitude, frequency 0.5

In a square simulation the reference at a 50% of amplitude and 50% of frequency, the results obtained (figure 17 and 18) were not satisfactory at any case, the FIS could not have a perfect control at this reference change and it was because of the active rules at each sample; the range of the fuzzy rules were not able to reach a change of a square reference.

5 Conclusions

Natural intelligence is the product of millions of years of biological evolution. Simulation of complex biological evolutionary processes may lead us to discover how evolution propels living systems toward higher-level intelligence. Greater attention it thus being paid to evolutionary computing techniques such genetic algorithms and ant colony systems.

The use of fuzzy controllers is very helpful when control issues we're talking about; fuzzy controller works to obtain a better performance reaching a goal faster than a non fuzzy controller. In this project the control problem that solved is the Ball and Beam model; where we tried to balance a ball on a beam, in every circumstance a good controller could maintain the system working perfectly. It's very known that finding good parameters for a controller to its well functioning is very hard and it means lost of time. Evolutionary methods that are used to optimize in this project are Genetic Algorithms (GA) and Ant Colony Optimization (ACO). GA is a paradigm that has proved to be a unique approach for solving various mathematical problems which other gradient type of mathematical optimizers have failed to reach; ACO has been applied successfully to a large number of difficult combinatorial optimization problems.

For the case of optimizing the ball and beam system, is fine to say that both methods performed with good quality; when parameters are adequate and the reference is set, the controller may reach the goal but when is also used to follow another reference things may not be clear, so the controller could lose sequence most of the time depending on its efficiency. Both methods have advantages and disadvantages; during this project the performance of these methods were kind of similar but once you have experimented individually is notable how a method could be more tolerable at changes, the lapses of time, the error averages at the simulation area. ACO have more possibilities at this time in my opinion, the experience of this project aloud me to say that for the Ball and Beam system ACO is more quick than GA, ACO reached a lower error than GA did, it also optimize more the base of knowledge with good results than GA did, but is also important to mention that sometimes reducing things reduces are possibilities, so in this case a reduce number of fuzzy rules is not always the best result you can obtain. ACO is better than GA, even when GA obtained good results in this project.

References

[1] Man, K.F., Tang, K.S., Kwong, S.: Genetic Algorithms: concepts and designs, City University of Hong Kong. Springer, Heidelberg (1998)

[2] Holland, J.H.: Adaptation in natural and artificial systems. MIT Press, Cambridge (1995)

[3] David, L.: Handbook of genetic algorithms. Van Nostrand Reinhold (1991)

[4] Golderb, D.E.: Genetic algorithms in search, optimization and machine learning. Addison-Wesley, Reading (1989)

[5] Michalewicz, Z.: Genetic Algorithms + Data Structures = Evolutionary Program, 3rd edn. Springer, Heidelberg (1996)

[6] Beasly, D., Bull, D.R., Martin, R.R.: An overview of Genetic Algorithms: Part 1, fundamentals. University Computing 15(2), 58–69 (1993)

[7] Beasly, D., Bull, D.R., Martin, R.R.: An overview of Genetic Algorithms: Part 2, research topics. University Computing 15(4), 170–181 (1993)

[8] Man, K.F., Tang, K.S., Kwong, S.: Genetic Algorithms: concepts and applications. IEEE Trans. Industrial Electronics 43(5), 519–534 (1996)

[9] Srinivas, M., Patnaik, L.M.: Genetic algorithms: a survey. Computing,?June 17–26 (1994)

[10] Tang, K.S., Man, K.F., Kwong, S., He, Q.: Genetic Algorithms and their applications in signal processing. IEEE Signal Processing Magazine 13(6), 22–37 (1996)

[11] Whitley, D.: The GENITOR algorithm and Selection pressure: Why rank-based allocation of reproductive trails is best. In: Schatfer, J.D. (ed.) Proc. 3rd Int. Conf. Genetic Algorithms, pp. 116–121 (1989)

[12] Bonabeau, E., Dorigo, M., Theraulaz, G.: Swarm Intelligence: From Natural to Artificial Systems. Oxford Univ. Press, New York (1999)

[13] Bonabeau, E., Dorigo, M., Theraulaz, G.: Inspiration for optimization from social insect behavior. Nature 406, 39–42 (2000)

[14] Camazine, S., Deneubourg, J.L., Franks, N.R., Sneyd, J., Theraulaz, G., Bonabeau, E.: Self-Organization in Biological Systems. Princeton Univ. Press, Princeton (2001)

[15] Clark, P., Niblett, T.: The CN2 induction algorithm. Mach. Learn 3(4), 261–283 (1989)

[16] Dorigo, M., Bonabeau, E., Theraulaz, G.: Ant algorithms and stigmergy. Future Gener, Comput. Syst. 16(8), 851–871 (2000)

[17] Deneubourg, J.L., Aron, S., Goss, S., Pasteels, J.M.: The Self-organizing exploratory pattern of the Argentine ant. Journal of Insect Behavior 3, 159 (1990)

[18] Dorigo, M., Maniezzo, V., Colorni, A.: Possitive feedback as a search strategy, Dipartimento di Elettronica, Politecnico di Milano, Italy, Tech. Rep. 91-016 (1991)

[19] Dorigo, M.: Optimization, learning and natural algorithms (in Italian), Ph. D. dissertation, Dipartimento di Elettronica, Politecnico di Milano, Italy (1992)

[20] Dorigo, M., Maniezzo, V., Colorni, A.: Ant System: Optimization by a Colony of cooperating agents. IEEE Trans. On Systems, Man and Cibernetics Part B 26(1), 29–41 (1996)

[21] Dorigo, M., Di Caro, G.: The Ant Colony Optimization meta-heuristic. In: Corne, D., et al. (eds.) New ideas in Optimization, pp. 11–32. McGraw Hill, London (1999)

[22] Dorigo, M., Di Caro, G., Gambardella, L.M.: Ant algorithms for discrete optimization. Artificial life 5(2), 137–172 (1999)

[23] Porta-Garcia, M., Montiel, O., Sepúlveda, R., Castillo, O.: Path Planning for Autonomous Mobile Robot Navigation with Rerouting Capability in Dynamic Search Spaces using Ant Colony Optimization. CITEDI-IPN, Department of Computing Science, Tijuana Institute of Technology, Tijuana, Mexico

[24] Wang, W., Bridges, S.M.: Genetic Algorithm Optimization of Membership Functions for Mining Fuzzy Association Rules. Department of Computer Science, Mississipi State University, USA (2000)

[25] Alcacla, R., Cordon, O., Herrera, F.: Algoritmos Geneticos para el Ajuste de Parametrosy Seleccion de Reglas en el Control Difuso de un Sistema de Climatizacion HVAC para Grandes Edificios. Department of Computer Science, Jaen University, Jaen, Spain (2002)

[26] Casillas, J.,Cordon, O., Herrera, F., Villa, P.: Aprendizaje Hibrido de la base de conocimiento de un sistema basado en reglas difusas mediante algoritmos geneticos y colonia de hormigas. Department of Computer Science and Artificial Intelligence, University of Granada, Department of Informatics, University of Vigo, Spain (2003)

Part II

Pattern Recognition

Type-1 and Type-2 Fuzzy Inference Systems as Integration Methods in Modular Neural Networks for Multimodal Biometry and Its Optimization with Genetic Algorithms

Denisse Hidalgo, Oscar Castillo, and Patricia Melin

Tijuana Institute of Technology, Tijuana BC. México

Abstract. We describe in this paper a comparative study of Fuzzy Inference Systems as methods of integration in modular neural networks (MNN's) for multimodal biometry. These methods of integration are based on type-1 and type-2 fuzzy logic. Also, the fuzzy systems are optimized with simple genetic algorithms. First, we considered the use of type-1 fuzzy logic and later the approach with type-2 fuzzy logic. The fuzzy systems were developed using genetic algorithms to handle fuzzy inference systems with different membership functions, like the triangular, trapezoidal and Gaussian; since these algorithms can generate the fuzzy systems automatically. Then the response integration of the modular neural network was tested with the optimized fuzzy integration systems. The comparative study of type-1 and type-2 fuzzy inference systems was made to observe the behavior of the two different integration methods f modular neural networks for multimodal biometry.

1 Introduction

Biometry, is a discipline that studies the recognition of people through its physiological characteristics (fingerprint, face, retina...) or of behavior (voice, signature,...). The interest of the society to use biometric patterns to identify or verify the authenticity of the people has had a drastic increase, that it is reflected in the appearance of diverse practical applications like identity Passports that include biometrics characteristics. Biometry provides a true identification of people, since this technology is based on the recognition of unique corporal characteristics, reason why recognizes the people based on who they are. Only biometric identification can provide a really efficient and precise control of the people, since it is possible to know with a high degree of certainty that the person that went through this form of recognition is the recognized person. As it is habitual in many scientific disciplines, before making a search of the solutions to the problem, it is reasonable and preferable to stop a moment and to make an analysis of the problem. The result of this analysis will be a vision of the different parts that form the whole, having transformed the initial task, probably complex, in a set of more elementary subtasks, susceptible to be approached in a simpler and efficient way. Once this is done, the problem is transformed into the opposite: to integrate the obtained partial results of each of the subtasks and of generating the solution to the complete problem [27].

O. Castillo et al. (Eds.): Soft Computing for Hybrid Intel. Systems, SCI 154, pp. 89–114, 2008.
springerlink.com

The primary goal of this research was to implement type-1 fuzzy logic and type-2 fuzzy logic as methods to integrate the partial results of each of the modules by which the modular neural network was formed and to optimize the type-1 and type-2 fuzzy systems with genetic algorithms; this with the purpose of obtaining the optimal results in the recognition, and making a comparison between the different fuzzy logic methods.

2 Modular Neural Networks

As it is common in many scientific disciplines, before making the search of the solutions to the problem, it is reasonable and preferable to stop a moment and make an analysis of the problem. The result of this analysis will be a vision of the different parts that form the whole, having transformed the initial task, probably complex, in a set of more elementary subtask, able to be approached in a simpler and efficient way. Once this is done, the problem is transformed into the opposite: to integrate the obtained partial results of each of the subtasks and of generating the solution to the complete problem. The first step is to divide the task (problem) in to subtasks, and later to create and to organize in a suitably way the constructed subsystems to allow the communication among them and thus to integrate them as a whole, which provides the desired solution. The idea of modularity, as it was proposed in the origins of the connectionist computation, has been inspired in the biological models. A review of the physiological structures of the nervous system in vertebrate animals reveals the existence of a representation and hierarchical modular processing of the information [22]. Considering as a basis the biological indications, one of the first modular approaches to complex systems was proposed by Jacobs and Jordan that can use two types of different methods of learning:

Supervised Learning: During which an external teacher provides for each input the correct output. However, this teacher does not specify which module is the one that must learn the corresponding pair (input, desired output).

Unsupervised Learning: That basically consists of a competitive learning, in that different modules do competitive learning whit the presented example [34].

In general, a computational system can be considered as a modular structure if it is possible to divide it in two or more modules, in which each individual module can evaluate different or the same inputs without communicating with the others. The outputs of the modules are aggregated by an integrating unit, which decides:

   How the modules are combined to form the final output of the system.
   - How each module must learn the patterns.

It is important to mention that the use of the MNN to solve a problem in particular, requires ample knowledge of the problem to be able to make the subdivision of the problem, and to build the suitable modular architecture to solve it, in such a way that it is possible to train each of the modules independently, and later to integrate the knowledge learned by each module, in the global architecture [8].

2.1 Methods of Integration

According to the form in which the division of the tasks takes place, the integration method allows to integrate or to combine the results given by each of the constructed modules. Some of the commonly used methods of integration are: Average, Gating Network, Fuzzy Inference Systems, Mechanism of voting using softmax function, the winner takes all, among others [8].In this paper we will describe methods of integration based on Fuzzy Inference Systems, since it is one that is of interest to us in this research work. We show next the block diagram of the modular neural network (see Figure 1).

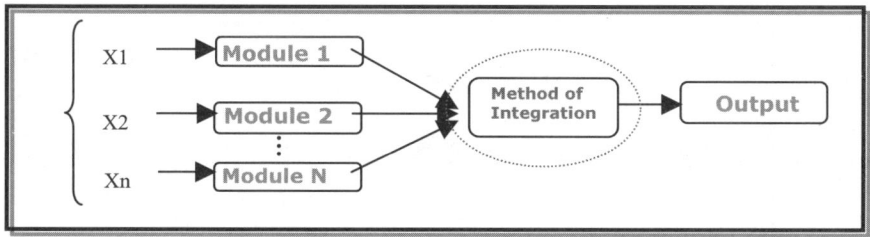

Fig. 1. Blocks diagram of a modular neural network

3 Type-1 Fuzzy Logic

Fuzzy Logic was created in 1965 with the publication of "Fuzzy Sets" [32] by Lofti A. Zadeh in the University of California in Berkeley for the Information an Control magazine, which was based on the work of J. Lukasiewicz [6] on multiple-valued logic. Once the foundations of fuzzy logic becomed firm, their applications have grown in number and diversity, and its influence within basic sciences, has become more visible and more substantial [33]. Fuzzy Logic creates mathematical approaches in the solution of certain kinds of problems. Fuzzy Logic produces exact results from vague data, thus it is particularly useful in electronic or computational applications.

Type-1 Fuzzy Inference System
The basic structure of fuzzy inference system consists of three conceptual components: a set of rules, which contains a selection of fuzzy rules; a data base (or dictionary), that defines the used membership functions in the rules; and a reasoning mechanism, that makes the inference procedure (usually fuzzy reasoning). The basic fuzzy inference system can take fuzzy or traditional inputs, but the outputs that are produced are always fuzzy sets. Some times it is necessary to have a traditional output, especially when a fuzzy inference system is used as a controller. Then, a "defuzzification" method is needed to extract the numerical value of output (see Figure 2).

A Fuzzy Inference System is a non linear mapping of its input space to its output space. This mapping is obtained by means of a set of fuzzy if-then rules, each of which describes the local behavior of the mapping. The basic structure of a type-1 fuzzy system is shown in Figure 3.

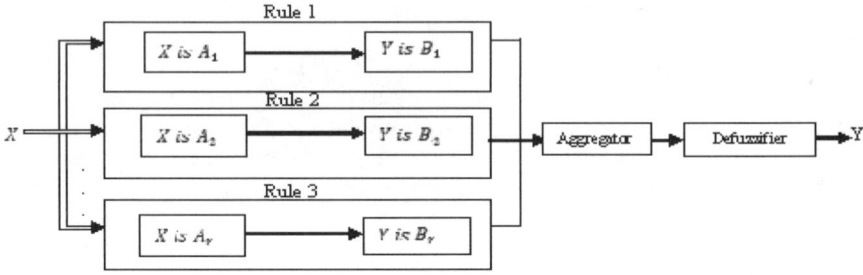

Fig. 2. Fuzzy Inference System

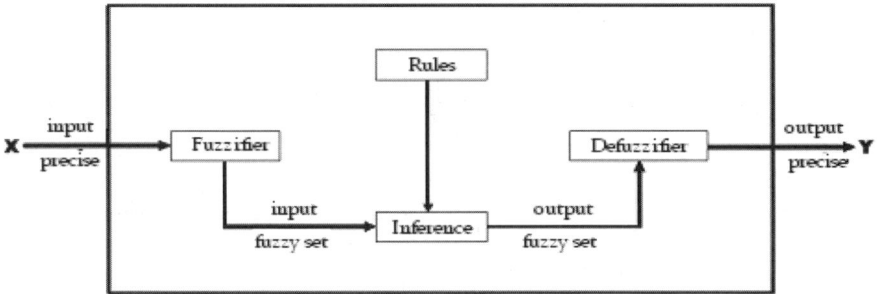

Fig. 3. Basic structure of Type-1Fuzzy Inference System

In this paper we study the integration method of modular neural networks. We use type-1 and type-2 fuzzy inference systems as integration methods and genetic algorithms are used to optimize the structure of the fuzzy system.

4   Type-2 Fuzzy Logic

The original theory of Fuzzy Logic (FL), was proposed by Lotfi Zadeh, more than 40 years ago, and this theory fully handle all the uncertainty present in real-world problems. "To handle," it is understood as "to model and to reduce to the minimum the effect of". That type-1 fuzzy logic cannot completely do this, sounds paradoxical because this has the uncertainty connotation. Type-2 Fuzzy Logic can handle uncertainty because it can model and reduce it to the minimum their effects. Also, if all the uncertainties disappear, type-2 fuzzy logic reduces to type-1 fuzzy logic, in the same way that, if the randomness disappears, the probability is reduced to the determinism [14]. Fuzzy sets and fuzzy logic are the foundation of fuzzy systems, and have been developed looking to model the form as the brain manipulates inexact information. Type-2 fuzzy sets are used to model uncertainty and imprecision; originally they were proposed by Zadeh in 1975 and they are essentially "fuzzy-fuzzy" sets in which the membership degrees are type-1 fuzzy sets.

Type-2 Fuzzy Inference System

A fuzzy Inference System is a system based on rules that uses fuzzy logic, instead of Boolean logic, to analyze data [5,10]. Its basic structure includes four main components:

- *Fuzzifier.* It translates inputs (real values) to fuzzy values.
- *Inference System.* Type-1 or type-2 applies a mechanism of fuzzy reasoning to obtain a fuzzy output.
- *Defuzzifier/Type Reducer.* The defuzzifier it translates an output to precise values; the type reducer transforms a fuzzy set of type-2 to type-1; and
- *Knowledge Base.* It contains a set of fuzzy rules, known as base rules, and a set of membership functions known as the data base.

In Figure 4 we can appreciate the basic structure of a type-2 Fuzzy Inference System.

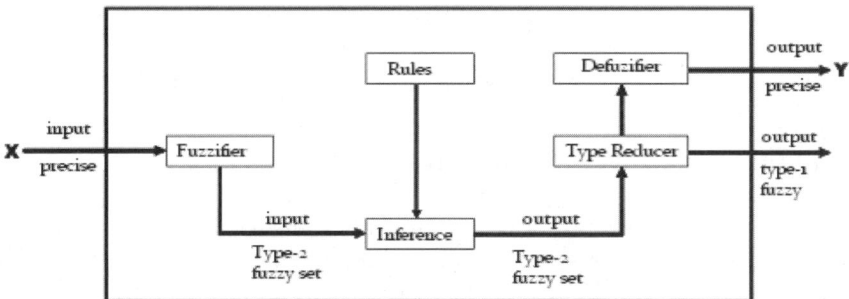

Fig. 4. Basic structure of Type-2 Fuzzy Inference System

Uncertainty is "the imperfection in the knowledge on the state or the processes of the nature". The statistical uncertainty is "the randomness or the originating error of several sources like described when using statistical methodology" [5].

5 Genetic Algorithms

The Genetic Algorithm (GA), is a search technique based on Darwin's theory of evolution, and has received tremendous popularity anywhere in the world during the past few years. They are adaptive methods that can be used to solve search and optimization problems. They are based on the genetic process of the living organisms. Throughout the generations, the populations evolve in the same form as in nature, with the principles of natural selection and the survival of fittest, postulated by Darwin. Simulating evolution, GA's are able to create solutions to problems of the real world. The evolution of these solutions towards optimal values of the problem depends largely on a suitable codification of the solutions [2]. The use of new

```
Procedure Genetic Algorithm

Begin
   t=o
   to initialize P(t)
   to evaluate P(t)
   While (the condition of shutdown is not fulfilled) do
        begin
             t=t+1
             to select P(t) from P(t-1)
             to apply crossover and mutation on P(t)
             to evaluate P(t)
        end
end
```

Fig. 5. Structure of a generational GA

representations and the construction of new operators to manipulate information have caused that the present conception of a GA is quite different and more general than the original idea. The basic structure of a GA is shown in Figure 5 [1].

6 Modular Architecture Description

In this paper we are making a comparison between the integration methods of a modular neural network; this comparison is done with the integration method as fuzzy systems, which used techniques of type-1 Fuzzy logic and type-2 Fuzzy Logic. These methods of integration were implemented in modular neural networks for biometry. In others words, the MNN´s were trained to make the recognition of persons using their face, fingerprint and voice. These features of persons are the 3 more important biometric measures in the field of pattern recognition, at the moment. The general architecture of the modular system is show in Figure 6.

Now we describe in more detail the pattern recognition system (Figure 7).

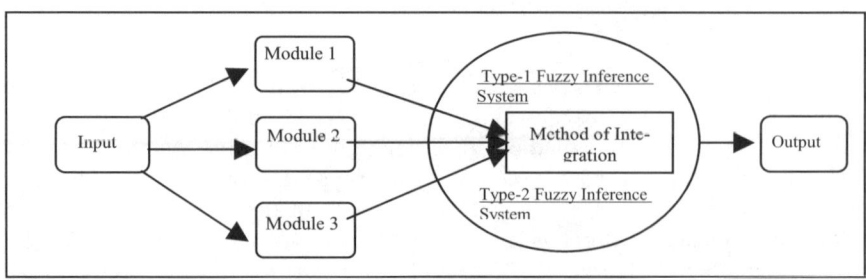

Fig. 6. General Architecture of the Modular System

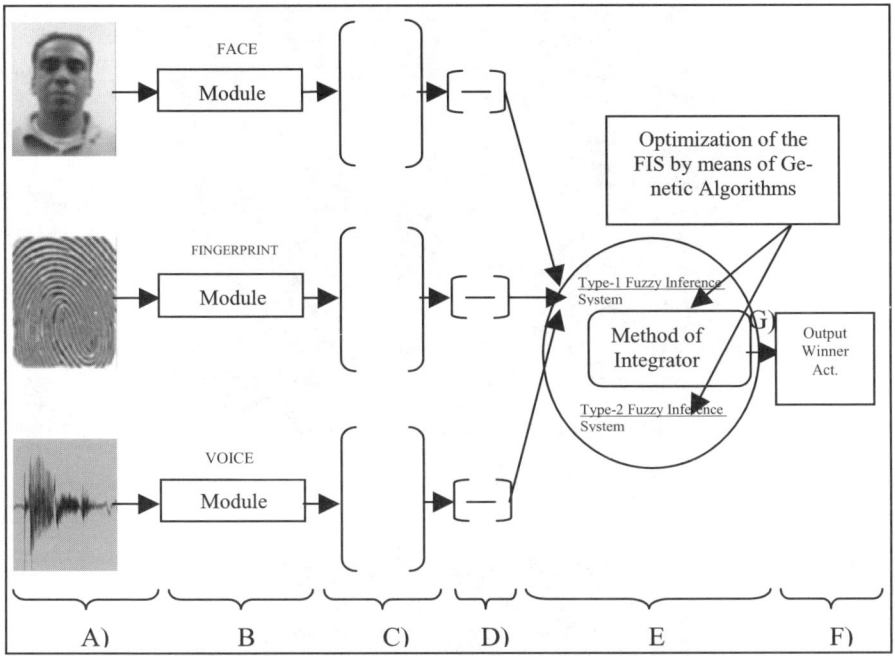

Fig. 7. General Scheme of the pattern recognition system

6.1 Description of the Problem and the Solution

In this section we describe in detail the different parts of the recognition system.

A) Input Data
We describe at this point the input data used in the modular architecture for pattern recognition:

Face: Images of the faces of 30 different people were used to make the training of the MNN without noise, we also use 30 images of the face of these same people but with different gestures, to use them in the training with noise. The used images were pre-processed with a wavelet function to obtain better results in the training, as given in [3]. These images were obtained from a group of students of Tijuana Institute of Technology, combined with some others of the ORL data base. The size of these images is of 268 x 338 pixels with extension .bmp. These are shown in Figure 8.

Fingerprint: Images of the fingerprints of 30 different people were used to make the training without noise. Then it was added random noise to the fingerprint use them in the training with noise. The used images were preprocessed with a wavelet function to obtain better results in the training, as given in [3]. These images were obtained from a group of students of Tijuana Institute of Technology, combined with some others of the ORL data base. The size of these images is of 268 x 338 pixels with extension .bmp. These are shown in Figure 9.

Fig. 8. Data base used for the training of faces

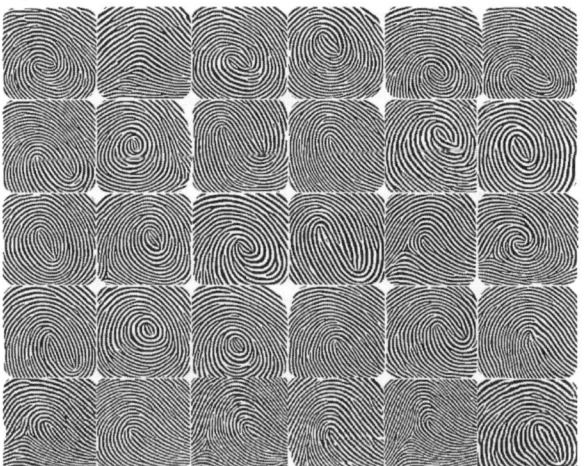

Fig. 9. Data base used for the training of fingerprints

Voice: For the training of module of voice was used word spoken by different persons with samples of 30 persons as with the face and fingerprint. We applied the Mel cepstrals coefficients [23], as preprocessing for the training in the MNN.

The used three Spanish words as follows:

- Accesar
- Hello
- Presentation

Where some people will say the words in Spanish "Accesar", others "Hello", and some other "Presentation".

We also have to mention that random noise was added to the voice signals to train the MNN with noisy signals.

B) Modules of the MNN for the Training Phase
In the modular neural network for the training phase we are considering three modules, one for the face, another one for the fingerprint and finally another for the voice, each of the three modules has three submodules. In others words, the architecture of the modules could be visualized as in Figure 10.

It is possible to mention that for each trained module and each submodule, different architectures were used, that is to say, different number of neurons, layers, etc., and different training methods.

C) Output of the Modular Neural Network (MNN)
The output of the MNN is a vector that is formed by 30 activations (in this case because the network has been trained with 30 different people).

D) Competition between Activations
The competition between the 30 winning activations is performed, and the final result from the MNN is obtained; where this is an activation by module, these activations were used as input to the fuzzy system, this data represented then the higher activation from a each module. In others words, three results were obtained; one for the module of the face, another one for module of the fingerpri0nt and another one for the module of the voice.

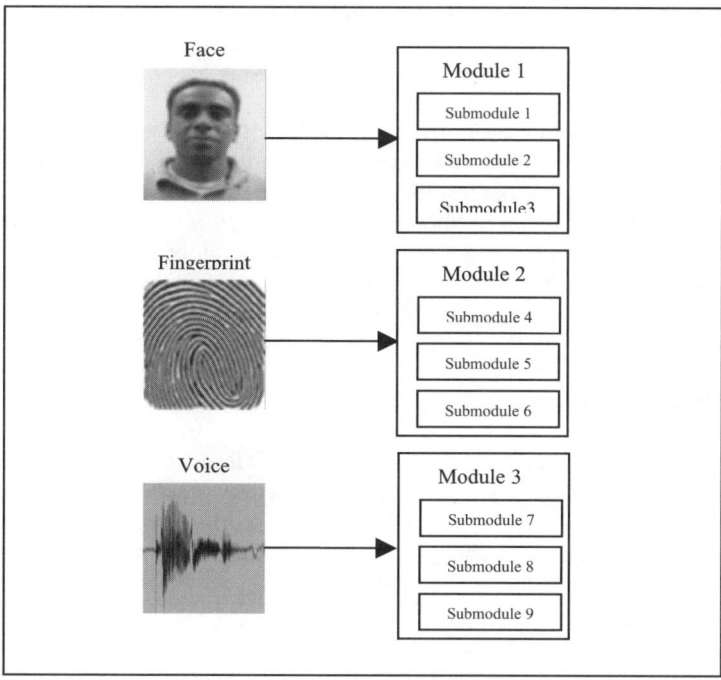

Fig. 10. Architecture of the modular neural network for the training

E) Integration of the MNN (with Type-1 and Type-2 Fuzzy Systems)
In this part, once the winning activations are obtained for each module, they are used as input to the fuzzy system, in which these activations are evaluated and depending on the characteristics of the fuzzy system, a result of final output is obtained, which will give us, the winning module.

F) Final Output (Winning Activation)
The result of the fuzzy system will give us to which module belongs, once this data is obtained we will know that a person has been recognized.

G) Optimization of the Fuzzy Systems using Genetic Algorithms
The optimization of the fuzzy systems consists of, as the name indicates it, optimizing the membership functions of the fuzzy systems, for type-1 and type-2 fuzzy logic, with genetic algorithms.

7 Simulation Results

The simulation results were obtained after several steps were carried out. First, a set of trainings with the MNN were performed. Second, several fuzzy systems (type-1 and type-2) were developed by using a genetic algorithm. The data used for training was obtained from previous research work and was also used to test the fuzzy integration modules.

Once we collected the data, the training phase of MNN with different architectures, was initiated, this with the purpose of being able to make comparisons between the different tests. We choose 5 of them (see figure 7) to test of integration module with the type-1 and type-2 fuzzy systems optimized with GA' s and to obtain a comparison of the results.

After we obtained the necessary trainings of the MNN´s, the genetic algorithm was used to obtain the type-1 fuzzy inference system with triangular membership functions and to use it then as integration method, therefore it was optimized with the genetic algorithm that allowed to obtain the type-1 fuzzy inference system with trapezoidal membership functions, and then with the genetic algorithm to obtain the type-1 fuzzy inference system with Gaussian membership functions and to test the different methods to integrate the results given by the MNN's. We show in table 1 the training of the modular neural networks for response integration.

7.1 Type-1 Fuzzy Inference System

As was mentioned previously several type-1 fuzzy inference systems were build, using Triangular, Trapezoidal and Gaussian membership functions; next we show the obtained fuzzy systems optimized by the GA. It is worth mentioning that not all the systems that were obtained are shown here because they were too many; therefore we show only one of each membership function.

7.1.1 Type-1 Fuzzy System
The type-1 fuzzy inference system shown in figure 11 have three inputs (activation of the face, activation of the fingerprint and activation of the voice), which are composed

Table 1. Training of the modular neural network

RNM	Método de Entrenamiento	No. de capas por submodulos	Neuronas por capa	Funcion de Rendimiento	Error Meta	Error Alcanzado	Épocas	# de pers. Recono-cidas por modulo	% de Reconocimiento	Tiempo de Entrenamiento
1	Trainscg	R: SubMod1: 2	198,160	MSE	0.01	0.00992082	4000	30	100	
		SubMod2: 2	287,136	MSE	0.01	0.00990072	4000	30	100	
		SubMod3: 2	285,95	MSE	0.01	0.00996058	4000	30	100	3 hrs
		H: SubMod4: 2	63,42	MSE	0.01	0.0200279	4000	19	63	2 min
		SubMod5: 2	70,50	MSE	0.01	0.00993908	4000	30	100	
		SubMod6: 2	60,42	MSE	0.01	0.0328742	4000	23	77	
		V: SubMod7: 2	35,45	MSE	0.001	0.000985456	3000	29	97	
		SubMod8: 2	31,35	MSE	0.001	0.000968815	3000	29	97	
		SubMod9: 2	22,38	MSE	0.001	0.000984716	3000	23	77	
RNM	Método de Entrenamiento	No. de capas por submodulos	Neuronas por capa	Funcion de Rendimiento	Error Meta	Error Alcanzado	Épocas	# de pers. Recono-cidas por modulo	% de Reconocimiento	Tiempo de Entrenamiento
2	Trainscg	R: SubMod1: 2	400,150	MSE	0.01	0.00999066	4000	30	100	
		SubMod2: 2	420,100	MSE	0.01	0.00996782	4000	30	100	
		SubMod3: 2	410,125	MSE	0.01	0.00997436	4000	30	100	3 hrs
		H: SubMod4: 2	350,130	MSE	0.01	0.00999281	4000	29	97	50 min
		SubMod5: 2	250,140	MSE	0.01	0.00999318	4000	29	97	
		SubMod6: 2	300,135	MSE	0.01	0.00999893	4000	30	100	
		V: SubMod7: 2	85,95	MSE	0.001	0.00476453	3000	7	23	
		SubMod8: 2	85,90	MSE	0.001	0.00079812	3000	28	93	
		SubMod9: 2	90,90	MSE	0.001	0.000987834	3000	28	93	
RNM	Método de Entrenamiento	No. de capas por submodulos	Neuronas por capa	Funcion de Rendimiento	Error Meta	Error Alcanzado	Épocas	# de pers. Recono-cidas por modulo	% de Reconocimiento	Tiempo de Entrenamiento
3	Trainscg	R: SubMod1: 2	200,160	MSE	0.01	0.00989252	4000	30	100	
		SubMod2: 2	300,150	MSE	0.01	0.00969909	4000	30	100	
		SubMod3: 2	320,85	MSE	0.01	0.00996425	4000	30	100	2 hrs
		H: SubMod4: 2	77,53	MSE	0.01	0.0295881	4000	20	67	57 min
		SubMod5: 2	72,61	MSE	0.01	0.0199445	4000	20	67	
		SubMod6: 2	74,53	MSE	0.01	0.00999007	4000	27	90	
		V: SubMod7: 2	40,55	MSE	0.001	0.000984098	3000	29	97	
		SubMod8: 2	38,58	MSE	0.001	0.00099729	3000	25	83	
		SubMod9: 2	42,56	MSE	0.001	0.000987742	3000	28	93	
RNM	Método de Entrenamiento	No. de capas por submodulos	Neuronas por capa	Funcion de Rendimiento	Error Meta	Error Alcanzado	Épocas	# de pers. Recono-cidas por modulo	% de Reconocimiento	Tiempo de Entrenamiento
4	Trainscg	R: SubMod1: 2	250,150	MSE	0.01	0.0099579	4000	30	100	
		SubMod2: 2	310,100	MSE	0.01	0.00996268	4000	30	100	
		SubMod3: 2	315,90	MSE	0.01	0.00984851	4000	30	100	1 hr
		H: SubMod4: 2	15,25	MSE	0.01	0.170638	4000	1	3	59 min
		SubMod5: 2	15,20	MSE	0.01	0.175748	4000	1	3	
		SubMod6: 2	15,30	MSE	0.01	0.0873482	4000	1	3	
		V: SubMod7: 2	45,50	MSE	0.001	0.000979182	3000	27	90	
		SubMod8: 2	47,55	MSE	0.001	0.000972833	3000	27	90	
		SubMod9: 2	25,30	MSE	0.001	0.00167415	3000	17	57	
RNM	Método de Entrenamiento	No. de capas por submodulos	Neuronas por capa	Funcion de Rendimiento	Error Meta	Error Alcanzado	Épocas	# de pers. Recono-cidas por modulo	% de Reconocimiento	Tiempo de Entrenamiento
5	Trainscg	R: SubMod1: 2	40,30	MSE	0.01	0.000999915	4000	26	87	
		SubMod2: 2	30,20	MSE	0.01	0.00116395	4000	3	10	
		SubMod3: 2	25,35	MSE	0.01	0.0318143	4000	1	3	3 hrs
		H: SubMod4: 2	23,38	MSE	0.01	0.0650265	4000	1	3	50 min
		SubMod5: 2	10,42	MSE	0.01	0.473676	4000	1	3	
		SubMod6: 2	15,33	MSE	0.01	0.653342	4000	6	20	
		V: SubMod7: 2	25,82	MSE	0.001	0.128213	3000	21	70	
		SubMod8: 2	5,200	MSE	0.001	0.0648562	3000	15	50	
		SubMod9: 2	10,8	MSE	0.001	0.0281692	3000	0	0	

by three membership functions each; and an output that defines the winning activation after going through the Mamdani inference machine.

As it was already mentioned different types of membership functions were used, therefore it is possible to observe next the different optimized systems created by the GA. It can be noticed that different values of the membership functions are obtained for each case; and this is true as well for the inputs as for the outputs of each fuzzy system.

7.1.1.1 Triangular Type-1 Fuzzy System. In figures 12, 13, 14 and 15, it is possible to appreciate the parameters values of the triangular membership function for each variable, as well as for the inputs and the output, for the type-1 fuzzy system. In this case the GA was used with a population of 55 individuals, a maximum of 100 generations, mutation of 0.001, one point crossover of 0.6; which lasted 2 minutes and it was stopped at generation 12, and the error was 0.000017082.

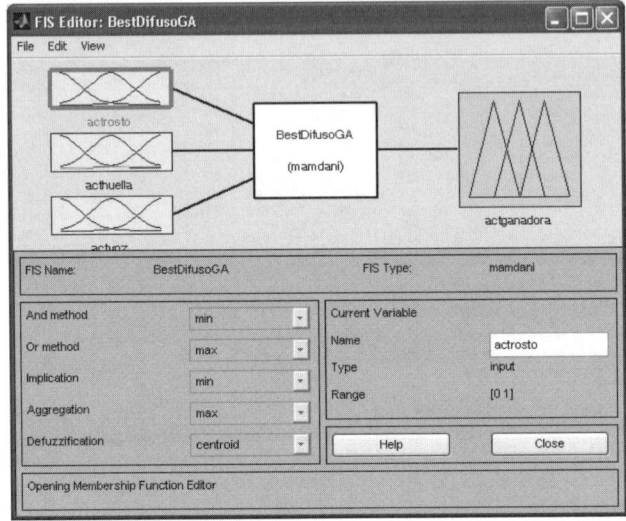

Fig. 11. Graphical representation of the Type-1 Fuzzy Inference System with its inputs and outputs

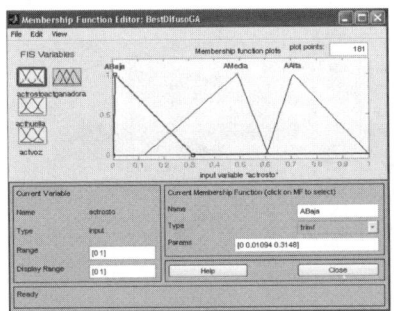

Fig. 12. First input variable (higher activation of the face)

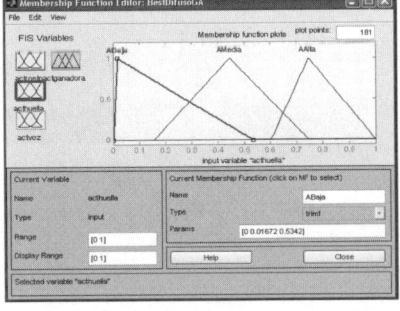

Fig. 13. Second input variable (higher activation of the fingerprint)

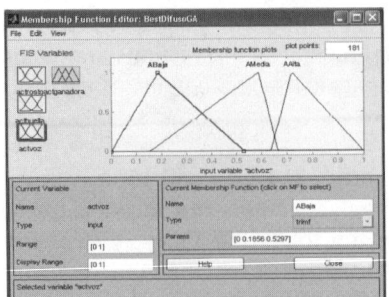

Fig. 14. Third input variable (higher activation of the voice)

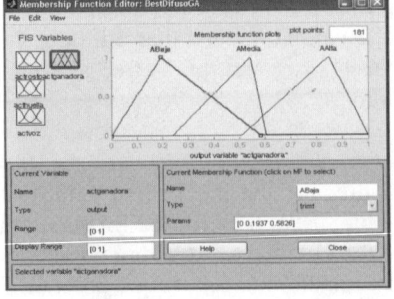

Fig. 15. Variable of output (winner activation)

7.1.1.2 Trapezoidal Type-1 Fuzzy System. In figures 16, 17, 18 and 19, it is possible to appreciate the parameters values of the trapezoidal membership function for each variable, as well as for the inputs and the output, for the type-1 fuzzy system. In this case the GA was used with a population of 85 individuals, a maximum of 100 generations, mutation of 0.0001, one point crossover of 0.25; which lasted of 12 minutes and it was stopped at generation 36, and the obtained error was 0.000046453.

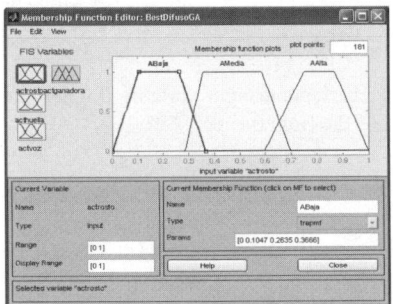

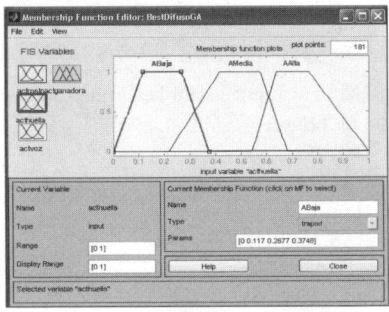

Fig. 16. First input variable (higher activation of the face)

Fig. 17. Second input variable (higher activation of the fingerprint)

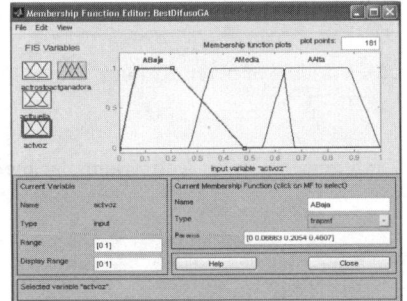

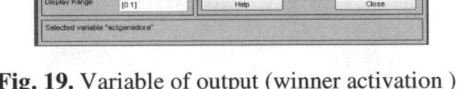

Fig. 18. Third input variable (higher activation of the voice)

Fig. 19. Variable of output (winner activation)

7.1.1.3 Gaussian Type-1 Fuzzy System. In figures 20, 21, 22 and 23, it is possible to appreciate the parameters values of the gaussian membership function for each variable, as well as for the inputs and the output, for the type-1 fuzzy system. In this case the GA was used with a population of 50 individuals, a maximum of 100 generations, mutation of 0.001, multipoint crossover of 0.65; which lasted of 4 minutes and it was stopped at generation 21, and the obtained error was 0.00011068.

We have to mention that the previously shown fuzzy systems are the ones with the smallest errors. In the following tables (2, 3 and 4) we show all the obtained results with the GA.

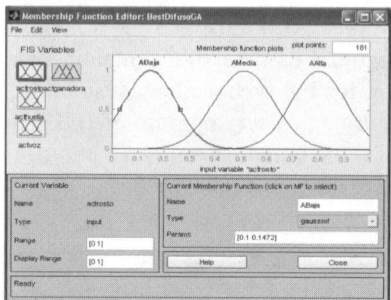

Fig. 20. First input variable (higher activation of the face)

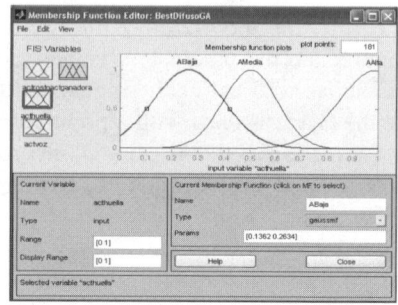

Fig. 21. Second input variable (higher activation of the fingerprint)

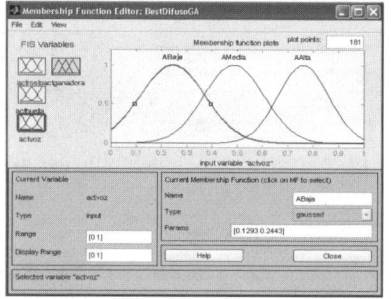

Fig. 22. Third input variable (higher activation of the voice)

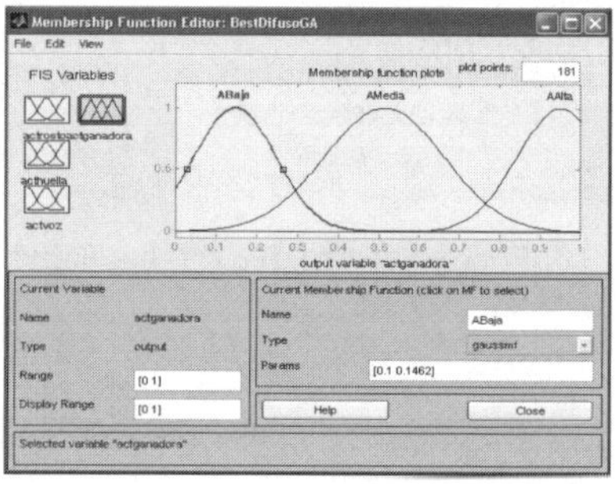

Fig. 23. Variable of output (winner activation)

Table 2. Results of the GA applied to Type-1 Fuzzy Inference Systems with Triangular membership functions

No.	Type MF's	Ind	Max Gen	Mutation	TypeMut	Crossover	TypeCross.	GGAP	Stopped Gen	Lasted of GA	Best Error
1	Triangular	100	150	0.0001	mutbga	0.1	xovmp	0.85	29	14 min	0.000673922
2	Triangular	90	100	0.0001	mutbga	0.2	xovmp	0.85	100	35 min	6.617518707
3	Triangular	80	100	0.0001	mutbga	0.3	xovmp	0.85	15	5 min	0.000038401
4	Triangular	70	100	0.0001	mutbga	0.5	xovmp	0.85	31	28 min	1.130730602
5	Triangular	60	100	0.0001	mutbga	0.45	xovmp	0.85	43	10 min	0.0004582
6	Triangular	50	100	0.001	mutbga	0.65	xovmp	0.85	100	20 min	0.00020151
7	Triangular	40	100	0.001	mutbga	0.7	xovsp	0.85	7	2 min	0.000075054
8	Triangular	30	100	0.01	mutbga	0.6	xovmp	0.85	100	10 min	0.0000014054
9	Triangular	20	100	0.01	mutbga	0.8	xovsp	0.85	10	1 min	0.00050029
10	Triangular	10	100	0.1	mutbga	0.9	xovsp	0.85	4	10 sec	0.00522077
11	Triangular	5	100	0.1	mutbga	1	xovsp	0.85	11	15 sec	0.00263167
12	Triangular	15	100	0.1	mutbga	0.95	xovsp	0.85	8	27 sec	0.00123392
13	Triangular	25	100	0.01	mutbga	0.85	xovsp	0.85	26	20 sec	0.000073374
14	Triangular	35	100	0.01	mutbga	0.65	xovmp	0.85	96	12 min	0.00040778
15	Triangular	45	100	0.001	mutbga	0.75	xovsp	0.85	19	3 min	0.00104095
16	**Triangular**	**55**	**100**	**0.001**	**mutbga**	**0.6**	**xovsp**	**0.85**	**12**	**2 min**	**0.000017082**
17	Triangular	65	100	0.001	mutbga	0.4	xovsp	0.85	16	4 min	0.0000412577
18	Triangular	75	100	0.0001	mutbga	0.35	xovsp	0.85	99	27 min	0.000085732
19	Triangular	85	100	0.0001	mutbga	0.25	xovsp	0.85	18	6 min	0.00119261
20	Triangular	95	100	0.0001	mutbga	0.15	xovsp	0.85	100	37 min	0.02178055

Table 3. Results of the GA applied to Type-1 Fuzzy Inference System with Trapezoidal membership functions

No.	Type MF's	Ind	Max Gen	Mutation	TypeMut	Crossover	Type Cross.	GGAP	Stopped Gen	Lasted of GA	Best Error
1	Trapezoidal	100	150	0.0001	mutbga	0.1	xovsp	0.85	42	16 min	0.00099803
2	Trapezoidal	90	100	0.0001	mutbga	0.2	xovmp	0.85	100	33 min	0.000164958
3	Trapezoidal	80	100	0.0001	mutbga	0.3	xovmp	0.85	100	31 min	0.00203463
4	Trapezoidal	70	100	0.0001	mutbga	0.5	xovmp	0.85	100	28 min	0.01164289
5	Trapezoidal	60	100	0.0001	mutbga	0.45	xovmp	0.85	100	23 min	0.00162315
6	Trapezoidal	50	100	0.001	mutbga	0.65	xovmp	0.85	15	3 min	0.00019062
7	Trapezoidal	40	100	0.001	mutbga	0.7	xovsp	0.85	13	2 min	0.00075258
8	Trapezoidal	30	100	0.01	mutbga	0.6	xovmp	0.85	13	1 min	0.00113649
9	Trapezoidal	20	100	0.01	mutbga	0.8	xovsp	0.85	100	8 min	0.00158774
10	Trapezoidal	10	100	0.1	mutbga	0.9	xovsp	0.85	10	1 min	0.0002004
11	Trapezoidal	5	100	0.1	mutbga	1	xovsp	0.85	100	2 min	4.66257156
12	Trapezoidal	15	100	0.1	mutbga	0.95	xovsp	0.85	100	6 min	0.00331437
13	Trapezoidal	25	100	0.01	mutbga	0.85	xovsp	0.85	100	10 min	0.01206595
14	Trapezoidal	35	100	0.01	mutbga	0.65	xovmp	0.85	100	13 min	0.00456134
15	Trapezoidal	45	100	0.001	mutbga	0.75	xovsp	0.85	100	6 min	0.00254746
16	Trapezoidal	55	100	0.001	mutbga	0.6	xovsp	0.85	100	21 min	0.00536313
17	Trapezoidal	65	100	0.001	mutbga	0.4	xovsp	0.85	100	25 min	1.243885488
18	Trapezoidal	75	100	0.0001	mutbga	0.35	xovsp	0.85	35	10 min	0.001008539
19	**Trapezoidal**	**85**	**100**	**0.0001**	**mutbga**	**0.25**	**xovsp**	**0.85**	**36**	**12 min**	**0.000046453**
20	Trapezoidal	95	100	0.0001	mutbga	0.15	xovsp	0.85	100	17 min	0.01042749

Table 4. Results of the GA applied to Type-1 Fuzzy Inference System with Gaussian membership functions

No.	Type MF's	Ind	Max Gen	Mutation	TypeMut	Crossover	Type Cross.	GGAP	Stopped Gen	Lasted of GA	Best Error
1	Gaussian	100	150	0.0001	mutbga	0.1	xovmp	0.85	3	1 min	0.000164788
2	Gaussian	90	100	0.0001	mutbga	0.2	xovmp	0.85	41	15 min	0.183854323
3	Gaussian	80	100	0.0001	mutbga	0.3	xovmp	0.85	5	2 min	0.00181081
4	Gaussian	70	100	0.0001	mutbga	0.5	xovmp	0.85	100	26 min	0.000615959
5	Gaussian	60	100	0.0001	mutbga	0.45	xovmp	0.85	10	4 min	0.02483356
6	**Gaussian**	**50**	**100**	**0.001**	**mutbga**	**0.65**	**xovmp**	**0.85**	**21**	**4 min**	**0.00011068**
7	Gaussian	40	100	0.001	mutbga	0.7	xovsp	0.85	53	9 min	0.06011866
8	Gaussian	30	100	0.01	mutbga	0.6	xovsp	0.85	5	34 sec	0.03585971
9	Gaussian	20	100	0.01	mutbga	0.8	xovsp	0.85	17	1 min	0.006970346
10	Gaussian	10	100	0.1	mutbga	0.9	xovsp	0.85	5	12 sec	0.04190672
11	Gaussian	5	100	0.1	mutbga	1	xovsp	0.85	3	5 sec	0.0045917
12	Gaussian	15	100	0.1	mutbga	0.95	xovsp	0.85	25	1 min	0.26194236
13	Gaussian	25	100	0.01	mutbga	0.85	xovsp	0.85	16	1 min	0.10640646
14	Gaussian	35	100	0.01	mutbga	0.65	xovmp	0.85	5	46 sec	0.00346566
15	Gaussian	45	100	0.001	mutbga	0.75	xovsp	0.85	3	30 sec	0.00036119
16	Gaussian	55	100	0.001	mutbga	0.6	xovsp	0.85	100	21 min	0.0474117
17	Gaussian	65	100	0.001	mutbga	0.4	xovsp	0.85	100	24 min	0.01250863
18	Gaussian	75	100	0.0001	mutbga	0.35	xovsp	0.85	2	36 sec	0.000164788
19	Gaussian	85	100	0.0001	mutbga	0.25	xovsp	0.85	11	4 min	0.00444927
20	Gaussian	95	100	0.0001	mutbga	0.15	xovsp	0.85	100	37 min	0.03247838

7.2 Type-2 Fuzzy Inference System

As was mentioned previously several type-2 fuzzy inference systems were build, using Triangular, Trapezoidal and Gaussian membership functions; next we show the obtained fuzzy systems optimized by the GA. It is worth mentioning that not all the systems that were obtained are shown here because they were too many; therefore we show only one of each membership function.

7.2.1 Type-2 Fuzzy System

The type-2 fuzzy inference system shown in figure 24 have three inputs (activation of the face, activation of the fingerprint and activation of the voice), which are composed by three membership functions each; and an output that defines the winning activation after going through the Mamdani inference machine.

As it was already mentioned different types of membership functions were used, therefore it is possible to observe next the different optimized systems created by the GA. It can be noticed that different values of the membership functions are obtained for each type; and this is true well as for the inputs as for the outputs of each fuzzy system.

7.2.1.1 Triangular Type-2 Fuzzy System. In the previous figures (25, 26, 27 and 28), it is possible to appreciate the parameters values of the triangular membership function for each variable, as well as for the inputs and the output, for the type-2 fuzzy system. In this case the GA was used with a population of 40 individuals, a maximum of 100 generations, mutation of 0.001, one point crossover of 0.7; which lasted of 19 minutes and it was stopped at generation 51, and the obtained error was 0.000011385.

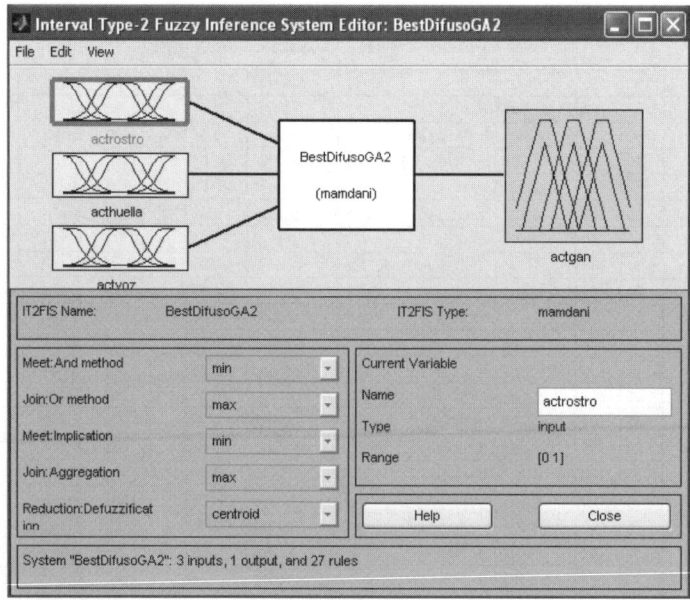

Fig. 24. Graphical representation of the Type-2 Fuzzy Inference System with its inputs and outputs

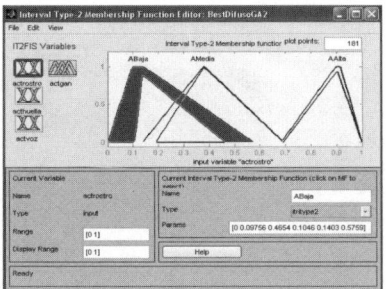

Fig. 25. First input variable (higher activation of the face)

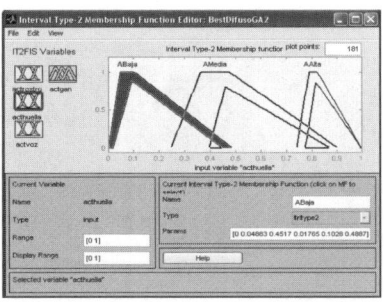

Fig. 26. Second input variable (higher activation of the fingerprint)

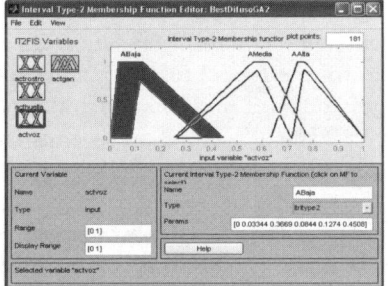

Fig. 27. Third input variable (higher activation of the voice)

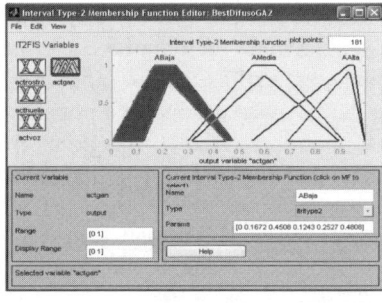

Fig. 28. Variable of output (winner activation)

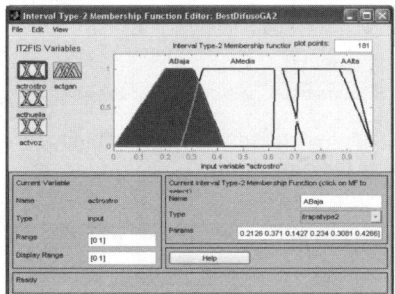

Fig. 29. First input variable (higher activation of the face)

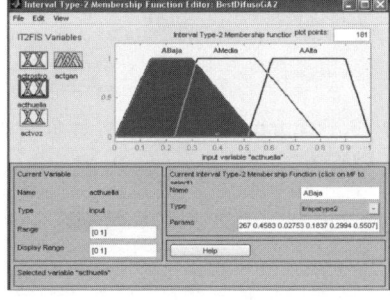

Fig. 30. Second input variable (higher activation of the fingerprint)

7.2.1.2   Trapezoidal Type-2 Fuzzy System. In the figures (29, 30, 31 and 32), it is possible to appreciate the parameters values of the trapezoidal membership function for each variable, as well as for the inputs and the output, for the type-2 fuzzy system. In this case the GA was used with a population of 80 individuals, a maximum of 100 generations, mutation of 0.0001, multipoint crossover of 0.3; which lasted of 13 minutes and it was stopped at generation 10, and the obtained error was 0.0247836.

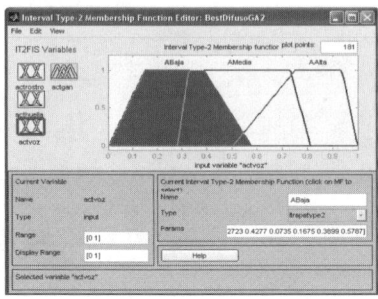 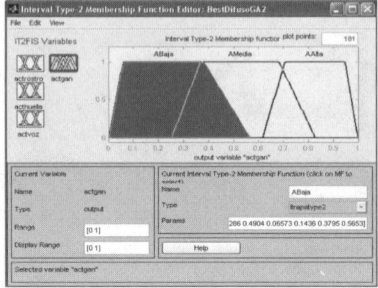

Fig. 31. Third input variable (higher activation of the voice)

Fig. 32. Variable of output (winner activation)

7.2.1.3   Gaussian Type-2 Fuzzy System. In the figures (33, 34, 35 and 36), it is possible to appreciate the parameters values of the gaussian membership function for each variable, as well as for the inputs and the output, for the type-2 fuzzy system. In this case the GA was used with a population of 55 individuals, a maximum of 100 generations, mutation of 0.001, one point crossover of 0.6; which lasted of 50 seconds and it was stopped at generation 4, and the obtained error was 0.0000103302.

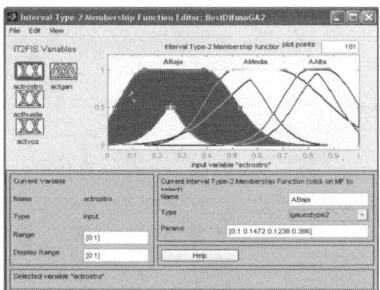 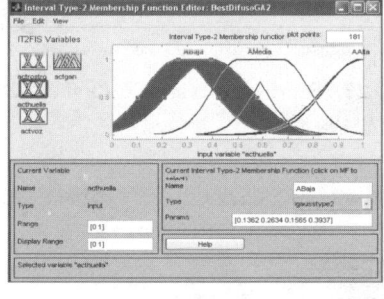

Fig. 33. First input variable (higher activation of the face)

Fig. 34. Second input variable (higher activation of the fingerprint)

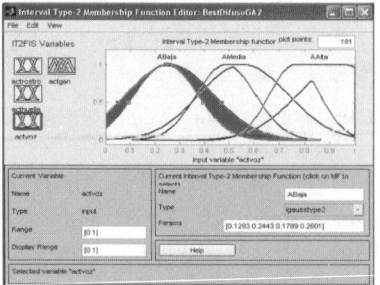 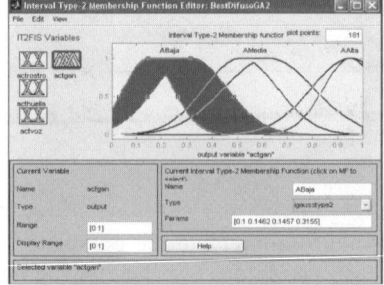

Fig. 35. Third input variable (higher activation of the voice)

Fig. 36. Variable of output (winner activation)

We have to mention that the previously shown fuzzy systems are the ones with the smallest errors. In the following tables (5, 6 and 7) we show all the obtained results with the GA.

Table 5. Results of the GA applied to Type-2 Fuzzy Inference System with Triangular membership functions

No.	Type MF's	Ind	Max Gen	Mutation	TypeMut	Crossover	TypeCross.	GGAP	Stopped Gen	Lasted of GA	Best Error
1	Triangular	100	150	0.0001	mutbga	0.1	xovmp	0.85	150	1 hr	0.243493825
2	Triangular	90	100	0.0001	mutbga	0.2	xovmp	0.85	14	4 min	0.001130642
3	Triangular	80	100	0.0001	mutbga	0.3	xovmp	0.85	100	30 min	0.42196264
4	Triangular	70	100	0.0001	mutbga	0.5	xovmp	0.85	100	29 min	0.441835768
5	Triangular	60	100	0.0001	mutbga	0.45	xovmp	0.85	100	22 min	0.22838559
6	Triangular	50	100	0.001	mutbga	0.65	xovmp	0.85	100	20 min	0.41049033
7	**Triangular**	**40**	**100**	**0.001**	**mutbga**	**0.7**	**xovsp**	**0.85**	**51**	**19 min**	**0.000011385**
8	Triangular	30	100	0.01	mutbga	0.6	xovmp	0.85	100	8 min	0.53426876
9	Triangular	20	100	0.01	mutbga	0.8	xovsp	0.85	100	5 min	0.240057
10	Triangular	10	100	0.1	mutbga	0.9	xovsp	0.85	100	4 min	0.60376824
11	Triangular	5	100	0.1	mutbga	1	xovsp	0.85	44	1 min	0.2344504
12	Triangular	15	100	0.1	mutbga	0.95	xovsp	0.85	100	6 min	0.37926694
13	Triangular	25	100	0.01	mutbga	0.85	xovsp	0.85	100	10 min	0.04789974
14	Triangular	35	100	0.01	mutbga	0.65	xovmp	0.85	8	1 min	0.00900691
15	Triangular	45	100	0.001	mutbga	0.75	xovsp	0.85	100	17 min	0.42580788
16	Triangular	55	100	0.001	mutbga	0.6	xovsp	0.85	100	21 min	0.01624613
17	Triangular	65	100	0.001	mutbga	0.4	xovsp	0.85	88	25 min	0.00232443
18	Triangular	75	100	0.0001	mutbga	0.35	xovsp	0.85	100	28 min	0.559013461
19	Triangular	85	100	0.0001	mutbga	0.25	xovsp	0.85	100	31 min	0.21063925
20	Triangular	95	100	0.0001	mutbga	0.15	xovsp	0.85	100	39 min	0.38952551

Table 6. Results of the GA applied to Type-2 Fuzzy Inference System with Trapezoidal membership functions

No.	Type MF's	Ind	Max Gen	Mutation	TypeMut	Crossover	TypeCross.	GGAP	Stopped Gen	Lasted of GA	Best Error
1	Trapezoidal	100	150	0.0001	mutbga	0.1	xovmp	0.85	150	5 hrs 58 min	0.542291089
2	Trapezoidal	90	100	0.0001	mutbga	0.2	xovmp	0.85	100	3 hrs 49 min	2.325
3	**Trapezoidal**	**80**	**100**	**0.0001**	**mutbga**	**0.3**	**xovmp**	**0.85**	**10**	**13 min**	**0.0247836**
4	Trapezoidal	70	100	0.0001	mutbga	0.5	xovmp	0.85	100	29 min	0.682264153
5	Trapezoidal	60	100	0.0001	mutbga	0.45	xovmp	0.85	100	24 min	0.4822949
6	Trapezoidal	50	100	0.001	mutbga	0.65	xovmp	0.85	100	17 min	0.73576041
7	Trapezoidal	40	100	0.001	mutbga	0.7	xovsp	0.85	36	34 min	0.035355
8	Trapezoidal	30	100	0.01	mutbga	0.6	xovmp	0.85	100	12 min	0.73576041
9	Trapezoidal	20	100	0.01	mutbga	0.8	xovsp	0.85	100	5 min	0.76300909
10	Trapezoidal	10	100	0.1	mutbga	0.9	xovsp	0.85	100	3 min	0.65602506
11	Trapezoidal	5	100	0.1	mutbga	1	xovsp	0.85	100	2 min	0.70884404
12	Trapezoidal	15	100	0.1	mutbga	0.95	xovsp	0.85	100	4 min	0.60458736
13	Trapezoidal	25	100	0.01	mutbga	0.85	xovsp	0.85	100	7 min	0.32737116
14	Trapezoidal	35	100	0.01	mutbga	0.65	xovmp	0.85	100	9 min	0.90410197
15	Trapezoidal	45	100	0.001	mutbga	0.75	xovsp	0.85	100	13 min	0.60458736
16	Trapezoidal	55	100	0.001	mutbga	0.6	xovsp	0.85	100	15 min	0.60458736
17	Trapezoidal	65	100	0.001	mutbga	0.4	xovsp	0.85	100	16 min	0.76300909
18	Trapezoidal	75	100	0.0001	mutbga	0.35	xovsp	0.85	100	29 min	0.875249036
19	Trapezoidal	85	100	0.0001	mutbga	0.25	xovsp	0.85	100	33 min	0.81848755
20	Trapezoidal	95	100	0.0001	mutbga	0.15	xovsp	0.85	100	38 min	0.57939829

Table 7. Results of the GA applied to Type-2 Fuzzy Inference System with Gaussians membership functions

No.	Type MF's	Ind	Max Gen	Mutation	TypeMut	Crossover	TypeCross.	GGAP	Stopped Gen	Lasted of GA	Best Error
1	Gaussian	100	150	0.0001	mutbga	0.1	xovmp	0.85	3	1 min	0.003115353
2	Gaussian	90	100	0.0001	mutbga	0.2	xovmp	0.85	6	2 min	0.001610318
3	Gaussian	80	100	0.0001	mutbga	0.3	xovmp	0.85	9	3 min	0.00294546
4	Gaussian	70	100	0.0001	mutbga	0.5	xovmp	0.85	100	25 min	0.0000787919
5	Gaussian	60	100	0.0001	mutbga	0.45	xovmp	0.85	100	21 min	0.00115187
6	Gaussian	50	100	0.001	mutbga	0.65	xovmp	0.85	100	18 min	0.0001781
7	Gaussian	40	100	0.001	mutbga	0.7	xovsp	0.85	4	37 sec	0.0000218
8	Gaussian	30	100	0.01	mutbga	0.6	xovmp	0.85	11	2 min	0.00298097
9	Gaussian	20	100	0.01	mutbga	0.8	xovsp	0.85	78	6 min	0.00239668
10	Gaussian	10	100	0.1	mutbga	0.9	xovsp	0.85	3	8 sec	0.00681549
11	Gaussian	5	100	0.1	mutbga	1	xovsp	0.85	2	3 sec	0.00031891
12	Gaussian	15	100	0.1	mutbga	0.95	xovsp	0.85	100	5 min	0.01700403
13	Gaussian	25	100	0.01	mutbga	0.85	xovsp	0.85	36	4 min	0.00047306
14	Gaussian	35	100	0.01	mutbga	0.65	xovmp	0.85	15	2 min	0.0000915214
15	Gaussian	45	100	0.001	mutbga	0.75	xovsp	0.85	100	16 min	0.00089849
16	**Gaussian**	**55**	**100**	**0.001**	**mutbga**	**0.6**	**xovsp**	**0.85**	**4**	**50 sec**	**0.0000103302**
17	Gaussian	65	100	0.001	mutbga	0.4	xovsp	0.85	20	5 min	0.00135865
18	Gaussian	75	100	0.0001	mutbga	0.35	xovsp	0.85	25	7 min	0.001623653
19	Gaussian	85	100	0.0001	mutbga	0.25	xovsp	0.85	8	3 min	0.00373201
20	Gaussian	95	100	0.0001	mutbga	0.15	xovsp	0.85	14	5 min	0.00031598

7.3 Results of the Type-1 Fuzzy Logic Integration

We show in table 8 and 9 the results of type-1 Fuzzy Logic Integration.

Table 8. Results of the response integration of the MNN's with Type-1 FIS for the best training

Best Training the MNN's			
Fuzzy System	% of Recognition for the FIS with Triangular MF's	% of Recognition for the FIS with Trapezoidal MF's	% of Recognition for the FIS with Trapezoidal MF's
1	100% (30/30)	100% (30/30)	100% (30/30)
2	100% (30/30)	100% (30/30)	100% (30/30)
3	100% (30/30)	100% (30/30)	100% (30/30)
4	100% (30/30)	100% (30/30)	100% (30/30)
5	100% (30/30)	100% (30/30)	100% (30/30)

Table 9. Results of the response integration of the MNN's with Type-1 FIS for the worse training

Worse Training the MNN's			
Fuzzy System	% of Recognition for the FIS with Triangular MF's	% of Recognition for the FIS with Trapezoidal MF's	% of Recognition for the FIS with Trapezoidal MF's
1	0% (0/30)	0% (0/30)	0% (0/30)
2	0% (0/30)	0% (0/30)	0% (0/30)
3	0% (0/30)	0% (0/30)	0% (0/30)
4	0% (0/30)	0% (0/30)	0% (0/30)
5	0% (0/30)	0% (0/30)	0% (0/30)

7.4   Results of the Type-2 Fuzzy Logic Integration

We show in table 10 and 11 the results of type-2 Fuzzy Logic Integration.

Table 10. Results of the response integration of the MNN's with Type-2 FIS for the best training

Best Training the MNN's			
Fuzzy System	% of Recognition for the FIS with Triangular MF's	% of Recognition for the FIS with Trapezoidal MF's	% of Recognition for the FIS with Trapezoidal MF's
1	100% (30/30)	100% (30/30)	100% (30/30)
2	100% (30/30)	100% (30/30)	100% (30/30)
3	100% (30/30)	100% (30/30)	100% (30/30)
4	100% (30/30)	100% (30/30)	100% (30/30)
5	100% (30/30)	100% (30/30)	100% (30/30)

Table 11. Results of the response integration of the MNN's with Type-2 FIS for the worse training

Worse Training the MNN's			
Fuzzy System	% of Recognition for the FIS with Triangular MF's	% of Recognition for the FIS with Trapezoidal MF's	% of Recognition for the FIS with Trapezoidal MF's
1	7% (2/30)	7% (2/30)	7% (2/30)
2	7% (2/30)	7% (2/30)	7% (2/30)
3	7% (2/30)	7% (2/30)	7% (2/30)
4	7% (2/30)	7% (2/30)	7% (2/30)
5	7% (2/30)	7% (2/30)	7% (2/30)

In the following graphic (see Figure 37) can appreciate the comparison of the percentage obtained for the best and worse training, the trainings show in the table 1.

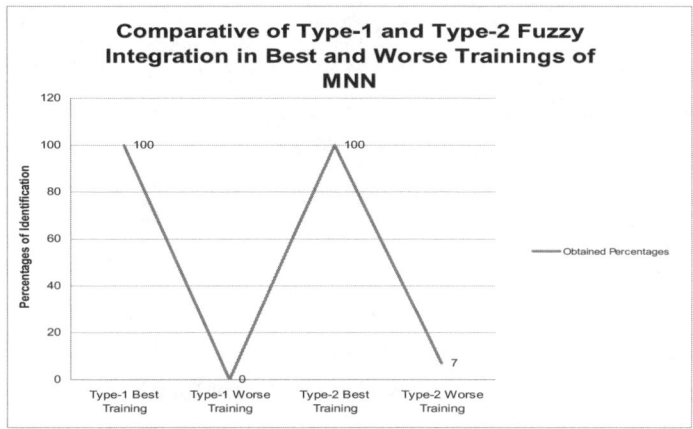

Fig. 37. Comparison of Type-1 and Type-2 Fuzzy Integration in Best and Worse Trainings of MNN

Table 12. Average Errors for the Fuzzy Inference Systems with Triangular membership functions

Type-1	Type-2
Average Error	Average Error
0.389196189	0.269979242

Table 13. Average Errors for the Fuzzy Inference Systems with Trapezoidal membership functions

Type-1	Type-2
Average Error	Average Error
0.298306163	0.688638346

Table 14. Average Errors for the Fuzzy Inference Systems with Gaussian membership functions

Type-1	Type-2
Average Error	Average Error
0.041501285	0.002356073

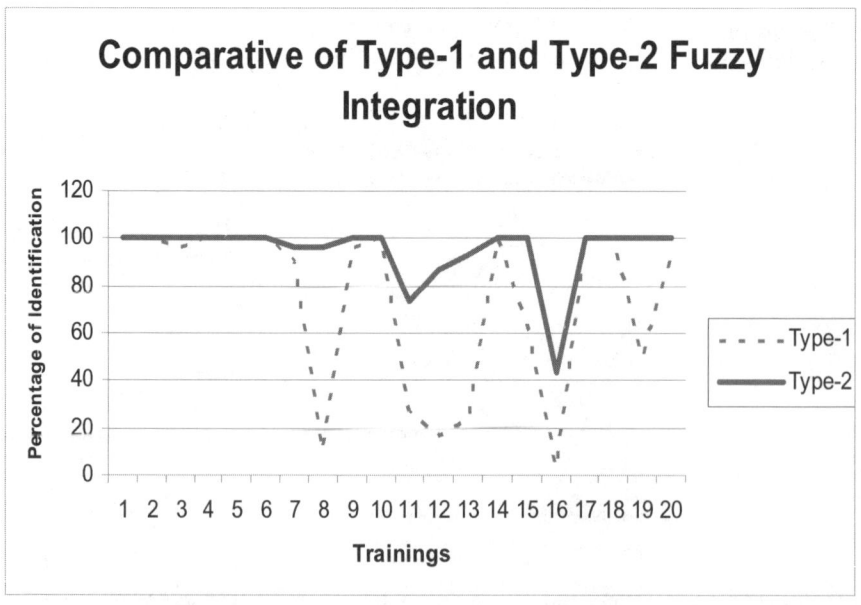

Fig. 38. Comparison of Integration with type-1 and type-2 Fuzzy Systems

Table 15. Comparison of Integration with type-1 and type-2 Fuzzy Systems

Percentages of identification obtained by test Type-1 and Type-2 Fuzzy Systems			
# of Training	Architecture of the MNN (By layer neurons for each submodule)	Average% for Type-1	Average % for Type-2
1	ModuleFace: 65,50; 75,73; 80,85 ModuleFingerprint: 110,90; 115,95; 120,100 ModuleVoice: 30,35; 35,43; 40,32	100%	100%
2	ModuleFace: 112,98; 112,98; 112,98 ModuleFingerprint: 121,93; 121,93; 121,93 ModuleVoice: 52,36; 52,36; 52,36	100%	100%
3	ModuleFace: 96,78; 96,78; 96,78 ModuleFingerprint: 113,95; 113,95; 113,95 ModuleVoice: 56,44; 56,44; 56,44	96.67%	100%
4	ModuleFace: 102,88; 102,88; 102,88 ModuleFingerprint: 111,83; 111,83; 111,83 ModuleVoice: 62,46; 62,46; 62,46	100%	100%
5	ModuleFace: 99,87; 99,87; 99,87 ModuleFingerprint: 118,96; 118,96; 118,96 ModuleVoice: 78,53; 78,53; 78,53	100%	100%
6	ModuleFace: 85,70; 85,70; 85,70 ModuleFingerprint: 90,80; 90,80; 90,80 ModuleVoice: 60,70; 60,70; 60,70	100%	100%
7	ModuleFace: 60,40; 95,80; 90,75 ModuleFingerprint: 150,100; 150,100; 150,100 ModuleVoice: 50,25; 50,25; 50,25	90%	96.67%
8	ModuleFace: 64,32; 64,32; 64,32 ModuleFingerprint: 120,60; 120,60; 120,60 ModuleVoice: 80,40; 80,40; 80,40	10%	96.67%
9	ModuleFace: 100,80; 100,80; 100,80 ModuleFingerprint: 100,100; 100,100; 100,100 ModuleVoice: 65, 50; 65, 50; 65, 50	96.67%	100%
10	ModuleFace: 50,70; 50,70; 50,70 ModuleFingerprint: 90,70; 90,70; 90,70 ModuleVoice: 75,35; 75,35; 75,35	100%	100%
11	ModuleFace: 100,90; 100,90; 100,90 ModuleFingerprint: 50,25; 50,25; 50,25 ModuleVoice: 60,60; 60,60; 60,60	26.67%	73.33%
12	ModuleFace: 110,80; 110,80; 110,80 ModuleFingerprint: 40,60; 40,60; 40,60 ModuleVoice: 80,70; 80,70; 80,70	16.67%	86.67%
13	ModuleFace: 120,85; 120,85; 120,85 ModuleFingerprint: 20,30; 20,30; 20,30 ModuleVoice: 60,50; 60,50; 60,50	23.33%	93.33%
14	ModuleFace: 130,90; 130,90; 130,90 ModuleFingerprint: 50,15; 60,20; 65,25 ModuleVoice: 68,32; 68,33; 68,34	100%	100%
15	ModuleFace: 125,90; 125,90; 125,90 ModuleFingerprint: 100,150; 100,150; 100,150 ModuleVoice: 70,65; 70,65; 70,65	63.33%	100%
16	ModuleFace: 120,85; 110,80; 125,90 ModuleFingerprint: 70,40; 60,35; 65,45 ModuleVoice: 70,30; 65,45; 75,35	3.33%	43.33%
17	ModuleFace: 135,100; 135,100; 135,100 ModuleFingerprint: 200,100; 200,100; 200,100 ModuleVoice: 80,50; 80,50; 80,50	100%	100%
18	ModuleFace: 150,100; 150,100; 150,100 ModuleFingerprint: 200,150; 200,150; 200,150 ModuleVoice: 80,50; 80,50; 80,50	100%	100%
19	ModuleFace: 128,72; 128,85; 128,92 ModuleFingerprint: 95,100; 100,85; 90,78 ModuleVoice: 55,43; 55,43; 55,43	50%	100%
20	ModuleFace: 165,100; 165,100; 165,100 ModuleFingerprint: 10,15; 30,50; 50,85 ModuleVoice: 75,30; 75,30; 75,30	93.33%	100%

7.5 Comparison Errors of GA between Type-1 and Type-2 Fuzzy Inference Systems

We show in tables 12, 13 and 14the comparison of errors of GA between the type-1 and type-2 Fuzzy Inference Systems for Triangular, Trapezoidal and Gaussian membership functions.

7.6 Comparative Integration with Type-1 and Type-2 Fuzzy Inference Systems

After the modular neural network trainings were obtained, we make the integration of the modules with the type-1 and type-2 optimized fuzzy systems. Next we show type-1 and type-2 graphics with the 20 modular neural network new trainings and the percentage of the identification (See Figure 38 and Table 15). We can appreciate that type-2 is better.

7.7 Average Percentages for Pattern Recognition

Table 16 shows the average percentages integration with type-1 and type-2 Fuzzy Inference Systems that we tested on the last experiment.

Table 16. Comparative table of average percentage integration with type-1 and type-2 fuzzy inference systems

Type-1	Type-2
Average percentage of identification	Average percentage of identification
73.50 %	94.50 %

8 Conclusions

In this paper a study of fuzzy integration methods for MNN's is presented. Type-1 and type-2 fuzzy system are considered as integration methods for a MNN's in biometry applications. The fuzzy systems were optimized using GA to be able to make an accurate comparison of type-1 and type-2 fuzzy logic as methods of integration. A comparison with simulation results for pattern recognition was made. Type-2 Fuzzy Logic is shown to be a superior method for integration of responses in MNN's.

References

[1] Algoritmos genéticos en la construcción de funciones de membresía difsa (February 2007),
    http://wotan.liu.edu/docis/dbl/iariia/2003_18_25_AGELCD.htm
[2] Algoritmos Genéticos, http://eddyalfaro.galeon.com/geneticos.html
[3] Alvarado-Verdugo, J.M.: Reconocimiento de la persona por medio de su rostro y huella utilizando redes neuronales modulares y la transformada wavelet, Instituto Tecnológico de Tijuana (2006)

[4] Bronstein, M.: Biolynx, Biometría, National Geographic Instituto de Inves-tigación Tecnológica de Georgia (April 2007),
http://www.tecnociencia.es/monografico/biometria/biometria.htm

[5] Castro, J.R.: Tutorial Type-2 Fuzzy Logic: theory and applications, Univer-sidad Autónoma de Baja California-Instituto Tecnológico de Tijuana (October 9, 2006),
http://www.hafsamx.org/cis-chmexico/seminar06/tutorial.pdf

[6] Chen, G., Phan, T.T.: Introduction to Fuzzy Sets, Fuzzy Logic and Fuzzy Control Systems. CRC Press, EEUU (2000)

[7] http://sci2s.ugr.es/publications/ficheros/estylf-2004-1-8.pdf (February 2007)

[8] INSYS, Gpo.: Soluciones Biométricas, Torreón Coah. MEXICO(February 2007),
http://www.insys.com.mx/biometria/biometria.htm

[9] Jang, J.-S.R., Sun, C.-T., Mizutani, E.: Neuro-Fuzzy and Soft Computing, A Computational Approach to Learning and Machine Intelligence. Prentice Hall, Englewood Cliffs (1997)

[10] Karnik, N., Mendel, J.M.: Operations on type-2 fuzzy sets. In: Signal and Image Processing Institute, Department of Electrical Engineering-Systems, University of Southern California, Los Angeles, CA, USA (May 11, 2000)

[11] Karnik, N., Mendel, J.M., Liang, Q.: Type-2 Fuzzy Logic Systems. IEEE Transactions on Fuzzy Systems 7(6) (December 1999)

[12] Man, K.F., Tang, K.S., Kwong, S.: Genetic Algorithms, Concepts and Designs. Springer, Heidelberg (1999)

[13] Mendel, J.M.: UNCERTAIN Rule-Based Fuzzy Logic Systems. In: Introduction and New Directions. Prentice Hall, Englewood Cliffs (2001)

[14] Mendel, J.M.: Why We Need Type-2 Fuzzy Logic Systems? Article is provided courtesy of Prentice Hall, By Jerry Mendel (May 11, 2001),
http://www.informit.com/articles/article.asp?p=21312&rl=1

[15] Mendel, J.M.: Uncertainty: General Discussions, Article is provided courtesy of Prentice Hall, By Jerry Mendel (May 11, 2001),
http://www.informit.com/articles/article.asp?p=21313

[16] Mendel, J.M., Bob-John, R.I.: Type-2 Fuzzy Sets Made Simple. IEEE Transactions on Fuzzy Systems 10(2), 117 (2002)

[17] Melin, P., Castillo, O., Gómez, E., Kacprzyk, J., Pedrycz, W.: Analysis and Design of Intelligent Systems Using Soft Computing Techniques. Advances in Soft Computing, vol. 41. Springer, Heidelberg (2007)

[18] Melin, P., Castillo, O.: Hybrid Intelligent Systems for Pattern Recognition Using Soft Computing. In: An Evolutionary Approach for Neural Networks and Fuzzy Systems. Studies in Fuzziness and Soft Computing, (Hardcover - April 29) (2005)

[19] Melin, P., Castillo, O., Kacprzyk, J., Pedrycz, W.: Hybrid Intelligent Systems. Studies in Fuzziness and Soft Computing, (Hardcover - December 20) (2006)

[20] Melin, P., Castillo, O., Gómez, E., Kacprzyk, J.: Analysis and Design of Intelligent Systems using Soft Computing Techniques. Advances in Soft Computing, (Hardcover - July 11) (2007)

[21] Mendoza, O., Melin, P., Castillo, O., Licea, P.: Type-2 Fuzzy Logic for Improving Training Data and Response Integration in Modular Neural Networks for Image Recognition. In: Melin, P., Castillo, O., Aguilar, L.T., Kacprzyk, J., Pedrycz, W. (eds.) IFSA 2007. LNCS (LNAI), vol. 4529, pp. 604–612. Springer, Heidelberg (2007)

[22] Quiliano, I.: Sistemas Modulares, Mezcla de Expertos y Sistemas Híbridos, Spain (February 2007), http://lisisu02.usal.es/~airene/capit7.pdf
[23] Ramos-Gaxiola, J.: Redes Neuronales Aplicadas a la Identificación de Locutor Mediante Voz Utilizando Extracción de Características, Instituto Tecno-lógico de Tijuana (2006)
[24] Romero, L.A.: Aplicaciones e Implementaciones de las Redes Neuronales en reconocimiento de Patrones (AIRENE) 30/06/99, cyted.html; Spain (March 2007), http://lisisu02.usal.es/~airene/airene.html
[25] Sigüenza, J., Tapiador, M.: Tecnologías Biométricas Aplicadas a la Seguri-dad (Rama) (March 2007), http://www.agapea.com/Tecnologias-biometricas-aplicadas-a-la-seguridad-n214440i.htm
[26] The 2007 International Joint Conference on Neural Networks. In: IJCNN 2007 Conference Proceedings. Orlando, Florida, USA, August 12-17, 2007. IEEE Catalog Number: 07CH37922C; ISBN: 1-4244-1380-X, ISSN: 1098-7576, ©2007 IEEE
[27] Urías, J.: Desarrollo de un nuevo Método de Integración utilizando Lógica Difusa Tipo-2 para Sistemas Biométricos, Instituto Tecnológico de Tijuana (2006)
[28] Urias, J., Melin, P., Castillo, O.: A Method for Response Integration in Modular Neural Networks using Interval Type-2 Fuzzy Logic. In: FUZZ-IEEE 2007. FUZZ, vol. 1, pp. 247–252. IEEE, London (2007)
[29] Urias, J., Hidalgo, D., Melin, P., Castillo, O.: A Method for Response Integration in Modular Neural Networks with Type-2 Fuzzy Logic for Biometric Systems. In: Patricia, M., et al. (eds.) Analysis and Design of Intelligent Systems using Soft Computing Techniques. Studies in Fuzziness and Soft Computing, vol. 1, pp. 5–15. Springer, Germany (2007)
[30] Urias, J., Hidalgo, D., Melin, P., Castillo, O.: A New Method for Response Integration in Modular Neural Networks Using Type-2 Fuzzy Logic for Biometric Systems. In: Proc. IJCNN-IEEE 2007, IEEE, Orlando (2007)
[31] Zadeh, L.A.: Knowledge representation in Fuzzy Logic. IEEE Transactions on knowledge data engineering 1, 89 (1989)
[32] Zadeh, L.A.: Fuzzy Sets. Information and Control 8 (1975)
[33] Zadeh, L.A.: Fuzzy Logic = Computing with Words. IEEE Transactions on Fuzzy Systems 4(2), 103 (1996)
[34] Zadeh, L.A.: Fuzzy Logic. Computer 1(4), 83–93 (1998)

Interval Type-2 Fuzzy Logic for Module Relevance Estimation in Sugeno Integration of Modular Neural Networks

Olivia Mendoza[1], Patricia Melín[2], and Guillermo Licea[1]

[1] Universidad Autónoma de Baja California, [2] Tijuana Institute of Technology
México
omendozad@uabc.mx, epmelin@hafsamx.org, glicea@uabc.mx

Abstract. In this paper we show the performance of an Interval Type-2 Fuzzy Inference System as a method to estimate the relevance of each module in a Modular Neural Network for images recognition. The aggregation operator used to make the integration of the simulation matrices is Sugeno Integral, and the output of the inference system are the fuzzy densities to calculate the fuzzy λ measures. Although this integration method was tested for image recognition, is possible to adapt it for distinct applications, which need information fusion of sources with uncertain relevance.

1 Introduction

Aggregation has the purpose of making simultaneous use of different pieces of information provided by several sources in order to come to a conclusion or a decision. The aggregation operators are mathematical objects that have the function of reducing a set of numbers into a unique representative number, and any aggregation or fusion process done with a computer underlies numerical aggregation [1].

Most aggregation operators use some kind of parameterization to express additional information about the objects that take part in the aggregation process; then the parameters are used to represent the background knowledge. Among all the existing types of parameters, the fuzzy measures are a rich and important family. They are of interest because they are used for aggregation purposes in conjunction with fuzzy integrals like Choquet and Sugeno Integrals [2].

In this paper we describe an image recognition method using Modular Neural Networks combined with the Sugeno Integral. The information to be combined is the simulation output of the 3 modules trained to recognize a different part of the image [3]. The modular architecture consists in dividing each image in 3 parts after the edges detection process, and using each part as training data for 3 monolithic neural networks [4]. Then the problem becomes how to combine the simulation of the three modules in order to recognize the maximum number of images possible.

We make the final decision using the Sugeno Integral, which is used to combine the simulation vectors into only one vector, then at the end of the method the system decides on the best choice of recognition in the same manner than made with only one monolithic neural network, but with the problem of complexity resolved with modularity [5].

O. Castillo et al. (Eds.): Soft Computing for Hybrid Intel. Systems, SCI 154, pp. 115–127, 2008.
springerlink.com   © Springer-Verlag Berlin Heidelberg 2008

Then the problem is to find the ideal input parameters for the Sugeno Integral, which means, the values of the fuzzy density for each module, to rank its relevance in the decision process. This problem was solved by building a FIS to estimate the fuzzy densities using only the simulation vectors as input variables. This step was implemented with two fuzzy logic systems, namely using Type-1 and Interval Type-2 Fuzzy Logic respectively [6].

2 Modular Neural Networks

In this section the modular neural networks are described.

2.1 Modular Structure

The modular structure was designed specifically for a database of images, like the Olivetty Research Laboratory database of faces (ORL) [7], but it is not limited to this data. To measure the recognition rate in an objective form, we trained the modular neural networks with the set of images in a random ordered fashion, this process is called a random permutation.

The design of the Modular Neural Network consists of 3 monolithic feed-forward neural networks, each one trained with a supervised method with the first 7 samples of the 40 images of the ORL database.

The edges vector for each image is accumulated into a matrix, as shown in the scheme of figure 1. Then the complete matrix of images is divided into 3 parts, each module is trained with a corresponding part, with the some rows for overlapping [8][9].

The target to the supervised training method consists of one identity matrix with dimensions 40x40 for each sample, building one matrix with total dimensions (40x40x7), as shown in figure 2 [10].

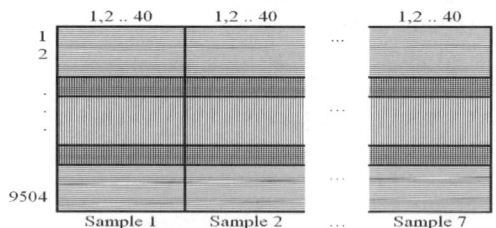

Fig. 1. Input: Seven images for each person

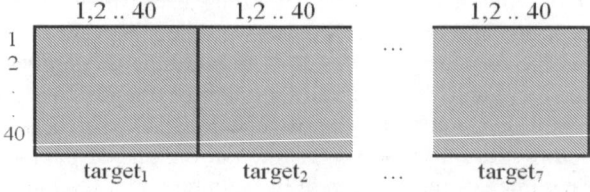

Fig. 2. Target: One identity matrix with dimensions 40x40 for each sample

Each monolithic neural network has the same structure of figure 3 and was trained under the same conditions [11]:

- Three hidden layers with 200 neurons and *tansig* transfer functions.
- The output layer with 40 neurons and *purelin* transfer functions.
- The training function is gradient descent with momentum and adaptive learning rate back-propagation (*traingdx*).

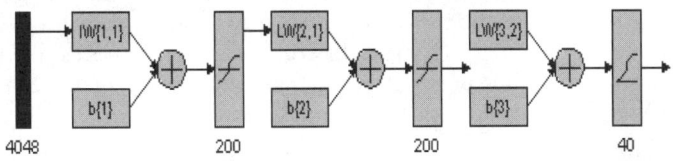

Fig. 3. Structure of each monolithic neural network

2.2 Training of the Modules

The next code segment was created to train each module, where *newff* is a Matlab Neural Network Toolbox function to create a feed-forward back-propagation network with the parameters specified as follows [12].

```
layer1=200; layer2=200; layer3=40;
net=newff(minmax(p),[layer1,layer2,layer3],{'tansig','tansig','log
sig'},'traingdx');
net.trainParam.goal=1e-5;
net.trainParam.epochs=1000;
```

2.3 Modules Simulation

A program was developed in Matlab that simulates each module with the 400 images of the ORL database, building a matrix with the results of the simulation of each module, as it is shown in figure 4. These matrices are stored in the file "mod.mat" to be analyzed later for the combination of results [13].

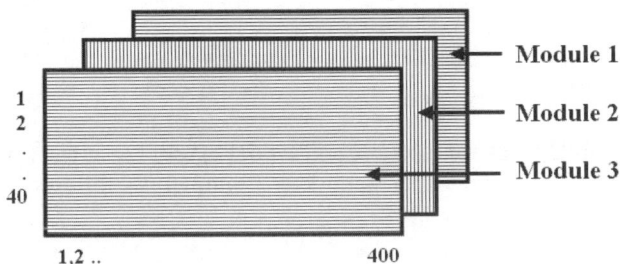

Fig. 4. Scheme of simulation matrices for the three modules

In the simulation matrix, columns corresponding to the images in the training data, with a value near one, are always selected correctly. However, some outputs of the training data have very low values in all positions, reason why it is very important to have a good combination method to recognize more images [13].

3   Sugeno Integral for Modules Fusion

For the recognition of one image we have to divide it in three parts, and then simulate each one using the corresponding module. Then for each image for recognition we have three simulation vectors. To make the final decision is necessary the fusion of the three vectors, using an aggregation operator like the Sugeno Integral. In this section we provide some basic concepts about Sugeno Measures and Sugeno Integral.

3.1   Sugeno λ-Meausures

The Sugeno measures (λ-measures) are monotonic measures, μ that are characterized by the following requirement: For all A, B ∈ P (X), if A ∩ B = Ø, then:

$$\mu(A \cup B) = \mu(A) + \mu(B) + \lambda\mu(A)\mu(B) \tag{1}$$

Where λ>-1 is a parameter by which different λ-measures are distinguished. Equation (1) is usually called λ-rule [14].

When X is a finite set and values μ({x}) (called fuzzy densities) are given for all x∈ X, then the value μ(A) for any A ∈ P (X), can be determined from these values on singletons by a repeated application of the λ-rule. This value can be expressed as:

$$\mu(A) = \left[\prod_{x \in S} (1 + \lambda\mu(\{x\})) \right] / \lambda \tag{2}$$

Observe that, given values μ({x}) for all x∈ X, the values of λ can be determined by the requirement that μ({X})=1. Applying this requirement to the equation (2) results in the equation (3):

$$\lambda + 1 = \prod_{i=1}^{n} (1 + \lambda\mu(\{x_i\})) \tag{3}$$

for λ. This equation determines the parameter uniquely under the conditions stated in the following theorem [14]:

Theorem 1. Let μ({x}) <1 for all x∈ X and let μ({x}) >0 for at least two elements on X. Then equation (3) determines the parameter λ uniquely as follows:

$$\sum_{x \in X} \mu(\{x\}) < 1$$

If the above expression is valid, then λ is equal to the unique root of the equation in the interval $(0, \infty)$, which means that μ qualifies as a lower probability, λ>0.

$$\sum_{x \in X} \mu(\{x\}) = 1$$

If the above expression is valid, then $\lambda=0$, which is the only root of the equation, which means that μ is a classical probability measure, $\lambda=0$.

$$\sum_{x \in X} \mu(\{x\}) > 1$$

If the above expression is valid, then λ is equal to the unique root of the equation in the interval $(-1,0)$, which means that μ qualifies as an upper probability, $\lambda<0$.

Once the value of λ is calculated using some numerical method for zero finding, the Sugeno measures can be calculated using equations (4) and (5), after descendent ordered of the sets X and $\mu(\{x\})$, respect to the elements of set X [15].

$$\mu(A_1) = \mu(x_1) \tag{4}$$

$$\mu(A_i) = \mu(x_i) + \mu(A_{i-1}) + \lambda\mu(x_i)\mu(A_{i-1}) \tag{5}$$

3.2 The Sugeno Integral

The Sugeno Integral can be interpreted in a way similar to the weighted maximum, but using the measures instead of possibility distributions. The difference is that now each value is weighted according to the weight (the measure) of all the sources that support the value (6) [16] .

$$h(\sigma_1,...,\sigma_n) = \max{}_{i=1}^{n}(\min(\sigma_i, \mu(A_i))) \tag{6}$$

Where $\sigma i = \sigma(xi)$ y $0 \le \sigma 1 \le ... \le \sigma n \le 1$

The Sugeno Integral generalizes both weighted minimum and weighted maximum.

A weighted maximum with a possibilistic weighting vector u is equivalent to a Sugeno integral with fuzzy measure

$$\mu_u^{w\,max}(A) = \max_{a_i \in A} u_i \tag{7}$$

A weighted minimum with a possibilistic weighting vector u is equivalent to a Sugeno integral with fuzzy measure

$$\mu_u^{w\,min}(A) = 1 - \max_{a_i \notin A} u_i \tag{8}$$

4 Fuzzy Logic for Density Estimation

After the simulation of an image in the Neural Network, the simulation value is the only known parameter to make a decision, then to estimate the fuzzy density of each module this is the only available information. For this reason we analyze the simulation matrix in many tests and decide that each of the inputs to the FIS corresponds to the maximum value of each column corresponding to the simulation of each module of the 400 images [17]. Then m1, m2 and m3 correspond to the simulation values to combine, max1, max2 and max3 correspond to the maximum values of the simulation vector, and d1, d2 and d3 the fuzzy densities for each module, as shown in figure 5 [18].

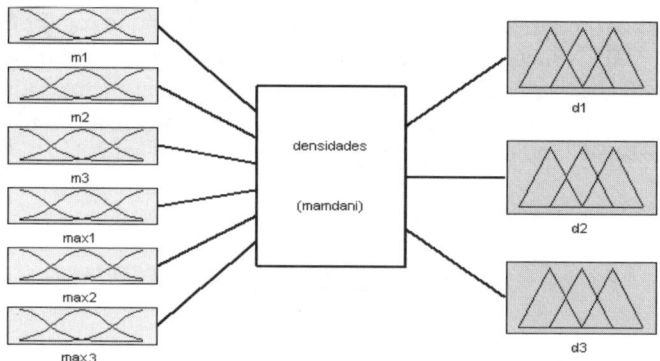

Fig. 5. Variables for the FIS to find the fuzzy densities

Each output corresponds to one fuzzy density, to be applied to each module to perform the fusion of results later with the Sugeno Integral [19]. The inference rules calculate fuzzy densities near 1 when de maximum value in the simulation is between 0.5 and 1, and near 0 when the maximum value in the simulation is near 0. The fuzzy rules are shown next.

1.If (m1 is LOW) and (max1 is LOW) then (d1 is LOW)
2.If (m1 is MEDIUM) and (max1 is MEDIUM) then (d1 is MEDIUM)
3.If (m1 is HIGH) and (max1 is HIGH) then (d1 is HIGH)
4.If (m2 is LOW) and (max2 is LOW) then (d2 is LOW)
5.If (m2 is MEDIUM) and (max2 is MEDIUM) then (d2 is MEDIUM)
6.If (m2 is HIGH) and (max2 is HIGH) then (d2 is HIGH)
7.If (m3 is LOW) and (max3 is LOW) then (d3 is LOW)
8.If (m3 is MEDIUM) and (max3 is MEDIUM) then (d3 is MEDIUM)
9.If (m3 is HIGH) and (max3 is HIGH) then (d3 is HIGH)
10.If (m1 is LOW) and (max1 is MEDIUM) then (d1 is LOW)
11.If (m1 is MEDIUM) and (max1 is HIGH) then (d1 is MEDIUM)
12.If (m2 is LOW) and (max2 is MEDIUM) then (d2 is LOW)
13.If (m2 is MEDIUM) and (max2 is HIGH) then (d2 is MEDIUM)
14.If (m3 is LOW) and (max3 is MEDIUM) then (d3 is LOW)
15.If (m3 is MEDIUM) and (max3 is HIGH) then (d3 is MEDIUM)
16.If (m1 is LOW) and (max1 is HIGH) then (d1 is LOW)
17.If (m2 is LOW) and (max2 is HIGH) then (d2 is LOW)
18.If (m3 is LOW) and (max3 is HIGH) then (d3 is LOW)

According to exhaustive tests made in the simulation matrices, we know that recognition of the images that were used for the training the neural networks is 100%. Therefore the interest is focused on the recognition of the samples that do not belong to the training set, in other words samples 8,9 and 10.

The parameters for the Sugeno Fuzzy Integral that will be inferred will be the Fuzzy Densities, with values between 0 and 1 for each module, which determines the relevance rate for each module.

4.1 FIS-1 for Estimate Fuzzy Densities

The membership functions in figure 6 and the solution surface in figure 7 show the system's behavior.

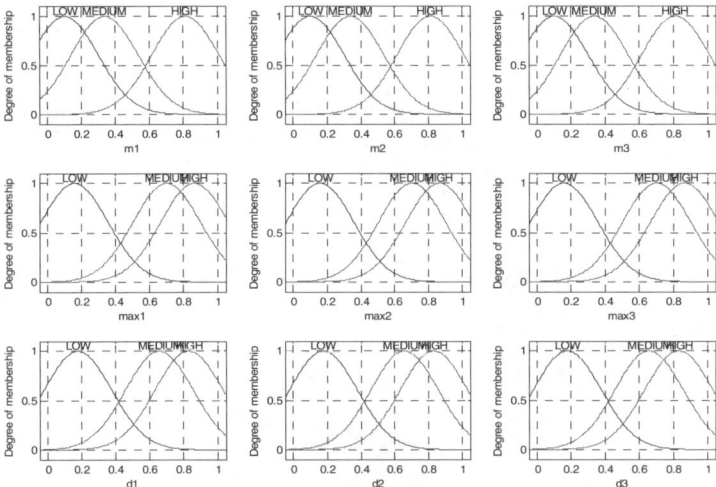

Fig. 6. Variables for FIS-1 to find fuzzy densities, before optimization

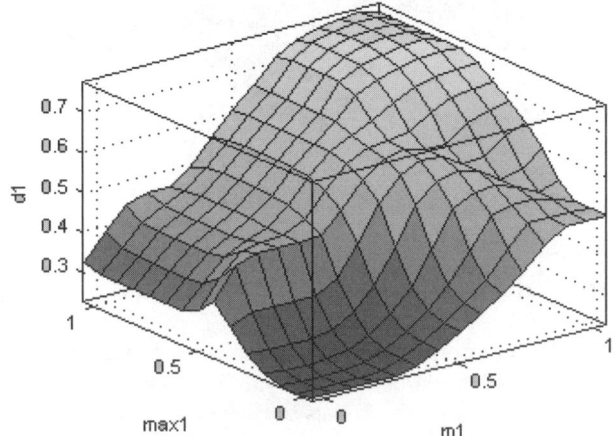

Fig. 7. Solution surface for max1, m1 and d1 in FIS-1 to find fuzzy densities

4.2   FIS-2 for Estimate Fuzzy Densities

To compare the results FIS-1 and FIS-2 to estimate the fuzzy densities, we added a FOU=0.2, to the same fuzzy variables in FIS-1, as we can see in figure 8 [20][21].

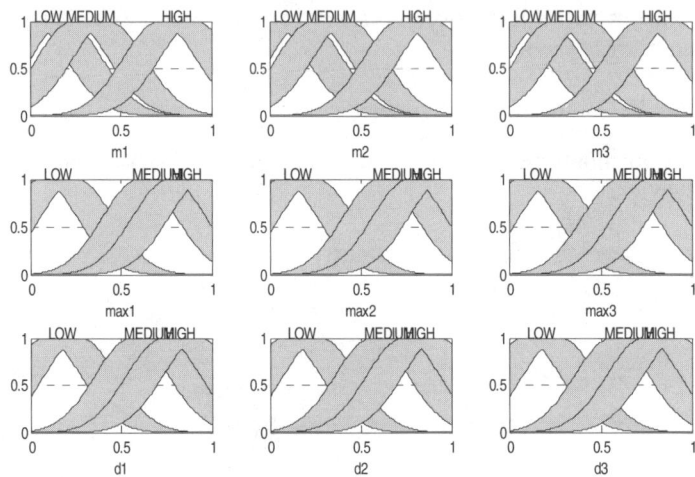

Fig. 8. Variables for the FIS-2 to estimate fuzzy densities, using the same centers of the FIS-1 and FOU=0.2

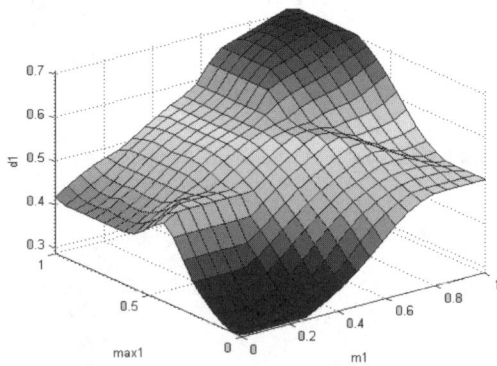

Fig. 9. Solution surface for max1, m1 and d1 in FIS-2 to find fuzzy densities

5   Sugeno Integral for Information Fusion

Then, after the simulation of one image divided in three modules we have three simulation vectors of length 40 to combine. The value for each element of the resulting vector is the Sugeno Integral for the values corresponding to the same position in the 3 vectors to combine [22].

Consider the following values corresponding to the simulation of sample 8 of person number 13, this sample is not on the training data.

	m1		m2		m3
1					
2					
3					
:					
13	0.1575		0.0094		0.0247
14					
:					
40					
	max1		max2		max3
	0.1575		0.0286		0.0574

The fuzzy densities were inferred by a FIS as follows: First the FIS calculates the fuzzy densities for each module.

d1= 0.3069
d2= 0.1788
d3= 0.1890

Then we need to solve the function f (λ)=$(1+0.3069\lambda)(1+0.1788\lambda)(1+0.1890\lambda)$ – $(1+\lambda)$ as we can see in figure 10, according with equation (11).

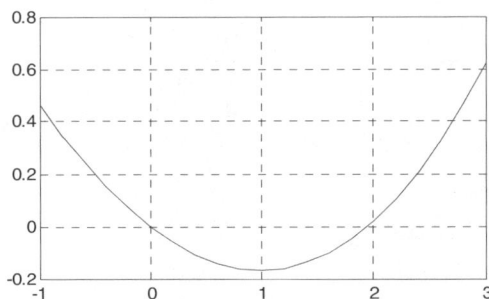

Fig. 10. Plot for f (λ)=$(1+0.3069\lambda)(1+0.1788\lambda)(1+0.1890\lambda)$ –$(1+\lambda)$

The value of λ can be calculated by equation (11), using a numerical method such as Newton-Raphson or bisection, to find the root of f (λ) shown in figure 10. To solve this equation we used the Matlab function 'fzero' that find the root of continuous function of one variable. The solution for λ= 1.9492 in this example, and the Sugeno Measures can be constructed using the recursive formulas (12) and (13), but before this calculation the fuzzy densities must be sorted descendent by σ (xi), that is shown in Table 1.

Table 1. The Values Sorted Descendent By X Before Normalization

x_i	$\sigma\,(x_i)$=trapmf([0 max(i) max(i) 2],xi)	$\mu(x_i)$
0.1575	0.9989	0.3069
0.0247	0.0492	0.1890
0.0094	0.0249	0.1788

$\mu(A_1)=\mu(x_1)$
$\mu(A_i)=\mu(x_i)+\mu(A_{i-1})+\lambda\mu(x_i)\mu(A_{i-1})$

$\mu(A_1)= 0.3069$
$\mu(A_2)= 0.1890+0.3069+(1.9492)(0.1890)(0.3069)= 0.6090$
$\mu(A_3)= 0.1788+0.6090+(1.9492)(0.1788)(0.6090)= 1$

The Fuzzy Sugeno Integral, can now be calculated using (6):
h(0.9989,0.0492,0.0249)=
max(
min(0.9989,0.3069),
min(0.0492,0.6090),
min(0.0249,1))
h(0.9989,0.0492,0.0249)=max(0.3089,0.0492,0.0249)=0.3089

As can be seen in figure 11, for person number 13 it was obtained the highest value of the Sugeno Integral and is correctly selected.

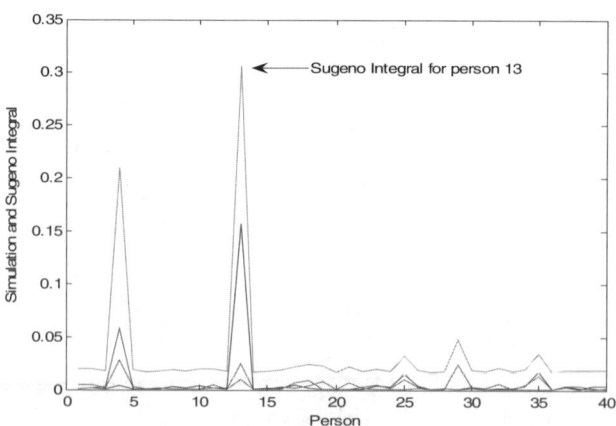

Fig. 11. Simulation of the Modules and the Sugeno Integral

In order to measure in an objective form the final results, we developed a method of random permutation, which rearranges the samples of each person before the training. Once a permutation is made, the modular neural networks are trained 5 times and the net weights and permutation order are saved for posterior tests.

6 Simulation Results

Some of the images don't reach a sufficient value in the simulation of the three modules, in these cases, there isn't enough information to select an image at the modules combination, and the image can be wrongly selected. But, in most of the cases the images are correctly selected. In tables 1 and 2 we show the recognition rates for the FIS-1 and FIS-2 methods.

Table 2. Recognition rates with FIS-1 Fuzzy Densities Estimation

P	Image Recognition (%)				
	Train 1	Train 2	Train 3	Avg	Max
1	94.00	95.75	94.50	94.75	95.75
2	94.25	94.75	94.25	94.41	94.75
3	94.25	94.25	95.25	94.58	95.25
4	94.00	93.25	93.50	93.58	94.00
5	94.75	94.75	94.00	94.36	94.75
				94.36	95.75

Table 3. Recognition rates with FIS-2 Fuzzy Densities Estimation

P	Image Recognition (%)				
	Train 1	Train 2	Train 3	Avg	Max
1	97.25	96.25	95.00	96.17	97.25
2	94.75	95.25	95.75	95.25	95.75
3	95.50	97.50	96.00	96.33	97.50
4	95.25	95.00	95.50	95.25	95.50
5	96.50	97.00	96.00	96.50	97.00
				95.90	97.50

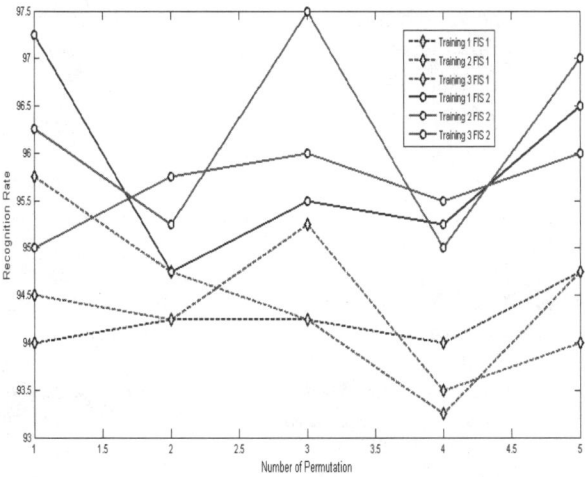

Fig. 12. Recognition Rates for FIS-1 and FIS-2 fuzzy densities estimation

To appreciate the performance of each fuzzy system more clearly, in figure 12 we show the recognition rates for the simulation of the Modular Neural Network, for 3 different trainings of type-1 and type-2 fuzzy logic.

In figure 13 we show the best recognition rates for both types of fuzzy systems. As we can see in this comparison, the best recognition rates using the Type-2 Fuzzy System are better than with the Type-1 Fuzzy System.

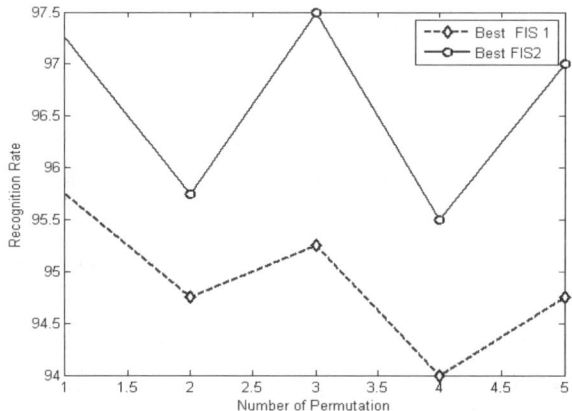

Fig. 13. The best Recognition Rates for the FIS-1 and FIS-2 fuzzy densities estimation

7 Conclusions

In image recognition with Modular Neural Networks, when an image is divided into several parts, each module is an expert to recognize only one part. Then, in the simulation phase if one of the modules doesn't reach the target values, is a module with low relevance for the recognition.

The Fuzzy density is the rate of the relevance of each module, and the estimation of its values improves the recognition rates on Modular Neural Networks combined with Sugeno Integral. Fuzzy densities estimated with the Interval Type-2 Fuzzy System produces better recognition rates than the Type-1 Fuzzy System.

References

1. Detyniecki, M.: Mathematical Aggregation Operators and Their Application to Video Querying, PHD Artificial Intelligence, Pierre and Marie Curie University, Paris (2000), http://www-poleia.lip6.fr/~marcin/publications.htm
2. Melín, P., Castillo, O.: An intelligent hybrid approach for industrial quality control combining neural networks. fuzzy logic and fractal theory, Journal of Information Science 77(7), 1543–1557 (2007)
3. Melín, P., Mancilla, A., Lopez, M., Mendoza, O.: A Hybrid Modular Neural Network Architecture with Fuzzy Sugeno Integration for Time Series Forecasting. Applied Soft Computing Journal 7(4), 1217–1226 (2007)
4. Sharkey, A.J.C.: Ensemble and Modular Multi-Net Systems. In: Combining Artificial Neural Nets, pp. 1–25. Springer, Germany (1999)

5. Melin, P., Mendoza, O., Soto, M., Gutierrez, M., Solano, D.: An Intelligent System for Pattern Recognition and Time Series Prediction Using Modular Neural Networks. In: IEEE World Congress on Computational Intelligence, pp. 8197–8193 (2006)
6. Mendel, J.M.: Advances in type-2 sets and systems. Information Sciences 177, 84–110 (2007)
7. AT&T Laboratories Cambridge, The ORL database of faces, http://www.cl.cam.ac.uk/research/dtg/attarchive/facedatabase.html
8. Mendoza, O., Melín, P.: Sistemas de Inferencia Difusos Tipo-1 y Tipo-2 Aplicados a la Detección de Bordes en Imágenes Digitales. In: International Seminar on Computational Intelligence 2006 IEEECIS, Tijuana, Mexico, pp. 117–123 (2006)
9. Mendoza, O., Melín, P.: The Fuzzy Sugeno Integral as a Decision Operator in the Recognition of Images with Modular Neural Networks. In: International Conference on Fuzzy Systems, Neural Networks and Genetic Algorithms, Tijuana, México, pp. 105–116 (2005)
10. Salinas, R., Larraguibel, L.: Red Neuronal de Arquitectura Paramétrica en Reconocimiento de Rostros, Divulgación Electrónica de las Ciencias, vol.17 (2002), http://cabierta.uchile.cl/revista/17/articulos/articulos.html
11. Mendoza, O., Melin, P., Licea, G.: Type-2 Fuzzy Systems as a Method for Improving Training Data and Decision Making in Modular Neural Networks for Pattern Recognition. International Journal of Information Technology And Intelligent Computing, Academy of Humanities And Economics (Wshe), Polonia, Int. J. It&Ic 1(4) (in prees, 2007)
12. Mendoza, O., Melin, P., Castillo, O., Licea, G.: Type-2 Fuzzy Logic for Improving Training Data and Response Integration in Modular Neural Networks for Image Recognition. In: Foundations of Fuzzy Logic And Soft Computing, pp. 604–612. Springer, Germany (2007)
13. Mendoza, O., Melin, P., Licea, G.: Modular Neural Networks and Type-2 Fuzzy Logic for Face Recognition. In: NAFIPS 2007 International Conference, San Diego, E.U.A, pp. 622–627 (2007)
14. Torra, V., Narukawa, Y.: Modeling Decisions, Information Fusion and Aggregation Operators. Springer, Germany (2007)
15. Mendoza, O., Melín, P., Licea, G.: Type-2 Fuzzy Systems for Improving Training Data and Decision Making in Modular Neural Networks for Image Recognition. In: International Joint Conference on Neural Networks, Orlando, E.U.A, pp. 1766–1770 (2007)
16. Klir, G.: Uncertainty and Information. Wiley & Sons, Inc., E.U.A (2005)
17. Mendoza, O., Melín, P.: The Fuzzy Sugeno Integral as a Decision Operator in the Recognition of Images with Modular Neural Networks. In: Hybrid Intelligent Systems:Design and Analysis, pp. 299–310. Springer, Germany (2007)
18. Melin, P., Castillo, O.: Hybrid Intelligent Systems for Pattern Recognition. Springer, Heidelberg (2005)
19. The MathWorks, Inc., Neural Network Toolbox (1994-2006), http://www.mathworks.com/products/neuralnet/
20. Castillo, O., Melin, P., Kacprzyk, J., Pedrycz, W.: Hybrid Intelligent Systems: Design and Analysis. Springer, Germany (2007)
21. Mendel, J.: Uncertain Rule-Based Fuzzy Logic Systems: Introduction and New Directions. Prentice-Hall, U.S.A (2001)
22. Castro, J.R., Castillo, O., Martínez, L.G.: Tool Box para Lógica Difusa Tipo-2 por Intervalos. In: International Seminar on Computational Intelligence 2006 IEEE - CIS Mexico Chapter, Tijuana, Mexico, pp. 100–108 (2006)
23. Mendoza, O., Torres, G., Melin, P.: An Intelligent System for Modeling and Simulation with Modular Neural Networks. In: International Conference on Artificial Intelligence, Las Vegas, E.U.A, pp. 406–411 (2005)

Optimization of Response Integration with Fuzzy Logic in Ensemble Neural Networks Using Genetic Algorithms

Miguel Lopez[1,2], Patricia Melin[2], and Oscar Castillo[2]

[1] PhD Student of Computer Science in the Universidad Autonoma de Baja California,
Tijuana, B.C., México
danym23@aol.com
[2] Computer Science in the Graduate Division Tijuana,
Institute of Technology Tijuana, B.C., Mexico
pmelin@tectijuana.mx, ocastillo@tectijuana.mx

Abstract. We describe in this paper a new method for response integration in ensemble neural networks with Type-1 Fuzzy Logic and Type-2 Fuzzy Logic using Genetic Algorithms (GA's) for optimization. In this paper we consider pattern recognition with ensemble neural networks for the case of fingerprints. An ensemble neural network of three modules is used. Each module is a local expert on person recognition based on their biometric measure (Pattern recognition for fingerprints). The Response Integration method of the ensemble neural networks has the goal of combining the responses of the modules to improve the recognition rate of the individual modules. Using GA's to optimize the Membership Functions of The Type-1 Fuzzy System and Type-2 Fuzzy System we can improve the results of the fuzzy systems. We show in this paper the results of a type-2 approach for response integration that improves performance over the type-1 logic approaches.

1 Introduction

At the moment, a variety of methods and techniques are available to determine the unique identity of the person, the most common being fingerprint, voice, face and iris recognition [12]. Of these, fingerprint and iris offer a very high level of certainty as to a person's identity, while the others are less accurate. The four primary methods of biometric authentication in widespread use today are face, voice, fingerprint, and iris recognition. In this paper, we consider pattern recognition with ensemble neural networks for the case of fingerprints.

Fingerprint Recognition. The process of authenticating people based on their fingerprints can be divided into three tasks. First, you must collect an image of a fingerprint, second, you must determine the key elements of the fingerprint for confirmation of identity, and third, the set of identified features must be compared with a previously-enrolled set for authentication. The system should be never expected to see a complete 1:1 match between these two sets of data. In general, you could expect to couple any collection device with any algorithm, although in practice most vendors offer proprietary, linked solutions.

O. Castillo et al. (Eds.): Soft Computing for Hybrid Intel. Systems, SCI 154, pp. 129–150, 2008.
springerlink.com © Springer-Verlag Berlin Heidelberg 2008

A number of fingerprint image collection techniques have been developed. The earliest method developed was optical: using a camera-like device to collect a high resolution image of a fingerprint. Later developments turned to silicon-based sensors to collect an impression by a number of methods, including surface capacitance, thermal imaging, pseudo-optical on silicon, and electronic field imaging.

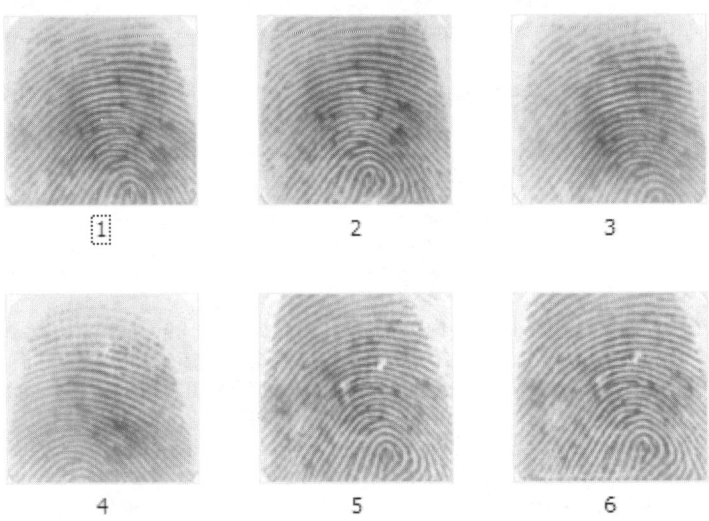

Fig. 1. Sample images from FCV200 database; each row shows different impressions of the same finger

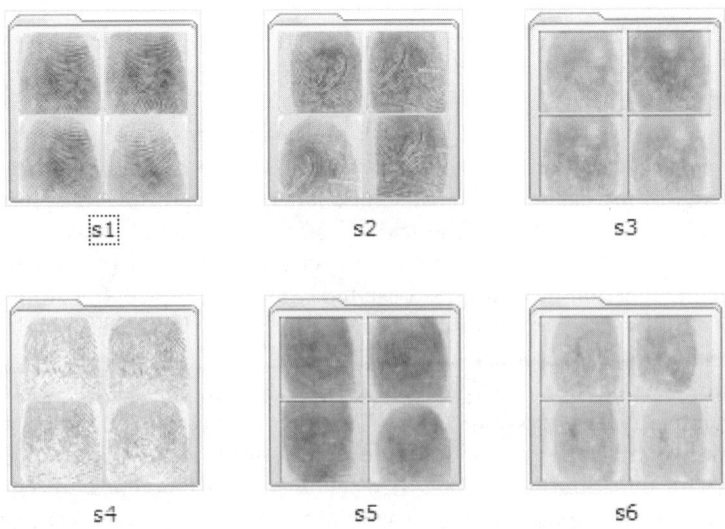

Fig. 2. Images from FCV200 database; all the samples are from different fingers and are ordered by persons

As discussed, a variety of fingerprint detection and analysis method exist, each with their own strengths and weaknesses. Consequently, researchers vary widely on their claimed (and achieved) false accept and false reject rates. The poorest systems offer a false accept rate of around 1:1,000, while the best are approaching 1:1,000,000. False reject rates for the same vendors are around 1:100 to 1:1000.

In the experiments performed in this research work, we used the database of the Fingerprint Verification Competition FCV2000 [8], [9]; the image size is 300 pixels wide and 300 pixels high with a resolution of 500 ppi, and representation of a gray scale. The fingerprints were acquired by using a low-cost optical sensor; up to four fingers were collected for each volunteer (forefinger and middle finger of both the hands). The database is 10 fingers wide (w) and 8 impressions per finger deep (d) (80 fingerprints in all); the acquired fingerprints were manually analyzed to assure that the maximum rotation is approximately in the range [-15°, 15°] and that each pair of impressions of the same finger have a non-null overlapping area.

Sample images from the FCV200 database are shown in figures 1 and 2; each row shows different impressions of the same finger:

2 Design of the Genetic Algorithms for Optimization

GA's are well-known robust methods of optimization. The GA deals with coding the problem using chromosomes to optimize the membership function of the type-1 and type-2 fuzzy logic system, used as response integration methods in ensemble neural networks. The Characteristics of the GA are the following:

1   GA's are parallel-search procedures that can be implemented on parallel processing machines for massively speeding up their operations.
2   GA's are applicable to both continuous and discrete optimization problems.
3   GA's are stochastic and less likely to get trapped in local minima, which inevitably are present in any practical optimization application.
4   GA's have flexibility in both structure and parameter identification in complex models such as neural networks and fuzzy inference systems.

In each generation, the GA's build a new population using genetic operators such as crossover and mutation, that is, members with higher fitness values are more likely to survive and participate in mating crossover operations. After a number of generations, the populations contains members with better fitness values in which this is analogous to Darwinian models of evolution by random mutation and natural selection. GA's are sometimes referred to as methods of population based optimization that improves performance by upgrading entire populations rather than individual members.

In this paper, the GA's are used to optimize the membership functions of the type-1 and type-2 fuzzy logic system as response integration methods in ensemble neural networks [3], [4]. The GA deals with coding of the problem using chromosomes for optimize the membership function of the type-1 and type-2 fuzzy logic system. The Chromosome Structure used to optimize the membership function of the type-1 fuzzy system as response integration method in ensemble neural networks shown in the figure 3.

Chromosome Type-1 Fuzzy Logic

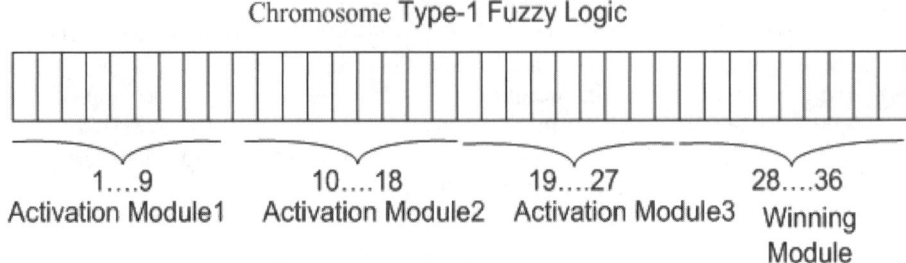

1....9	10....18	19....27	28....36
Activation Module1	Activation Module2	Activation Module3	Winning Module

Fig. 3. The Chromosome Structure of Type-1 Fuzzy Logic to optimize the input and output membership functions

Chromosome for Type-1 Fuzzy Logic. The first 9 bits represent the activation of the membership functions for Module1. Each Module has three input linguistic variables, which are Activation Low, Activation Medium and Activation High. In the same way the bits from 10 to the 18 for Module2 and the bits from the 19 to the 27 are for Module3, and the bits from the 28 to the 36 represent to the output variable Winning Module for three Membership functions, WinnningModule1, WinningModule2 and WinningModule3.

The Chromosome Structure used to optimize the membership function of the type-2 fuzzy logic system, as response integration method, in ensemble neural network is shown in the figure 4.

Chromosome Type-2 Fuzzy Logic

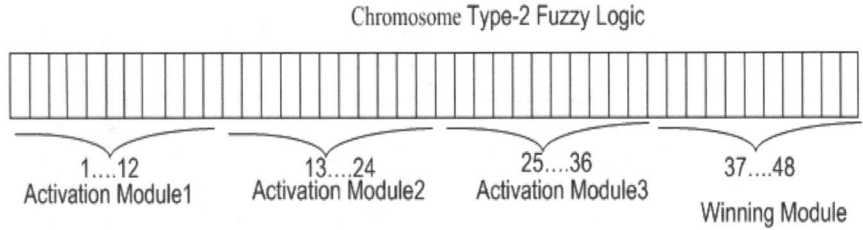

1....12	13....24	25....36	37....48
Activation Module1	Activation Module2	Activation Module3	Winning Module

Fig. 4. The Chromosome Structure Type-1 Fuzzy Logic for optimizes the input and output membership function

Chromosome for Type-2 Fuzzy Logic. The first 12 bits represent the activation of the membership functions for Module1. Each Module has three input linguistic variables, which are Activation Low, Activation Medium and Activation High. In the same way the bits from 13 to 24 are for the Module2 and the bits from 25 to the 36 are for the Module3, the bits from the 37 at the 48 represents to the output variable Winning Module for three Membership functions, WinnningModule1, WinningModule2 and WinningModule3.

3 Proposed Architecture for Fingerprint Recognition

The proposed architecture in this paper consists of three main modules, in which each module in turn consists of a set of neural networks trained with the same data

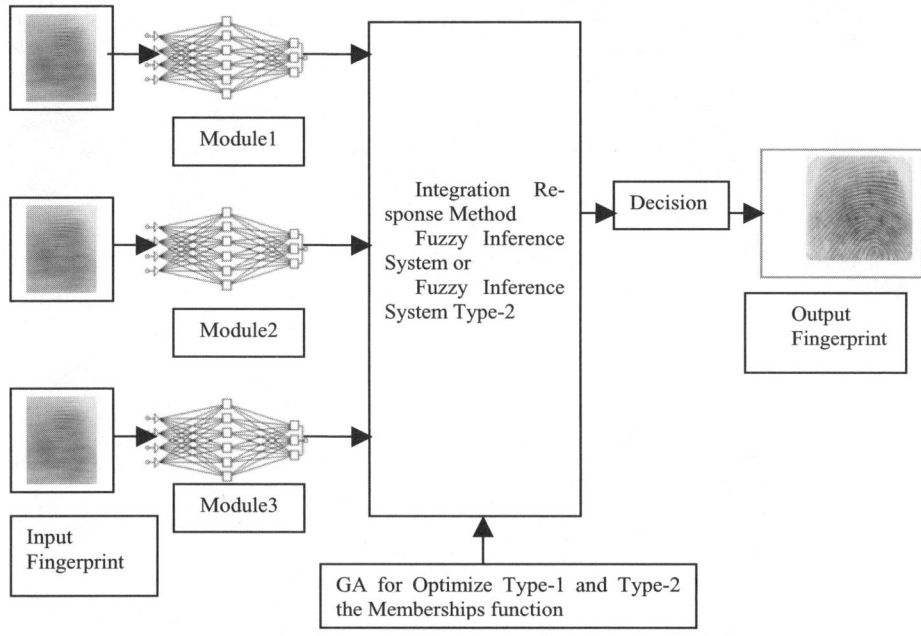

Fig. 5. Architecture of the Proposed Ensemble Neural Network for Fingerprint Recognition using GA's for Optimize Type-1 and Type-2 Membership Function

(fingerprints), which provides the modular architecture that is shown in Figure 5. The integration of responses is performed with a fuzzy system that is optimized using a GA.

4   Response Integration with Type-1 Fuzzy Logic

Over the past decade, fuzzy systems have displaced conventional technology in different scientific and system engineering applications, especially in pattern recognition and control systems. The same fuzzy technology, in approximate reasoning form, is resurging also in the information technology, where it is now giving support to decision making and expert systems with powerful reasoning capacity and a limited quantity of rules [25]. For the case of modular networks, a fuzzy system can be used as an integrator of results [24].

The Type-1 Fuzzy Inference System, as method of response integration, of the ensemble neural network output, is considering as input three linguistic variables, Activation Low, Activation Medium, and Activation High, and one output linguistic variable, Winning Activation of the three modules.

Three membership functions were used for each linguistic variable (input and output) of the triangular type, to be considered in a range from 0 to 1.

We show in figures 6, 7, 8, and 9, the membership functions designed using the editor of the fuzzy logic toolbox of MATLAB [10], for all the linguistic variable of the fuzzy system.

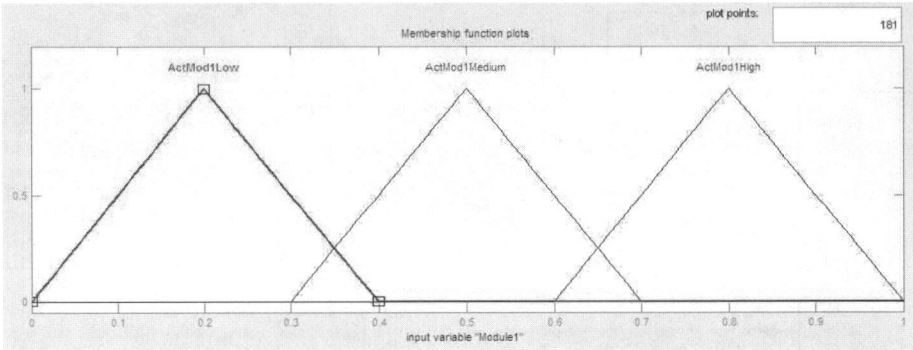

Fig. 6. Membership Functions of Input Module 1

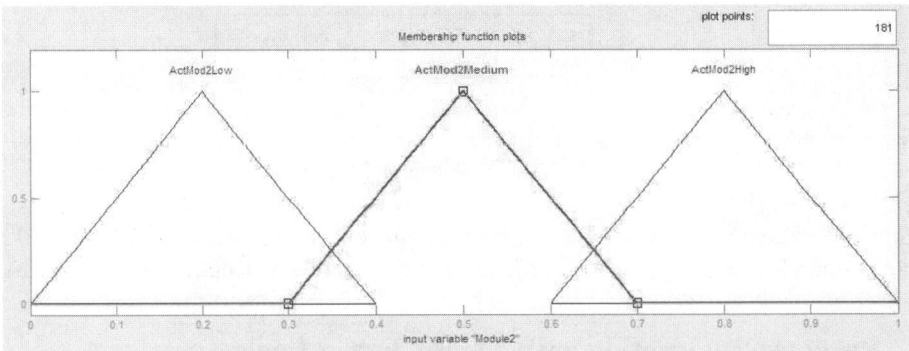

Fig. 7. Membership Functions of Input Module 2

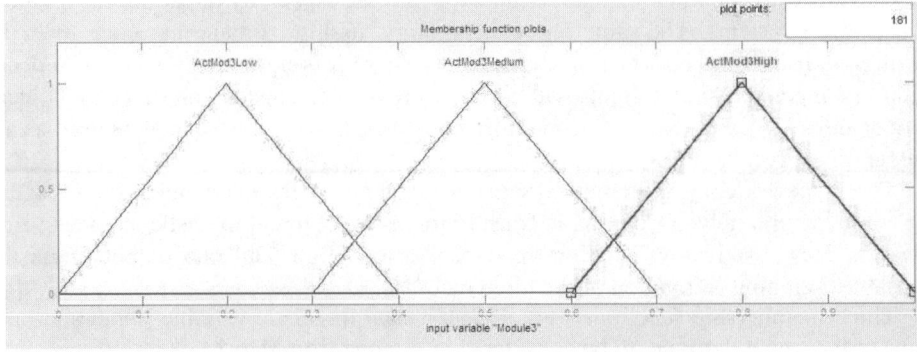

Fig. 8. Membership Functions of Input Module 3

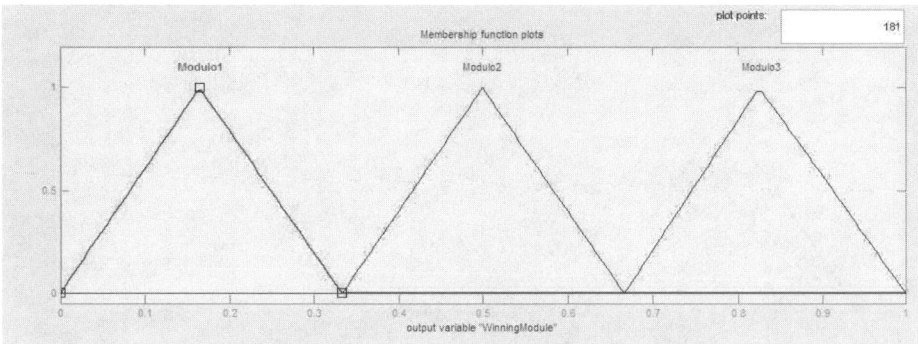

Fig. 9. Membership Functions of Fuzzy System Output

5   Response Integration with Type-1 Fuzzy Logic Using the GA

We used a genetic algorithm to optimize the triangular membership functions of the type-1 fuzzy system, with the following GA parameters: population of 5 individuals, crossing rate of 85%, mutation rate of 10%, roulette wheel selection, Stopping Conditions for the Algorithm are a maximum of 100 generations or if the error is zero between the type-1 fuzzy system base (figs. 6 to 9) and the type-1 fuzzy system obtained with the genetic algorithm (fig.10). We show in figures 11, 12, 13, and 14, the membership functions obtained with the genetic algorithm using the editor of the fuzzy logic toolbox of MATLAB [10]. The results of the optimized type-1 fuzzy system will be presented later.

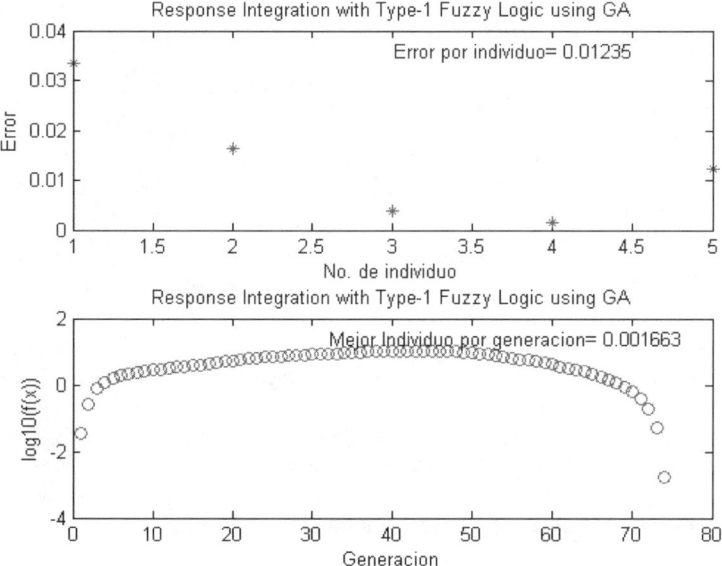

Fig. 10. Results of convergence of the genetic algorithm for Response Integration Type-1

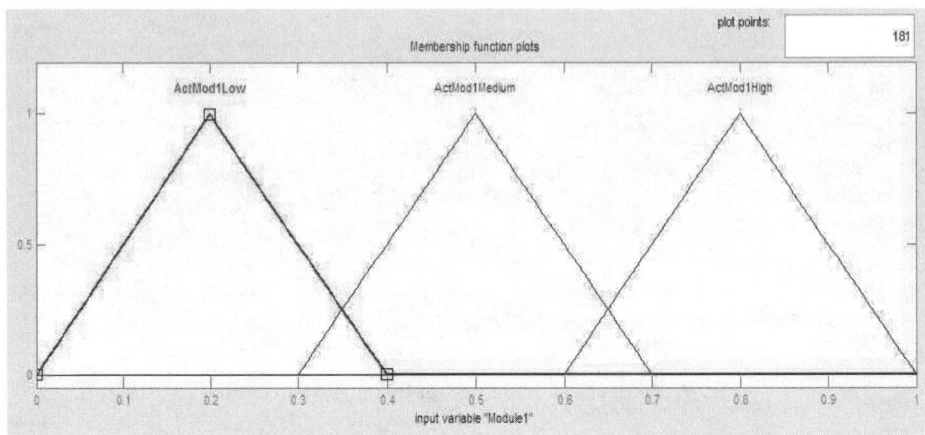

Fig. 11. Membership Functions of Input Module 1 using the GA

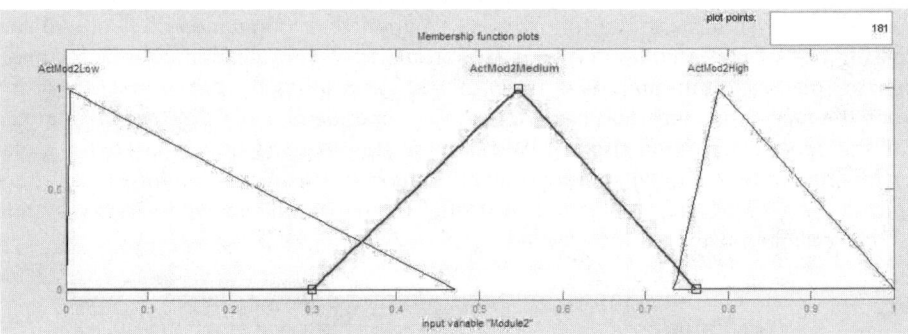

Fig. 12. Membership Functions of Input Module 2 using the GA

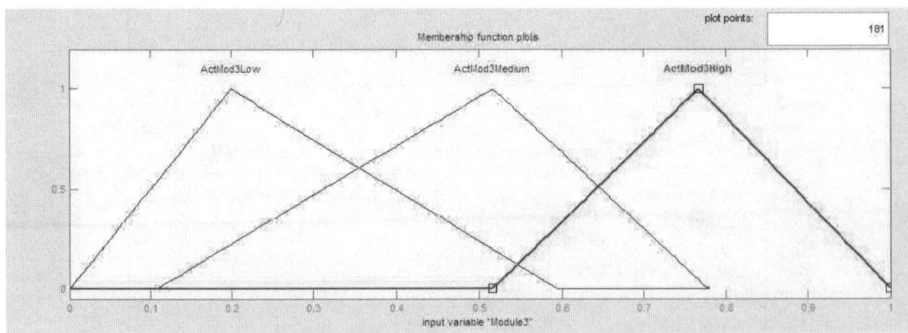

Fig. 13. Membership Functions of Input Module 3 using the GA

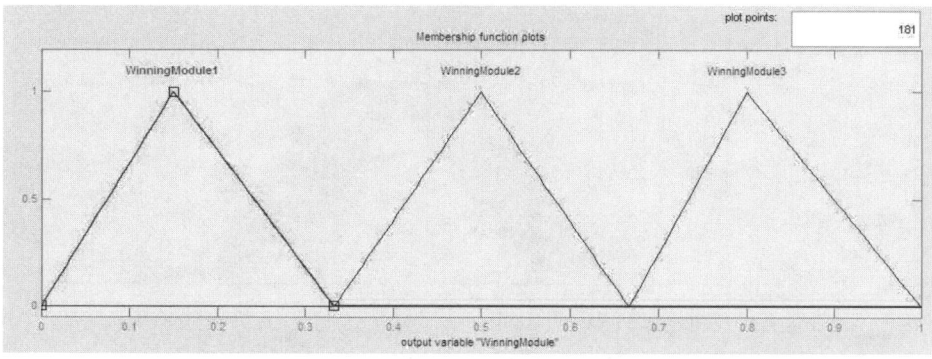

Fig. 14. Membership Functions of Fuzzy System Output using the GA

6 Response Integration with Type-2 Fuzzy Logic

The Type-2 Fuzzy Inference System, as method of Response integration, of the ensemble neural network output, is defined considering as input three linguistic variables, Activation Low, Activation Medium, and Activation High, and one output linguistic variable, Winning Activation of the three modules.

Three membership functions were used for each linguistic variable of input and output of the Gaussian type, to be managed in a range from 0 to 1.

We show in figures 15, 16, 17, and 18, the membership functions designed using the editor of IT2FUZZY fuzzy logic toolbox, which was developed by our group [1].

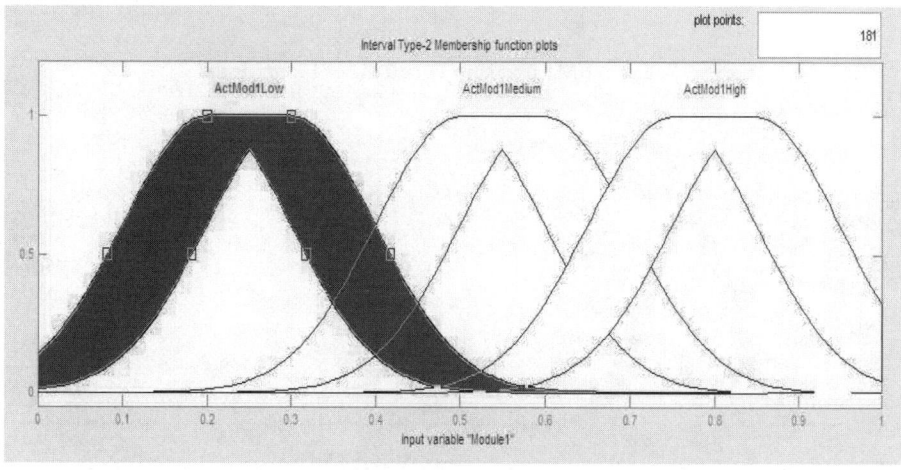

Fig. 15. Type-2 Membership Functions of Input Module 1

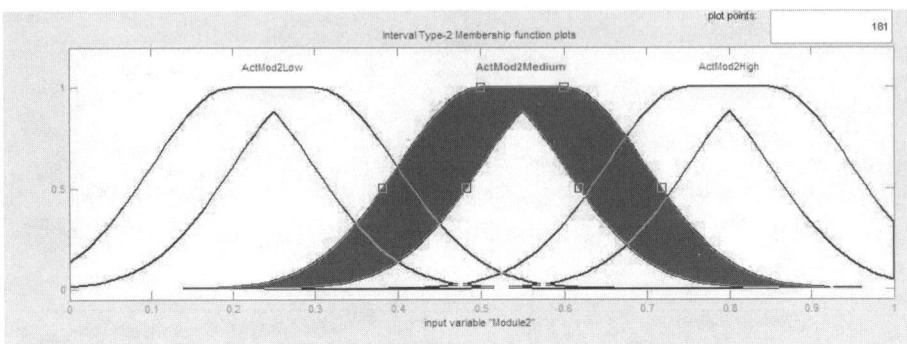

Fig 16. Type-2Membership Functions of Input Module 2

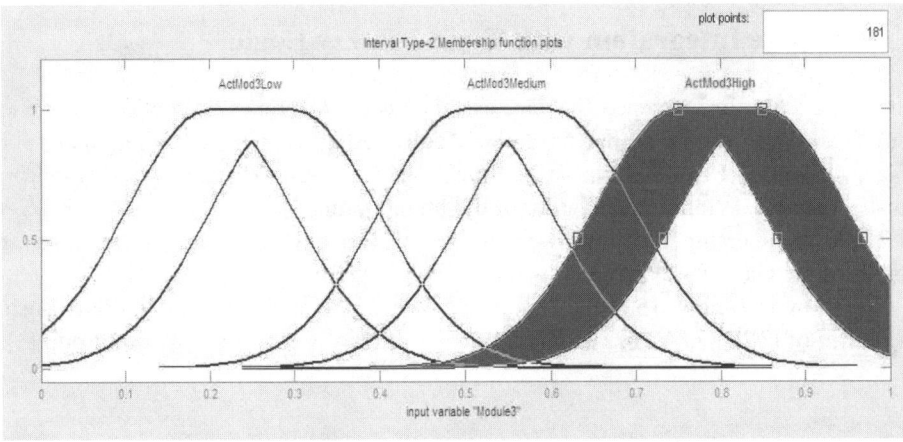

Fig. 17. Type-2 Membership Functions of Input Module 3

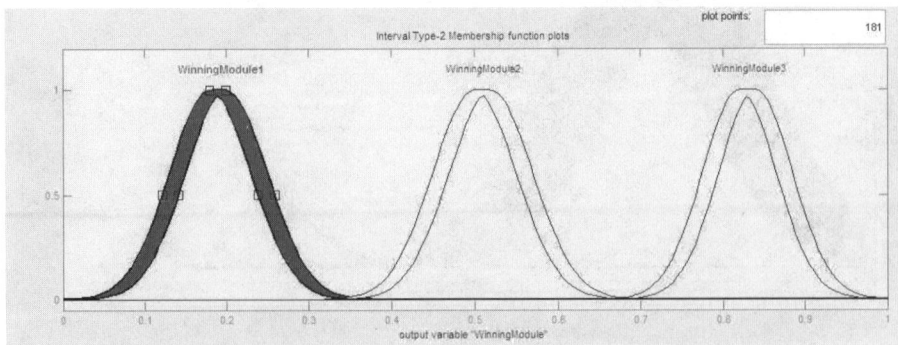

Fig. 18. Type-2 Membership Functions of Type-2 Fuzzy System Output

7   Response Integration with Type-2 Fuzzy Logic and GA

We used a genetic algorithm to optimize the Gaussian membership functions of the type-2 fuzzy s, with the following parameters: population of 5 individuals, crossing rate of 85%, mutation rate of 10%, roulette wheel selection, Stopping Conditions for the Algorithm are a maximum of 100 generations or if the error is zero between the type-1 fuzzy system base (figs. 15 to 18) and the type-2 fuzzy system obtained for the

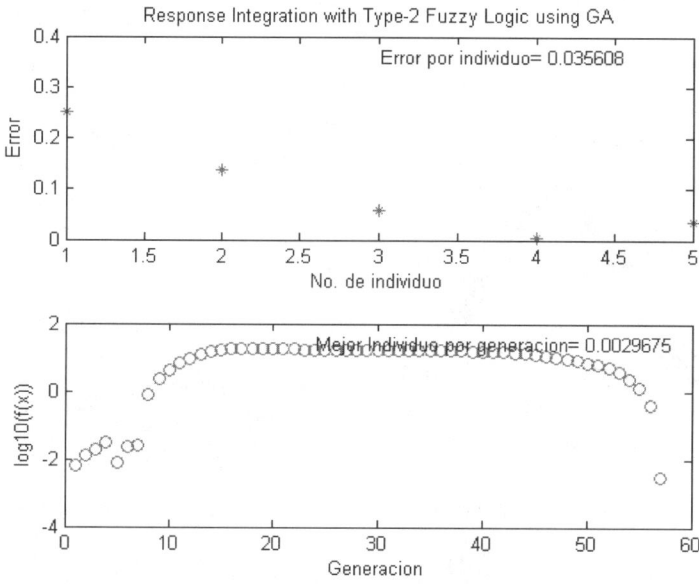

Fig. 19. Results of convergence of the genetic algorithm

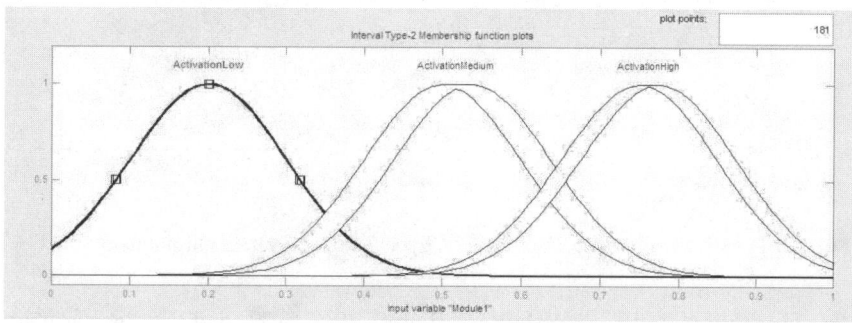

Fig. 20. Type-2 Membership Functions of Input Module 1 using the GA

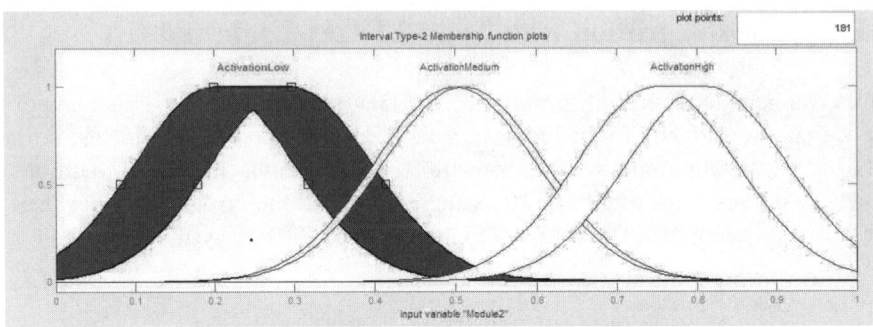

Fig. 21. Type-2 Membership Functions of Input Module 2 using the GA

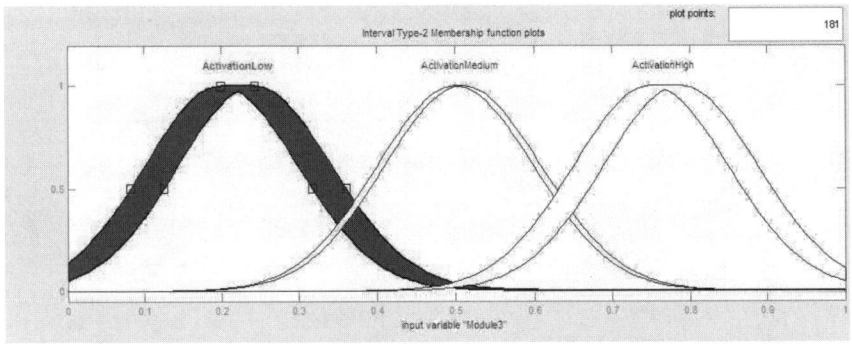

Fig. 22. Type-2 Membership Functions of Input Module 3 using the GA

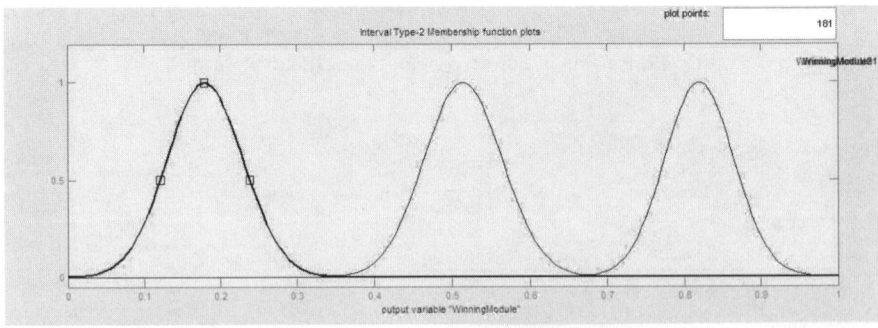

Fig. 23. Type-2 Membership Functions of Type-2 Fuzzy System Output using the GA

genetic algorithm (fig. 19). We show in figures 20, 21, 22, and 23, the membership functions obtained with the genetic algorithm using the editor of IT2FUZZY fuzzy logic toolbox [1].

8   Simulation Results with Blur Motion Noise Using GA's

Once the Ensemble Neural Network is trained, the fuzzy inference system integrates the outputs of the modules. We used the same 80 people's images to which we had applied different levels of noise with blur motion, both the type-1 and type-2 fuzzy

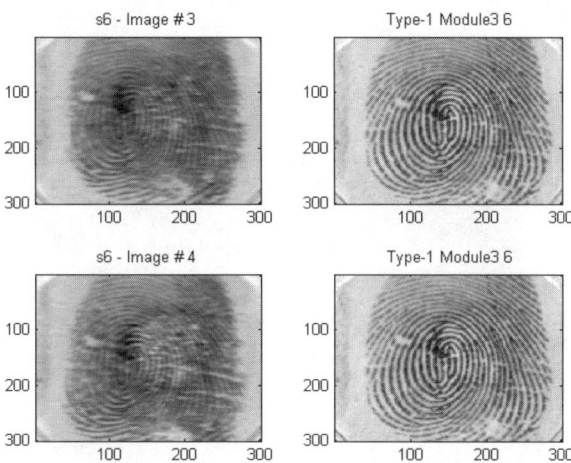

Fig. 24. Experimental results of the fingerprints using the type-1 Fuzzy Inference System (blur motion 10 displacement pixels)

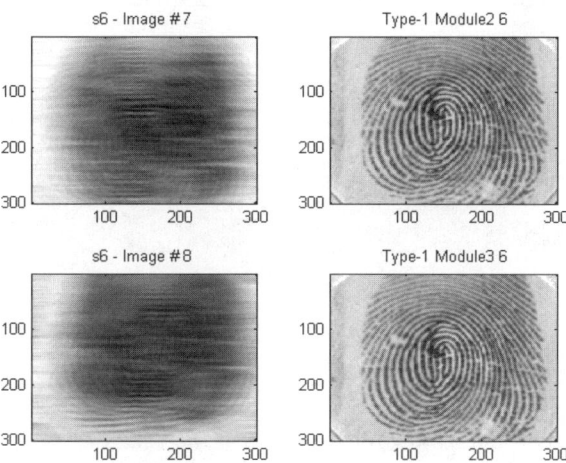

Fig. 25. Experimental results of the fingerprints using the type-1 Fuzzy Inference System (blur motion 50 displacement pixels)

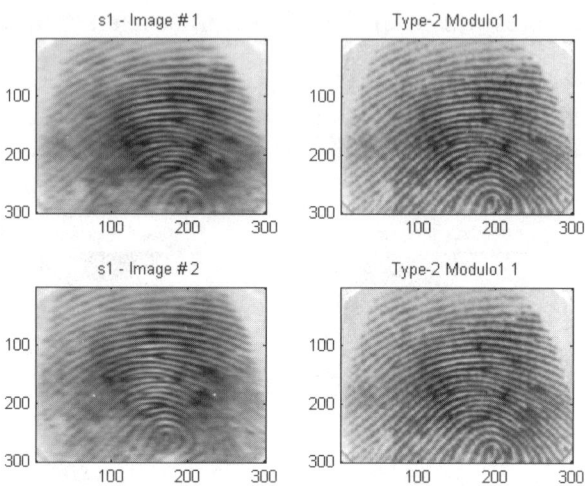

Fig. 26. Experimental results of the fingerprints using the type-2 Fuzzy Inference System (blur motion 10 displacement pixels)

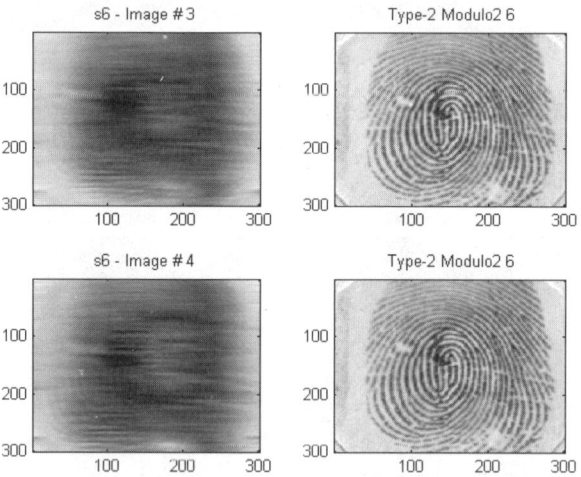

Fig. 27. Experimental results of the fingerprints using the type-2 Fuzzy Inference System (blur motion 50 displacement pixels)

inference system gives an answer for the stage of the final decision, and show the result if the fingerprint input was recognized. We show in Figures 24 and 25 the experimental results using the type-1 fuzzy inference system and in Figures 26 and 27 the experimental results using the type-2 fuzzy inference system obtained for the genetic algorithm.

9   Simulation Results with Blur Radial Noise Using GAs

Once the Ensemble Neural Network is trained, the fuzzy inference system integrates the answers of the modules. We used the same 80 people's images to which we had applied different levels of noise with blur radial, both the type-1 and type-2 fuzzy inference systems give an answer for the stage of the final decision, and shows the result if the fingerprint input was recognized. We show in Figures 28 and 29 the experimental results

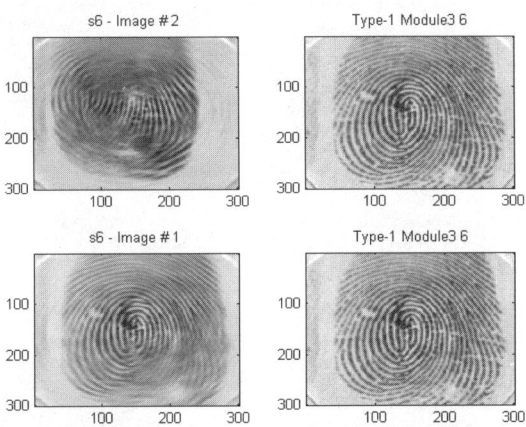

Fig. 28. Experimental results of the fingerprints using the type-1 Fuzzy Inference System (blur radial 10 displacement pixels)

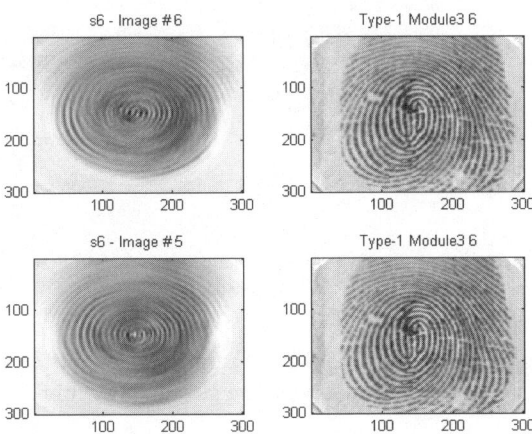

Fig. 29. Experimental results of the fingerprints using the type-1 Fuzzy Inference System (blur radial 50 displacement pixels)

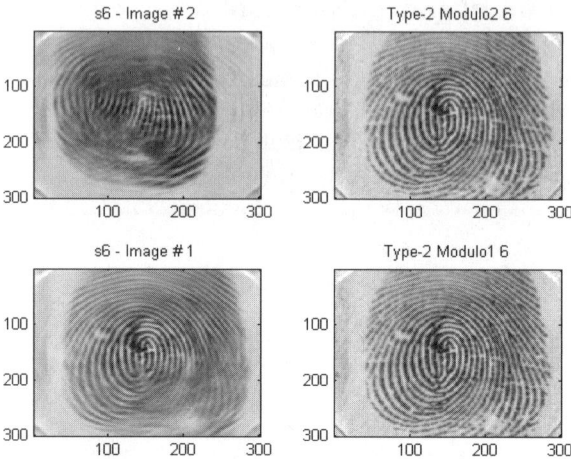

Fig. 30. Experimental results of the fingerprints using the type-2 Fuzzy Inference System (blur radial 10 displacement pixels)

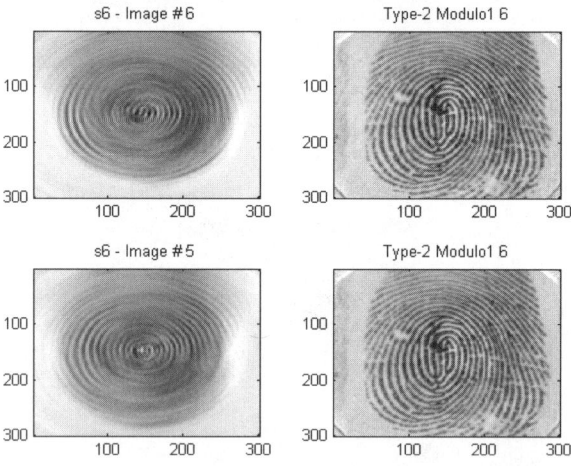

Fig. 31. Experimental results of the fingerprints using the type-2 Fuzzy Inference System (blur radial 50 displacement pixels)

using the type-1 fuzzy inference system and Figures 30 and 31 the experimental results using the type-2 fuzzy system obtained with the genetic algorithm.

10   Simulation Results with Blur Gaussian Noise Using GAs

Once the Ensemble Neural Network is trained, the fuzzy inference system integrates the answers of the modules. We used the same 80 people's images to which we had applied different levels of noise with blur Gaussian, both the type-1 and type-2 fuzzy inference system gives an answer for the stage of the final decision, and show the result if the fingerprint input was recognized. We show in Figures 32, and 33, the experimental results using the type-1 fuzzy inference system and Figures 33, and 34, the experimental results using the type-2 fuzzy inference system obtained with the genetic algorithm.

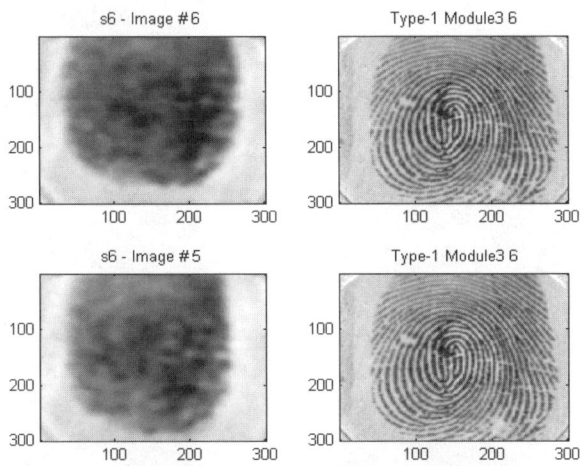

Fig. 32. Experimental results of the fingerprints using the type-1 Fuzzy Inference System (blur Gaussian 5 radius pixels)

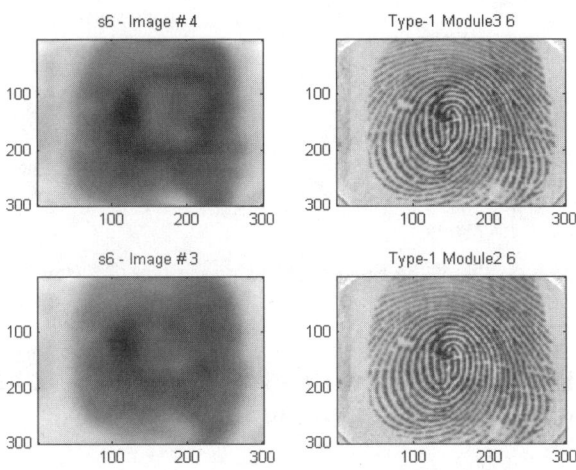

Fig. 33. Experimental results of the fingerprints using the type-1 Fuzzy Inference System (blur radial 10 radius pixels)

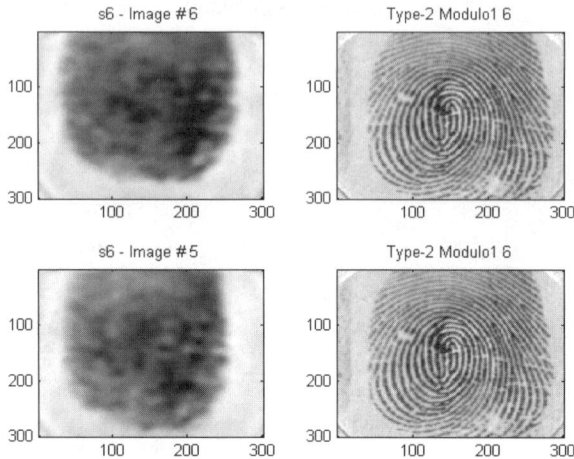

Fig. 34. Experimental results of the fingerprints using the type-2 Fuzzy Inference System (blur radial 5 displacement pixels)

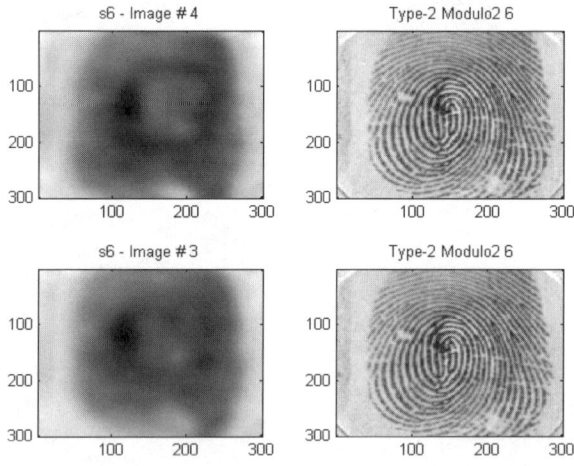

Fig. 35. Experimental results of the fingerprints using the type-2 Fuzzy Inference System (blur radial 10 radius pixels)

11   Comparison of Results with Type-1 and Type-2 Fuzzy Logic as Response Integration Methods

We performed 30 trials of the fingerprint recognition with an average of 100% recognition without noise for all methods. For the case in which noise was added to the fingerprint input the obtained results are as follows. The Identification Rate with blur motion of 10 displacement pixels is 98.75 % for type-1 (not optimized) and also for

Table 1. Comparison between Type-1 and Type-2 using GA's with blur motion noise

Response Integration Method	Recognition Rate	Identification Rate				
		Blur Motion Noise Distance Pixels				
		10	20	30	40	50
Type-1 (not optimized)	80/80 100%	79/80 98.75%	77/80	72/80	69/80	66/80 82.5%
Type-1 using GA's	80/80 100%	79/80 98.75%	78/80	73/80	70/80	68/80 85%
Type-2 (not optimized)	80/80 100%	77/80 96.25%	73/80	71/80	69/80	59/80 73.75%
Type-2 using GA's	80/80 100%	78/80 97.5%	74/80	72/80	71/80	61/80 76.25%

Table 2. Comparison between Type-1 and Type-2 using GA's with blur radial noise

Response Integration Method	Recognition Rate	Identification Rate				
		Blur Radial Noise Distance Pixels				
		10	20	30	40	50
Type-1 (not optimized)	80/80 100%	73/80 91.75%	73/80	71/80	68/80	59/80 73.75%
Type-1 using GA's	80/80 100%	74/80 92.5%	74/80	73/80	72/80	60/80 75%
Type-2 (not optimized)	80/80 100%	78/80 97.5%	73/80	70/80	63/80	63/80 78.75%
Type-2 using GA's	80/80 100%	79/80 98.75%	74/80	71/80	65/80	64/80 80%

type-1 using GA's, for type-2 (not optimized) is 96.25 % and for type-2 using GA's is 97.5%. The Identification rate with blur radial noise of 10 displacement pixels is 91.75% for type-1 (not optimized) and is 92.5% for type-1 using GA's, for type-2 not optimized is 97.5 % and for type-2 using GA's is 98.75%. The Identification rate with blur Gaussian noise of 5 radius pixels is 82.5% for type-1 (not optimized) and 83.75% for type-1 using GA's, for type-2 not optimized is 75 % and for type-2 using GA's 76.25%. In tables 1, 2 and 3 we show a detailed comparison between the type-1 and the type-2 fuzzy systems using genetic algorithms with the levels of noise mentioned. We can appreciate from Table 2 that type-2 fuzzy logic is superior than type-1 in the case of blur radial noise.

We used the same 80 people's images to which we had applied noise with blur motion, blur radial, and blur Gaussian noise, and the type-1 and type-2 fuzzy inference system gives an answer for the stage of the final decision.

Table 3. Comparison between Type-1 and Type-2 using GA's with blur Gaussian noise

Response Integration Method	Recognition Rate	Identification Rate	
		Blur Gaussian Noise Radius Pixels 5	10
Type-1 (not optimized)	80/80 100%	73/80 98.75%	66/80 82.5%
Type-1 using GA's	80/80 100%	74/80 98.75%	67/80 83.75%
Type-2 (not optimized)	80/80 100%	68/80 96.25%	60/80 75%
Type-2 using GA's	80/80 100%	69/80 97.5%	61/80 76.25%

In the figure 35 we shown the Comparison of Results with Type-1 and Type-2 Fuzzy Logic as Response Integration methods in Ensemble Neural Networks using GA's.

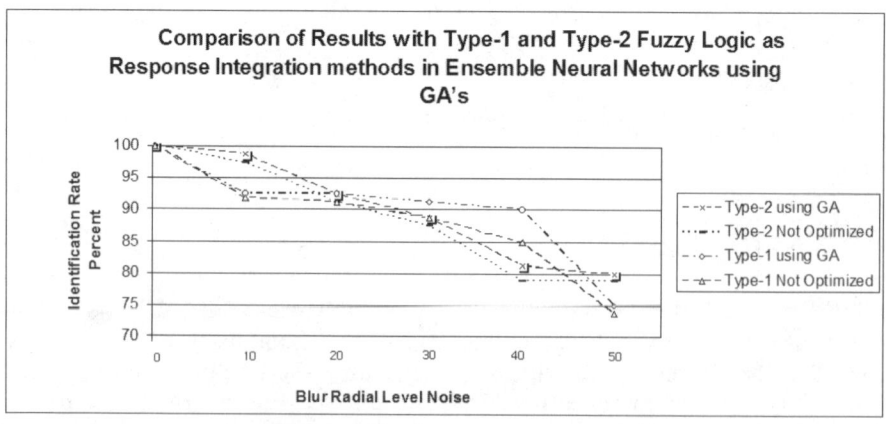

Fig. 35. Comparison between Type-1, and Type-2 using GA's. Trainscg, 2 Layers, Neurons by layer (36,18), Goal Error=.001, MSE, 80 Input Fingerprint, 10 Persons and 8 samples by persons, 80 input fingerprints.

12   Conclusions

Based on the experimental results, we can conclude that using a genetic algorithm to optimize the membership function of the type-1 and type-2 fuzzy logic, as response integration of the output ensemble neural networks for the fingerprints is a good choice. It is necessary to make more tests, to validate the proposed architecture of ensemble neural networks for fingerprints.

For the case of the type-1 fuzzy and type-2 fuzzy inference systems for response integration of ensemble neural networks, we can conclude that the behavior can be improved. We think that there is an advantage in using a type-2 fuzzy inference system to manage the uncertainty of the knowledge base in pattern recognition problems.

Future work will include, testing with more kinds of noise, using wavelets like feature extraction, and other methods of the image compression, with the goal of improving the identification rate. Also using genetic algorithms to optimize the fuzzy rules of the type-1 and type-2 fuzzy logic.

Acknowledgments

We would like to express our gratitude to the CONACYT, Universidad Autonoma de Baja California and Tijuana Institute of Technology for the facilities and resources granted for development of this research project.

References

[1] Castro, J.R., Castillo, O., Melin, P., Martinez, L.G., Escobar, S., Camacho, I.: Building Fuzzy Inference Systems with Interval Type-2 Fuzzy Logic Toolbox, 1st edn. Number 1 in Studies in Fuzziness and Soft Computing, vol. 6, pp. 53–62. Springer, Germany (2007)
[2] Castro, J.R., Castillo, O., Melin, P.: An Interval Type-2 Fuzzy Logic Toolbox for Control Applications. In: Proc. FUZZ-IEEE 2007 (2007)
[3] Chang, S., Greenberg, S.: Fuzzy Measures and Integrals: Theory and Applications, pp. 415–434. Physica-Verlag, NY (2003)
[4] Cunningham, P.: Overfitting and Diversity in Classification Ensembles based on Feature Selection, TCD Computer Science Technical Report, TCD-CS-2000-07.
[5] Grabisch, M., Murofushi, T., Sugeno, M.: Fuzzy Measures and Integrals: Theory and Applications, pp. 348–373. Physica-Verlag, NY (1989)
[6] Grabisch, M.: A new algorithm for identifying fuzzy measures and its application to pattern recognition. In: Proc. of 4th IEEE Int. Conf. on Fuzzy Systems, Yokohama, Japan, pp. 145–150 (1995)
[7] Gutta, S., Huang, J., Takacs, B., Wechsler H.: Face Recognition Using Ensembles of Netrworks. In: 13th International Conference on Pattern Recognition (ICPR 1996), Vienna, Austria, vol. 4, p. 50 (1996)
[8] Maio, D., Maltoni, D., Cappelli, R., Wayman, J.L., Jain, A.K.: FVC 2004: Third Fingerprint Verification Competition. In: Zhang, D., Jain, A.K. (eds.) ICBA 2004. LNCS, vol. 3072, pp. 1–7. Springer, Heidelberg (2004)

[9] Maltoni, D., Maio, D., Jain, A.K., Prabhakar, S.: The full FVC2000 and FVC2002 data-
     bases are available in the DVD included. In: Handbook of Fingerprint Recognition,
     Springer, New York (2003)
[10] MATLAB Trade Marks, by the MathWorks, Inc. © (1994-2007)
[11] Mostafa Abd Allah, M.: Artificial Neural Networks Fingerprints Authentication with
     Clusters Algorithm. Informatica 29, 303–307 (2005)
[12] Melin, P., Castillo, O.: Hybrid Intelligent Systems for Pattern Recognition Using Soft
     Computing. Springer, Heidelberg (2005)
[13] Melin, P., Castillo, O.: Hybrid Intelligent Systems for Pattern Recognition using Soft
     Computing. Springer, Berlin (2005)
[14] Melín, P., González, F., MartínezG.: Pattern Recognition Using Modular Neural Net-
     works and Genetic Algorithms. In: IC-AI 2004, pp 77–83 (2004)
[15] Melín, P., Mancilla, A., Lopez, M., Solano, D., Soto, M., Castillo, O.: Pattern Recogni-
     tion for Industrial Security using the Fuzzy Sugeno Integral and Modular Neural Net-
     works. In: WSC11 11th Online World Conference on Soft Computing in Industrial Ap-
     plications (September 18- October 6) (2006)
[16] Melín, P., Urias, J., Solano, D., Soto, M., Lopez, M., Castillo, O.: Voice Recognition with
     Neural Networks, Type-2 Fuzzy Logic and Genetic Algorithms. Engineering Let-
     ters 13(2), 108–116 (2006)
[17] Mendoza, O., Melin, P., Licea, G.: Modular Neural Networks and Type-2 Fuzzy Logic
     for Face Recognition. In: Reformat, M. (ed.) Proceedings of NAFIPS 2007, San Diego,
     vol. 1 (June 2007); CD Rom
[18] Nemmour, H., Chibani, Y.: Neural Network Combination by Fuzzy Integral for Robust
     Change Detection in Remotely Sensed Imagery. EURASIP Journal on Applied Signal
     Processing 14, 2187–2195 (2005)
[19] Opitz, D.W.: Feature Selection for Ensembles. In: Sixteenth National Conference on Arti-
     ficial/ Intelligence (AAAI), Orlando, FL, pp. 379–384 (1999)
[20] Opitz, D., Maclin, R.: Popular Ensemble Methods: An Empirical Study. Journal of Artifi-
     cial Intelligence Research 11, 169–198 (1999)
[21] Opitz, D.W., Shavlik, J.W.: Generating accurate and diverse members of a neural net-
     work ensemble. In: Touretzky, D.S., Mozer, M., Hasselmo, M. (eds.) Advances in Neural
     Information Processing Systems, vol. 8, pp. 535–541. MIT Press, Cambridge (1996)
[22] Sharkey, A.C.: Modularity, combining and artificial neural nets. Connection Science 8,
     299–313 (1996)
[23] Sharkey, A.: On combining artificial neural nets. Connection Science 8, 299–313 (1996)
[24] Urias, J., Solano, D., Soto, M., Lopez, M., Melín, P.: Type-2 Fuzzy Logic as a Method of
     Response Integration in Modular Neural Networks. In: IC-AI 2006, pp. 584-590 (2006)
[25] Zadeh, L.A.: Fuzzy Logic. Computer 1(4), 83–93 (1998)

Optimization of Modular Neural Network, Using Genetic Algorithms: The Case of Face and Voice Recognition

José M. Villegas, Alejandra Mancilla, and Patricia Melin

Department of Computers Science, Tijuana Institute of Technology,
Tijuana, México
pmelin@tectijuana.mx

Abstract. This paper deals with two optimization problems as the architecture (modules, layers and neurons) and the best training of an artificial neural network (ANN). For that matter is used a Hierarchical Genetic Algorithm, which theorically has the capacity to bring the optimal architecture and the training result of the ANN, for a particular task; in this case the recognition of an individual is via voice and face.

1 Introduction

Biometrics is the study of automated methods for recognizing humans based only on one or more intrinsic physical or behavioral features. The term derives from the Greek words "bios" life and "metrics" measurement [2].

That's why the current systems increasingly require a solution-oriented use of security systems based on biometrics because of the constant identity theft and fraud to increase in all types of organizations.

The Applied Biometrics can provide solutions needed to protect companies in these constant attacks ensuring the user, employee or customer who is requesting access it is really who they say they are the process involves verification of their identity through reviewing their biometric features as their official documents, ensuring that these are not false.

Security is a very important issue for all people and even more for businesses, which is why each time the are emerging new techniques to improve it.

One of these techniques as mention is: biometrics of voice or speech recognition, which is the process of automatically recognizes who is talking on the basis of the information contained in wave's voice [3] . Other biometric technologies are widely used identification of the face, fingerprint, hand geometry, iris, retina and signing among others.

To accomplish biometrics there are several ways to do Artificial Neural Networks (ANN) is a paradigm of learning and automatic processing inspired by the way in which the nervous system of animals, so the pattern recognition through Artificial neural networks has proved to give good results in this type of application, which can be very complex to be dealt with neural networks monolithic therefore recommend the use of neural networks Modular which divide the complex problem into several problems smaller.

O. Castillo et al. (Eds.): Soft Computing for Hybrid Intel. Systems, SCI 154, pp. 151–169, 2008.
springerlink.com

It's hard to find a neural network architecture for modular, with successful results, that is why this research is focused on the optimization of neural networks through Genetic Algorithms, these are systematic methods for troubleshooting search and optimization, are based on the concept of natural selection and evolutionary processes of living beings (selection based on the population, mating and mutation. These are systematic methods for troubleshooting search and optimization, are based on the concept of natural selection and evolutionary processes of living beings (selection based on the population, mating and mutation.

To decide who is the person to recognize through his face and voice there is a technique called Fuzzy Logic and has been proven to be particularly useful in expert systems and other applications of artificial intelligence.

The present paper describes the methods used during the development of this work both for face and voice. Also presents the results obtained after developing the system and doing some tests. And finally describes the conclusions it arrived.

2 Basic Theoretical Concepts

2.1 Artificial Neural Networks

It is a system consisting of a large number of basic elements (Artificial Neurons), grouped into layers, and who are highly interconnected Synapses); his structure has several inputs and outputs; these will be trained to respond in a manner desired, to input stimuli.

These systems emulate, in a certain way, the human brain. They need to learn how to behave (Learning) and someone should be responsible for teaching (Training), based on previous knowledge of the environment problem. [6]

2.2 Modular Artificial Neural Networks

A complex problem can be divided into a number of sub-problems simpler than can be solved efficiently by smaller networks. From a point of view neurocomputation, Modularity is an essential feature of the human brain.

The advantage is that if the model admits in a natural way A breakdown in functions simpler, The use of a modular network translates into a learning faster and easier.

Each module can be constructed differently, to fit the requirements of each sub-task.

2.3 Genetic Algorithms

They were introduced by John Holland in 1970 inspired by the process observed in the natural evolution of living beings.

They are systematic methods for troubleshooting search and optimization, are based on the concept of natural selection and evolutionary processes of living organisms (Selection based on the population, mating and mutation).

These algorithms make evolve a population individuals Subjecting random actions Similar to those operating in biological evolution (Mutations and genetic recombination), as well as a selection according to some criteria, Depending on which decides which individuals are more adapted to survive, and which are less suitable, which are discarded.

3   Development and Methods Used

For the development of this research was necessary to divide it into three parts: Voice Recognition, Face Recognition and Decision making. We describe below each of these parts.

3.1   Part One: Voice Recognition

For training Voice of the Modular Neural Network tube is a base pattern of 36 samples and 12 samples to train without training, which is comprised of six alumni of the institution, taken from the reference [4] . Everyone read his name on a slow, normal and fast.

Each of the words were engraved on individual files, for the recording of these jobs are code and MatLab Wav Floor to add noise to the original samples.

The signals were acquired through MatLab code, using a standard PC microphone for this. The signals are captured with the following characteristics:

- Speed Transmission   88 kbps.
- Sample Size   8 bits.
- Channels   mono.
- Speed   11025 MPS.
- Duration   1 sec. [4]

To add noise to sound signals are using the program Wave Flow 5.6, the noise level was 0.5 gaussian noises [4].

The signal from the word Daniel said of the three forms shown in Figure 1.

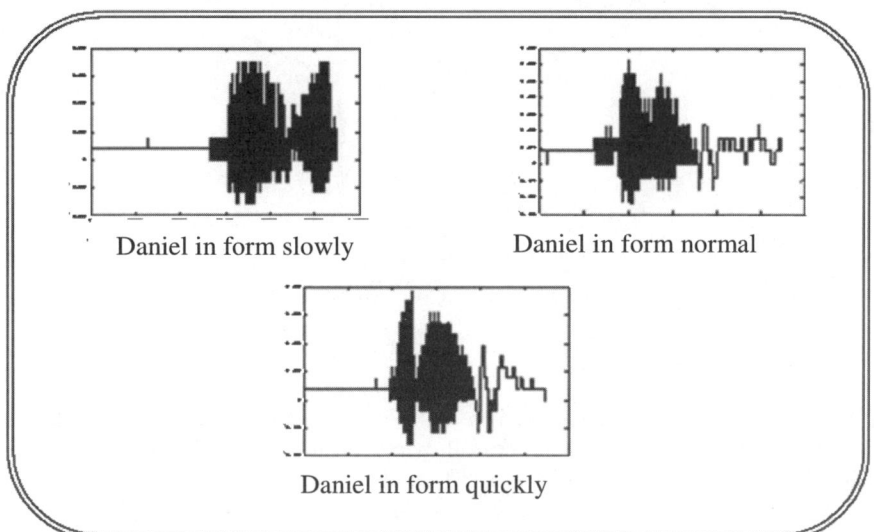

Daniel in form slowly   Daniel in form normal

Daniel in form quickly

Fig. 1. Signal Of The Word Daniel

The architecture chosen for particular Artificial neural networks can determine the success or failure of using them in a given application, that is why in this research was proposed the use of the Method of Genetic Algorithms Hierarchical (HGA) with a view that the search and optimization of the architecture is carried out automatically and in a way that the HGA we projected as a result an architecture appropriate to solve the problem silver.

3.1.1 Proposed Architecture

To define the chromosome of Genetic Algorithms Hierarchical, must be taken into account a Architecture initial Modular Neural Network so that it will decrement and So that it will decrement at the same time optimizing the parameters of the modular neural network with the genetic algorithm, the objective of the optimization is to modulate the neural network has the lowest number of modules in each of the modules will have the lowest number of layers and neurons for each layer; further find the best training, this with the intention that the neural network can learn the best, This is why it proposes a modular neural network with 4 modules as maximum as shown in Figure 2 To solve the problem of voice recognition.

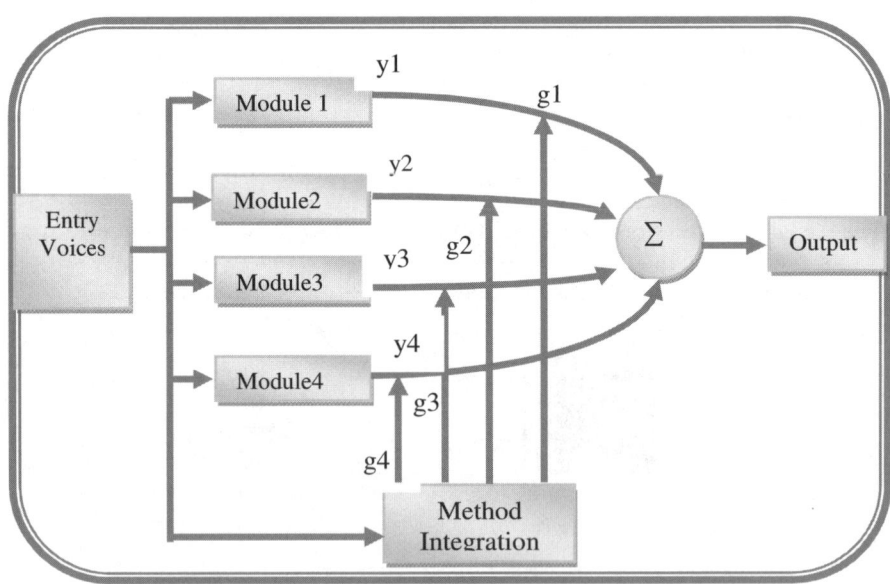

Fig. 2. Proposed Modular Architecture for Voice

3.1.1.1 For the Voice Recognition Modular Neural Network. The entries of the modules are set out in Table 1 and by the following formula (ns ÷ nm), (ns where are the sounds that number is equal to 36 nm and are the number of modules, where 1 <nm ≤ 4). For example, the Genetic Algorithm Hierarchical have to find less than 4 modules and minimum two modules, in the event that ns outside two modules each module will feed samples of 18 words or sound, that is the first module with the trains first 18 words and the second module with the following 18 words.

Table 1. Number Of Sounds ÷ Number Of Modules

Module	Number Voice
1	36
2	18
3	12
4	9

3.1.2 Method of Integration Modules Voice

The method was performed to make the integration of the modules of the voice was: "Gating Network", why was chosen this technique integration was because it is one of the best and also that by giving the entries to the modules, there are modules that are not fed with the same sounds or voices as shown in Table 5.1, the sounds are different in each module.

3.1.3 Chromosome the Genetic Algorithm for Hierarchical Voice

The architecture of the Modular Neural Network will form the chromosome of Genetic Algorithm Hierarchical as follows:

- 4 Modules.
- 3 Layers For each module (12 layers in total).
- 90 Neurons for each layer (1080).

Based on these data, will proceed to form Hierarchical chromosome binary of two levels, which possesses genes control and genes parameters; With a length of 1096 bits, The first 4 bits symbolize the second level of genetic algorithm Hierarchical and represent 4 modules Modular Neural Network, the following 12 genes represent the first level of genetic algorithm Hierarchical and symbolize the number of layers that exist for each of the modules of the Modular Neural Network (3 layers for module) and 1080 bits which represent genes parameters Genetic Algorithm Hierarchical, symbolizing the neurons for each of the layers of the Modular Neural Network.

In Figure 3, we can see how you can represent the architecture of the chromosome of Genetic Algorithm Hierarchical we explained above.

The optimization process of the neural network modular took place as follows, first generates a random group of chromosomes which takes architecture to train the neural network system, then provides an assessment of each individual and select the most suitable and to select the most suitable for operations crossing and mutation in order to give new life to individuals, after that is done recombination in which individuals are least able replaced by the children and after this takes place again training the neural network system, this process can be seen in Figure 4.

Depending on the result given by the Hierarchical Genetic Algorithm relative to the number of modules that must have the modular neural network, is the division between the number of voices and the number modules, for example: if the Hierarchical Genetic Algorithm result is that the best architecture consists of 3 modules, which are

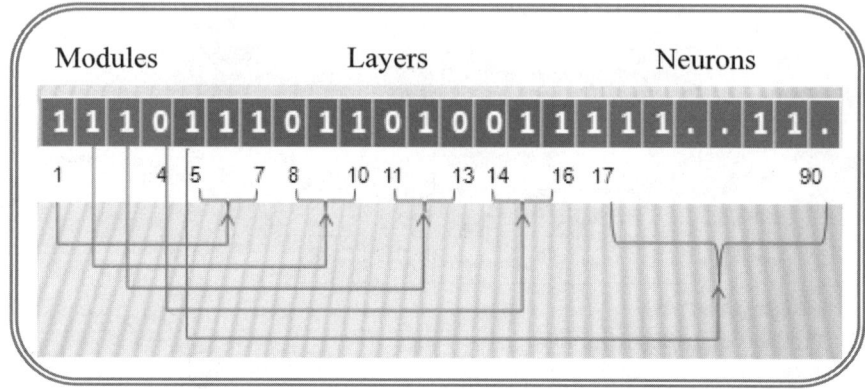

Fig. 3. Architecture Binary Chromosome From Hierarchical Genetic Algorithm

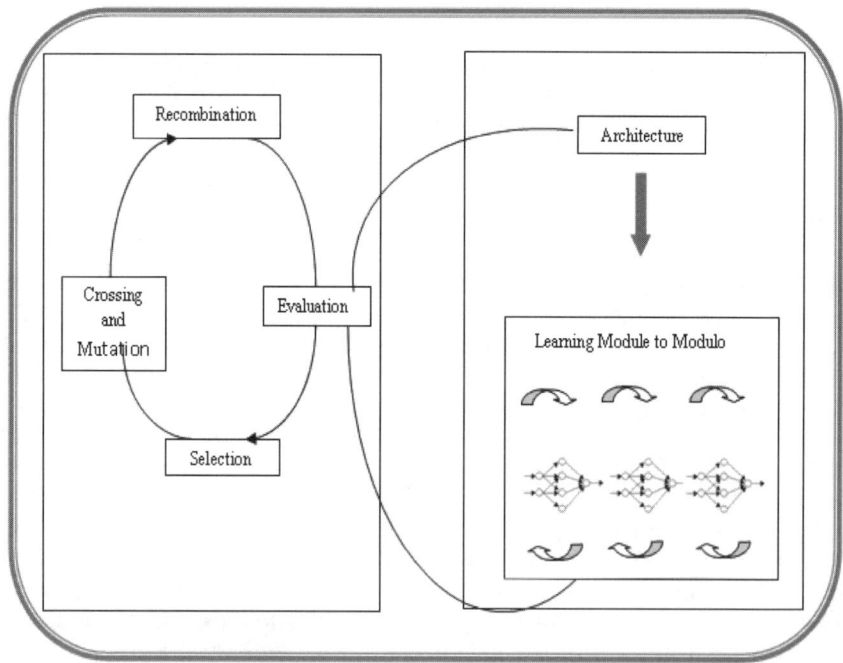

Fig. 4. Training Process Of A Modular Neural Network With HGA

36 sounds is divided between 3, which will serve as data entry for each of the modules of the Modular Neural Network; 12 for the first module, 12 for the second module and another 12 for the third module. In Figure 5 you can appreciate the architecture of a number N of words for 3 modules.

The Hierarchical Genetic Algorithm scheduled for this never produces a single module and the minimum numbers of modules are 2 and 4 the highest. The use of neural networks provides a modular increase in the speed of learning, since each module expert has to learn a smaller number of data and is responsible for a subtask that is easier to complete the entire task, as if it were a modular ensemble, it is easier to understand the task that has been responsible for each module.

In this figure we see they are giving 12 different voices, they are giving 12 different voices as input to each of the modules, subsequently employ a method of integration and finally get out that the person would recognize.

Table 2. Division Sounds Of The Number Of Modules

Number of Modules	Division
1 Modules	36 Samples of sound
2 Modules	1-18 Samples of sound
	19-36 Samples of sound
3 Modules	1-18 Samples of sound
	1-18 Samples of sound
	1-18 Samples of sound
4 Modules	1-18 Samples of sound
	1-18 Samples of sound
	1-18 Samples of sound
	1-18 Samples of sound

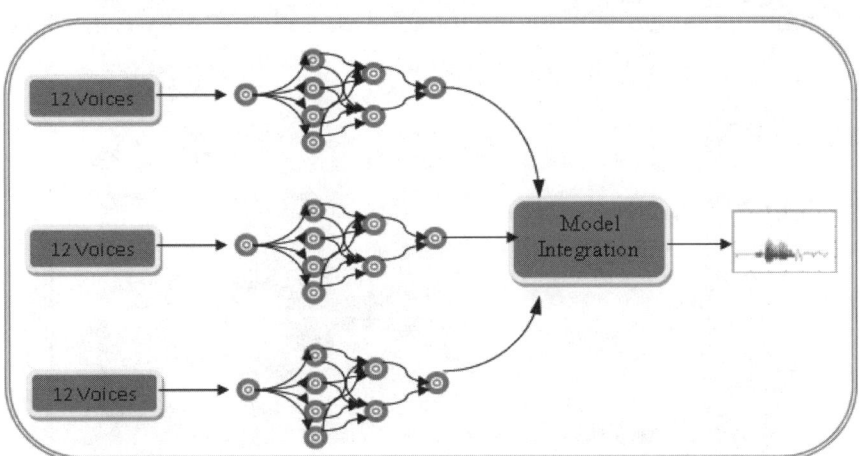

Fig. 5. Example Of The Neural Network Architecture For 3 Modules

The training method that was used for training the neural network is the Trainscg (*Scaled Conjugate Gradient*), with an error goal of 0.001 and between 150 and 300 times, we use this method because it is one of the best training methods for the pattern recognition [3] and [4].

3.2 Part Two: Face Recognition

For training Face of the Modular Neural Network tube is a base pattern of 48 samples and 12 samples to train without training, which is comprised of six alumni of the institution, taken from the reference [4] .

The first images were taken with them was going to work for training the neural network, these images were taken with a digital camera, will use a white background and were taken with a resolution of 4 Megapixels, then cut the face with a paint program and through code in MatLab was given the format of 90 * 100 pixels and 16-bit.

Were used 30 photographs normal for training, Figure 3.6 shows the images used.

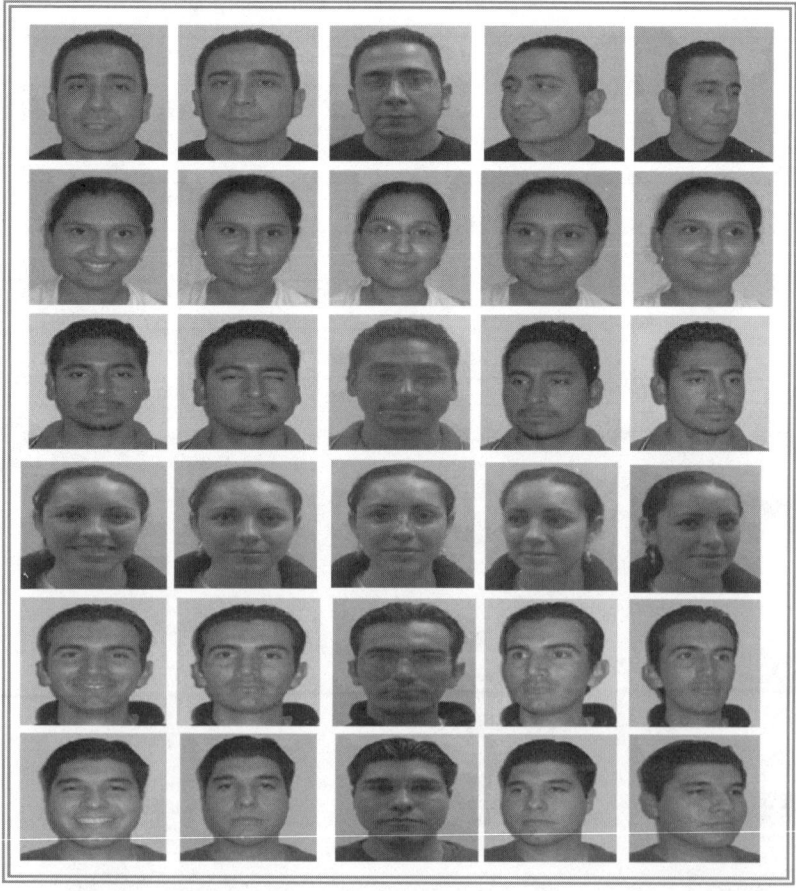

Fig. 6. The Faces Used For Training

At the same time these same images were used to be included in training with noise, was used to add noise, the function of MatLab addnoise .

The noise was added 0.02 salt and pepper, Figure 7 shows the images with noise.

Fig. 7. The Images With Noise added

3.2.1 Proposed Architecture for Face Recognition

Like on the first part, it is for voice recognition, will be, avoided training network a trial and error and I choose to use Hierarchical Genetic Algorithm; since it is necessary to solve two types of optimization problems, such as: architecture (modules, layers and neurons) and the search for better training of an artificial neural network modules.

To define the chromosome of a Hierarchical Genetic Algorithms are must taken into account initial Architecture Neural Network so that it will decrement and at the same time optimizing the parameters of Modular neural network with the genetic algorithm. The figure 8 shows this architecture similar to the architecture for voice.

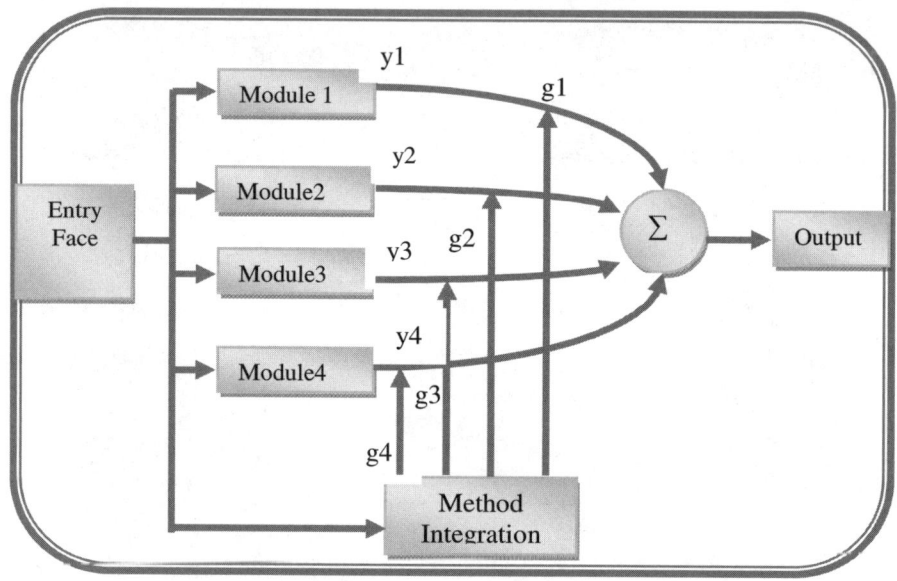

Fig. 8. Modular Architecture Proposals to Face

Table 3. Images by module

Number of Modules	Images by module
1 Modules	48 Sample Face
2 Modules	1-24 Sample Face
	25-48 Sample Face
3 Modules	1-16 Sample Face
	17-32 Sample Face
	33-48 Sample Face
4 Modules	1-12 Sample Face
	13-24 Sample Face
	25-36 Sample Face
	37-48 Sample Face

3.2.1.1 Modular Neural Network For Face Recognition. For the development of this investigation for face, took place in the same way that the voice; the number of pictures divided by the number of modules.

The number of modules is determined by the Genetic Algorithm Hierarchical, there may be minimum 2 modules and maximum 4 modules.

The entries of the modules are set out in Table 3.

3.2.2 Method of Integration Modules in Face
As in voice, this is by modulating number of images between the number of modules, and the method of integration is the same as that of the voice, as it does not feed modules with the same images.

3.2.3 Chromosome of Hierarchical Genetic Algorithm for the Face
It was proposed to create a chromosome similar to Figure 3.3 to 1096 bit, which in Table 4 describes the parameters of chromosome proposed Face.

Table 4. Criteria Chromosome Nominated To Face

Levels Of Chromosome	Name	# BITS in the Chromosome	Descripsion
Second Level Control	Modules	4 bits	4 Modules
First Level Control	Layers On Module	12 bits	3 Layers For Module
Nivel De Parámetros	Neurons For Layer	1080 bits	90 Neurons For Layer

3.3 Objective Function

When it is intended to optimize a neural network, the objective function of a Genetic Algorithm is to minimize both the precision (f1) and complexity (f2) of the network which is defined by the number of active connections on the network; this is using the equation (1). [1]

$$f1 = 1/N \sum_{i=1}^{N} (\tilde{y}i - yi) \tag{1}$$

Where N is the size of the vector, $\tilde{y}i$ y yi, are output the network and the desired output for the sample i^{th} a vector test respectively.

The fitness determines always the Chromosomes to be reproduced which will eliminate, the ability of chromosome hierarchical   genetic algorithm for voice and face; for optimizing neural network is determined by the function (2).

$$f(z) = \alpha \cdot rank\ [f1\ (z)] + \beta \cdot f2\ (z) \tag{2}$$

Where α is the coefficient of the accuracy of weights; β is the ratio of the complexity of weights and rank [f1 (z)] ε Z $^{+}$ value ranking.

The equation 2, was used to evaluate individuals of each generation in hierarchy genetic algorithm.

The function is multiple goal, why they want to search for the best training **[f1 (z)]** and minimize architecture **f2 (z),** then: **α** has to be a number larger than β to be able to find a good training, because it is the alpha rank error; which case results are shown below; this find the Better training more beta number nodes by the f2 (z).

3.4   Part Three: Fuzzy System

It proposed a thesis as to allow recognition of individuals, it proposed a thesis as to allow the recognition of people so united by integrating information 2 important biometric measures, such as: Voice and Face, to perform such recognition using artificial neural networks, In a manner that got a Modular Neural Network for each of the biometric measures.

Once they obtained the exits of the modules in each of the neural networks, was the integration of results using Gating Network, this allowed us to determine independently, that person belongs to the voice or face, because it gave us 2 results (one for each Measure Biometric), will tube that integrate those results to determine the identity of the person, for this integration will developed a system of decision, it used Fuzzy Logic to take into account the uncertainty involves the decision-making process.

In the diagram that shown in Figure 3, are displayed and more clearly, which is the system for the recognition of people through a system of decision diffuse.

This phase consisted of a system that would allow diffuse integrate the results Gating Network, to determine the identity of the person. Figure 10 shows the fuzzy system that was conducted for this research.

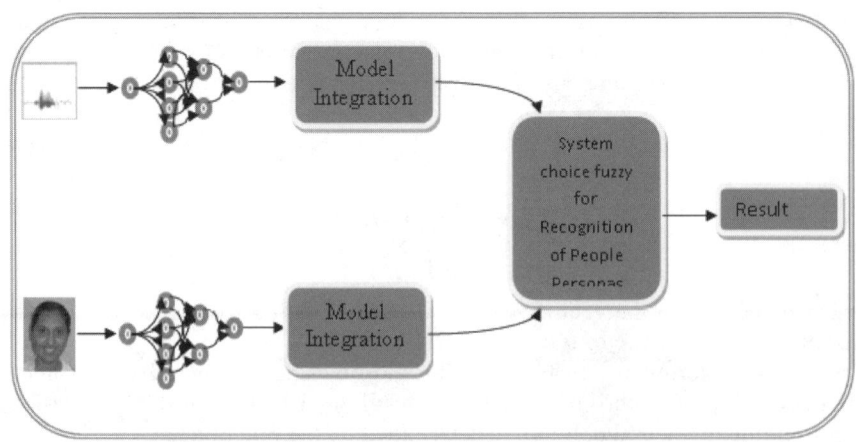

Fig. 9. System Architecture Recognition of Persons

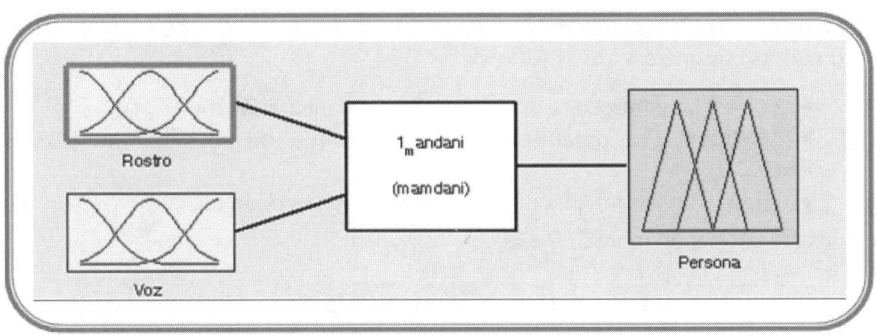

Fig. 10. Variable Input and Output System fuzzy

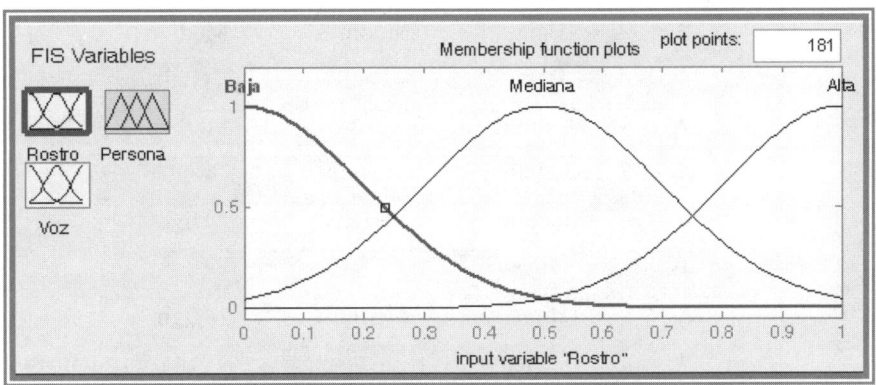

Fig. 11. Functions Gaussians Membership of the Input Variables

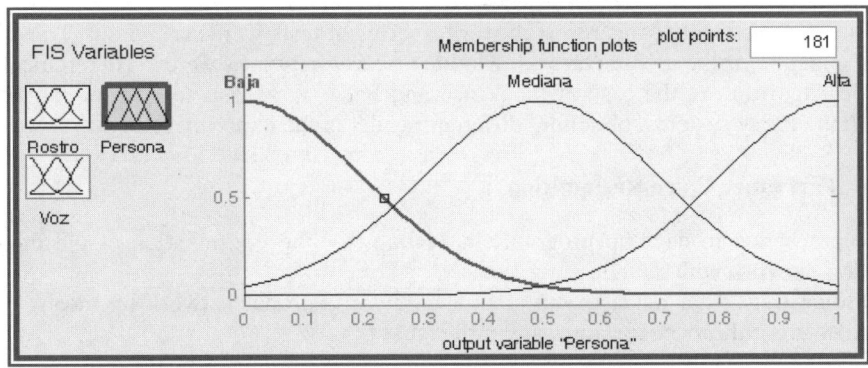

Fig. 12. Gaussians Membership of the Output Variables

This fuzzy system was thought to have input 2 variable these are: Face and the Voice, and a variable exit which is: Person and the method of fuzzy inference Mamdani.

The membership functions of this fuzzy system are Gaussians, both for the entry to the exit and its parameters are as follows:

- Low: The parameters of this function membership are [0.2 0].
- Medium: The parameters of this function membership are [0.2 0.5]. [0.2 0].
- High: The parameters of this function membership are [0.2 1].

The rules are shown on table 5.

Table 5. Rules Fuzzy System

Reglas	Face	Voice	Result
1	Low	Low	Low
2	Baja	Medium	Medium
3	Baja	High	High
4	Medium	Low	Medium
5	Medium	Medium	Media
6	Medium	High	High
7	High	Low	High
8	High	Medium	High
9	Alta	Alta	Alta

4   Results

In the development of this research work a series of tests were carried out, To achieve optimization of the architectures of Modular neural networks we use Hierarchical Genetic Algorithm in the case of the Voice and Face, as well as tests were carried out with the fuzzy systems, obtaining different results in the experiments.

4.1   Part One: Voice Recognition

We proceeded to develop programs in Matlab, for the optimization of the modular neural network with genetic algorithm.

Some tests were made to optimize a modular neuronal network by means of the genetic algorithm whose values are in the table 6:

4.2   Part Two: Face Recognition

In this second phase the goal is to develop a Hierarchical Genetic Algorithm to seek the best training and minimizing the architecture neural network for modular to Face. The results of these experiments can be found in Table 8.

Table 8 Results optimized with GA with 48 images for training, and 12 to identify.

Table 6. Parameters of the GA and RNA for Voice

GA			RNA		
Crossing	**Mutation**	**GGap**	**M. E**	**Meta error**	**Times**
0.5	0.1	0.5	trainscg	0.001	150

GA= Genetic Algorithm, GGap = Number of new individuals to create.
RNA= Data of the Artificial Neural Network, M.E= Method of training.

Simulation results are shown in Table 7. Figure 13 shows the convergence of the genetic algorithm.

Table 7. Optimized Results with Genetic Algorithms

Test	GA	Arq	Time	Recognizes	Identifies
1	Gen 20	44-38	2 hrs 41 min 27 seg	35/36	12/12
	Ind 10	48-45		97.22%	100%
2	Gen 15	45-43	3 hrs 12 min 28 seg	31/36	10/12
	Ind 13	41		86.10%	83%
3	Gen 10	44-37	1 hrs 45 min 9 seg	36/36	12/12
	Ind 5	44-44-46		100%	100%
4	Gen 40	52-50	6 hrs 17 min 3seg	28/36	08/12
	Ind 20	35		77.70%	66%
5	Gen 12	39-36	2 hrs 5 min 43 seg	36/36	12/12
	Ind 12	44-45		100%	100%
6	Gen 30	44-51	3 hrs 8min 33 seg	28/36	09/12
	Ind 10	41		77.70%	75%
7	Gen 15	39-40	3 hrs 15 min 56 seg	35/36	11/12
	Ind 10	42-52-34		97.20%	91.60%
8	Gen 10	48-46-45	1hr 17 min 7 seg	36/36	12/12
	Ind 10	41-48		100%	100%
9	Gen 15	45-38	3 hr 1 min 30 seg	34/36	11/12
	Ind 5	47		94.40%	91.60%
		37-43			

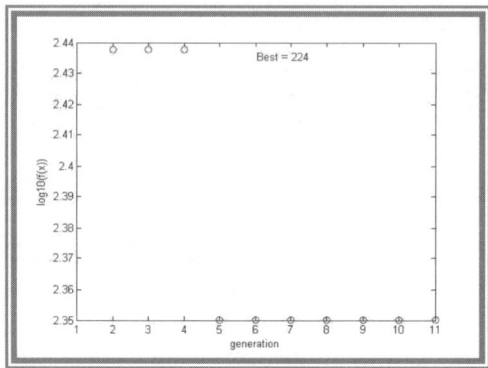

Fig. 13. Convergence of the algorithm genetic

Prueba	GA	Cruce	Mut	Arquitectura	Tiempo	Reconoce	Identifica
1	Gen 7 Ind 5	0.3	0.5	47-51 39	5 hrs 51 min 14 seg	33/48 68.75%	9/12 75%
2	Gen 8 Ind 6	0.3	0.5	45-43 41	6 hrs 50 min 53 seg	24/48 50%	6/12 50%
3	Gen 5 Ind 4	0.3	0.5	43 47-47	2 hrs 53 min 10 seg	31/48 70.83%	8/12 66.66%
4	Gen 3 Ind 3	0.3	0.1	45-48 53-44	1hr 15 min 24 seg	34/48 64.58%	8/12 66.66%
5	Gen 5 Ind 4	0.1	0.7	42-52 41-43	3 hrs 30 min	41/48 85.42%	9/12 75%

Traincsg was used to train the network with a goal error of 0.001 and executed 500 times.

4.3 Part three: Fuzzy System

This section explains the results of the experiments of the part the Fuzzy System; where, the architecture was shown in Figure 9.

Several Modular neural networks were trained whose architecture was suggested by the genetic algorithm in which it holds a good share of recognition in the case of voice, and not very successful in the case of Face, as well as training in which is not achievement get a high percentage of recognition, and this could help us to test the system later decision diffuse.

As we look in Table 7 are training for Voice, where there are very good results and others not so good, just as it also happens in Table 8 in the case of Face.

In Table 9 shows the results obtained with the system diffuse type mamdani to identify the person, whose rules can be found in Table 5 with the best results of training for Modular Neural Network Voice and Face obtained with the Hierarchical Algorithm Genetic.

Table 9. Results of the fuzzy system to identify the people

File voice	File Face	AnswerR	AnswerV	Sis. Dif	Person
1-6a	1-6ª	1	1	1	Daniel
1-6a	1-6b	6	1	1	Daniel
9-16a	7-12ª	2	2	2	Jazzyni
9-16b	7-12b	2	2	2	Jazzyni
17-24a	13-18ª	3	3	3	Karim
17-24b	13-18b	3	3	3	Karim
25-32a	19-24ª	4	4	4	Maribel
25-32b	19-24b	3	4	4	Maribel
33-40a	25-30ª	5	5	5	Mike
33-40a	25-30b	5	5	5	Mike
41-48a	31-36ª	2	6	6	Omar
41-48b	31-36b	6	6	6	Omar

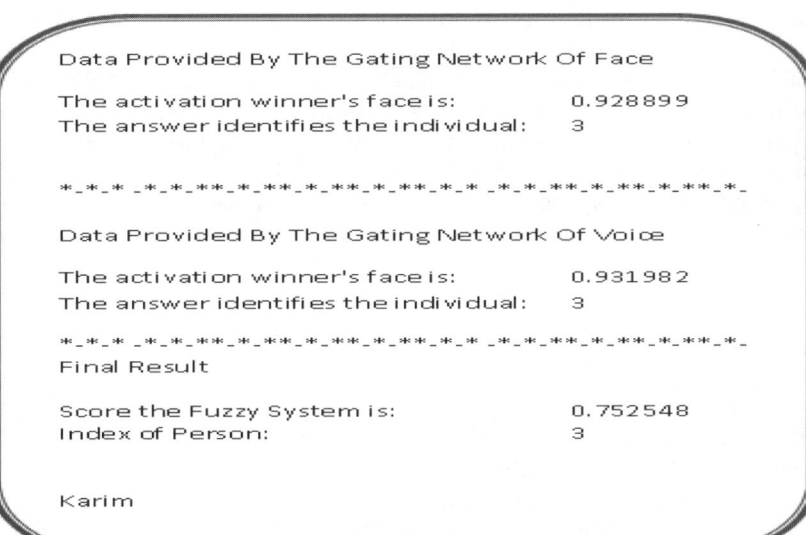

Fig. 14. Results of the fuzzy system to identify the people

Table 10. Percent identification of the person fuzzy system

Good Trainings	Trainings Regular	Trainings Ill
100 %	83 %	16.6%

In Figure 14 you can see the result, the method of integration. "The Winner takes all "(Gating Network), face and voice. And the result of the identification of the person. There may be the case where the module of face says it is a person diferent that of the voice, this is why the fuzzy system decision was used.

The activations (AnswerR and AnswerV) are the input variables for the fuzzy system.

And the Exit fuzzy system is the result of the identification of the individual, with his two biometric measures (Face and Voice).

5 Conclusions

At the beginning of the research work it was uncertain about how the chromosomes should be used, it was thought to segment the voice and face to find the modules, used in previous works. This led us to: instead of segmenting, Why we not divide the number of samples between the number of modules and in that way train the modular neuronal network in an automatically way by the GA .

The best Modular Neural Network is obtained to develop modules in accordance with mistaken identity and the complexity of the modules, the best architecture consists of two modules of the four modules proposed initially, for both Voice and Face.

The results of the problem of Speech Recognition are good and show the feasibility of Hierarchical Genetic Algorithms for the optimization of a neural network topology modular, however, the GAs can require hundreds and thousands of evaluations which may take some time computer huge this could mean hours or days in order to obtain an optimum result, but also relies heavily parameters of both neural networks and the GA.

References

[1] Schmidt, A.: Manchester Metropolitan University, Department of Computing, A Modular Neural Network Architecture with Additional Generalization Abilities for High Dimensional Input Vectors, (September 1996)

[2] Biometría Aplicada, S.A. De C.V. (2006-2007),
     http://www.biometriaaplicada.com

[3] Hernandez, D.S.: Reconocimiento De Personas Por Medio De Su Huella, Rostro Y Voz, Mediante El Uso De Redes Neuronales Modulares Y Logica Difusa, Instituto Tecnológico de Tijuana, Maestría en Ciencias en Ciencias de la Computación, (Octubre de 2006)

[4] Martínez, G.E.: Optimización de Módulos y Métodos de Integración de una Red Neuronal Modular usando Algoritmos Genéticos Jerárquicos Aplicado al reconocimiento de Voz, Instituto Tecnológico de Tijuana, Maestría en Ciencias en Ciencias de la Computación. (Febrero de 2005).

[5] Castro, M.G.S.: Reconocimiento de Personas por medio de su Rostro y Voz en tiempo Real utilizando Redes Neuronales Artificiales, Instituto Tecnológico de Tijuana, Maestría en Ciencias en Ciencias de la Computación. (Octubre de 2006).

[6] Melin, P., Castillo, O.: Hybrid Intelligent Systems for Pattern Recognition Using Soft Computing. Springer, Heidelberg (2005)

A New Biometric Recognition Technique Based on Hand Geometry and Voice Using Neural Networks and Fuzzy Logic

Pedro Antonio Salazar-Tejeda, Patricia Melin, and Oscar Castillo

Tijuana Institute of Technology, Tijuana BC. México

Abstract. In this paper, we describe an application of biometric recognition that is structured basically with three inputs: the hand geometry, voice and image. The hand geometry is given by an image of "the palm" of the hand with a 480x640 size which is preprocessed with a feature extraction that uses computer vision techniques and with certain features we recognize the individual. After that we preprocessed the image and get some variables as the fingers, palm, wrist, also a segment of the palm; they appear to be from a with a fuzzy system that will tell us how much they seemed to a certain person, comparing each variable given by the preprocessing of the image according to the data base that its already stored (all the images of the individuals, voice, etc.).

1 Introduction

Since ancient times, man has been concerned with security, and has laid hands on physical characteristics to achieve this end, today the subject to recognition and identification has become essential for achieving solve security problems that have arisen the solution to like, in recognizing a vehicle, a car registration, identification of a person. Has been inspired by the mechanics of how a living being interacts with its environment to make decisions, as an example, one could say that a human being is able to recognize a person through one of your senses, listening to his tone of voice, to identify any odor, even when viewing a person in a place with little light, is able to know who is, thanks to the senses that owns and pattern recognition analysis. The pattern recognition, also called reading patterns, consist of identifying shapes and shape recognition, pattern recognition signals. Not only is a field of computer science, is a fundamental process that is found in almost all human actions [4].

2 The Problem

The problem was solved by developing an implementation in Matlab 7.1 based on the analysis of voice, measurements of the geometry of the hand and similarity of the palms of the individual; techniques using hand geometry, voice and computer vision extraction of features, neural networks and fuzzy logic.

Neural Networks were applied to a database of 29 persons which consisted of a sample of original sound ("which was the name") and 3 more variants which were the

O. Castillo et al. (Eds.): Soft Computing for Hybrid Intel. Systems, SCI 154, pp. 171–186, 2008.
springerlink.com © Springer-Verlag Berlin Heidelberg 2008

same sounds original (116 samples in total), with the difference that these samples they had been implemented variations in the signal with respect to the original sample, this would be spelled later. On the other hand is working with a cut made to the palms which a neural network assessment to obtain a result of identification of an individual, this cut in the image were applied different techniques for extracting features regarding the image with the intention of starting from the general to the particular with regard to the image data to work with the neural network. Another key point for the application of neural networks is the need to find suitable architectures; it is noteworthy that these were determined to trial and error. With regard to the geometry of the hand, was conducted in Matlab an algorithm which its purpose is to find some measurements on the image that happens to analyze. Finally, we developed two fuzzy systems, the first of which is responsible for assessing the results of the analysis resulting geometry of the hand, and give an outcome, the second fuzzy system is responsible for evaluating the final results of each analysis, i.e. (the result of the geometry of the hand result of neural network + voice + result of the neural network palms = identification of individuals), see figure 1.

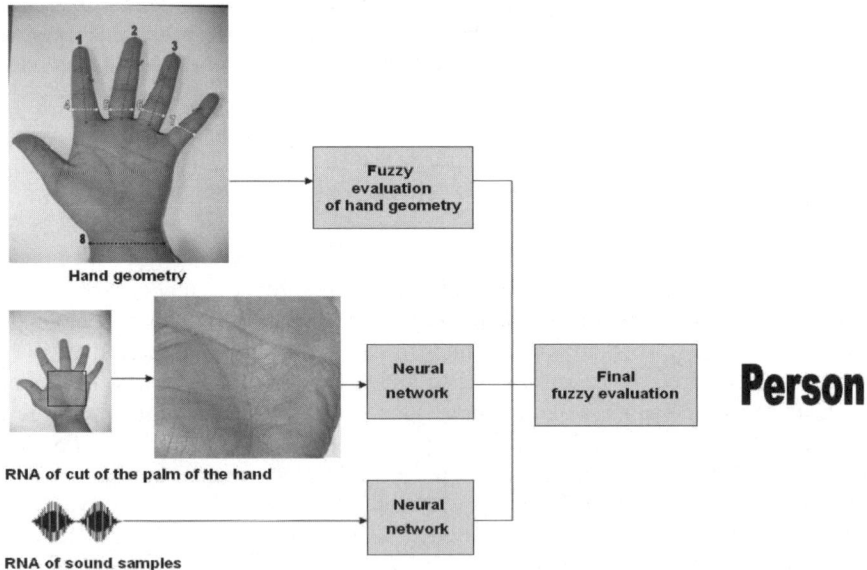

Fig. 1. Proposed architecture for recognition

3 Database

3.1 Database of Sounds

There is a database that consists in 116 sound of samples, of 29 different people, which recorded its name; these samples are divided in the following way, See the figure 2.

29 original files (feed network for training).
29 with noise option # 1.
29 with noise option # 2.
29 with noise option # 3.

Characteristic of the sound file

Transmission speed   ....: 64kbps
Size of the sample of sound   ....: 8bits
Channel   ....: mono
Speed of sample of sound   ....: 8 KHz
Format of audio   ....: PCM
Size   ....: 7.85KB

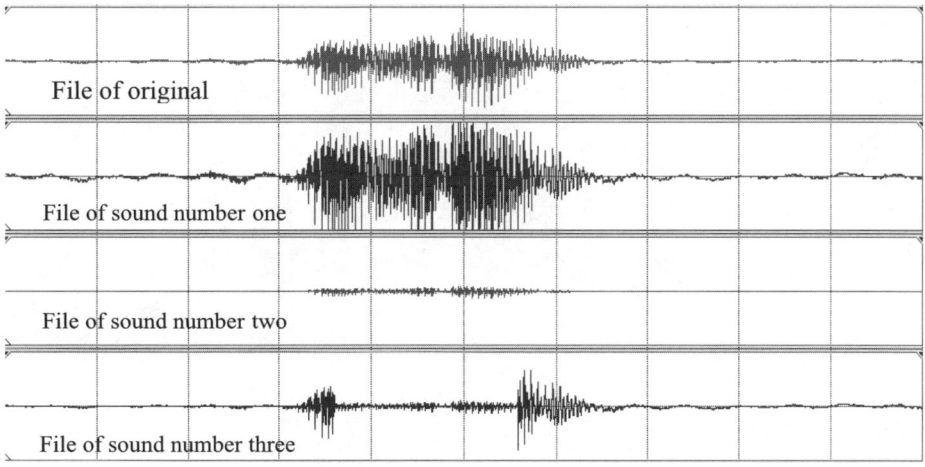

Fig. 2. Example of the database of sounds

3.2 Database of Hand Geometry

There is a database that consists in 42 images of samples, three images for person, and 14 people in total, of which in the case of geometry of the hand, was elected as one of three reference images to identify the two remaining images of each person, that is to say the database of 42 images of the 14 people alone were registered 14 images and 28 images remaining were used to identify the individual, as is observed in the figure 3.

3.3 Database of Part of the Hand Palm

For the case of the neural network of part of the hand palm, of the same mentioned previous database new images of the process of part of the hand palm were obtained, 42 samples in total of same 14 people, see figure 4.

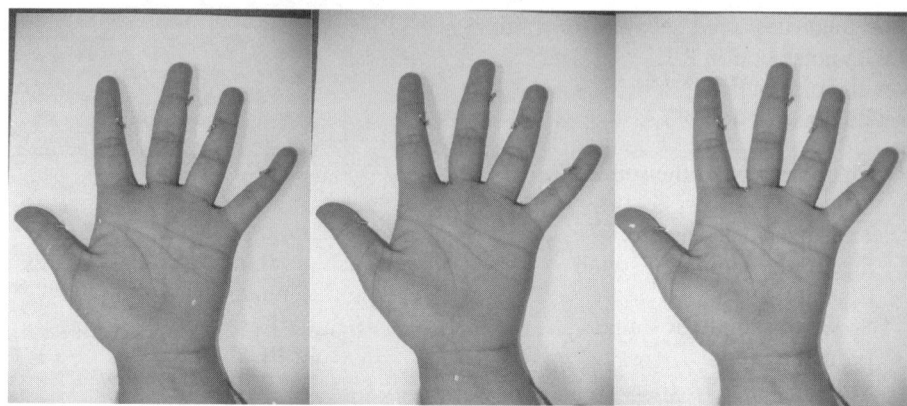

Fig. 3. Example of three samples images, each image is of different taking. Size of the image is 480 x 640 of format jpg.

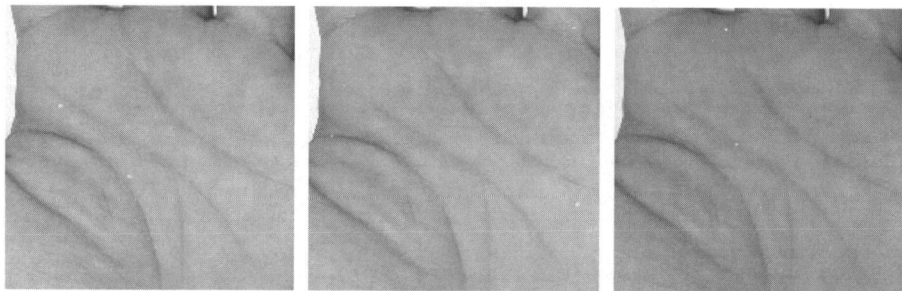

Fig. 4. Example of three samples of images, each image is of different taking. Size of the image is 201 x 201 of format jpg.

4 General Theory

We describe in this section some basic concepts relevant to the paper.

Biometrics: The term biometrics applies generally to science that is dedicated to the statistical study of the quantitative characteristics of living beings: weight, length, etc.. But in more recent times this term is also used to refer to the automatic methods that analyze certain human characteristics to identify and authenticate individuals [14].

Pattern Recognition: It is the science that is responsible for the description and classi-fication (recognition) of objects, people, signs, representations, and so on. While the margin of applications is very wide, the most important are related to vision and hearing by a machine, in a manner similar to humans. Some examples of its application are:

- Character Recognition
- Industrial Applications

- Mapping medical applications
- Guided Vehicle
- Automatic Speech Recognition
- Recognition biometric people

Neural networks: They aim to mimic the shape of tiny scale operation of the neurons that make up the human brain. Any development of the neural network has a lot to do with the neuro-physiology, it is not in vain to imitate a human neuron as accurately as possible, Figure 5.

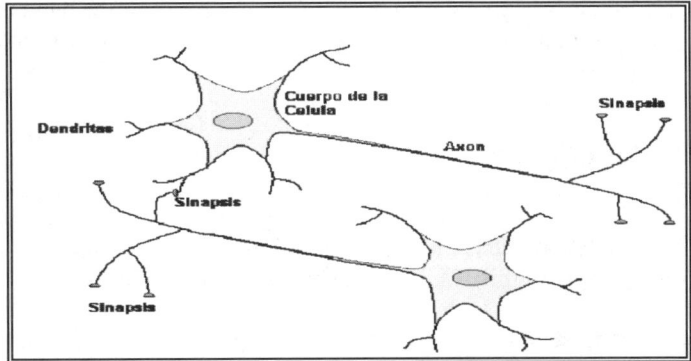

Fig. 5. Structure of a biological neuron

The ANNs (Artificial Neural Networks) were originally an abstract simulation of biological nervous systems, composed of units called neurons "or" nodes "connected with each other. These connections have a great resemblance to the dendrites and axons in the biological nervous systems [22] [15], see figure 6.

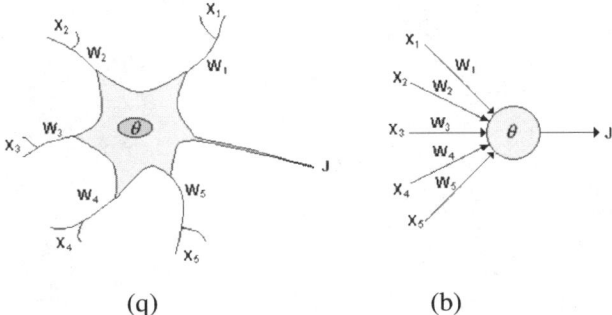

(q)   (b)

Fig. 6. (a) Biological neuron (b) Artificial neuron

Fuzzy Logic: It has two different meanings. In a narrow sense, is a fuzzy logic system can be considered an extension of the multiple – valued logic. But in a broad sense is dominant, fuzzy logic is almost synonymous with the fuzzy set theory, a theory that relates to object classes with undefined borders where membership is a matter of degree. The basic concept fundamental fuzzy logic is the linguistic variable which is a variable whose values are words rather than numbers. While the words are inherently less accurate than the numbers, their use is closest to human intuition. Calculating with words rather than numbers lowers the cost of the solution.

Geometry of the hand: The use of the geometry of various body parts to identify people, started as study [7] [14], at the time of the ancient Egyptians, although closest to our times can be found in the Bertillon system of the late nineteenth century. The authentication systems based on the analysis of the geometry of the hand are undoubtedly the fastest within the biometric: with a low probability of error in most cases, in approximately one second can determine whether a person is who is said.

5 Identification Process

The first step is the analysis of the complete image of the hand process based on the technique of the hand geometry, this analysis will give a result that later can be evaluated with a Fuzzy System, this result that they can be (Identified, resemblance or different). See figure 7.

The second step is the analysis of part of the image of the hand palm, by means of a neural network, the result is if the image that was given from entrance to the net corresponds the individual required, if this is certain will tell us identified if not, not identified.

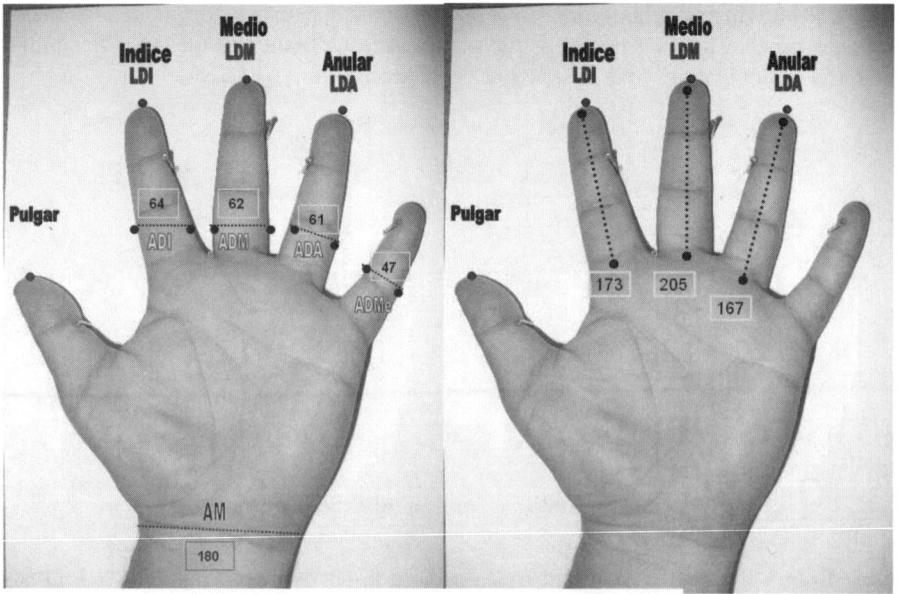

Fig. 7. Result of method of hand geometry

Step three, is the analysis of the sounds samples by means of a neural network, the result of the neural network is if the sound that was given from entrance to the net corresponds the individual required, if this is certain will tell us identified if not, not identified.

Step four, for finish a fuzzy system evaluates the final results of the voice analyses, image of the hand palm and measures of geometry of the hand, and the system decides if the individual is identified, resemblance or different, as a result final, as is observed in the figure 1.

6 Fuzzy System

The fuzzy system presented in the figure 8, is the one that evaluates the result of the analysis of the geometry of the hand.

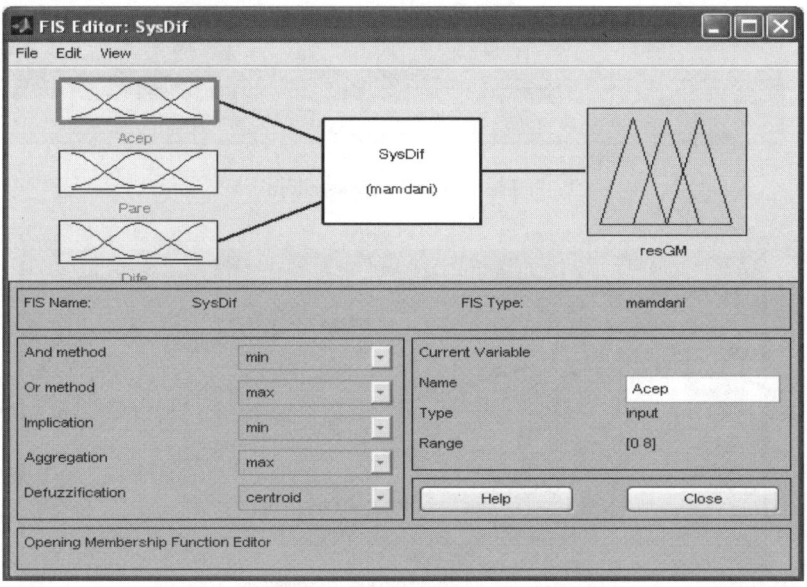

Fig. 8. Fuzzy system of hand geometry

The fuzzy system presented in the figure 9, is the one that evaluates the final results of the analysis of the geometry of the hand, voice and part of hand palm.

The fuzzy system work in the following way has three of entrance variables that correspond to the final results of the analysis of the geometry of the hand, voice and part of palm of the hand. The variables of entrance netsound and netimg, will receive the number 0 when it is not identified (membership D) and 1 when it is identified (membership R), see figure 10 y 11. The i variable GM (Hand Geometry), they will receive the number 0 when it is not identified (membership D), number 0.5 when it is not seemed (membership P), and 1 when it is identified (membership R), see figure 12.

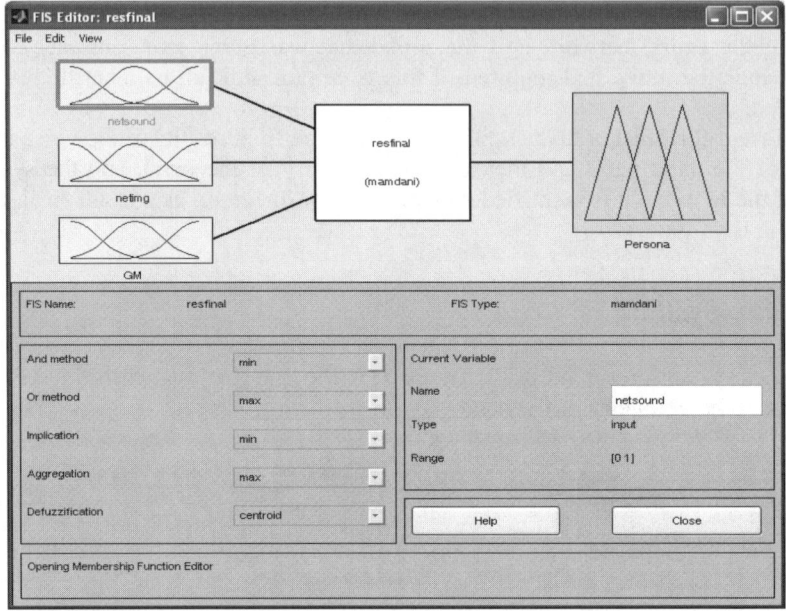

Fig. 9. Final fuzzy system

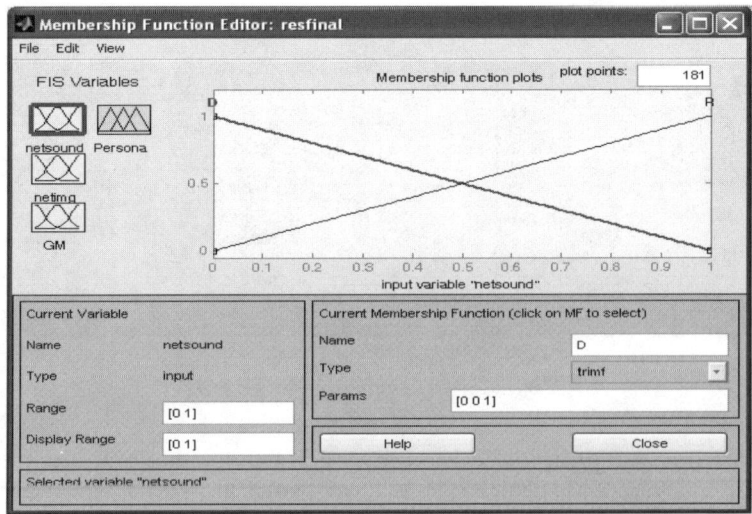

Fig. 10. Functions of membership of the variable netsound

The variable of outp person, will receive values from 1 to 3, the value 1 mean that alone one of the three analyses identifies to a person and the value 3 that the three analyses identified the person, see figure 13.

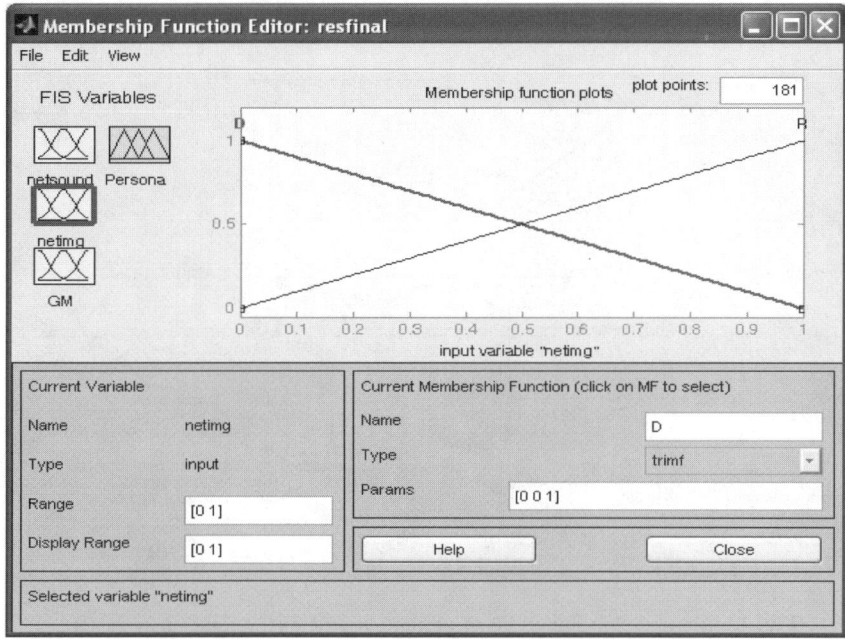

Fig. 11. Membership functions of the variable netimg

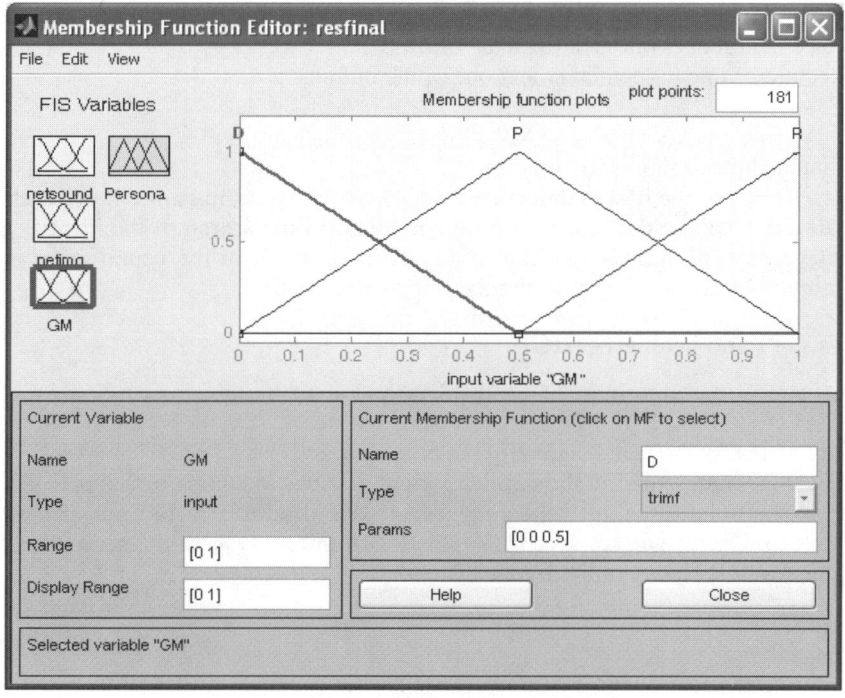

Fig. 12. Functions of membership of the variable GM (Hand Geometry)

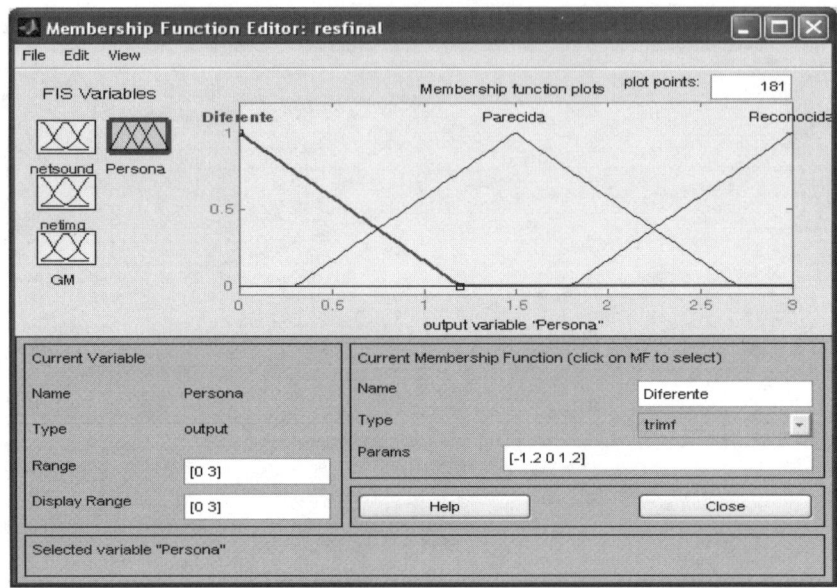

Fig. 13. Membership Functions of output variable of the final fuzzy system

7 Results

The voice neural network is defined as follows.
Architecture of one hidden layer and one of the output.
S1=58;
net = newff(minmax(sonido),[S1 S2],{'logsig' 'purelin'},'trainrp');
Resilient backpropagation **'trainrp'**
   In the search of the best architecture for an ANN of voice, the nine shown architectures are the result of the best nine trainings found in the search of the identification percentage most high possible. The results of the training of the neural network are shown in the figure 14; the result number nine is the best.

Analysis of hand geometry (HG)

When running the algorithm of hand geometry, eight variables are obtained, ADI, ADM, ADA, ADMe, LDI, LDM, LDA y AM, see figure 7. In the figure 16, all the results of the analysis of hand geometry are observed.
   The values represented in the figure 15 are the values to compare, that is to say, referring to the point three, of the 42 images 14 were used to register each person and the 28 remaining images are used for evaluation of the analysis of the geometry of the hand, the result is shown in the figure 17.

Example
Original value.
Person: 1 ADI= 66 ADM= 57 ADA= 65 ADMe= 52 LDI= 149 LDM= 197 LDA= 155 AM= 193, see figure 15, number 1.

Value obtained to compare with the original.
Person: 1 ADI= 63 ADM= 57 ADA= 61 ADMe= 48 LDI= 146 LDM= 194 LDA= 151 AM= 184 see figure 16, person 1 image 2.

				Recognition percentage				
#Training	neu-rons	Time of training	Architecture	Normal voice	Voice noise #1	Voice noise#2	Voice noise#3	Method of training
1	58	38seg	1 hidden layer	100%	100%	93%	100%	Scg
2	58	23seg	1 hidden layer	100%	100%	97%	97%	Scg
3	50	27seg	1 hidden layer	100%	100%	97%	100%	Scg
4	58	01:22seg	1 hidden layer	100%	100%	97%	100%	Gdx
5	50	01:13 seg	1 hidden layer	100%	100%	97%	100%	Gdx
6	50	36seg	1 hidden layer	100%	100%	93%	97%	Gdx
7	58	39seg	1 hidden layer	100%	100%	97%	100%	Gdx
8	58	18seg	1 hidden layer	100%	100%	100%	97%	Rp
9	58	15seg	1 hidden layer	100%	100%	100%	100%	Rp

trainscg   Scaled conjugate gradient backpropagation
traingdx   Gradient descent with momentum and adaptive learning rate backpropagation
trainrp   Resilient backpropagation

Fig. 14. Is observed that the best training was the finish

	ADI	ADM	ADA	ADMe	LDI	LDM	LDA	AM
1	66	57	65	52	149	197	155	193
2	66	63	62	49	175	207	170	178
3	63	54	60	46	164	187	147	171
4	62	55	57	39	148	180	137	164
5	60	49	51	22	129	141	97	160
6	75	62	58	47	191	231	172	183
7	58	53	58	43	154	193	145	178
8	51	47	48	39	139	163	132	154
9	53	50	48	44	153	190	158	161
10	48	44	44	42	137	166	136	147
11	55	51	50	45	165	187	141	155
12	57	51	50	38	147	168	131	147
13	69	60	58	43	135	182	151	179
14	56	52	53	42	157	187	137	163

Fig. 15. Sample fourteen results of the 42 images analyzed for hand geometry

Person	IMG	LDI	LDM	LDA	ADI	ADM	ADA	ADMe	AM
Alma Isabel	1	149	197	155	66	57	65	52	193
1	2	146	194	151	63	57	61	48	184
	3	147	195	153	65	57	63	50	186
alma martinez	1	175	207	170	66	63	62	49	178
2	2	173	205	167	64	62	61	47	180
	3	171	204	168	64	61	60	47	182
cinthya	1	164	187	147	63	54	60	46	171
3	2	161	183	143	63	53	60	47	172
	3	161	183	144	62	54	60	46	170
erika	1	148	180	137	62	55	57	39	164
4	2	147	180	139	62	55	57	39	164
	3	147	180	138	61	54	57	39	167
jaime	1	129	141	97	60	49	51	22	160
5	2	130	146	106	58	46	50	33	152
	3	129	145	103	58	46	51	30	152
jose	1	191	231	172	75	62	58	47	183
6	2	190	233	174	74	63	57	47	199
	3	190	232	174	75	63	58	49	208
magda	1	154	193	145	58	53	58	43	178
7	2	153	191	142	58	53	57	42	178
	3	153	190	144	58	53	59	43	178
manuel	1	139	163	132	51	47	48	39	154
8	2	133	160	133	50	47	47	42	158
	3	132	160	133	49	46	47	41	159
mirsa	1	153	190	158	53	50	48	44	161
9	2	150	188	157	51	50	47	43	162
	3	147	183	152	50	48	46	42	151
monica	1	137	166	136	48	44	44	42	147
10	2	136	163	132	47	44	45	40	140
	3	133	161	131	46	44	44	39	138
patricia	1	165	187	141	55	51	50	45	155
11	2	163	184	139	55	48	49	42	155
	3	162	180	135	56	47	49	44	146
ray	1	147	168	131	57	51	50	38	147
12	2	146	166	129	57	50	48	38	147
	3	144	163	125	56	49	48	38	141
Ricardo_Marro	1	135	182	151	69	60	58	43	179
13	2	135	184	153	70	61	59	44	179
	3	134	183	153	69	60	58	43	179
Victor_Manuel	1	157	187	137	56	52	53	42	163
14	2	157	190	145	55	52	53	42	161
	3	160	191	146	55	51	53	41	160

Fig. 16. Results of the analysis of hand geometry of fourteen people with their 3 images to compare

Person	Result HG	Acep	Pare	Dife	Person	Result HG	Acep	Pare	Dife
Person 1 image 1	R	8	0	0	Person 8 image 1	R	8	0	0
Person 1 image 2	P	4	3	1	Person 8 image 2	R	6	2	0
Person 1 image 3	P	4	3	1	Person 8 image 3	R	6	1	1
Person 2 image 1	R	8	0	0	Person 9 image 1	R	8	0	0
Person 2 imagen 2	R	8	0	0	Person 9 image 2	R	8	0	0
Person 2 imagen 3	R	6	2	0	Person 9 image 3	P	4	2	2
Person 3 imagen 1	R	8	0	0	Person 10 image 1	R	8	0	0
Person 3 imagen 2	R	6	2	0	Person 10 image 2	R	6	1	1
Person 3 image 3	R	7	1	0	Person 10 image 3	P	4	3	1
Person 4 image 1	R	8	0	0	Person 11 image 1	R	8	0	0
Person 4 image 2	R	8	0	0	Person 11 image 2	R	8	0	0
Person 4 image 3	R	8	0	0	Person 11 image 3	P	4	2	2
Person 5 image 1	R	8	0	0	Person 12 image 1	R	8	0	0
Person 5 image 2	P	4	1	3	Person 12 image 2	R	8	0	0
Person 5 image 3	P	4	2	2	Person 12 image 3	P	5	3	0
Person 6 image 1	R	8	0	0	Person 13 image 1	R	8	0	0
Person 6 image 2	R	7	0	1	Person 13 image 2	R	8	0	0
Person 6 image 3	R	7	0	1	Person 13 image 3	R	8	0	0
Person 7 image 1	R	8	0	0	Person 14 image 1	R	8	0	0
Person 7 image 2	R	8	0	0	Person 14 image 2	R	7	0	1
Person 7 image 3	R	8	0	0	Person 14 image 3	R	6	1	1

Fig. 17. Are observed all the results of the analysis of the hand geometry

Differences with regard to the original values.

Person: 1 ADI= 3 ADM= 0 ADA= 4 ADMe= 4 LDI= 3 LDM= 3 LDA= 4 AM= 9

These variables are labeled in three ways, Acep (for variables accepted), Pare (for similar variables), Dife (for different variables).

X can have any value of (1...8), for the eight variables that one obtains of the analysis of the hand geometry.

```
if (valorX >= 0)&&(valor1 <= 3)
   Acep=Acep+1;
end
if (valorX >= 4)&&(valor1 <= 6)
   Pare=Pare+1;
end
if (valorX >= 7)
   Dife=Dife+1;
end
```

The result: Acep=4.; Pare=3; Dife = 1.

This result evaluated it the fuzzy system of hand geometry, see figure 8.

Result = P (similar person). See figure 17, person 1 image 2.

The neural network of the part of the hand palm

Architecture of one hidden layer and one of exit.

S1 = 56;
net = newff(minmax(imagen),[S1 S2],{'tansig' 'logsig'},'trainscg');

Scaled conjugate gradient backpropagation **'trainscg'**

The network was fed with 28 of 42 images of 14 people, as mentioned in section three, see figure 4, and the 14 remaining images were used to identify. In figure 3 some results of the network are shown. The one selected was based on speed of training and recognition percentage and identification is the one in row number 6 of Table 1.

Table 1. Some results of the neural network of images that were observed

Time	S1	Recognition	Identification	Method	Error
1.8seg	62	28 de 28	14 de 14	trainscg	0.00001
8seg	27	18 de 28	5 de 14	trainscg	0.00001
1.3seg	54	28 de 28	11 de 14	trainscg	0.00001
1.6	56	26 de 28	10 de 14	traingdx	0.00001
1.4	40	22 de 28	7 de 14	traingdx	0.00001
1.4seg	56	28 de 28	14 de 14	trainscg	0.00001

8 Conclusions

In this paper, we have described an application of biometric recognition that is structured basically with three inputs: the hand geometry, the voice and an image of the hand palm. The hand geometry is given by an image of "the palm" of the hand with a 480x640 size, which is preprocessed with a feature extraction that uses computer vision techniques and with certain features we recognize the individual. Simulation results show the feasibility of the proposed approach.

References

[1] Bishop, C.M.: Neuronal networks for pattern recognition. Oxford Press (1995)
[2] Chen, G., Phan, T.T.: Introduction to Fuzzy Sets. In: Fuzzy Logic and Fuzzy Control Systems. CRC Press, USA (2000)
[3] Lecanda, E., Irigoyen, E.: Sistemas biométricos, Universidad del País Vasco, Escuela Superior de Ingenieros (2004)
[4] Meyer, E.A.: Glosario de términos técnicos, Entrada P, pattern recognition. Grupo de Informática Aplicada al Inglés Técnico, Argentina, bajo la licencia de documentación libre GNU (1995)
[5] Gran enciclopedia Larousse Planeta S.A. Barcelona, Spain (1969)
[6] Gray, H.: Anatomy descriptive and surgical. Gramercy Books, Nueva York (1977)
[7] Lee, H.C., Gaensslen, R.E., et al.: Advances in Fingerprint Technology Elsevier, 1991. CRC Press LLC., USA (1994)
[8] Jan, J., Sun, J., Mizutani, E.: Neuro-Fuzzy and soft computing: a computational approach to learning and machine Intelligence. Prentice Hall, Englewood Cliffs (1997)
[9] Woodward, J.D.: Biometrics: Privacy's Friend. Proc. IEEE Special Issue on Automated biometrics 85(9), 1480–1492 (1997)
[10] Jang, J-s.R.: Neuro-Fuzzy and Soft Computing: a computational approach to learning and machine intelligence
[11] Mendel, J.M.: UNCERTAIN Rule-Based Fuzzy Logic Systems. In: Introduction and New Directions. Prentice Hall, Englewood Cliffs (2001)
[12] Fukunaga, K.: Statistical Pattern recognition, 2nd edn. Academic Press, London (1990)
[13] Acosta, M.I., Salazar, H., Zuluaga, C.A.: Tutorial de Redes Neuronales, Universidad Tecnológica de Pereira, Facultad de Ingeniería Eléctrica (2000),
http://ohm.utp.edu.co/paginas/docencia/neuronales/
[14] Mateos, M.T. Juan, A., Pizarro, S.: Tecnologías biométricas aplicadas a la seguridad, Editorial Alfa omega, ISBN 970-15-1128-X
[15] Soto Castro, M.G.: Thesis Reconocimiento de rostro y voz en tiempo real utilizando redes neuronales artificiales, Instituto Tecnológico de Tijuana (2006)
[16] Barley, N.: El antropologo inocente. In: Notas desde una choza de barro, Anagrama (1989)
[17] Peter Pacheco.: Parallel Programming with MPI. (Deiciembre de 2005),
http://www.cs.usfca.edu/mpi.html
[18] Melin, P., Castillo, O.: Hybrid Intelligent Systems for Pattern Recognition (2005)
[19] Clarke, R.: Human identification for information System: Management Challenges and Public Policy Issues. Info. Technol. People 7(4), 6–37 (1994)
[20] Schalkoff, R.: Pattern recognition: Statistical, structural and neuronal approaches. John Wiley & Sons, Chichester (1992)
[21] Seix, M. Bernando: Estudio de redes neuronales modulares para el modelado de sistemas dinámicos no lineales, Universidad UPC (febrero 2006),
http://www.tdc.cesca.es/TDX-0726101-162142/
[22] Shorlemmer, M.: Tutorial básico sobre redes neuronales. In: Artificial Intelligence Research Institute, Barcelona, Spain (1995)

[23] Nordstrom, T.: Highly Parallel Computers for Artificial Neural Networks. PhD Thesis. Division of Computer Science & Engineering. Luleå University of Technology, Sweden
[24] Universidad de Antoquia, Facultad de ingeniería, Redes Neuronales (2004), http://ingenieria.udea.edu.co/investigacion/mecatronica/mectronics/redes.htm
[25] Universidad Politécnica de Madrid E.S.T.I. de telecomunicaciones (2005)
[26] Zadeh, L.A.: Knowledge representation in Fuzzy Logic. IEEE Transactions on knowledge data engineering 1, 89 (1989)
[27] Zadeh, L.A.: Fuzzy sets, Information & Control, Fuzzy logic. IEEE Computer 8 (1975)
[28] Zadeh, L.A.: Fuzzy Logic = Computing with Words. IEEE Transactions on Fuzzy Systems 4(2), 103 (1996)

Part III

Intelligent Agents and Social Systems

A Hybrid Model Based on a Cellular Automata and Fuzzy Logic to Simulate the Population Dynamics

Cecilia Leal Ramírez[1] and Oscar Castillo[2]

[1] Facultad de Ciencias Químicas e Ingeniería, Universidad Autónoma de Baja California, Tijuana B.C. México, División de Estudios de Postgrado e Investigación
[2] Instituto Tecnológico de Tijuana B.C. México

Abstract. At present time, new advances in the generation of computational models can be applied to improve tasks in different areas of research. The hybrid computational models can be considered as new advances in science. In the present work a hybrid model has been proposed on the basis of a cellular automata and fuzzy logic to simulate, in space and time, the dynamics of a population structured by ages and where the changes in the levels of the biomass are induced by a stochastic variation of the environment. The model can be used as computational tool in the area of the Biology to describe and quantify the changes that continuously occurs in the population, knowing not only their size and its structure, but the form and the intensity in which it changes and renews.

Keywords: hybrid model, cellular automata, fuzzy logic, biomass.

1 Introduction

The satisfaction of human needs of natural resources implies creating a balance between the needs and the resources, in other words, the resources required by a human society that increase at certain moment must exceed the supply (Malthus 1798). In order of handle this balance is necessary to have information about the current state of the environment and use efficient tools to predict possible changes of this state, due to diverse effects (Bulla and Rácz 2004). This implies knowing, in the area of the ecology, the population dynamics to negotiate the biological resources and to evaluate the environmental consequences of the human actions.

The population dynamics takes care of the study of the changes that the biological populations suffer as for size, physical dimensions of their members, age structure and sex and other parameters that define them, as well as of the factors that cause these changes and the mechanisms that produce them. The development and analysis of the mathematical models lead to a better and deeper interpretation of the dynamic processes that are produced in the populations. The formulation of the models is focused to the acquisition of new knowledge from the mathematical development or the population processes when allowing to draw greater conclusions and to obtain more complete results. In the last two hundred years a great variety of mathematical models on populations dynamics have been proposed (Molina 2004). In 1966, von Neumann introduced the Cellular Automata (AC), these mathematical models are broadly used

O. Castillo et al. (Eds.): Soft Computing for Hybrid Intel. Systems, SCI 154, pp. 189–203, 2008.

by scientists to model complex ecological systems and well known due to its mathematical simplicity when not using differential equations (Wolfram (1986), Gutowitz (1991), Moreno et al. (2002), Molofsky and Bever (2004), Rohde (2005)).

In the present work we constructed a model composed of a cellular automata and a fuzzy logic system that we call a 'hybrid model' to model the population dynamics structured by ages. The attributes or characteristic that are studied in all the populations, are the mortality, reproduction and migration. The transition function is defined in terms of these characteristics, which are induced by the environment stochastic variation and controlled by a fuzzy logic system. The join between cellular automata and the fuzzy logic system is given by the transition function definition, which determines the state of a cellular space's cell in a time unit.

From the point of view of Zadeh (1975, 1988), founder of the fuzzy sets theory, any field can be represented by fuzzy sets wherein the benefits increase the ability to model problems of the real world. The fuzzy approach as a possible road to handle uncertainty is particularly useful to process uncertainty or imprecise data of the environment. These can be defined as fuzzy sets that reflect better the continuous character of the nature.

Several models have been constructed combining cellular automata and fuzzy sets applied in different problems in the ecological investigation. For example, Di Stefano (2000) developed a model based on this combination for analysis of spread of epidemics, and Mandelas et al. (2006) developed a shell for modeling urban growth by using the same combination. However all they have differences in their construction, because their purposes are different. Our purpose is to represent environment variables by using fuzzy sets, whose combination is the result of the factors that determine the growth model of the population dynamics.

In the second section of this paper a detailed description of the architecture of the hybrid model is presented. In the third section an implementation is made using a simulation program. We show trajectories described by the population dynamics behavior under different biological interpretations represented by means of the fuzzy rules. In the fourth section we finished with the conclusions about the applicability of the hybrid model.

2   The Hybrid Model

A model is a representation of a particular thing, idea or condition. The models can be very simple, such as models of logistical growth for species or extremely complex such as models based on individuals. In this study, a hybrid model is presented which is composed by a cellular automata and a fuzzy logic system wherein the union of both architectures is given by the transition function. The transition function is defined in terms of the general characteristics of any study of populations' dynamics. The fuzzy logic system controls the environment factors that induce the population behaviour variation.

2.1   Cellular Automata

The Cellular Automatas (CA) are flexible and particularly useful for the investigation in space-temporary modeling of natural processes (Wolfram (1986), Czárán (1998)).

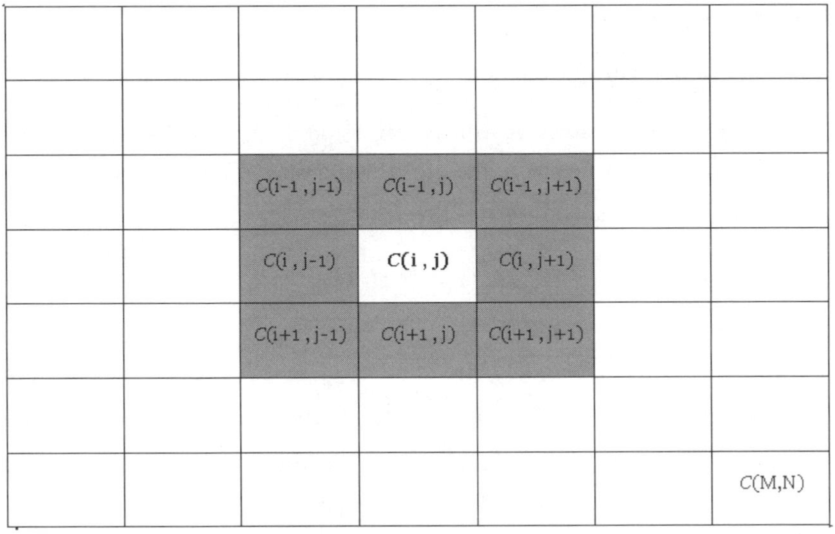

		C(i-1,j-1)	C(i-1,j)	C(i-1,j+1)		
		C(i,j-1)	C(i,j)	C(i,j+1)		
		C(i+1,j-1)	C(i+1,j)	C(i+1,j+1)		
						C(M,N)

Fig. 1. The cellular automata architecture with neighborhood $r = 1$ for each cell

The concept of a CA is based on the original Von Neumann's idea (1966). A CA is a formal model composed of a rectangular region of MxN cells in which the evolution of each cell depends on its present state and the state of its immediate neighboring cells (figure 1). All of the cells, at the same time, pass to the following generation according to a transition function, being this same one for all the cells. Let C the representation of a cell, then $C(i, j)$ is the cell that is centered in the point of coordinates where $1 \le i \le M$ and $1 \le j \le N$.

Then, we defined the union of all the cells of the cellular automata by

$$\Omega(M, N) = \bigcup_{ij} \{C(i, j)\}.$$

Each cell interacts with the others one within a finite neighborhood according to a local rule (gray blocks in figure 1). The cell neighborhood $C(i, j)$ in terms of a radius r is represented by means of $\Pi_r(i, j)$, which is defined as

$$\Pi_r(i, j) = \left\{ C(k, l) \left| \begin{array}{l} \max\{|k - i|, |l - j|\} \le r \\ 1 \le k \le M; 1 \le l \le N \end{array} \right. \right.,$$

where r is a positive integer number and k, l are the coordinates of another cell where the magnitude of the difference between i, k and j, l does not exceed the value r for any case. The defined CA for this study considers the kind of Moore neighborhood ($r = 1$) with absorbent edges, wherein the cells of the edges do not have neighboring cells beyond of the grid limits (figure 1).

2.2 Population Dynamics

Each CA cell is a fundamental element for the population change, in each cell exists a population of individuals structured in ages, which reproduce, die and emigrate towards neighboring cells, reason why the contribution of all the cells that constitute the cellular space $\Omega(M, N)$ results in the global dynamics of the population. Then total number of individuals in $\Omega(M, N)$ is given by

$$N(t) = \sum_k n_k(t).$$

For each cell, let $n_k(C(i, j), t)$ the number of individuals of the age class k^{th} that at the same time t is over the cell $C(i, j)$. Therefore, at time t the total number $n_k(t)$ of individuals of the age class k^{th} distributed in $\Omega(M, N)$ is expressed by

$$n_k(t) = \sum_i \sum_j n_k(C(i, j), t),$$

and the number of individuals that are distributed in the cell $C(i, j)$ at time t is expressed by means of

$$n(C(i, j), t) = \sum_k n_k(C(i, j), t).$$

Every population is constantly under the effect of opposed factors that at the same time tend to increase and diminish it. The size of the population depends at any moment on the existing balance among these factors. Therefore the transition function that calculates the number of individuals in a cell at time $t + 1$ is expressed in terms of the reproduction rate, mortality rate and emigration, wherein these factors are controlled by a fuzzy logic system.

Let μ_1, μ_2 and μ_3 the mortality rates of the individuals of the age class one, two and three respectively, therefore the number of surviving individuals in each age class at time t is expressed by

$$n_1(C(i, j), t) - \mu_1(n_1(C(i, j), t)), \tag{1}$$

$$n_2(C(i, j), t) - \mu_2(n_1(C(i, j), t)), \tag{2}$$

$$n_3(C(i, j), t) - \mu_3(n_1(C(i, j), t)), \tag{3}$$

and let b_1, b_2 y b_3 the reproduction rates of the individuals of the age class one, two and three respectively

$$b_1[n_1(C(i, j), t) - \mu_1(n_1(C(i, j), t))] \tag{4}$$

$$b_2[n_2(C(i, j), t) - \mu_2(n_1(C(i, j), t))] \tag{5}$$

$$b_3[n_3(C(i, j), t) - \mu_3(n_1(C(i, j), t))] \tag{6}$$

All the individuals generated by the reproduction belong to the individuals of the first age class at time $t+1$. The individuals that time t survive in the first class transfer to time $t+1$ to the second class. In the same way it occurs with the individuals of the second class, the survivors to time t pass at time $t+1$ to the third class. The survivors in the third age class to time t remain on it to time $t+1$. This is expressed of the following way by taking the expressions (1), (2), (3), (4), (5) and (6)

$$n_1(C(i,j),t+1)=b_1[n_1(C(i,j),t)-\mu_1(n_1(C(i,j),t))]+b_2[n_2(C(i,j),t)-\mu_2(n_1(C(i,j),t))]+$$
$$b_3[n_3(C(i,j),t)-\mu_3(n_1(C(i,j),t))] \tag{7}$$

$$n_2(C(i,j),t+1)=n_1(C(i,j),t)-\mu_1(n_1(C(i,j),t)) \tag{8}$$

$$n_3(C(i,j),t+1)=n_2(C(i,j),t)-\mu_2(n_1(C(i,j),t))+n_3(C(i,j),t)-\mu_3(n_1(C(i,j),t)) \tag{9}$$

All the individuals have optimal environment conditions under which they can be developed better and even though all the conditions are not the optimal ones, each population will always try to be located in environment conditions where each one of these conditions is within tolerable limits that allow their normal development. In this context the function that measures the intensity of the individual's emigration process that are in cell $C(i,j)$ towards the cells of their neighborhood $\Pi_r(i,j)$ is defined by means of the following expression

$$E_k(\Pi_r(i,j),t) \; \forall \; k=1,2,3 \tag{10}$$

The emigration of the individuals from the $C(i,j)$ towards $C(k,l)$ occur when the cell $C(k,l)$ has better relative environmental factor than the cell $C(i,j)$, where this factor is obtained by means of

$$F_{ij}(t)=\frac{f_{ij}(t)}{\sum\limits_{N_r(i,j)} f_{ij}(t)},$$

Where in $f_{ij}(t)$ is absolute environmental factor of cell $C(i,j)$. We supposed that the amount of individuals $e^k_{ij\to kl}(t)$ of age class k^{th} that transfer themselves from the cell $C(i,j)$ towards the cell $C(k,l)$ is given by the expression

$$e^k_{ij\to kl}(t)=\varepsilon^{kl}_{ij}(t)n_k(C(i,j),t),$$

where $\varepsilon^{kl}_{ij}(t)$ is a function of the gradient of relative environmental factors between cells $C(i,j)$ and $C(k,l)$, it is to say

$$\varepsilon^{kl}_{ij}(t)=\begin{cases} \alpha^{kl}_{ij}(t)\left|F_{kl}(t)-F_{ij}(t)\right| & si \;\; F_{kl}(t)>F_{ij}(t) \\ 0 \;\; si \;\; F_{ij}(t)>F_{kl}(t) \end{cases}$$

where $\alpha_{ij}^{kl}(t)$ is a random number distributed uniformly between 0 and 1. Therefore the transition function that calculates the number of individuals for the following generation of the cell $C(i, j)$ at time t is expressed in terms of the equations (7), (8), (9) and (10) and given by

$$n_1(C(i,j),t+1) = b_1[n_1(C(i,j),t) - \mu_1(n_1(C(i,j),t))] + b_2[n_2(C(i,j),t) - \mu_2(n_1(C(i,j),t))] +$$

$$b_3[n_3(C(i,j),t) - \mu_3(n_1(C(i,j),t))] + E_1(\Pi_r(i,j),t)$$

$$n_2(C(i,j),t+1) = n_1(C(i,j),t) - \mu_1(n_1(C(i,j),t)) + E_2(\Pi_r(i,j),t)$$

$$n_3(C(i,j),t+1) = n_2(C(i,j),t) - \mu_2(n_1(C(i,j),t)) + n_3(C(i,j),t) - \mu_3(n_1(C(i,j),t)) +$$

$$E_3(\Pi_r(i,j),t) \tag{11}$$

2.3 Fuzzy Logic System

Fuzzy logic (FL) is a area of soft computing that handles the fuzzy sets. These sets are defined by their membership functions, which indicate the degree of membership (a value between 0 and 1) of an element to given set (Anderson 1983).

A fuzzy logic system is composed by five elements: fuzzifier, rules, inference mechanism and defuzzifier (figure 2). The system starts with the fuzzyfier, which maps the input values on the fuzzy sets that characterize to both, antecedents and consequences. These sets are used as inputs to the inference mechanism through the rules. These rules are combined in the inference mechanism to produce a fuzzy exit (fuzzy sets), which it is mapped to a numerical value, by means of the defuzzyfier, to produce an output.

The fuzzy logic system used in the transition function is made up of two antecedents and two consequent ones, whose relation controls the reproduction and mortality rates of a population structured in ages.

In the hybrid model the population distributed in the CA cells is considered as population of fish, whose pattern of growth describes a sigmoidal curve type (figure 3). In the very early stages of the life of the fish the increment in weight is very slow. But later, the growth goes accelerating until reaching approximately 1/3 of its maximum weight. After this, an inflection appears and the growth becomes more and more slow, and the fish goes coming closer asymptotically to his maximum weight (von Bertalanffy (1938)).

The fish population growth pattern is taken as reference to structure the population in three fuzzy sets that represent three edge classes: Small, Young and Adult (figure 4). The fuzzy sets are defined by means of membership functions, which assign the membership that input values have on the fuzzy sets, being these values the rank of the weights that a fish is acquiring during all the stage of its life (figure 3).

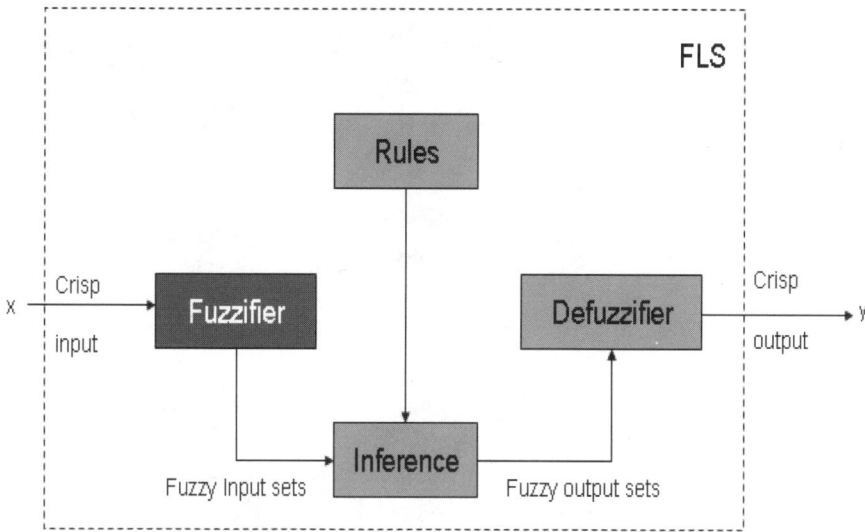

Fig. 2. The fuzzy logic system architecture

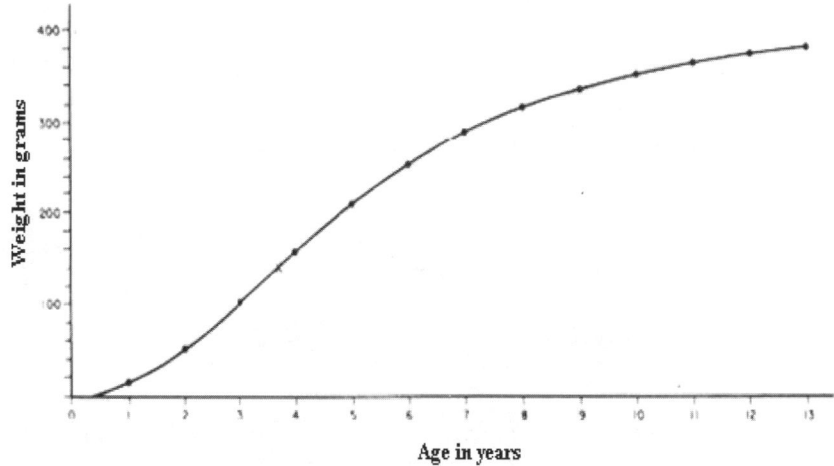

Fig. 3. Growth curve in weight of a fish population

For example, a fish with a weight of 150 grams belongs to but of a fuzzy set being a fish that this stopping being small to happen to the youthful age, which means that the fish has a membership degree to the fuzzy set Small less than that membership degree that it has to the Young fuzzy set. That's why it is deduced that the fish has a membership degree equal to zero on the Adult diffuse set. The membership function that defines the class of fish Adult concludes with the pattern of growth of the fish described in figure 3, wherein it shows as the growth is become slower approaching asintotically to its maximum weight, in which it will remain until aging or dying.

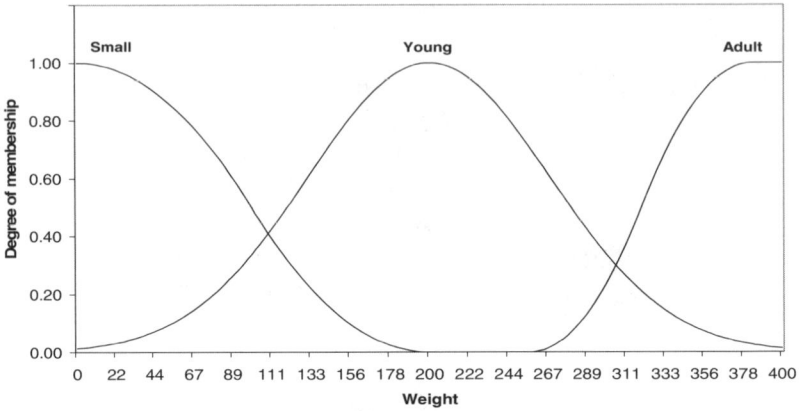

Fig. 4. Fuzzy sets defined to the structure of ages

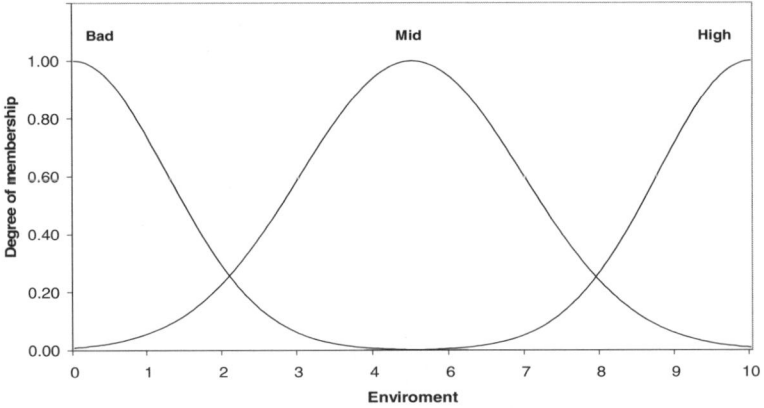

Fig. 5. Fuzzy sets defined to the environment factor

The environment is another factor that has great influence in the development and the state of the fish populations. The conditions can be several that alter positively or negatively the size and the population dynamic balance, for example: temperature, salinity, amount of oxygen, direction and wind force and currents, amount and quality of the organic material in suspension, structure of the land, etc. The environment must be considered like result of several variables that independently are affecting the population, causing some changes in its growth density and growth speed, as well as in any of the population parameters, changes that are evidently not related to the size of the same population.

In the fuzzy logic system, the environment factor is considered a value that represents the result of several environmental variables, which means that these variables are not handled of independent way in this system. Consequently the environment factor is represented by fuzzy sets: Bad, Mid, High (figure 5). The rank of the input

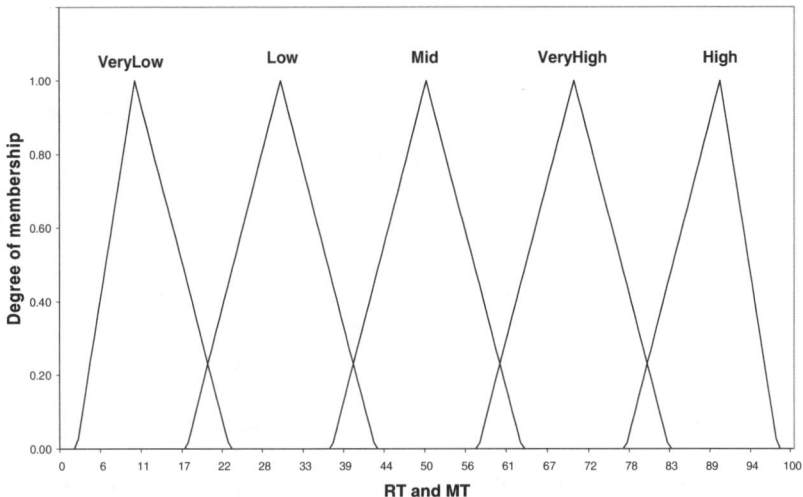

Fig. 6. Fuzzy sets defined to the reproduction and mortality rates

values for the membership function that define such fuzzy sets are values between 0 and 10, nevertheless this rank can be changed in terms of others units of measurement in accordance with any variable that is considered.

In the fuzzy logic system, the operations among sets are considered in terms of their membership functions and the results of these operations are the reproduction and mortality rates, which cause the changes in the population of each age class.

The reproduction and mortality rates are characterized by the same fuzzy set, the figure 6 show in a graph the states in which the both variables can appear as result of the environment effects, but in the design of the fuzzy system these are managed in an independent way. The fuzzy logic system considers two antecedents 'WEIGHT' and 'ENVIRONMENT', and two consequences 'RT' and 'MT'. These are characterized by fuzzy sets defined for age structure and environment factor, and the consequences ones are characterized by the fuzzy sets defined to reproduction and mortality rates. The results of the implication depends on the composition of the decision rules, these must be adapted to the type behavior desired. In the present work the decision rules composition is presented according to population behavior characteristics considered for hybrid model

- The young fish class is more able of reproducing than the adult fish class.
- Therefore, the biggest reproduction is centered in the young fish class.
- The small fish class is considered as unable of reproducing.
- The small fish class reduces its mortality under good environmental conditions.
- The small fish class has the highest index in mortality under environment low conditions.
- The young fish class resists more the environment unfavorable changes that the others classes.

Decision Rules

1. IF (Environment is High) AND (Weight is Adult) THEN (RT is High) AND (MT is Mid)
2. IF (Environment is Mid) AND (Weight is Adult) THEN (RT is Mid) AND (MT is High)
3. IF (Environment is Bad) AND (Weight is Adult) THEN (RT is VeryLow) AND (MT is VeryHigh)
4. IF (Environment is High) AND (Weight is Young) THEN (RT is VeryHigh) AND (MT is VeryLow)
5. IF (Environment is Mid) AND (Weight is Young) THEN (RT is High) AND (MT is Mid)
6. IF (Environment is Bad) AND (Weight is Young) THEN (RT is Low) AND (MT is Mid)
7. IF (Environment is High) AND (Weight is Small) THEN (MT is Low)
8. IF (Environment is Mid) AND (Weight is Small) THEN (MT is Mid)
9. IF (Environment is Bad) AND (Weight is Small) THEN (MT is VeryHigh)

The fuzzy logic system Integra in the transition function redefining itself the reproduction and mortality rates like functions that depend on the factor 'ENVIRONMENT' and of the weight of 'FISH', whose evaluation is given by the fuzzy logic system. The redefined function is given by

$$n_1(C(i,j),t+1) = b_1(environment, weight)[n_1(C(i,j),t) - \mu_1(environment, weight)n_1(C(i,j),t)] +$$

$$b_3(enviroment, weight)[n_3(C(i,j),t) - \mu_3(environment, weight)n_3(C(i,j),t)] +$$

$$b_2(environment, weight)[n_2(C(i,j),t) - \mu_2(environment, weight)n_2(C(i,j),t)] +$$

$$E_1(\Pi_r(i,j),t)$$

$$n_2(C(i,j),t+1) = n_1(C(i,j),t) - \mu_1(environment, weight)n_1(C(i,j),t) +$$

$$E_2(\Pi_r(i,j),t)$$

$$n_3(C(i,j),t+1) = n_2(C(i,j),t) - \mu_2(environment, weight)n_2(C(i,j),t) +$$

$$n_3(C(i,j),t) - \mu_3(environment, weight)n_3(C(i,j),t) +$$

$$E_3(\Pi_r(i,j),t) \tag{12}$$

3   Simulation Results

In the hybrid model, the transition function is defined as a pattern of the population behaviour affected by environment (equation 12), whose objective is to generate the states (changes) of population in space and time (figure 7).

En el fuzzy logic system, the fuzzy rules are combined, according to effects that the environment causes to population. The model considers that the environment must

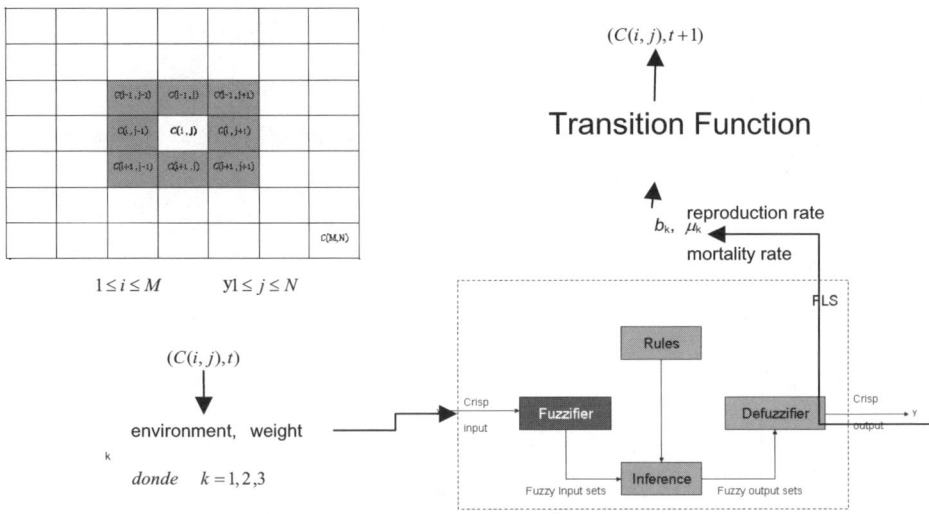

Fig. 7. The hybrid model architecture

vary at the time and in the cellular space, and must vary the effect among the different age classes. The transition function is applied synchronously to each cell in the cellular space; this opens an opportunity for the approach in the field of situations of local colonization and extinction.

With the objective of seeing the feasibility of the model to describe the population dynamics, we implement in a simulation program the hybrid model and used Matlab 7.1 to program it. The obtained results are divided in three cases of study. In the first of them, considered that the reproduction and mortality rates do not represent the result of a relation between the age classes and the environment (a controller does not exist in the transition function), are simply stochastic values uniformly distributed. This is, the fish reproduces or die without concerning the effect of environment on themselves (figure 8).

In figure 8 we can observe that exist too much variation in the density of the population of the three classes, without a clear tendency in the trajectories. The changes are given by the main characteristics in any study on population dynamics: the reproduction, mortality and emigration. Although the emigration is given by the environmental favorability, the reproduction and the mortality not, these change without an order, its variation is stochastic without concerning the environment effect on each population. Leal (2004) made a detailed study of the variability and stability of the trajectories in a population structured by ages under the same characteristics of this case.

In the following cases we introduce the fuzzy logic system, and defined through the fuzzy rules, the effect that the environment could have on the individuals. Therefore in a second case, the biggest reproduction is centered in the class of young fish, which means that this population resists more the unfavorable changes of the environment that the other classes. The results of this case are shown in figure 9, wherein we can observe, how the adult fish population increases quickly, because the young

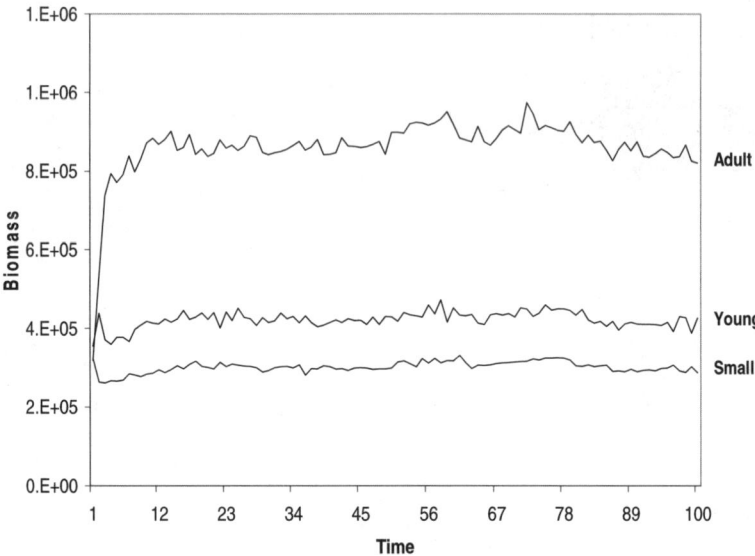

Fig 8. Density of population generated by the first case of study

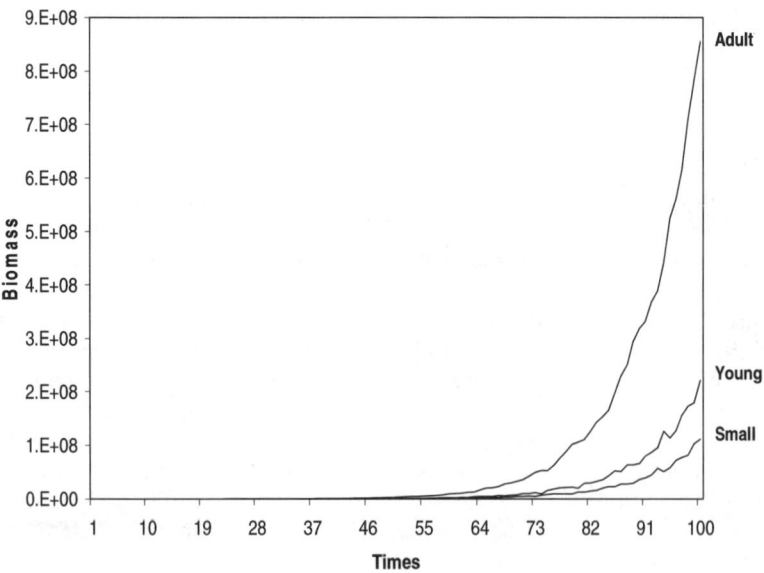

Fig. 9. Density of the population generated by the second case of study

fish population go to the class of adult fish in more proportion, with a clear tendency towards a model of exponential growth in the three classes.

In a third case, the rules change so the biggest reproduction effort is centered in the adult fish population, therefore the rules are combined by obtaining

1. IF (Environment is High) AND (Weight is Adult) THEN (RT is VeryHigh) AND (MT is Low)
2. IF (Environment is Mid) AND (Weight is Adult)   THEN (RT is Mid) AND (MT is Mid)
3. IF (Environment is Bad) AND (Weight is Adult)   THEN (RT is VeryLow) AND (MT is VeryHigh)
4. IF (Environment is High) AND (Weight is Young) THEN (RT is Mid) AND (MT is VeryLow)
5. IF (Environment is Mid) AND (Weight is Young) THEN (RT is Low) AND (MT is Mid)
6. IF (Environment is Bad) AND (Weight is Young) THEN (RT is VeryLow) AND (MT is High)
7. IF (Environment is High) AND (Weight is Small) THEN (MT is Low)
8. IF (Environment is Mid) AND (Weight is Small) THEN (MT is Mid)
9. IF (Environment is Bad) AND (Weight is Small) THEN (MT is VeryHigh)

In the figure 10 show that the class of young fish diminishes because they have a reproduction rate smaller than the class of adult fish and of the last example. Then, the young fish go to the class of adult fish in minus proportion. In addition, the class of adult fish has a mortality rate higher than the last example.

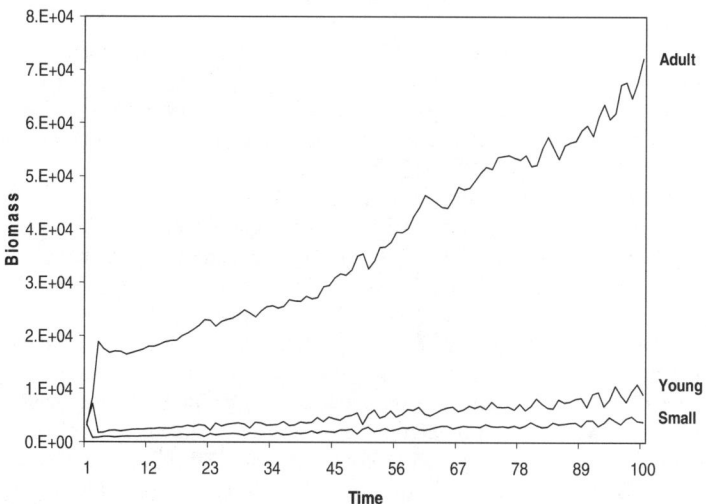

Fig. 10. Density of the population generated by the third case of study

Also, in figure 10 only the class of adult fish tends to grow exponentially, the others two classes present a growth mode different. Those classes could maintain their density in a low level of threshold.

4 Conclusions

The combination of fuzzy logic with cellular automata allowed us to model the dynamics of a population, wherein we controlled the effects that the environment could cause to a population, according to different interpretations defined by the fuzzy sets and the combination from the fuzzy rules in each case of study. Nevertheless the interpretations that we can represent with fuzzy logic system are limited, firstly because the environment is considered as a representative value of several variables that interacts with the environment, which also cause changes in the population dynamics, and the interpretations that can be formed from the fuzzy sets that characterize to environment and to population are few, this must to that only some of the combinations can be used because the representations must be similar to the reality. The tendencies to the exponential growth presented in figures 8 and 9, do not represent the generalization for all interpretation of the environmental effects, can be expected another type of tendencies, more when are added more variables to the fuzzy logic system that allows to establish conditions that regulate the growth by different way. A right combination of the fuzzy rules could represent the real interpretation of population behaviour in space and time.

References

1. Anderson, J.R.: The architecture of cognition. In: Cognitive sciences series. Harvard University Press, Cambridge (1983)
2. Czárán, T.: Spatiotemporal Models of Population and Community Dynamics. Chapman & Hall, London (1998)
3. Di Stefano, B., Fuks, H., Lawniczak, A.T.: Application of fuzzy logic in CA / LGCA models as a way of dealing with imprecise and vague data. Electrical and Computer Engineering, Canadian Conference 1, 212–217 (2000)
4. Fortuna, L., Rizzotto, G., Lavorgna, M., NunnarI, G., Xibili, M.G., Capponetto, R.: Soft Computing. In: New trends and Applications. Springer, Heidelberg (2001)
5. Gutiérrez, J.D., Riss, W., Ospina, R.: Lógica difusa como herramienta para la bioindicación de la calidad del agua con macroinvertebrados acuáticos en la sabana de bogotá – colombia Caldasia. 26(1), 161–172 (2004)
6. Mandelas, E.A., Hatzichristos, T., Prastacos, P.: A fuzzy cellular automata based shell for modelling urban growth – a pilot application mesogia area. In: 10th AGILE International Conference on Geographic Information Science Alborg University, Denmark (2007)
7. Molina-Becerra, M.: Análisis de algunos modelos de dinámica de poblaciones estructurados por edades con y sin difusión, PhD Thesis, España Universidad de Sevilla (2004)
8. Molofsky, J., Bever, J.: A new kind of ecology? BioScience 54(5), 440–446 (2004)
9. Moreno, N., Ablan, M., Tonella, G.: SpaSim: A software to Simulate Spatial Models. Integrated Assessment and Decision Support. In: Proceedings of the First International Environmental Modeling and Soft-ware Conference, Vol 3. pp. 348–358. June 24-27. Lugano, Switzerland (2002); ISBN 8890078707

10. Leal, R.C.: Desarrollo de un simulador, basado en un autómata celular, para la generación de dinámica poblacional inducida por gradientes de favorabilidad ambiental, Ms. Thesis, Universidad Autónoma de Baja California (2004)
11. Rohde, K.: Cellular Automata and Ecology, Zoology, University of New England, Armidale NSW 2351, Australia (2005)
12. Von Bertalanffy, L.: A quantitative theory of organic growth. Hum. Biol. 10(2), 181–213 (1938)
13. Von Neumann, J.: Theory of Self-reproducing Automata. University of Illinois Press, Urbana (1966)
14. Wolfram, S.: Theory and Applications of Cellular Automata. World Scientific, Singapore (1986)
15. Zadeh, L.A.: The concept of a linguistic variable and its application to approximate reasoning, Parts 1, 2, and 3, Information Sciences (1975)
16. Zadeh, L.A.: Fuzzy Logic. Computer 1(4), 83–93 (1988)

Soft Margin Training for Associative Memories: Application to Fault Diagnosis in Fossil Electric Power Plants

Jose A. Ruz-Hernandez[1], Edgar N. Sanchez[2], and Dionisio A. Suarez[3]

[1] Universidad Autonoma del Carmen, Av. 56 # 4 X Av. Concordia,
Col. Aviación, C.P. 24180, Cd. del Carmen, Campeche, Mexico
jruz@pampano.unacar.mx
[2] CINVESTAV, Unidad Guadalajara, Apartado Postal 31-430, Plaza La Luna, C.P. 45091,
Guadalajara, Jalisco, Mexico, on sabbatical leave at CUCEI, Universidad de Guadalajara
sanchez@gdl.cinvestav.mx
[3] Instituto de Investigaciones Electricas, Calle Reforma # 113, Col. Palmira,
C.P. 62490, Cuernavaca, Morelos, Mexico
suarez@iie.org.mx

Abstract. In this paper, the authors discuss a new synthesis approach to train associative memories, based on recurrent neural networks. They propose to use soft margin training for associative memories, which is efficient when training patterns are not all linearly separable. On the basis of the soft margin algorithm used to train support vector machines, the new algorithm is developed in order to improve the obtained results via optimal training algorithm also innovated by the authors, which are not fully satisfactory due to that some times the training patterns are not all linearly separable. This new algorithm is used for the synthesis of an associative memory implemented by a recurrent neural network with the connection matrix having upper bounds on the diagonal elements to reduce the total number of spurious memory. The scheme is evaluated via a full scale simulator to diagnose the main faults occurred in fossil electric power plants and taking into account three different cases.

1 Introduction

The implementation of associative memories via recurrent neural networks is discussed in [1], where a synthesis approach is developed based on the perceptron training algorithm. The goal of associative memories is to store a set of desired patterns as stable memories such that a stored pattern can be retrieved when the input pattern (or the initial pattern) contains sufficient information about that stored pattern. In practice the desired memory patterns are usually represented by bipolar vectors (or binary vectors). There are several well-known methods available in the literature, which solve the synthesis problem of RNNs for associative memories, including the *outer product method*, the *projection learning rule* and the *eigenstructure method*, [2].

Due to their high generalization performance, Support Vector Machines (SVMs) have attracted great attention for pattern recognition, machine learning, neural networks and so on, [3]. Learning of a SVM leads to a quadratic programming (QP) problem, which can be solved by many techniques [4]. Furthermore, the relation

O. Castillo et al. (Eds.): Soft Computing for Hybrid Intel. Systems, SCI 154, pp. 205–230, 2008.
springerlink.com

existing between SVMs and the design of associative memories based on the generalized brain-state-in-a-box (GSB) neural model is formulated in [5].

On the basis of the optimal hyperplane algorithm used to train SVMs, an optimal training algorithm for associative memories has been developed and applied by the authors in [6] and [7]. The proof related to the convergence properties for this algorithm when patterns are linearly separable, and the corresponding proof for the design of an associative memory via RNNs using constraints in the diagonal elements of connection matrix can be reviewed in [6] and [8].

This paper proposes a new soft margin training for associative memories implemented by RNNs. On the basis of the soft margin algorithm used to train SVMs as described in [3], the new algorithm is developed in order to improve the obtained results via optimal training algorithm when the training patterns are not all linearly separable.

2 Preliminaries

This section introduces useful preliminaries about associative memories implemented by RNNs, the perceptron training algorithm and a synthesis for RNNs based on this algorithm, which is proposed in [1]. The class of RNNs considered is described by equations of the form

$$\frac{dx}{dt} = -Ax + T\text{sat}(x) + I \quad ,$$

$$y = \text{sat}(x)$$

(1)

where x is the state vector, $y \in D^n = \{x \in R^n \mid -1 \leq x_i \leq 1\}$ is the output vector, $A = \text{diag}[a_1, a_2, ..., a_n]$ with $a_i > 0$ for $i = 1, 2, ..., n$, $T \in R^{nxn}$ is the connection matrix with elements $T_{ij} \in R$, $I = [I_1, I_2, ..., I_n]^T$ is a bias vector, and $\text{sat}(x) = [\text{sat}(x_1), ..., \text{sat}(x_n)]^T$ represents the activation function, where

$$\text{sat}(x_i) = \begin{cases} 1 & x_i > 1 \\ x_i & -1 \leq x_i \leq 1 \\ -1 & x_i < -1 \end{cases} .$$

(2)

It is assumed that the initial states of (1) satisfy $\mid x_i(0) \mid \leq 1$ for $i = 1, 2, ..., n$. System (1) is a variant of the analog Hopfield model with activation function sat($\bullet$).

2.1 Synthesis Problem for Associative Memories Implemented by RNNs

For the sake of completeness, the following results are taken from [1] and included in this section. A vector α will be called a (stable) memory vector (or simply, a memory) of system (1) if $\alpha = \text{sat}(\beta)$ and if β is an asymptotically stable equilibrium point of system (1). In the following lemma, B^n is defined as a set of n-dimensional bipolar vectors $B^n = \{x \in R^n \mid x_i = 1 \text{ or } -1, i = 1, 2, ..., n\}$. For $\alpha = [\alpha_1, \alpha_2, ..., \alpha_n]^T \in B^n$ define $C(\alpha) = \{x \in R^n \mid x_i\alpha_i > 1, i = 1, 2, ..., n\}$.

Lemma 2.1. If $\alpha \in B^n$ and if

$$\beta = A^{-1}(T\alpha + I) \in C(\alpha),$$ (3)

then (α, β) is a pair of stable memory vector and an asymptotically stable equilibrium point of (1). The proof of this result can be reviewed in [1].

The following synthesis problem concerns the design of (1) for associative memories.

Synthesis Problem: Given m vectors $\alpha^1, \alpha^2, ..., \alpha^m$ in the set of n-dimensional bipolar vectors, B^n, choose $\{A, T, I\}$ in such a manner that:

1. $\alpha^1, \alpha^2, ..., \alpha^m$ are stable memory vectors of system (1);
2. the total number of spurious memory vectors (i.e., memory vectors which are not desired) is as small as possible, and the domain (or basin) of attraction of each desired memory vectors is as large as possible.

Item 1) of the synthesis problem can be guaranteed by choosing the $\{A, T, I\}$ such that every α^j satisfies condition 3 of Lemma 2.1. Item 2) can be partly ensured by constraining the diagonal elements of the connection matrix.

In order to solve the synthesis problem, one needs to determine A, T and I from (3) with $\alpha = \alpha^k$, $k = 1, 2,..., m$.

Condition given in (3) can be equivalently written as

$$\begin{cases} T_i\alpha^k + I_i > a_i & \text{if } \alpha_i^k = 1 \\ T_i\alpha^k + I_i < -a_i & \text{if } \alpha_i^k = -1 \end{cases},$$ (4)

for $k = 1, 2,..., m$ and $i = 1, 2,..., n$ where T_i represents the ith row of T, I_i denotes the ith element of I, a_i is the i-th diagonal element of A, and α_i^k is the i-th entry of α^k.

3 New Approach: Soft Margin Training for Associative Memories

This section contains our principal contribution. First, we describe soft margin algorithm used for SVMs when training patterns are not all linearly separable. We propose a new soft margin training for associative memories implemented by RNN (1).

3.1 Soft Margin Algorithm

Consider the case where the training data can not be separated without error by an SVM. In this case one may want to separate the training set with a minimal number of errors. To express this formally let us introduce some non-negative variables $\xi_i \geq 0$, $i = 1, ..., l$.

We can now minimize the functional

$$\phi(\xi) = \sum_{i=1}^{l} \xi_i^\sigma$$ (5)

for small $\sigma > 0$, subject to the constraints

$$y_i(W \bullet X_i + b) \geq 1 - \xi_i, \quad i = 1, ..., l. \tag{6}$$

$$\xi_i \geq 0, \quad i = 1, ..., l. \tag{7}$$

For sufficiently small σ the functional (5) describes the number of the training errors. Minimizing (3) one finds some minimal subset of training errors

$$(y_{i1}, X_{i1}), ..., (y_{ik}, X_{ik}). \tag{8}$$

If these data are excluded from the training set one can separate the remaining part of the training set without errors. To separate the remaining part of the training data one can construct an optimal separating hyperplane. To construct a soft margin hyperplane we maximize the functional

$$\Phi = \frac{1}{2}W \bullet W + C(\sum_{i=1}^{l} \xi_i^\sigma) \tag{9}$$

subject to constraints (6) and (7), where C is a constant. The solution to the optimization problem under the constraint (6) is given by the saddle point of the Lagrangian, as obtained in [3]

$$L(W, \xi, b, \Lambda, R) = \frac{1}{2}W \bullet W + C(\sum_{i=1}^{l} \xi_i) - \sum_{i=1}^{l} \alpha_i[y_i(x_i \bullet W + b) - 1 + \xi_i] - \sum_{i=1}^{l} r_i \xi_i \tag{10}$$

where $R^T = [r_1, r_2, r_3, r_4, r_5]$ enforces the constraint (6). In [3] is described that vector W can be written as a linear combination of support vectors when optimal hyperplane algorithm is used for training. To find the vector $\Lambda^T = [\lambda_1, ..., \lambda_l]$ one has to solve the dual quadratic programming problem of maximizing

$$W(\Lambda, \delta) = \Lambda^T P - \frac{1}{2}[\Lambda^T Q \Lambda + \frac{\delta^2}{C}] \tag{11}$$

subject to the constraints

$$\Lambda^T Y = 0 \tag{12}$$

$$\delta \geq 0 \tag{13}$$

$$0 \leq \Lambda \leq \delta P \tag{14}$$

where P is an l-dimensional unit vector, $\Lambda^T = [\lambda_1, \lambda_2, ..., \lambda_l]$ contains Lagrange multipliers, Y contains the entries, and Q is a symmetric matrix are the same elements as used in the optimization problem for constructing an optimal hyperplane, δ is a scalar, and (14) describes coordinate-wise inequalities.

Note that (14) implies that the smallest admissible value δ in the functional (11) is

$$\delta = \lambda_{max} = \max[\lambda_1, \lambda_2, ..., \lambda_l] \tag{15}$$

Therefore to find a soft margin classifier one has to find a vector Λ that maximizes

$$W(\Lambda) = \Lambda^T P - \frac{1}{2}[\Lambda^T Q A + \frac{\lambda_{max}^2}{C}] \tag{16}$$

under the constraints $\Lambda \geq 0$ and (14). This problem differs from the problem of constructing an optimal margin classifier only by the additional term with λ_{max} in the functional (11). Due to this term the solution to the problem of constructing the soft margin classifier is unique and exist for any data set.

3.2 New Synthesis Approach

Considering the soft margin algorithm, we propose the soft margin training for associative memories implemented by RNNs which is described as follows.

Synthesis Algorithm 3.1: Given m training patterns α^k, $k = 1, 2,..., m$ which are known to belong to class X_1 (corresponding to $Z = 1$) or X_2 (corresponding to $Z = -1$), the weight vector W can be determined by means of the following algorithm.

1. Start out by solving the quadratic programming problem given by

$$F(\Lambda^i) = (\Lambda^i)^T P - \frac{1}{2}[(\Lambda^i)^T Q \Lambda^i + \frac{\lambda_{max}^2}{C}], \tag{17}$$

under the constraints

$$(\Lambda^i)^T Y^i = 0, \text{ where } Y^i = \begin{bmatrix} \alpha_i^1 & \alpha_i^2 & \cdots & \alpha_i^m \end{bmatrix},$$

$$\delta \geq 0,$$

$$0 \leq \Lambda \leq \delta Q$$

to obtain $\Lambda^i = \begin{bmatrix} \lambda_1^i, \lambda_2^i,..., \lambda_n^i \end{bmatrix}$.

2. Compute the weight vector

$$W^i = \sum_{k=1}^m \lambda_i^k \alpha_j^k \alpha^{-k} = \begin{bmatrix} w_1^i, w_2^i,..., w_n^i, w_{n+1}^i \end{bmatrix}, \tag{18}$$

$i = 1, 2, ..., n$, such that

$$\begin{aligned} W^i \bullet \alpha^{-k} + b \geq 1 - \xi_i & \text{ if } \alpha_i^k = 1, \\ W^i \bullet \alpha^{-k} + b \leq -1 + \xi_i & \text{ if } \alpha_i^k = -1, \end{aligned} \tag{19}$$

and for $k = 1, 2, ..., m$ where

$$\alpha^{-k} = \begin{bmatrix} \alpha^k \\ ... \\ 1 \end{bmatrix}. \tag{20}$$

Choose $A = \text{diag } [a_1, a_2, \ldots, a_n]$ with $a_i > 0$. For $i, j = 1,2,\ldots,n$ choose $T_{ij} = w_j^i$ if $i \neq j$, $T_{ij} = w_j^i + a_i \mu_i - 1$ if $i=j$, with $\mu_i > 1$ e $I_i = w_{n+1}^i + b_i$.

4 Application to Fault Diagnosis in Fossil Electric Power Plants

In order to illustrate the applicability of the above proposed optimal procedure to train associative memories, based on recurrent neural networks, we discuss its application to fault diagnosis in fossil electric power plants.

Fault diagnosis can be performed by a three steps algorithm [9]. First, one or several signals are generated which reflect faults in the process behavior. These signals are called residuals. For the second step, the residual are evaluated. A decision has to be taken in order to determine time and location of possible faults from the residuals. Finally the nature and the cause of the fault is analyzed by the relations between the symptoms and their physical causes.

In order to describe the fault free nominal behavior of the process under supervision, a model (mathematics or heuristic) is employed, giving to this concept the name of model-based fault diagnosis. Model-based approaches have dominated the fault diagnosis research [10], [11].

Employing measurements of the process under normal operation, if possible, or with the help of a simulator as realistic as possible, a suitable neural network can be trained to learn the process input-output behaviour [12].

This section presents a neural network scheme for fault diagnosis. It uses for residual generation a predictor which consists of a bank of recurrent multilayer perceptron neural network models. Fault diagnosis is carried out by an associative memory, which is based on a recurrent neural network, and trained with the proposed soft margin learning algorithm.

4.1 Problem Description

Fault diagnosis in fossil electric power plants is a task carried out by an expert operator. This operator recognizes typical faults via supervision of key variables evolution. Adequate fault detection and diagnosis aids will help the human operator in order to take the right decisions to maintain the required electric energy production, avoiding failures and even accident risky to humans and the environment [13].

For this kind of plants, the main faults can be clustered as: faults related to temperature control of the superheated and reheated steam, faults related to combustion control and faults related to the steam generator drum water level. In order to understand the first group of faults, it is helpful to briefly describe the steam generator and superheated and reheated steam system. A simplified scheme is presented in Fig. 1, which illustrates the main components of a typical steam generator and superheating/reheating system.

The feedwater from the economizer enters the steam drum, and by forced circulation, the drum water flows down the downcomers and rises through the furnace wall tubes to generate steam by means of the hot combustion gases in the furnace. The water and steam in the drum are separated by steam separators and the steam becomes

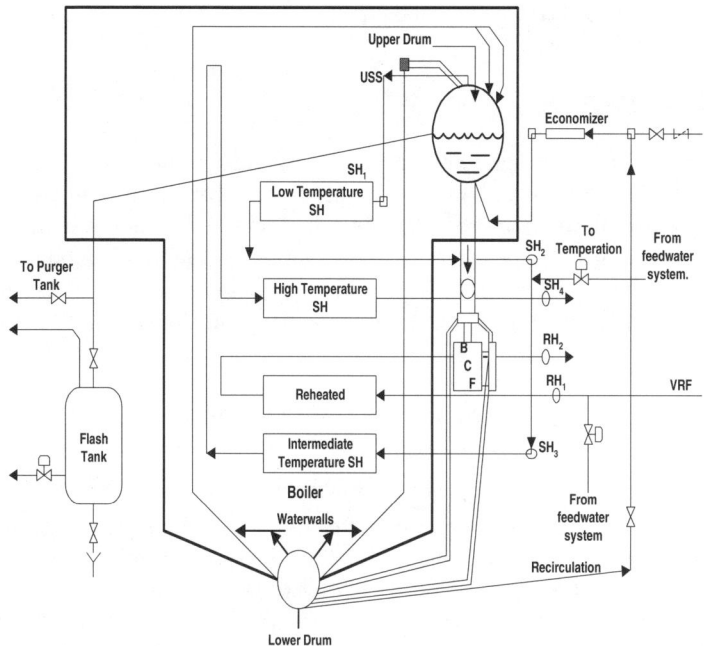

Fig. 1. Steam Generator and Reheating / Superheating System

superheated as it passes through various superheaters. The turbine exhaust steam is again superheated in the reheater before generating power in the intermediate and low pressure turbines. For this system, the main automatic control loops are the main steam temperature control and the reheated steam temperature control. The first one is controlled by the spray attemperator, and the second one is controlled by the burners inclination angle, as well as by other spray attemperator.

As an example, we discuss a typical fault: *waterwall tubes breaking*, which is part of the above first fault group. It could be due to inadequate design, selection of materials, and/or unsuitable start-up operations [14]. In presence of this fault, the combustion gases do not circulate properly and the waterwall tubes are not suitable cold. Additionally, the water level on the steam generator drum goes down and the level control tries to keep it by means of varying the feedwater flow. However if the maximum value of this flow is reached, and the water level continues to decrease, the low level monitoring orders the steam generator out of operation. If this order takes a long time to be executed or if it is not performed, the waterwalls tubes operating normally will suffer strong damages.

This fault also diminishes the steam generator drum pressure, causing reductions on the superheated and reheated steam pressures. The combustion control tries to correct this situation by increasing the air flow and the fossil oil flow; these actions could increase the steam generator pressure beyond the allowed limit, and as a consequence the steam generator would be taken out of operation. If the human operator, in presence of this fault, does not take the adequate corrective actions, the healthy waterwalls tubes could be damaged due to thermal stress. The turbine will also suffer from thermal and mechanical stress.

4.2 Scheme for Fault Diagnosis

This scheme has two components: residual generation and fault diagnosis. The scheme is displayed in Fig. 2. The first component is based on comparison between the measurements coming from the plant and the predicted values generated by a neural network predictor. The predictor is based on neural network models, which are trained using healthy data from the plant. The differences between these two values, named as residuals, constitute a good indicator for fault detection. The residuals are calculated as

$$r_i(k) = x_i(k) - \hat{x}_i(k), \; i = 1, 2, \ldots, n. \tag{21}$$

where $x_i(k)$ are the plant measures and $\hat{x}_i(k)$ are the predictions. The residuals should be independent of the system operating state under nominal plant operating conditions. In absence of faults, the residuals are only due to noise and disturbance. When a fault occurs in the system, the residuals deviate from zero in characteristic ways.

For the second component, residuals are encoded in bipolar or binary vectors using thresholds to obtain fault patterns. These fault patterns are used to train an associative memory based on a recurrent neural network, which is employed to carry out the fault diagnosis. Our proposed soft margin algorithm is used to train this associative memory.

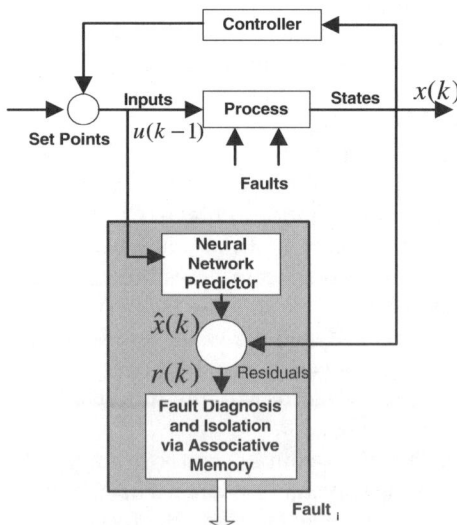

Fig. 2. Scheme for Fault Diagnosis

4.2.1 Residual Generation

For residual generation purposes the neural network replaces the analytical model describing the process under normal operation. The neural networks training is done using the series-parallel scheme [15]. After finishing the training, the neural networks can be applied for residual generation (Fig. 3); its weights are fixed and used as a parallel scheme to carry out predictions. The neural network predictor is designed using ten neural network models which were training via the Levenberg-Marquardt Learning Algorithm ([16], [17]). Each neural network is a recurrent multilayer perceptron. The networks have one hidden layer with hyperbolic tangent activation functions and a single neuron with a linear activation function as the output layer. The neural network models are obtained employing the toolbox NNSYSID [18], which runs in MATLAB[1].

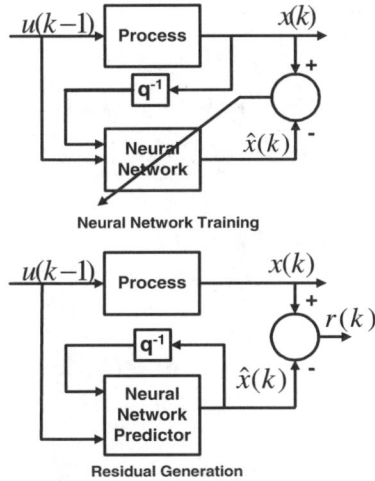

Fig. 3. Scheme for training and application of neural networks for residual generation

All the models have eight input variables and a one output variable with a NNARX structure as:

$$\hat{x}_1(k) = F_1[W_1, x_1(k-1), \cdots, x_1(k-6), u_1(k-1), \cdots, u_1(k-6),$$
$$\cdots, u_8(k-1), \cdots, u_8(k-6)] \tag{22}$$

$$\hat{x}_2(k) = F_2[W_2, x_2(k-1), \cdots, x_2(k-4), u_1(k-1), \cdots, u_1(k-4),$$
$$\cdots, u_8(k-1), \cdots, u_8(k-4)] \tag{23}$$

---

[1] MATLAB is a registered trademark of The Math Works, Inc.

$$\hat{x}_3(k) = F_3[W_3, x_3(k-1), \cdots, x_3(k-6), u_1(k-1), \cdots, u_1(k-6), \\ \cdots, u_8(k-1), \cdots, u_8(k-6)] \tag{24}$$

$$\hat{x}_4(k) = F_4[W_4, x_4(k-1), \cdots, x_4(k-6), u_1(k-1), \cdots, u_1(k-6), \\ \cdots, u_8(k-1), \cdots, u_8(k-6)] \tag{25}$$

$$\hat{x}_5(k) = F_5[W_5, x_5(k-1), \cdots, x_5(k-5), u_1(k-1), \cdots, u_1(k-5), \\ \cdots, u_8(k-1), \cdots, u_8(k-5)] \tag{26}$$

$$\hat{x}_6(k) = F_6[W_6, x_6(k-1), \cdots, x_6(k-3), u_1(k-1), \cdots, u_1(k-3), \\ \cdots, u_8(k-1), \cdots, u_8(k-3)] \tag{27}$$

$$\hat{x}_7(k) = F_7[W_7, x_7(k-1), \cdots, x_7(k-6), u_1(k-1), \cdots, u_1(k-6), \\ \cdots, u_8(k-1), \cdots, u_8(k-6)] \tag{28}$$

$$\hat{x}_8(k) = F_8[W_8, x_8(k-1), \cdots, x_8(k-6), u_1(k-1), \cdots, u_1(k-6), \\ \cdots, u_8(k-1), \cdots, u_8(k-6)] \tag{29}$$

$$\hat{x}_9(k) = F_9[W_9, x_9(k-1), \cdots, x_9(k-6), u_1(k-1), \cdots, u_1(k-6), \\ \cdots, u_8(k-1), \cdots, u_8(k-6)] \tag{30}$$

$$\hat{x}_{10}(k) = F_{10}[W_{10}, x_{10}(k-1), \cdots, x_{10}(k-6), u_1(k-1), \cdots, u_1(k-6), \\ \cdots, u_8(k-1), \cdots, u_8(k-6)] \tag{31}$$

where the input variables are

$u_1(.)$ = Fossil oil flow (%).
$u_2(.)$ = Air flow (%).
$u_3(.)$ = Condensed water flow (Litres per minute).
$u_4(.)$ = Water flow to attemperator (Kg/s).
$u_5(.)$ = Feedwater flow (T/H).
$u_6(.)$ = Replacement flow to condenser (Litres per second).
$u_7(.)$ = Steam water flow (Litres per minute).
$u_8(.)$ = Burner inclination angle (Degrees)

and the output variables are
$x_1(.)$ = Load power (MW).
$x_2(.)$ = Boiler pressure (Pa).

$x_3(.)$ = Drum level (m).
$x_4(.)$ = Reheated steam temperature (°K).
$x_5(.)$ = Superheated steam temperature (°K).
$x_6(.)$ = Reheated steam pressure (Pa).
$x_7(.)$ = Drum pressure (Pa).
$x_8(.)$ = Differential pressure (spray steam − fossil oil flow) (°K).
$x_9(.)$ = Fossil oil temperature to burners (°K).
$x_{10}(.)$ = Feedwater temperature (°K).

W_i represents the weights for each neural network model. The lag structure of each neural network model is determined using the same criterion as in [18]. Once neural networks have been trained, its weights are fixed and used as a parallel scheme for carry out the predictions. Neural networks models are validated with healthy fresh data. Prediction errors close to 1 % are obtained for each model. We display in Fig. 4 a validation test with neural network model given by equation (22) working as a parallel scheme to carry out predictions. This validation test considers load power changes, and it is assumed that initial condition for neural network model is different to the data acquired $x_1(0)$ from full scale simulator.

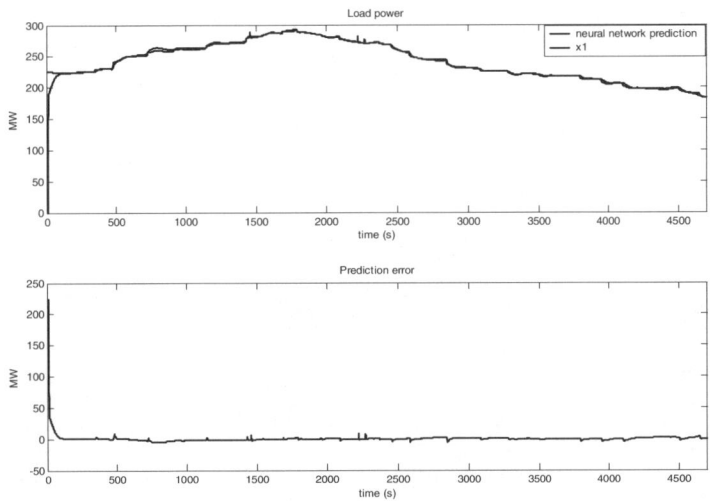

Fig. 4. Validation test for neural network model given by equation (32) working as a parallel scheme

The residual generation scheme is implemented to six faults: *waterwall tubes breaking, superheater tubes breaking, superheated steam temperature control fault, dirty regenerative preheater, velocity varier of feedwater pumps operating to maximum value* and *blocked fossil oil valve* named as *fault* 1 to *fault* 6, respectively. For faults 1 to 4 , data bases are acquired with a full scale simulator for 75% of initial load power (225 MW), 15 % of severity fault, 2 minutes for inception and 8 minutes of simulation time. Furthermore, for *fault* 5 and *fault* 6 the simulator has only available severity and inception which are chosen as 15 % and 2 minutes, respectively. For these two faults, data bases are acquired for 3 and 4 minutes of simulation time, respectively. The *fault 5* is very critical because it can shoot the drum level alarm and break out of operation the fossil electric power plant. It is clear that *fault 6* is visible when load power is changed by operator because the fossil oil valve does not work adequately. In six cases, residuals are close to zero during time for inception. After this interval, residuals deviate of zero in different ways. The residuals for *fault 1* and

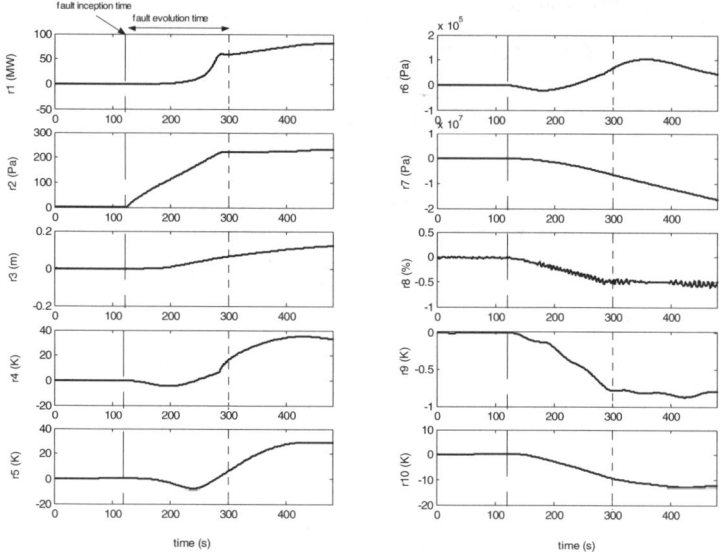

Fig. 5. Residuals for *fault 1*

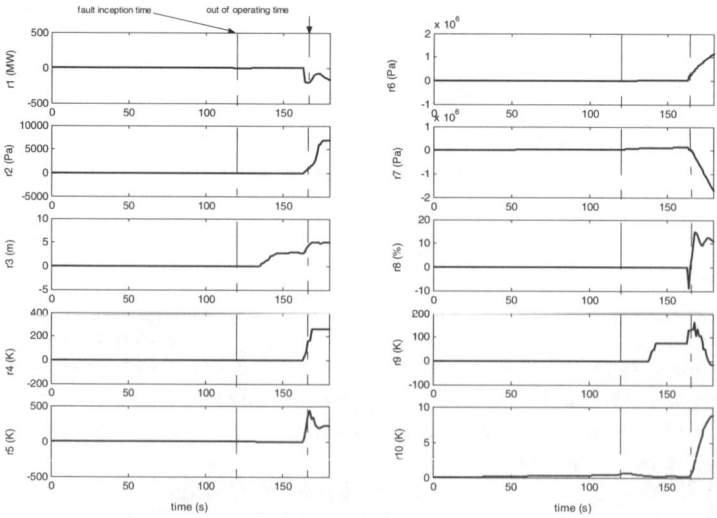

Fig. 6. Residuals for *fault 5*

fault 5 are displayed in Fig. 5 and Fig. 6. In Fig. 7 a load power change at 140 s can be seen; after this change, the load power decreases due to that the fossil oil valve is blocked. Fig. 8 displays residuals for *fault* 6 which are closed to zero before the valve is blocked. Once the valve is blocked, the fault is not detected until the operator carries out the load power change and the residuals starts its deviation from zero in distinctive ways to indicate the fault 6 is occurring.

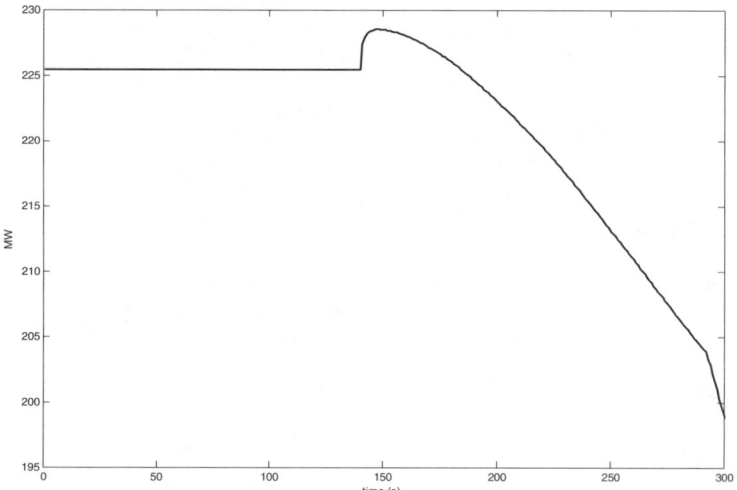

Fig. 7. Load power change made by the operator to detect *fault 6*

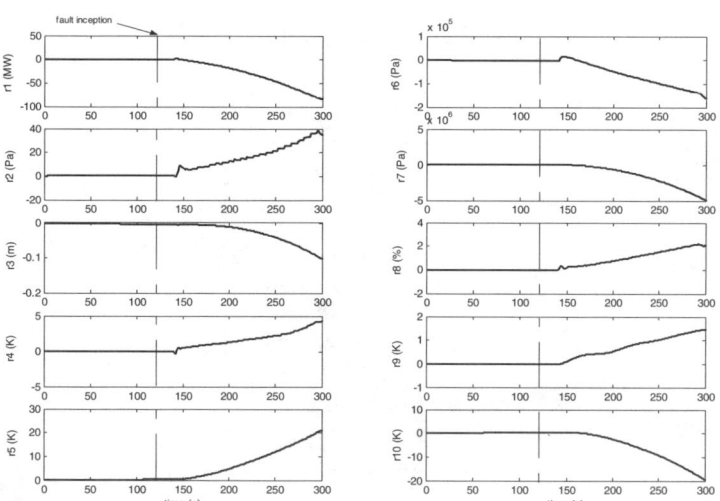

Fig. 8. Residuals for *fault 6*

4.2.2 Fault Diagnosis

Fig. 9 presents a scheme to carry out the fault diagnosis via this associative memory. The previous stage generates a residual vector with ten elements which are evaluated by detection thresholds. Detection thresholds are contained in Table 1. This evaluation provides a set of residuals encoded (bipolar vectors) $[s_1(k), s_2(k), ..., s_{10}(k)]^T$ where

$$s_i(k) = \begin{cases} -1 \text{ if } r_i < \tau_i \\ 1 \text{ if } r_i \geq \tau_i \end{cases}, \ i = 1, 2, ..., 10. \tag{32}$$

Detection thresholds are determined taking into account the following criteria:

1. They are selected bigger than the corresponding prediction errors.
2. In [13], it is explained how each fault evinces on steam generator operation variables. Based on this information, thresholds are selected by trial and error, in order to reproduce these behaviors.
3. Encoded residuals are all equal to -1 to indicate normal operating conditions.

Residuals are encoded on-line for every fault. Encoded residuals for *fault* 1, *fault* 5 and *fault* 6 are displayed in Fig. 10, Fig. 11 and Fig. 12. *Fault* 1 presents an evolution as indicated; encoded residuals values before inception (2 minutes) have elements

Table 1. Detection thresholds

i	1	2	3	4	5	6	7	8	9	10
τ_i	25 MW	30 Pa	0.022 m	4 K	10 K	20000 Pa	42000 Pa	0.85 %	4 K	10 K

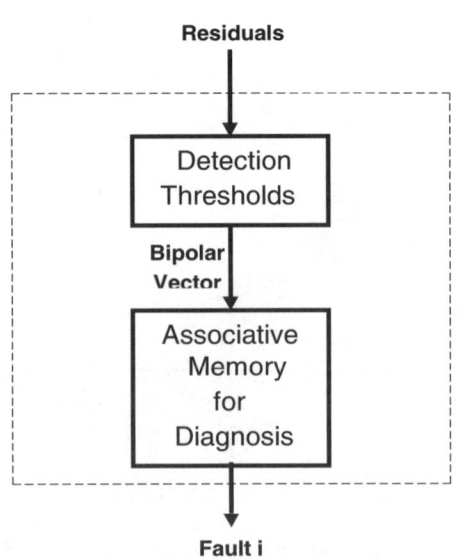

Fig. 9. Scheme for fault diagnosis

equals to −1 indicating normal operating condition. After this time, some components of encoded residuals starts to take values equal to +1 indicating that a fault is present. When fault evolution time is over the encoded residuals do not change anyway. *Fault 5* and *fault* 6 do not have evolution time. For these faults, some residuals present transient values, which are used for fault classification. Once residuals are encoded, it is necessary to analyze them to choose the fault patterns to store in the associative memory. This selection is done in order to discriminate adequately every fault, to reduce false alarms and to isolate fault as soon as it is possible.

The obtained patterns are used, based on the synthesis algorithm proposed by us to train the recurrent neural network and to design the respective associative memory as a way to isolate the faults. Fault patterns are contained in Table 2 where *fault* 0 pattern is included to denote a normal operating condition.

Table 2. Fault patterns to store in associative memory

α^0	α^1	α^2	α^3	α^4	α^5	α^6
-1	1	-1	-1	-1	-1	1
-1	1	-1	-1	-1	-1	-1
-1	1	1	1	1	1	1
-1	1	-1	1	-1	-1	-1
-1	1	1	1	-1	-1	1
-1	1	-1	1	-1	1	1
-1	1	1	1	1	-1	1
-1	-1	-1	-1	-1	-1	1
-1	1	-1	-1	-1	1	1
-1	1	-1	-1	-1	-1	1

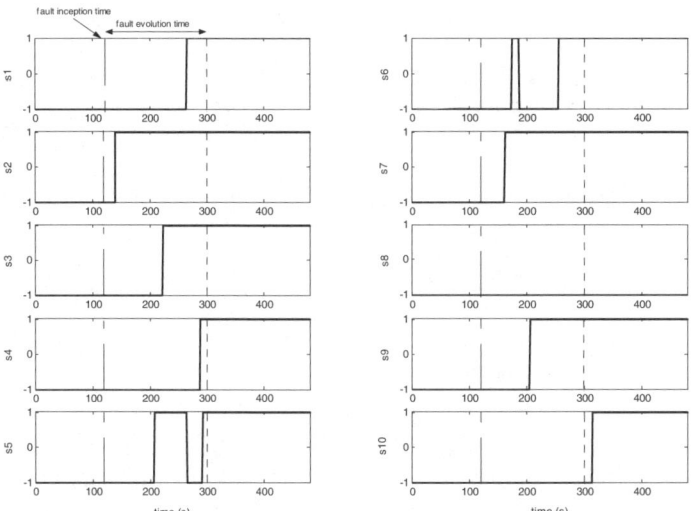

Fig. 10. Encoded residuals for *fault* 1

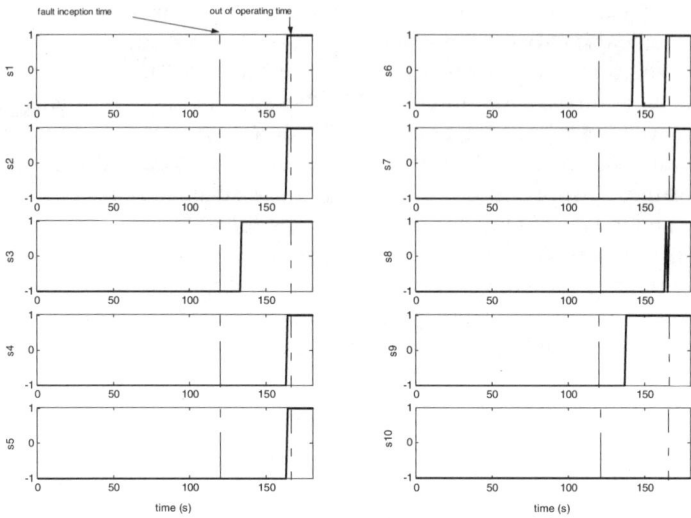

Fig. 11. Encoded residuals for *fault 5*

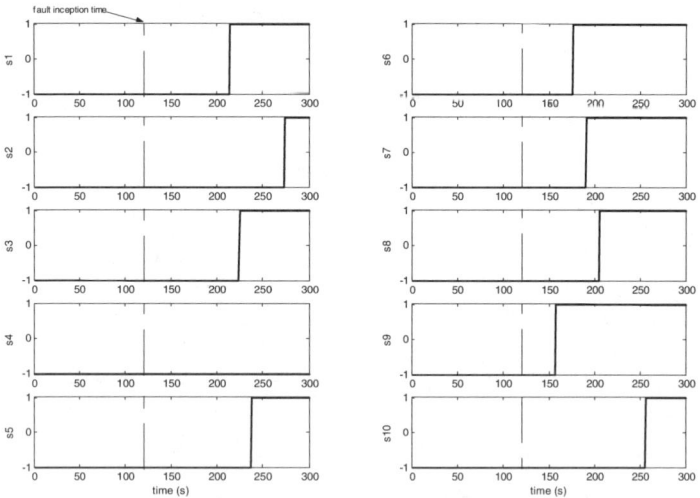

Fig. 12. Encoded residuals for *fault 6*

The soft margin training algorithm is encoded in MATLAB. The number of neurons is $n=10$ (fault pattern length) and the patterns are $m=7$ (number of fault patterns). The Lagrange multipliers matrix $LM= [\Lambda^1, \Lambda^2, ...,\Lambda^n]$, the weight matrix $WM=[W^1, W^2, ...,W^{n+1}]$ and the bias vector $BV=[b^1, b^2, ..., b^n]$ are obtained as in (33), (34) and (35). The matrices A, T and I are calculated as in (36), (37) and (38).

$$
LM = \begin{bmatrix}
0.00 & 0.00 & 0.10 & 0.00 & 0.02 & 0.05 & 0.10 & 0.00 & 0.05 & 0.00 \\
0.08 & 0.12 & 0.00 & 0.10 & 0.00 & 0.00 & 0.00 & 0.06 & 0.06 & 0.08 \\
0.00 & 0.00 & 0.00 & 0.08 & 0.10 & 0.10 & 0.03 & 0.02 & 0.00 & 0.00 \\
0.09 & 0.04 & 0.00 & 0.10 & 0.08 & 0.10 & 0.03 & 0.00 & 0.10 & 0.09 \\
0.00 & 0.00 & 0.06 & 0.00 & 0.10 & 0.08 & 0.10 & 0.00 & 0.04 & 0.00 \\
0.05 & 0.00 & 0.03 & 0.03 & 0.10 & 0.10 & 0.10 & 0.02 & 0.10 & 0.05 \\
0.07 & 0.06 & 0.00 & 0.08 & 0.04 & 0.03 & 0.03 & 0.10 & 0.02 & 0.07
\end{bmatrix} , \tag{33}
$$

$$
WM = \begin{bmatrix}
0.51 & 0.37 & -0.07 & 0.22 & 0.22 & 0.22 & 0.07 & 0.37 & 0.37 & 0.51 & -0.22 \\
0.30 & 0.42 & -0.18 & 0.68 & 0.30 & 0.06 & -0.06 & 0.18 & 0.18 & 0.30 & -0.30 \\
0.00 & 0.00 & 0.00 & 0.00 & 0.00 & 0.00 & 0.00 & 0.00 & 0.00 & 0.00 & 0.00 \\
0.26 & 0.43 & -0.08 & 0.64 & 0.61 & 0.26 & 0.08 & 0.08 & 0.08 & 0.26 & -0.26 \\
0.13 & 0.04 & 0.13 & 0.00 & 0.13 & 0.13 & 0.22 & 0.04 & 0.04 & 0.13 & 0.04 \\
0.10 & 0.03 & 0.10 & 0.05 & 0.10 & 0.23 & 0.03 & 0.03 & 0.16 & 0.10 & 0.03 \\
0.03 & -0.03 & 0.16 & 0.03 & 0.03 & 0.03 & 0.23 & -0.03 & -0.03 & 0.03 & 0.10 \\
0.50 & 0.30 & -0.30 & 0.22 & 0.10 & 0.10 & -0.10 & 0.70 & 0.30 & 0.50 & -0.50 \\
0.14 & 0.08 & 0.02 & 0.00 & 0.02 & 0.14 & -0.02 & 0.08 & 0.20 & 0.14 & -0.02 \\
0.51 & 0.37 & -0.07 & 0.22 & 0.22 & 0.22 & 0.07 & 0.37 & 0.37 & 0.51 & -0.22
\end{bmatrix} , \tag{34}
$$

$$
BV = [-0.07 \quad -0.80 \quad 0.80 \quad -0.52 \quad 0.16 \quad 0.20 \quad 0.26 \quad -0.88 \quad 0.05 \quad -0.07]^{T} , \tag{35}
$$

$$
A = Identity \ matrix \ of \ 10 \ X \ 10 \ , \tag{36}
$$

$$
T = \begin{bmatrix}
1.00 & 0.37 & -0.07 & 0.22 & 0.22 & 0.22 & 0.07 & 0.37 & 0.37 & 0.51 \\
0.30 & 1.00 & -0.18 & 0.30 & 0.06 & 0.06 & -0.06 & 0.18 & 0.18 & 0.30 \\
0.00 & 0.00 & 1.00 & 0.00 & 0.00 & 0.00 & 0.00 & 0.00 & 0.00 & 0.00 \\
0.26 & 0.43 & -0.08 & 1.00 & 0.26 & 0.26 & 0.08 & 0.08 & 0.08 & 0.26 \\
0.13 & 0.04 & 0.13 & 0.13 & 1.00 & 0.13 & 0.22 & 0.04 & 0.04 & 0.13 \\
0.10 & 0.03 & 0.10 & 0.10 & 0.00 & 1.00 & 0.03 & 0.03 & 0.16 & 0.10 \\
0.03 & -0.03 & 0.16 & 0.03 & 0.10 & 0.03 & 1.00 & -0.03 & -0.03 & 0.03 \\
0.50 & 0.30 & -0.30 & 0.10 & 0.16 & 0.10 & -0.10 & 1.00 & 0.30 & 0.50 \\
0.14 & 0.08 & 0.02 & 0.02 & 0.10 & 0.14 & -0.02 & 0.08 & 1.00 & 0.14 \\
0.51 & 0.37 & -0.07 & 0.22 & 0.02 & 0.22 & 0.07 & 0.37 & 0.37 & 1.00
\end{bmatrix} , \tag{37}
$$

$$
I = [-0.29 \quad -1.10 \quad 0.80 \quad -0.78 \quad 0.21 \quad 0.23 \quad 0.36 \quad -1.38 \quad 0.02 \quad -0.29]^{T} . \tag{38}
$$

It is worth to mention that all diagonal elements in matrix T satisfies the optimal constrainst $T_{ii} \leq a_i$ in order to reduce the total number of spurious memories as described in [1] and [6]. New soft margin training algorithm improves the obtained results via optimal training developed by the authors where $T_{88} = 2.59 > a_8$ and $w_{88} = 1.59 > 1$ does not satisfy the corresponding optimal constraint as obtained in [7 p. 60]. Using soft margin training, $T_{88} = a_8 = 1$ and $w_{88} = 0.7 < 1$ are obtained.

The associative memory is evaluated with these matrices fixed using encoded residuals as input bipolar vectors. According to encoded residuals analysis and selected fault patterns, if an encoded residual as an input bipolar vector contains sufficient information about stored pattern in associative memory then corresponding fault pattern is retrieved. Three cases are considered for fault diagnosis via this associative memory using full scale simulator as plant. These cases are described as follows.

Case 1: Fault diagnosis when load power is constant. Fig. 13 displays retrieved fault pattern by the associative memory when load power is constant and fault 1 appears. It is clear that fault pattern α^1 is retrieved when encoded residuals as illustrated in Fig. 10

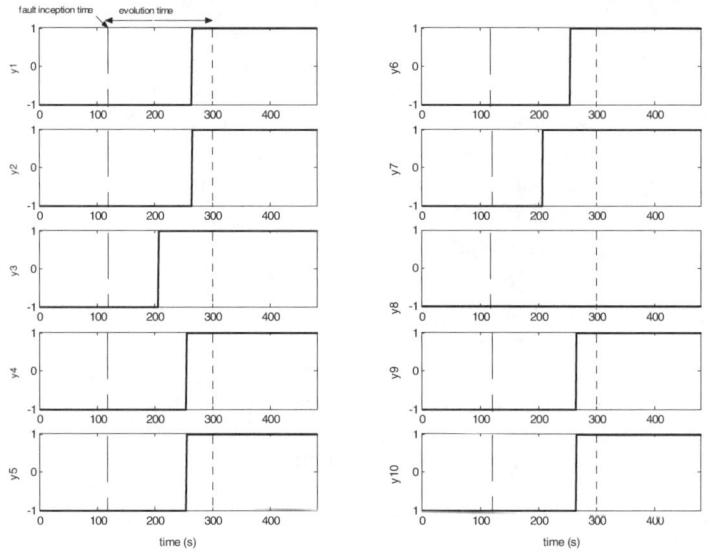

Fig. 13. Fault pattern retrieved by associative memory, *fault 1*

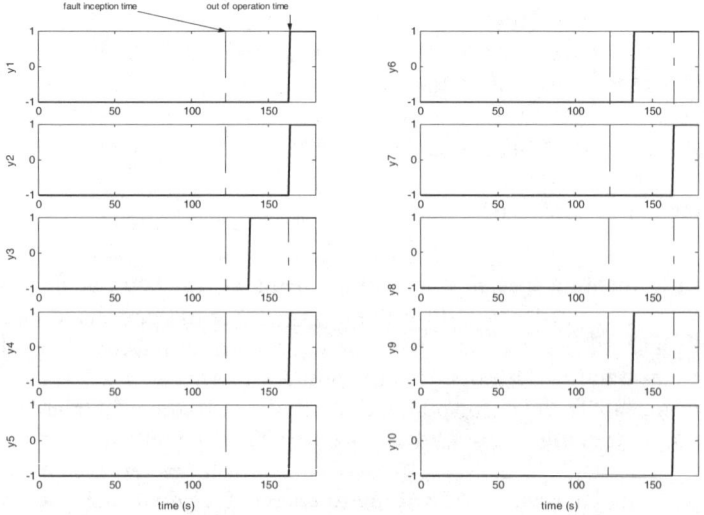

Fig. 14. Fault pattern retrieved by associative memory, *fault 5*

contain enough information about this fault pattern. Fault pattern α^1 is retrieved before fault evolution time is finished.

For this same case, Fig. 14 displays retrieved fault pattern by associative memory when *fault* 5 occurs. Fault pattern α^5 is retrieved before fossil electric power plant is forced by this fault to take out if operation. As explained for residual generation, *fault* 6 is not detected when load power holds constant and then *fault* 6 diagnosis is not considered in this case.

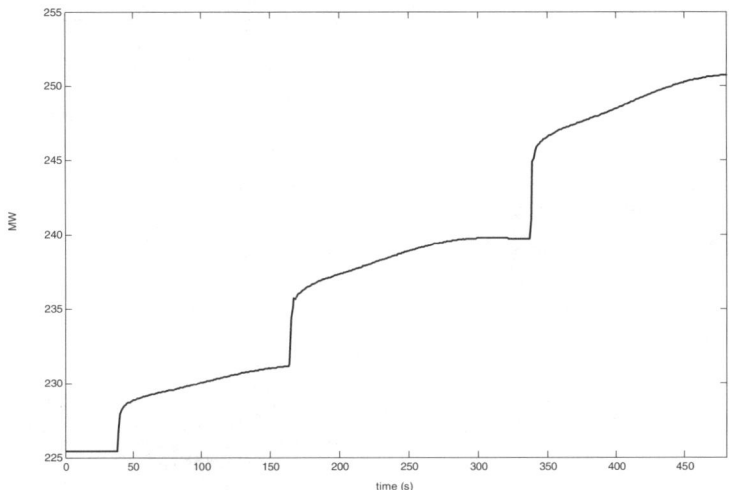

Fig. 15. Load power changes free of faults

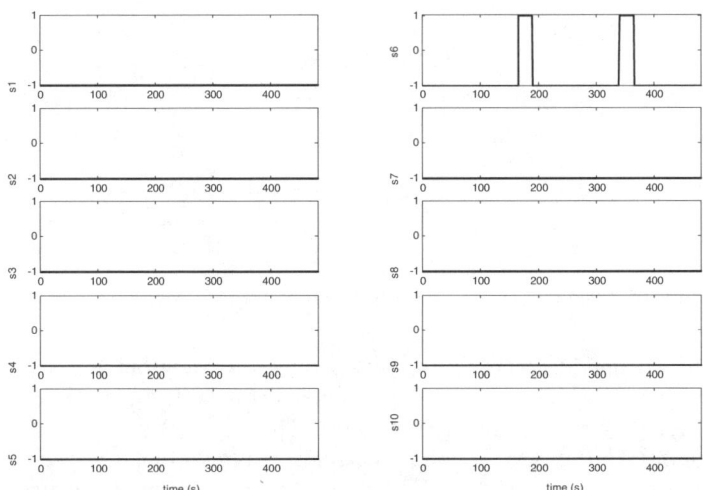

Fig. 16. Encoded residuals when load power changes are made by the operator

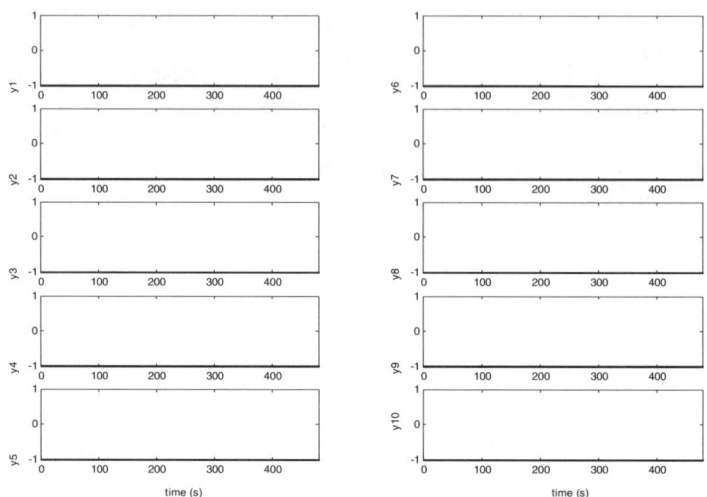

Fig. 17. Fault pattern retrieved by associative memory when load power changes are made by the operator

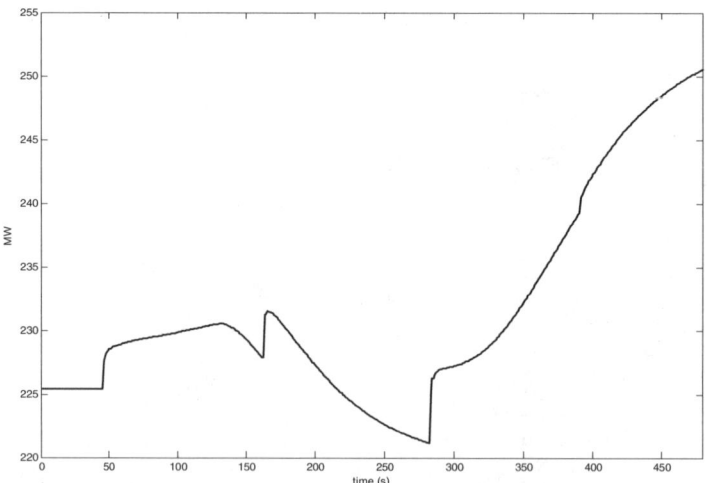

Fig. 18. Load power changes made by the operator when *fault* 1 occurs

Case 2: Normal operating conditions when operator carry out load power changes free of faults). In this case, the proposed scheme for fault diagnosis is evaluated in presence of load power changes free of faults. This changes are generally carried out by the operator (see Fig. 15). Encoded residuals as input bipolar vector to associative memory are displayed in Fig. 16 which shows that $s_6(k)$ take values between −1 and +1 during a short transient time. However, in Fig. 17 the associative memory retrieves α^0 indicating normal operating conditions.

Case 3: Fault diagnosis when the operator is carrying out load power changes. This case considers a fault appearing when the operator carries out load power changes to satisfy the electric power demand requirement by the users. As in the first case, faults have an inception time of 120 s. Fig. 18, 19 y 20 illustrate load power changes made by the

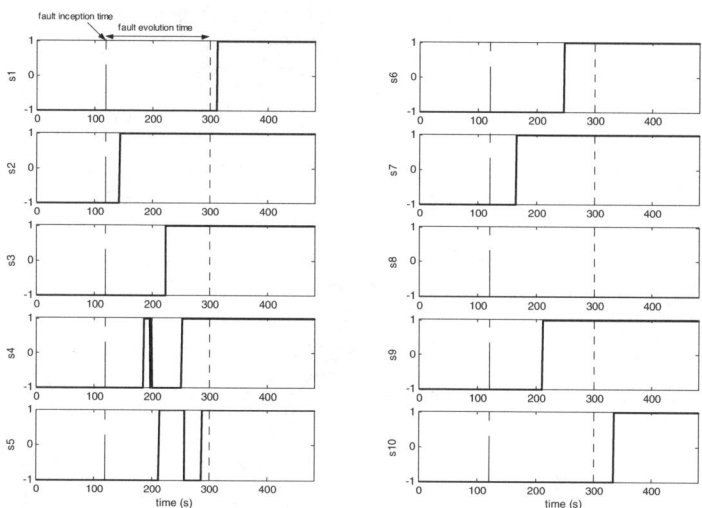

Fig. 19. Input bipolar vector to associative memory when load power changes are made by the operator and *fault* 1 occurs

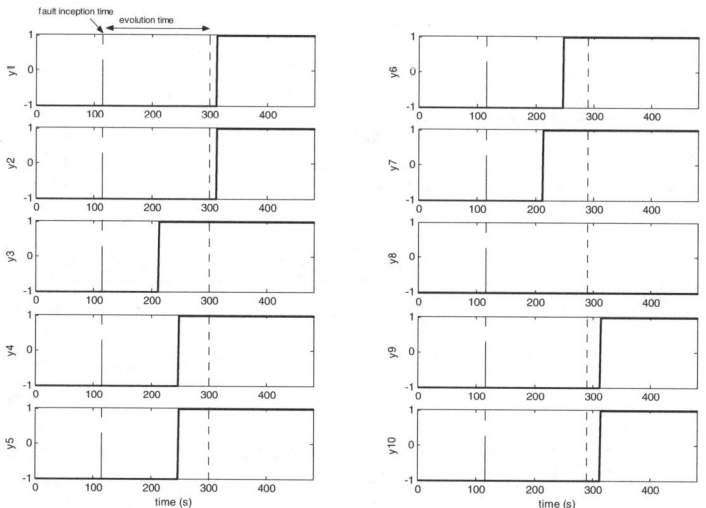

Fig. 20. Fault pattern retrieved by associative memory when load power changes made by the operator and *fault* 1 occurs

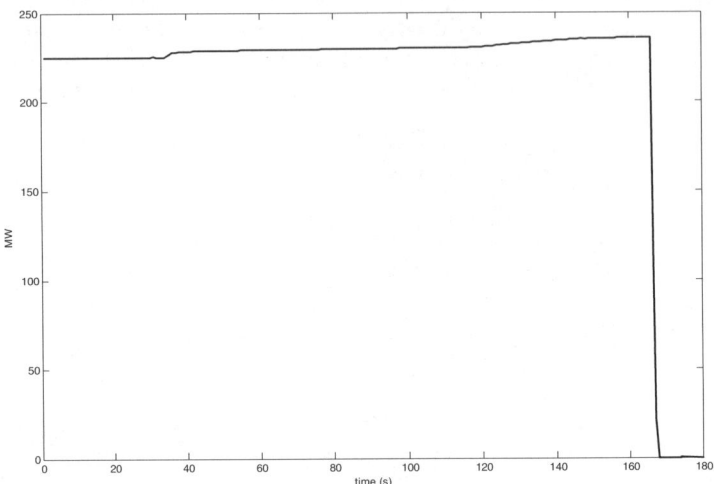

Fig. 21. Load power changes made by the operator when *fault* 5 occurs

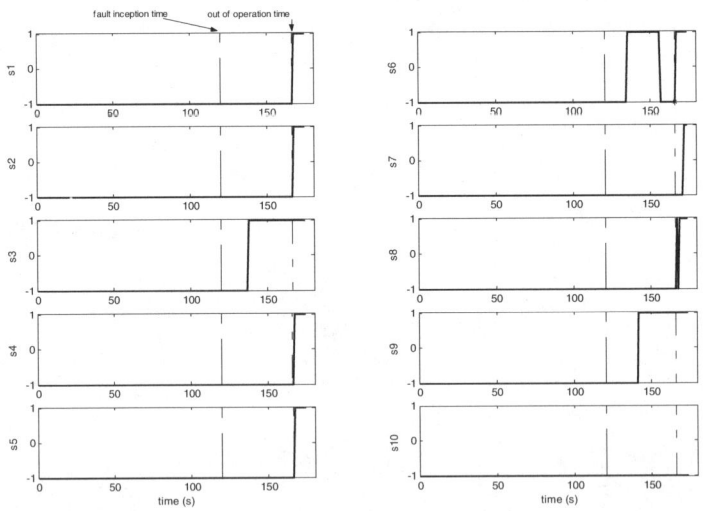

Fig. 22. Input bipolar vector to associative memory when load power changes made by the operator and *fault* 5 occurs

operator, encoded residuals used as input bipolar vector to associative memory and retrieved fault pattern when fault 1 occurs, respectively. As expected, fault pattern α^1 is retrieved.

Fig. 21 illustrates load power changes made by the operator when *fault* 5 has occurred. After 170 seconds, fossil electric power plant is out of operation. Fig. 22 illustrates encoded residuals used as an input bipolar vector to associative memory and Fig. 23 displays retrieved fault pattern which corresponds to α^5. Despite this fault is very critical, fault diagnosis is fast. For *fault* 6, retrieved fault pattern is only displayed in Fig. 24 due to that load power change used for fault detection and corresponding encoded residuals have been illustrated in Fig. 8 and Fig. 12, respectively. Retrieved fault pattern corresponds to α^6 to indicate that fossil oil valve is blocked.

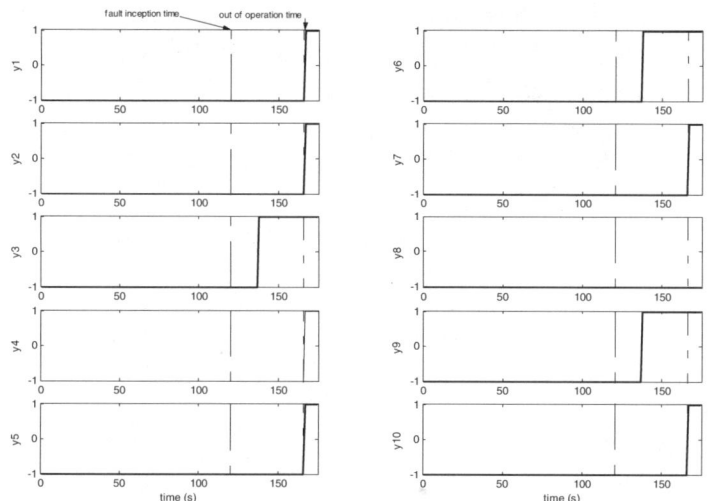

Fig. 23. Fault pattern retrieved by associative memory when load power changes are made by the operator and *fault* 5 occurs

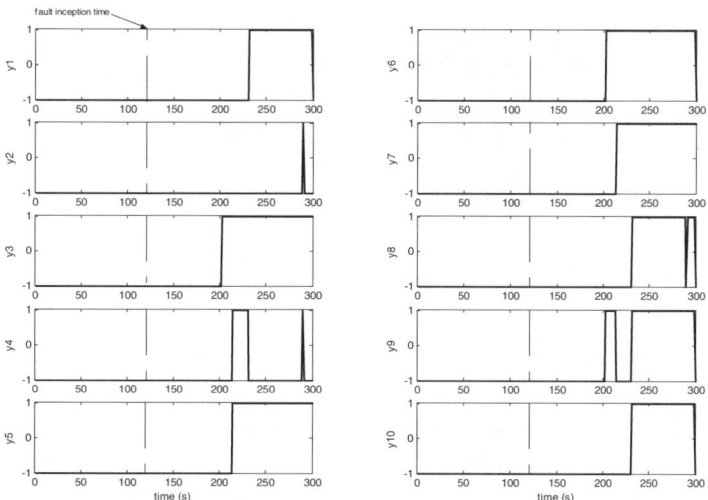

Fig. 24. Fault pattern retrieved by associative memory when load power changes are made by the operator and *fault* 6 occurs

4.2.3 Diagnosis Results

Diagnosis results based on associative memory must be useful to the operators. In order to this requirement in Fig. 25 and 26 are displayed the results for *fault* 1 diagnosis considering case 1 and case 3, respectively. Diagnosis results show that two logic values are possible for each fault indicator, the logic state placed as 1 indicates that *fault* 1 pattern has been retrieved by associative memory and the logic state placed as

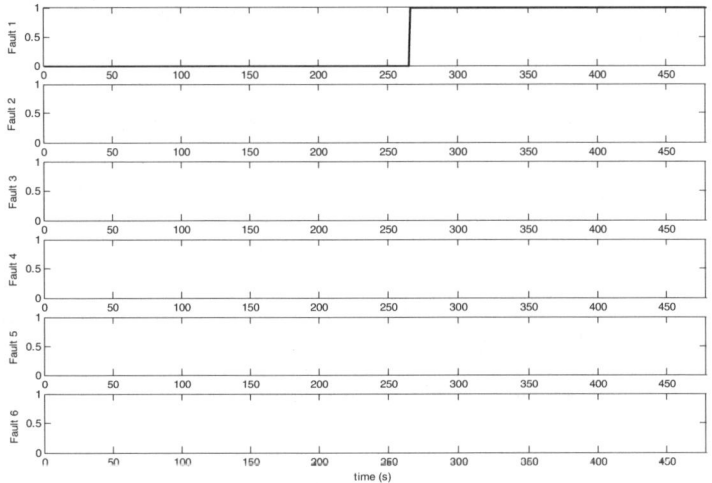

Fig. 25. Diagnosis results for *fault 1*, case 1

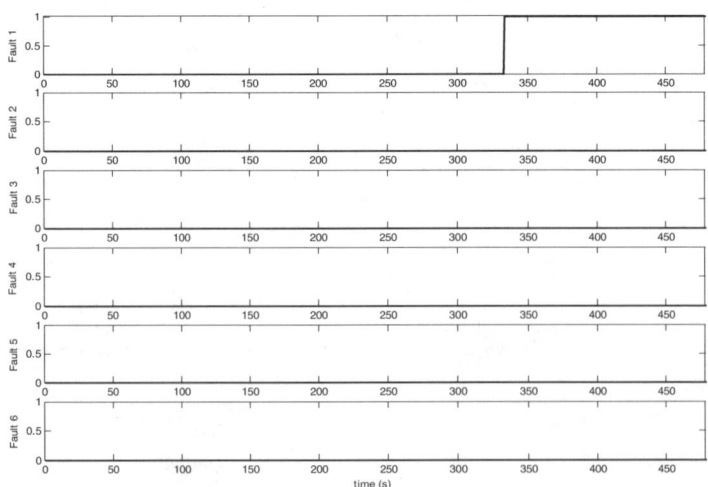

Fig. 26. Diagnosis results for *fault 1*, case 3

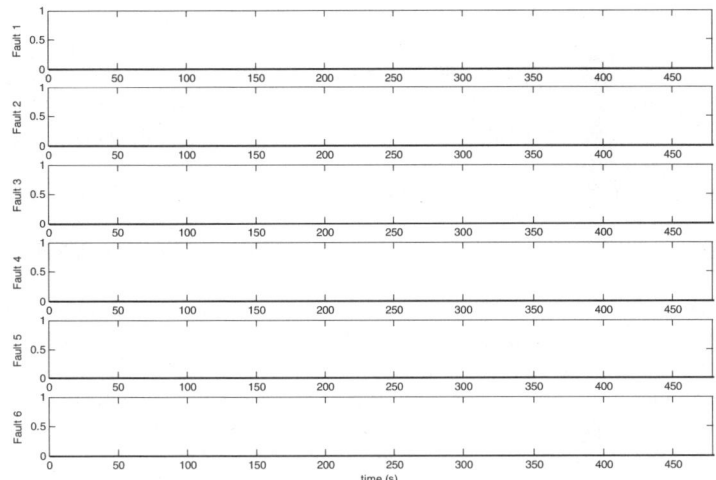

Fig. 27. Diagnosis results when the operator carries out load power changes free of faults, case 2

0 indicates that any other fault is not occurring. Similar diagnosis results are obtained when other faults are appearing.

Finally, Fig. 27 illustrates the corresponding results when load power changes free of faults are carried out by the operator. The logic state for each indicator is zero, then fossil electric power plant is operating on normal conditions. This information is very easy to be interpreted by the operator.

5 Conclusions

The obtained results illustrate that soft margin training proposed in this work is adequated to train associative memories based on RNNs. By means of this new approach, a n associative memory is designed and applied to fault diagnosis in fossil electric power plants. This application considers three cases to evaluate its robustness. Using soft margin training, all diagonal elements on connection matrix T are equals to diagonal elements in matrix A. This fact indicates that the total number of spurious memories is reduced. As a future work, it is necessary to analyze convergence properties for this new algorithm and it is necessary to establish the corresponding properties on connection matrix T.

Acknowledgment

The authors thank support of the CONACYT, Mexico on project 39866Y. Besides, authors also thank to the Process Supervision Department of IIE, Mexico, for allowing us to use its full scale simulator. The first author thanks support of UNACAR, Mexico on project PR/59/2006.

References

1. Liu, D., Lu, Z.: A new synthesis approach for feedback neural networks based on the perceptron training algorithm. IEEE Trans. Neural Networks 8, 1468–1482 (1997)
2. Michel, A.N., Farrel, J.A.: Associative memories via artificial neural networks. IEEE Contr. Syst. Mag. 10, 6–17 (1990)
3. Cortes, C., Vapnik, V.N.: Support Vector Networks. Machine Learning 20, 273–297 (1995)
4. Luemberger, D.: Linear and Non Linear Programming. Addison Wesley Publishing Company, USA (1984)
5. Casali, D., Constantini, G., Perfetti, R., Ricci, E.: Associative Memory Design Using Support Vector Machines. IEEE Transactions on Neural Networks 17, 1165–1174 (2006)
6. Ruz-Hernandez, J.A., Sanchez, E.N., Suarez, D.A.: Designing and associative memory via optimal training for fault diagnosis. In: Proceedings of International Joint Conference on Neural Networks, Vancouver, B. C., Canada, pp. 8771–8778 (2006)
7. Ruz-Hernandez, J.A., Sanchez, E.N., Suarez, D.A.: Optimal training for associative memories: application to fault diagnosis in fossil electric power plants, Book Chapter of Hybrid Intelligent Systems Analysis and Design. In: Castillo, O., Melin, P., Kacprzyc, J., Pedrycz, W. (eds.) International Series on Studies in Fuzzyness and Soft Computing, vol. 208, pp. 329–356 (2007); ISBN: 3-540-37419-1
8. Ruz-Hernandez, J.A.: Development and application of a neural network-based scheme for fault diagnosis in fossil electric power plants (In Spanish). Ph. D. Thesis, CINVESTAV, Guadalajara Campus (2006)
9. Frank, P.M.: Diagnostic procedure in the automatic control engineering. Automatic Control Engineering 2, 47–63 (1994)
10. Patton, R.J., Frank, P.M., Clark, R.N.: Fault Diagnosis in Dynamic Systems: Theory and Application. Prentice Hall, New York (1989)
11. Chen, J., Patton, R.J.: Robust Model Based Fault Diagnosis for Dynamic Systems. Kluwer Academic Publishers, Norwell (1999)
12. Köppen-Seliger, B., Frank, P.M.: Fault detection and isolation in technical processes with neural networks. In: Proceedings of the 34th Conference on Decision & Control, New Orleans, USA, pp. 2414–2419 (1995)
13. Comisión Federal de Electricidad, Manual del Centro de Adiestramiento de Operadores Ixtapantongo, Módulo III, Unidad 1, México (1997)
14. Ruz-Hernandez, J.A., Suarez, D.A., Shelomov, E., Villavicencio, A.: Predictive control based on an auto-regressive neuro-fuzzy model applied to the steam generator startup process at a fossil power plant. Revista de Computación y Sistemas 6(3), 204–212 (2003)
15. Noorgard, M., Ravn, O., Poulsen, N.K., Hansen, L.K.: Neural Networks for Modelling and Control of Dynamic Systems. Springer, London (2000)
16. Marquardt, D.: An algorithm for least-squares estimation of nonlinear parameters. SIAM Journal Appl. Mathematics 11(2) (1963)
17. Levenberg, K.: A method for solution of certain nonlinear problems in least squares. Quart. Appl. Mathematics 2, 164–168 (1944)
18. Nøorgard, M.: Neural Network based System Identification Toolbox (NNSYSID TOOLBOX), Tech. Report 97 E-851, Department of Automation, DTU, Lyngby, Denmark (1997)

Social Systems Simulation Person Modeling as Systemic Constructivist Approach

Manuel Castañón-Puga[1], Antonio Rodriguez-Diaz[1], Guillermo Licea[1], and Eugenio Dante Suarez[2]

[1] Baja California Autonomous University, Chemistry and Engineering Faculty,
Calzada Tecnológico 14418, Mesa de Otay, Tijuana,
Baja California, México, 22390
{puga,ardiaz,glicea}@uabc.mx
[2] Trinity University, Department of Business Administration, One Trinity Place,
San Antonio, TX, USA, 78212
esuarez@trinity.edu

Abstract. In recent years, social simulation has become one of the main tools for social research due to its ability to explore and validate social phenomena. While simulations traditionally consider populations as whole, and as such they tend to miss individual decisions and cultural reasons for actions. Our work focuses on the need for models that take into account personality for individuals, and its importance for characterizing virtual persons inside a simulation. We discuss different personality models and define a Systemic Constructivist Fuzzy Model that includes Transactional Analysis Theory as a basis for defining personality structure and behavior. An example is overviewed in a simulation for a typical social interaction case study. An autonomous intelligent agent is implemented, and the corresponding interactions with other agents inside a simulation are analyzed. Several diagrams in UML representation are given in order to discuss design and implementation features in Agent Oriented Paradigm. We conclude with a discussion on how this approach helps to a social scientist explore social processes and individual behavior in a more systematic way.

Keywords: Social Simulation, Personality, Virtual Persons, Multi Agents Systems, Transactional Analysis.

1 Introduction

Simulations as a research tool have gained more attention by researchers as a possibility for the study and understanding of social phenomena. Several disciplines have adopted it as a regular tool with success to generate data that closely resembles experimental results.

Traditionally, social sciences use statistical methods for creating and studying models that describe observed phenomena, but the advent of emergent systemic approach has made the possibility of creating new software is more and more appealing.

In artificial societies, one interesting and challenging task is to show interaction between individuals, in a process where the personality of the actors comes to light. Several proposals have been made on how to achieve this and the Multi-Agent Systems (MAS) paradigm—along with other cognitive model—seems to be a promising

O. Castillo et al. (Eds.): Soft Computing for Hybrid Intel. Systems, SCI 154, pp. 231–249, 2008.
springerlink.com

option to model and implement virtual persons in artificial worlds. On the one hand, agent technology adapts very well to model issues of real-life human behavior, such as adaptability, mobility, learning, reasoning, and personality. Moreover, cognitive models represent naturally what people think or know. Even though the MAS paradigm is not limited to any particular discipline, it is still not a fully developed technology. A growing literature is currently being developed; one that should leads us to further advance a modeling strategy of computer programs that behave and think more and more like human.

Getting closer to a virtual person model is an important challenge to undertake, since it is in the cross path of many different disciplines such as philosophy, psychology, cognitive sciences, sociology, communications, artificial intelligence, cybernetics, and others. With that in mind, what we are interested is in creating is artificial societies with distinct individuals, and claim to achieve that by following a MAS approach and psycho-cognitive theories in order to replicate human behavior by simulating internal processes that occur in human minds and those characteristics that psychologists consider as driven by personality.

The proposal of this document is to use autonomous intelligent agents and cognitive-psychological theories to simulate realistic persons. In particular, we focused on the topic of personality with the purpose of providing a tool that allows for agents to have a personality profile.

A well known technique used by therapists all over the world to identify personality profiles that describe communication conflicts between individuals in a very successful way is Transactional Analysis (TA). A tool that implements such a technique enables a social scientist to generate a whole spectrum of behaviors that will lead to finding explanations for the underlying mechanisms of human interaction.

An advantage of using this technique is that it is well documented, it was derived form practice and therapists experience, and that it has proven useful for explaining and solving real-life communication problems between individuals. Because its concepts are based on real life experience, TA has the advantage of being intuitive, easy to learn and understand by people that are not experts in psychology.

On the next section, we introduce the core concepts considered into this paper. Section 2 we introduce to Minsky´s theory of mind and in Section 3 we introduce to the theory of Zadeh concerning to perceptions. Section 4 we propose a person architecture and in Section 5 we show several agent design diagrams on UML. Section 7 discusses some positions about cognitions-action agents and Section 8 presents our conclusions. Finally section 9 concerns to future work.

1.1 Multi Agents Systems (MAS)

From a computational systems point of view, an agent is a computational process that implements autonomy (through internal decision strategies) and communication skills (through a set of symbols and semantics associated to those symbols). Through their actions, agents make a system functional. For FIPA (Foundation for Intelligent Physical Agents), concrete instances of this abstraction are key elements for implementing and application according to a given agent architecture [1].

MAS are built by multiple layers of interacting agents. Each of these instances is prepared for interchanging messages with other agents and show group behavior.

These agent's capacities for interacting as well as their capacities for autonomy and adaptability are determined by the type of research they are being used for, and generally have a life time and evolve inside the virtual world that contains them. MAS have evolved and the new tools seem able to escalate them in various ways, for example, every time more and more concurrent agents can be used at the same time, or the capacity to process more internal processes inside each agent is achieved.

1.1.1 Intelligent Autonomous Agents

Agents are normally defined as entities with attributes that are considered useful in a particular domain. This is the case of intelligent agents, where agents are seen as entities that emulate mental processes or simulate a rational behavior [2]; personal assistants, where agents are entities that help users with their tasks; mobile agents, where entities are capable of traveling inside a network in order to achieve their goals; information agents, where agents organize in a coherent way data gathered from different and sometimes unrelated places; and autonomous agents, where agents are capable of achieving tasks in an unsupervised way.

Flores-Mendez makes an interesting list of common agent attributes [3]:

- Adaptation: The capability for making internal changes through learning and experience.
- Autonomy: Reflecting important characteristics of the entity, which is goal oriented, proactive and the existence of a decision-making mechanism.
- Collaborative Behavior: refers to the ability to work with other agents for a common objective.
- Reasoning: Ability to infer new knowledge.
- Communication: Ability to communicate at the knowledge level.
- Mobility: Ability to migrate from one computer (container or world) to another.
- Personality: Ability to manifest behavioral attributes commonly ascribed to humans.
- Reactivity: Ability to "feel" its environment and act in a selective way.
- Temporal continuity: Identity and states persistence through long periods of time.
- So, we refer to [4]:

"A software agent is an interface that looks like a person, acts like a person and even appears to think like one" [4], and *"An agent has mental properties, such as knowledge, belief, intention and obligation. In addition, it may have mobility, rationality ..."* [4]

Based on the above description, our work is based on the idea that it is possible to achieve a "virtual person" based on agent architectures.

1.2 Adaptive Complex Systems

Adaptive Complex Systems are used to study natural and artificial systems generally defined by populations of adaptive agents that interact in a non-lineal way, and where an emergent property is created as a result of the interaction   [5]. Even though there are several approaches to address and analyze complex systems, one common approach involves investigating adaptive complex systems by building systems based on artificial intelligence [6] [7].

1.3 Personality

To talk about personality means to talk about people. Different theories have been developed over the years, but it is in the last decades that a systemic approach has improved those theories, allowing us to pursue the modeling of these theories with MAS.

1.3.1 Psychoanalysis

Freud's proposed Psychoanalysis Theory absolutely revolutionized the way in which we understand human behavior. Freud draws the attention to the idea that in human mind we find multiple processes, including dynamics that generate tension over motivations. In his seminal propositions, Freud introduced concepts such as libido, ego, superego, consciousness, etc., which can be thought of representing internal energies or sub-systems in tension. These potentially conflicting produce—as the resultant energy—observed behavior. Each system processes different types of information and represent different types of knowledge. All knowledge (learned or experienced) is taken into account by Freud [8], whether it was acquired from childhood or from the way in which people communicate with their environment, and in general referring to the way in which people handle internal and external drives. Freudian theories evolved in many different ways, and became a method for analyzing personality and therapeutic techniques.

1.3.2 Psychodynamics

Freud associated different systems that drive personality as an emergent behavior, since they are the result of the interaction of various dynamical systems (similar to a thermo dynamical system). The components of these systems potentially engage in competition among them, generating tensions throughout the general system. These tensions, if not liberated, can be responsible for observed pathological behavior [8].

1.3.3 Reflexive Knowledge Versus Embodied Knowledge

In relation with our research, it is worth mentioning the HUMAINE Group [9], as it is also involved in research intended to understand emotions and how they can be implemented in computer software. This group distinguishes between reflexive knowledge and embodied knowledge, in order to better analyze how emotions are related to perception-action and cognition-action processes. The former is proposed for tackling body related knowledge, albeit to that which is not necessarily related to language, whereas the latter is proposed for tackling the high-level knowledge which is represented by language.

1.3.4 Fuzzy Inference Systems

A Fuzzy Inference System (FIS) converts measures to linguistics variables and solves actions using a rule-based database. In this approach, we can apply Fuzzy Logic techniques to convert perceptions to linguistics variables using a Fuzzify method, and then solve the decision problem through FIS. Finally, we can apply a Defuzzify method to return perception like values. Figure 1 shows this technique, a Fuzzy Inference Engine (FIE) solves linguistic variables querying (production) rules from a data base.

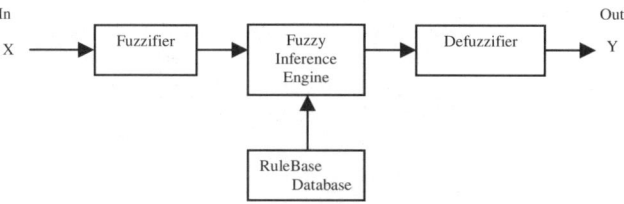

Fig. 1. Main components of a Fuzzy Inference System

2 Minsky´s Theory of Mind

In the process of creating a virtual person, an important theory to consider is Minsky's Society of Mind [10]. Inspired by his experiences in Artificial Intelligence and Cybernetics, as well as some ideas from psychological theories, Minsky proposes that the mind is best understood as a collection of multiple processes. Originally, he referred to these as agents, but later, in his last publication entitled "The Emotional Machine" [11], he referred to them as "resources," so they would not be mistaken with software agents, or with agents in the economical sense (as independent beings). The function of these resources is to solve concrete problems, which encapsulate inside their own representational system the knowledge and its processing. Complexity of the mind is then complexity of the interconnections between the different resources (agents) at different levels. These resource agents have the ability to associate and create agencies to solve more complex problems. Singh, [12] examines the different concepts proposed by Minsky in his Society of Mind Theory [10].

One of the interesting aspects of Minsky´s last book [11] is that it integrates Freudian Theory as an integral component of the analysis. Freud's idea of the mind as a kind of sandwich, connects very low-level knowledge (drives of the Id) with higher-level knowledge (ideals of the superego). In the middle part we find the ego, which handles the ways to settle conflicts between the two. Figure 2 reflects the overlap where Freud's and Minsky's ideas are integrated into Minsky's model.

Figure 3 shows Minsky's six intermediate levels, which can be used in an "increasing way of thinking". These intermediate levels were called "Common Sense Knowledge", representing a kind of knowledge that can be shared between several individuals, with several similar ways for interpreting and solving problems.

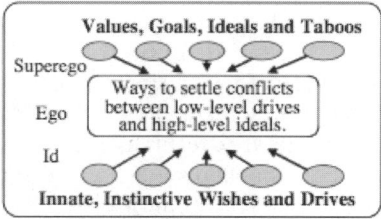

Fig. 2. Minsky´s illustration of Freud's idea of the mind as a "sandwich" with three major parts [11]

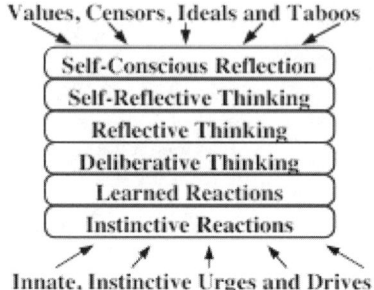

Fig. 3. Minsky´s sequence of levels at which we can use increasingly ways to think [11]

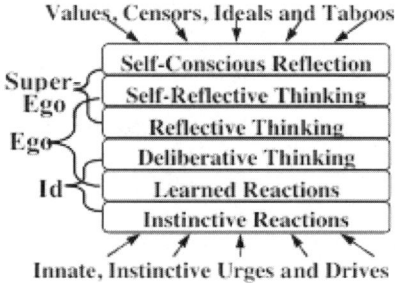

Fig. 4. Minsky´s illustration where Sigmund Freud's idea of the mind as a "sandwich" with three major parts can be directly compared to Minsky´s Model Six [11]

Furthermore, figure 4 shows that—while on the subject of central control—we should point out that Minsky´s Model Six is showing in terms of Sigmund Freud's idea of the mind as a "sandwich," with three major interacting and potentially competing parts [11].

3   From Computing with Numbers to Computing with Words

Perceptions that can be translated into linguistic variables and vice versa are discussed by Lofti Zadeh [13]. Following his theories of fuzzy sets, fuzzy logic and soft-computing, he suggests that—in a "counter-traditional" way of treating perceptions in a computer—perceptions can be converted primarily in measurements. Furthermore, these measurements are computed for decision-making in a high-knowledge level. However, once this decision has been taken, perceptions should be computed in order to be translated into measurements that result as a consequence of actions. Many of these decisions are taken based on knowledge expressed in IF-THEN rules, and formed by linguistic variables (discursive knowledge formed at the language level). As an example, consider IF *IS_HOT* THEN *MOVE_FAST*. In this example it can be seen that, even though body perceptions can be defined by continuous values, the final decision of moving is taken at another level, where words that can be used to

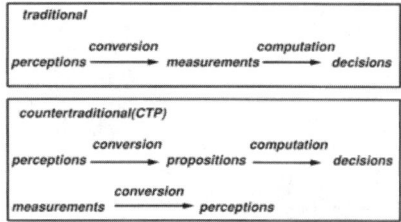

Fig. 5. Zadeh's illustration where a different approach to the conversion of measurements into perceptions. Traditionally, perceptions are converted into measurements [13].

evaluate and to take a decision, are ambiguous and with fuzzy meaning [13]. Figure 5 shows Zadeh's idea that perceptions are converted in measures so they can be computed and vice versa.

Most importantly, Zadeh's main idea is that perceptions with decision-making and its consequent actions can be understood as a fuzzy system whose problematic is to be approached with fuzzy inference systems (FIS).

4 Person Model

Taking all of the above into account, our goal is to build an autonomous intelligent agent model that implements personality profiles using a constructivist systemic approach. To complement the model, a psychology theory can be used to give a basis for personality. As an example, we combine Transactional Analysis and a Model of Mind to build our agent. Transactional Analysis is on its own a constructivist model and is suited for systemic approach [14].

4.1 Person Architecture

Our model consists of an agent that contains a "mind," which is in turn formed by a set of resources. The function of these resources is to represent ways of thinking related with knowledge-action. This agent represents a real person and can be formed by a set of subsystems which implement different ways of representing and processing information (knowledge). This agent will have a communication system and a dynamic system.

Each cognition-action agent encapsulates its own functionality, just as the "person" agent. This object will also contain a communication system and a dynamic system. Different ways of representing and processing knowledge can be implemented as part of the characteristics of these resources. This structure is shown in Figure 6.

Figure 7 shows how a person can use a resource, which in turn consults different sources of cognition-action. The resource's dynamic system will propose a set of possible actions. Figure 8 reflects the fact that different agents can form new agencies by communicating among them in order to collaborate in solving a given problem.

These intermediate agents will allow us to represent critical systems that help select the best way to solve a problem.

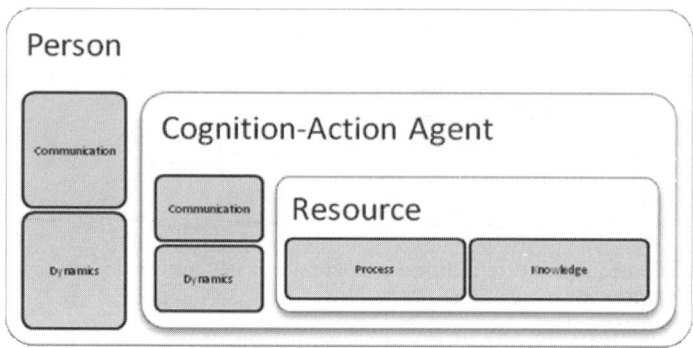

Fig. 6. Person architecture with communication, dynamics and cognition-action structure

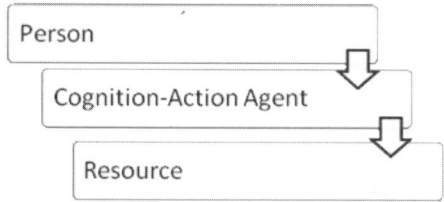

Fig. 7. Use of a resource by "person" and its link with a cognition-action structure

Fig. 8. Different cognition-action agents can collaborate to tackle a complex problem. The agents involved in this process implicitly create new forms of agency through their coordinated actions.

Fig. 9. Activation and process of person perception transfer dynamics to cognition-action

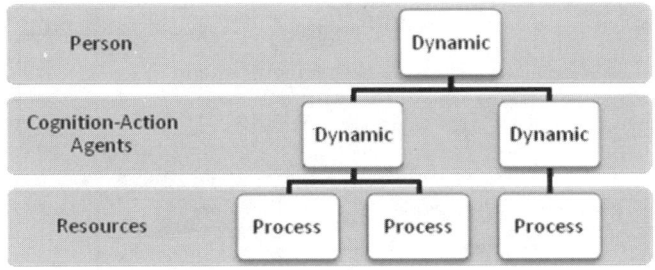

Fig. 10. Perception dynamics starts several agencies than activate different resources processes

Figure 9 shows how person perception and resource process are activated. It also describes transfer dynamics to cognition-action dynamics. Several resources can be activated to resolve different questions.

As we show in Figure 10, perception dynamics starts several resources that activate different cognition-action processes in a cascade fashion.

5   Person Agent Design Model

Through a computational approach, we propose the construction of a virtual person. To do so, we use as a basis for design the agent's specifications delineated by FIPA [1]. This organization establishes standards for agents with the capacity of executing a number of behaviors, to communicate with each other through messages codified in a standardized protocol, and to have mobility through a distributed system. The proposed designs are intended to adapt these characteristics to the proposed model of a virtual person.

5.1   Agent Oriented Design

We shall use one agent to represent a virtual person. This virtual person in turn encapsulates a set of additional subsystems. Each one of these subsystems is also implemented by computational agents, which have their own knowledge base and its own dynamical processes.

We present below a class diagram in the unified modeling language of UML [15], which represents a design oriented to agents that perform the proposed abstractions described in previous sections.

Figure 11 displays the implementation of a person agent and a cognition-action agent. A person agent can contain a number of cognition-action agents. The design is makes reference in the available library of JADE® [16], which is integrated in FIPA standards.

Figure 12 describes in further detail the relationship between the person agent and the cognition-action agent. In turn, the cognition-action agent can reference multiple resource objects.

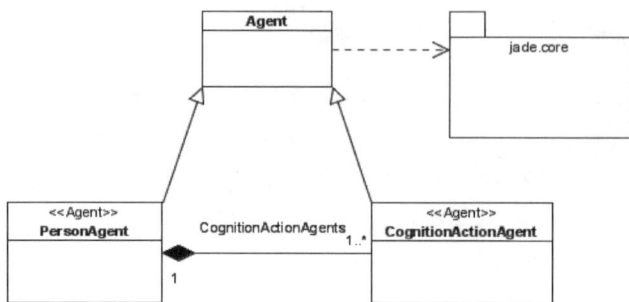

Fig. 11. A person agent contains a set of cognition-action agents

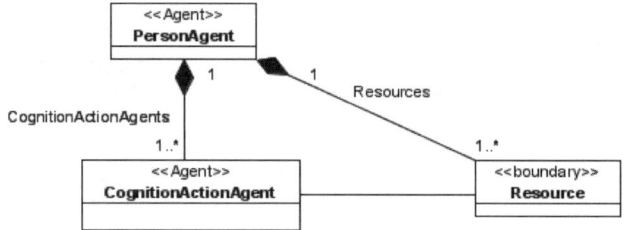

Fig. 12. Cognition-action agents reference different resource objects

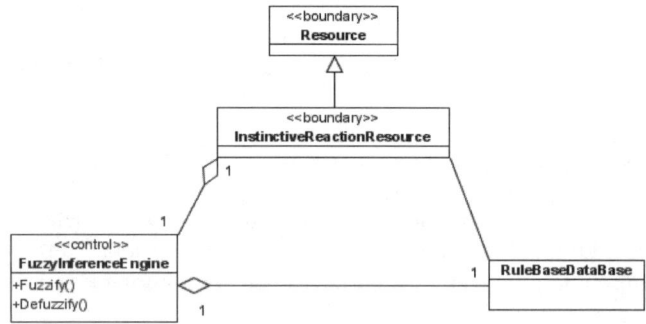

Fig. 13. One type of resource may be an instinctive reaction resource that could be implemented using a fuzzy engine with a rule-based database

Figure 13 shows an example of how we can specify the specialization of a type of resource. The type `InstinctiveReactionResource` encapsulates the dynamics and attributes related to knowledge at this level. This type of resource could, for example, contain a fuzzy inference engine that solves for certain numerical variables with linguistic values.

This simple design allows us to implement different subsystems that compose a person, by means of programming oriented to agents and to different ways in which to represent and process knowledge.

Many types of cognition-action agents can be implemented to tackle a given problem, just as many different techniques will produce similar results. The capacity of the agents to communicate through a language and a standardized communication protocol (ACL) is allowed by the establishment of an interface between them [17].

Through the application of techniques for collaboration amongst agents, nets can be formed in which a number of agents propose different solutions to a given problem.

New cognition-action agents should be created as the result of these collaborative interactions, establishing new societies that know how to solve problems for which they may already have previous experience.

6 Case Study

To exemplify our approach, we analyze a known pathological profile described with the previously discussed technique of Transactional Analysis. Our case study represents a viable option to describe the profile of an agent that is represented within the system. Different pathological profiles are available in the Psychology literature, and in many cases these profiles have also been expressed within the framework of Transactional Analysis.

6.1 Transactional Analysis

Transactional Analysis (TA) is a technique developed by Berne [18], who intended to disentangle personality based on the ideas described above. This procedure evolved out of a more general literature on personality. Berne proposes the identification of personality traits associated with the three profiles or states of the ego, which represent distinct characteristics of behavior, in turn stemming from different internal processes. He coined these three states as the father state, the adult state, and the kid state. He proposes main roles and attitudes too which the people adopt during interactions. Based on these states, roles and attitudes, he created a methodology for studying the way in which people communicate with each other; as well as how one can detect communication problems stemming from natural human interaction (one can find a thorough summary of the basic concepts of TA in [19]).

We can link each TA Ego State with each Freudian subsystem; child TA state represents in part the Id states of the Freudian model of personality, and adult state represents the Ego states and Parent the Superego states respectively. For all practical purposes, TA is a way to manage Freudian concepts in a friendly technique and easy to associate to a conventional person experience. The great advantage of TA is that it has grown since its origins out of therapeutic practice, and as such it is intended to help real people with real problems.

6.2 Cathexis Flux

Cathexis represents the psychodynamic processes that control the selection of different psychological states and activities at any given moment. This concept puts forward the idea that the combination of different energies (such as potential energy, kinetic energy and free energy) defines the emergent final action and the corresponding state of the ego at any given moment [20]. This energy is affected by internal and external stimuli, and is translated into the motivations held by a person as he or she relates to a specific activity, drawn form a set of possible activities.

6.3 Don Juan Syndrome

Novellino [21] describes "The Don Juan Syndrome" as a pathological profile of an individual who follows a specific script of behavior similar to that portrayed in the play *El Burlador de Sevilla*, by Tirso de Molina (1620), Moliere's play *Don Juan* (1665) and Mozart's opera *Don Giovanni* (1787).

The psychoanalytic interpretation of Don Juan is based on three key concepts:

1   *The constant need to prove his own sexual identity.*
2   *The constant search for a maternal element in women.*
3   *The constant resurfacing of a female element, which is then suppressed once seduction has been accomplished.* [21]

Our 'hero's' game unfolds in the sexual-emotional sphere, it represents a variation of the "Kick me" TA game [18], developing through the following moves:

1. Don Juan uses flattery and promises to present himself as the rescuer of a woman who is prey to her own need to be free and to feel appreciated (the Victim). This first move comprehends the con and the gimmick in Berne's (1972, p. 24) game formula.
2. The work of seduction continues until she capitulates (the response).
3. The moment our 'hero' reneges on any further demands for emotional closeness; the woman remains bewildered and shocked.
4. As soon as she realizes how gullible she has been, she turns into a persecutor seeking revenge, and Don Juan, in turn, becomes the victim of female voracity ready to start fresh anew as another woman's rescuer (this move is the switch).
5. The game's payoff is for Don Juan to prove once again how voracious women are and to feel "all set and raring to go" in a new attempt to win over the ideal woman; the woman's payoff is to confirm men are untrustworthy. [21]

This psychological profile will allow us to prove our methodology for developing a virtual person with such behavior.

6.4 Characterizing Don Juan

In order to develop the Don Juan syndrome we have the specifications described below. Figure 14 reflects the specifications for the relevant types: `PersonAgent`, `CognitionActionAgent` and resources with of the types `DonJuanAgent`, `EgoStatesSystem` and `PhsychologicalStateSystem`, respectively.

Figure 15 illustrates a representative interaction sequence among the environment's systems, Don Juan, and the internal system of his ego. The main idea is that the environment will interact with the `DonJuan` agent, and that he in turn interact with the `EgoSystem` agent to resolve how he would respond to multiple stimuli. The different states can be represented as a resource, in which the relationships among different states, as well as the rules that control the transitions from one state to another.

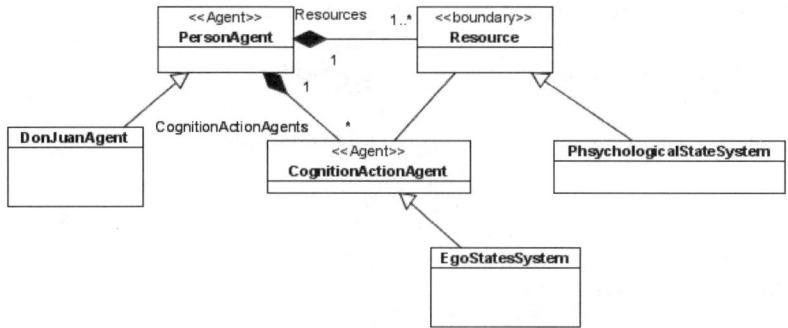

Fig. 14. DonJuanAgent and EgoStateSystem especification

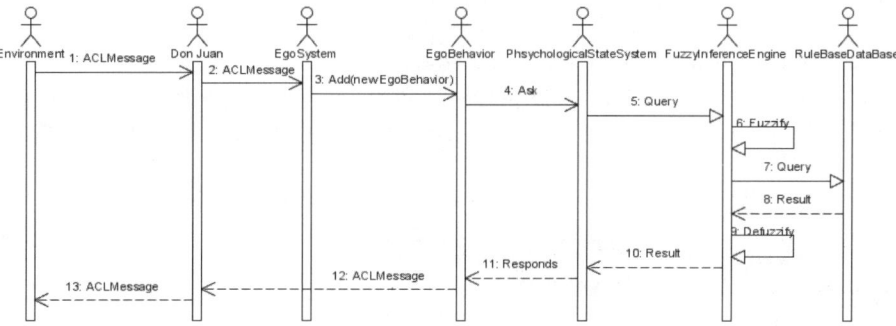

Fig. 15. Sequence diagrams of possible interactions between Environment, Don Juan and the internal Ego System

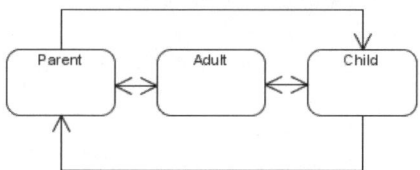

Fig. 16. State diagram expressing ego states transitions on AT

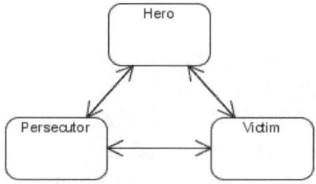

Fig. 17. State diagram expressing role switching transitions on AT

Figure 16 shows the different states and their possible transitions. Furthermore, Table 1 displays these transitions as expressed by rules.

Figure 17 shows the different roles and their possible transitions.

Table 1. Example representing AT ego states transitions with rule base database

```
IF parentState THEN adultState
IF parentState THEN childState
IF adultState THEN parentState
IF adultState THEN childState
IF childState THEN parentState
IF childState THEN adultState
```

Similarly, we can express activities as they relate to different states. For example, if the person were to be in a child state, then the corresponding action may be to play, while if the person were in an adult state, then the corresponding action would be to work, and if the person were in a father state, then the corresponding action may then be to rest. We can see in Table 2 how these conditions could be expressed as a rule set.

Table 2. Different states expressed as rules

```
IF parentState THEN restAction
IF adultState THEN workAction
IF childState THEN playAction
```

Other rules that are not as clearly defined can also be expressed. For example, if the person has been working for a significant amount of time, then it would necessarily have to rest. The degree of tiredness can be expressed in terms of energy levels, and as such the evaluation of this condition presents ambiguity of defining the state `tired`. For this reason, we opt for a different way in which we solve this condition, using a fuzzy methodology to evaluate the numerical variable as a linguistic variable, and afterwards take the corresponding action.

Table 3. State and action rule example

```
IF tired THEN restAction
```

For this type of decision-making, one would need for the system of the ego to include a fuzzy inference machine that transforms the numerical values into linguistic values and vice versa. Including these capacities, we can then establish new rules involving environment actions that are related to states of the ego. For example, to establish the rule of when the person is in the child state and at the same time it is very cold, and then the state is irrelevant.

Table 4. Fuzzy value and ego state example

```
IF veryCold && childState THEN restAction
IF veryCold && parentState THEN movingToAction
```

We can represent the script of behavior by following the Don Juan syndrome, and writing rules that establish the sequence of characteristic actions of this behavior such as, for example, by combining conditions like IF adultState && noCompromised THEN lookinForGirlAction.

Table 5. Don Juan Sript rules example

```
IF me.victimEgoRole THEN lookingForGirlAction
IF lookingForGirlAction THEN me.rescuerEgoRole
IF me.rescuerEgoRole && she.victimEgoRole THEN
me.flatering
IF me.flatering && she.victimEgoRole && she.capitulate
THEN me.inLove
IF me.inLove  && she.persecutorEgoRole THEN
me.victimEgoRole
```

The dynamics of the different actions can be controlled by a dynamical system in which each action could potentially contain different energy behaviors, and in a selection process in agreement with a resulting cathexis. This notwithstanding, a decision to change action could be taken if only the requisites of certain pre-established rules are met, such as in the previous examples.

A complementary paper to this work explains the system of cathexis flux that implements the dynamic part of this approximation [22]. Figure 18 displays an advance of this implementation.

In Figure 19 we can observe the different energies that compete for each state of the ego in the TA, and which determines what will be the dominant state of the ego.

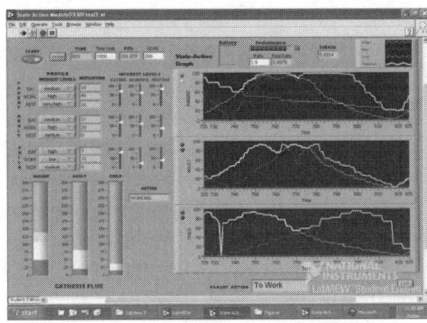

Fig. 18. Different functions determine the dynamic behavior of different actions

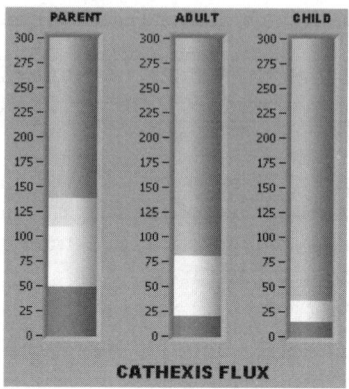

Fig. 19. Only the state with most energy will dominate all others and thus determining the current activity

The methodology of cathexis flux can be understood as energy containers that react to diverse internal and external stimuli (that is, internal motivations and environment stimuli), and create the illusion of a dynamic of motivations that make people pay attention to any given activity.

7 Discussion

In previous work, we start to build a first approach to virtual persons applying basics elements of agents [23]. In this approximation, we have considered a number of ideas abut how an actual person can be modeled. A good starting point for this bold task is Freud's psychoanalysis theory [8], since it represents the basis for modern psychology, and it includes many original concepts that are used today. By proposing that humans represent a collection of interacting subsystems, it prescribes abstractions that encapsulate much of what may occur inside the mind, and to some degree explains the nature of thought and behavior. Freud conceived of behavior as an emergent property of a dynamic system, which is also related to the systemic idea of homeostasis.

In another sense, one of the most relevant contemporary models of theory of mind is that of Minsky [10, 11], which without a doubt includes many of the advances in the area of artificial and cybernetic intelligence, in terms of how to reproduce the human mind. In his last publication, "the emotional machine" [11], he comprises important concepts of Freud [8], such as that which is traditionally called common sense [24, 25], the idea of multiple ways of thinking [11], of mental resources [10], and others. These ideas serve as a basis for future research in many different levels. For one, they allow the possibility of better understanding the human mind. Moreover, the theory of mind described by Minsky has the advantage of having been born out of experience building artificial individuals, and therefore is well aware of the problems that this area has faced.

At this point, something that is worth stressing is how Minsky sees emotions as emergent behavior stemming from the different and complex systems that form a human [11]. In contrast with many psychological theories of traits and types of personality [26], Minsky begins from Freudian theory [8], where personality is not programmed, but rather an emergent phenomenon of the multiple internal subsystems of a human being.

"There is not one unique way of thinking, but instead many and very complex ways of doing it, if any one way of solving a problem does no work, then the mind selects another one" says Minsky [11]. This vision is important because it offers us the necessary flexibility to tackle problems in multiple ways. In the same way that this theory describes how our minds work, we can just as well incorporate different approaches to the way in which our proposed virtual person can solve a problem. The agents that we build based on this methodology allow us these capabilities. This is because our virtual persons could be implemented by a number of agents that can collaborate in different ways in order to solve a problem; the subsystems have the facility to combine through a communications system based on messages. These messages provide the required flexibility needed to model different systems, whose combinations are defined by the communication language in which they are imbedded [23].

Furthermore, Zadeh [13] also in a way comes close to these same ideas by treating perceptions as lower-level knowledge, but that can relate to a higher-level knowledge through fuzzy logic techniques. The author is aware of this connection and using Fuzzify and Defuzzify methods proposes the conversion of numerical values to linguistic ones. This idea coincides with those of Minsky and Freud, by indicating the difference between different types of knowledge, as well as offering a way in which they can be translated from one onto the other.

Finally, transactional analysis is a methodology that has helped resolve communication problems between real people, and has the characteristic of systematically modeling people and their behavior. Even though the states of the ego described by transactional analysis do not perfectly overlap with Freud's concepts [8], it nonetheless resembles both psychoanalysis as well as Minsky's theory of the mind [10]. Therefore, even though there may be some who criticize methodology, one of its best features is that it can be easily understood. Cases such as the one we have developed can be extremely useful in the development of tools for social researchers, since the language used is accessible for many people. The concept of cathexis has also been criticized by some researchers, but by the same token, the idea of internal energies can be assimilated by lay people, as it helps to better visualize the proposed dynamics inside people. At the end of the day, the most important issue for the social researcher is to be able to assign behavioral profiles to individuals being studied, and through these means provide a coherent explanation to their behavior, as set within a community.

8 Conclusions

An approximation of virtual persons can be achieved through the use of systems based on agents. In psychology, the study of people leads us to the issue of personality. We propose the use of the concepts of transactional analysis as a basis for creating personality profiles in virtual persons. This methodology is well known, well documented, developed from practice with real individuals, and easily assimilated by lay people. We can with this methodology represent the different psychological states, playbooks, goals and actions, all set within a model of person from a systemic and constructivist point of view. With this we want to say that the concepts of TA have a systemic basis, and, furthermore, they refer to a cognitive where people represent their experiences as mental constructions. Different agents can represent different mental processes and knowledge. Agents can be considered for handling knowledge at different levels, and can use fuzzy logic for handling knowledge at different levels. In the same fashion, the fuzzy logic methodology can be used for solving the communication corporal sensations to the knowledge represented by linguistic variables. In terms of sensations, what we are referring to are low-level numerical values that are generated by a dynamical system imbedded in the agent. The emerging result of these interacting variables can be converted into linguistic variables by Fuzzify and Defuzzify methods, as well as a fuzzy inference machine that understands observed actions. Using the same technique, these actions can in turn be converted into low-level perceptions.

9 Future Work

As part of our future work, we are considering the advancement of our proponed methodology in several fronts. The first one referring to the refining the characteristics of the model so as to closer resemble the conceptual model proposed, that is, by incorporating more varied ways of representing and administering knowledge. Another route is to experiment with more examples related to transactional analysis, applying different profiles to populations of multi-agents. Another research avenue implies the incorporation of knowledge bases representing common sense, and applying them to different projects being studied. Yet another route has to do with scaling this person model to an organizational model. Following the ideas of Minsky, and incorporating into the mix the ideas of the Beer´s Viable Systems model [27], the research avenue would imply ascribing profiles to organizations.

References

1. FIPA, FIPA Abstract Architecture Specification, F.f.I.P. Agents, Editor (2002)
2. Hales, D., Edmonds, B.: Evolving social rationality for MAS using "tags". In: Proceedings of the second international joint conference on Autonomous agents and multiagent systems %@ 1-58113-683-8, pp. 497–503. ACM, Melbourne (2003)
3. Flores-Mendez, R.A.: Towards a Standardization of Multi-Agent System Frameworks. Crossroads 5(4), 18–24 (1999)
4. Wooldridge, M.J., Jennings, N.R.: Intelligent agents: Theory and Practice. The Knowledge Engineering Review 10, 115–152 (1995)
5. Brownlee, J.: Complex Adaptive Systems, Complex Intelligent Systems Laboratory, Centre for Information Technology Research, Faculty of Information Communication Technology, Swinburne University of Technology: Melbourne, Australia (2007)
6. Anderson, P.W.: Complexity: metaphors, models, and reality. In: George, A.C., David, P., David, M. (eds.) The Eightfold Way to the Theory of Complexity, p. 731. Perseus Books, USA (1999)

7. Mitchell, M., Newman, M.: Complex Systems Theory and Evolution. In: Pagel, M. (ed.) Encyclopedia of Evolution. Oxford University Press, New York (2002)

8. Freud, S.: A general introduction to psychoanalysis. Boni and Liveright, New York (1920)

9. Cañamero, L., et al.: Proposal for exemplars and work towards them: Emotion in Cognition and Action. In: Gelin, P. (ed.) Human-Machine Interaction Network on Emotions, University of Hertfordshire: Hatfield, Herts, United Kingdom, p. 46 (2005)

10. Minsky, M.: The Society of Mind. Simon and Schuster, New York (1986)

11. Minsky, M.: The Emotion Machine: Commonsense Thinking. Artificial Intelligence, and the Future of the Human Mind. Simon & Schuster (2006)

12. Singh, P.: Examining the Society of Mind. Computing and Informatics, 1001–1023 (2004)

13. Zadeh, L.A.: From Computing with Numbers to Computing with Words - From Manipulation of Measurements to Manipulation of Perceptions. International Journal Applications Math and Computer Sciences 12(3), 307–324 (2002)

14. Kreyenberg, J.: Transactional analysis in organizations as a systemic constructivist approach. Transactional Analysis Journal 35(4), 300–310 (2005)

15. Rumbaugh, J., Jacobson, I., Booch, G.: Unified Modeling Language Reference Manual, 2nd edn. Object Technology Series. Addison-Wesley, Reading (2004)

16. Jade. Java Agent DEvelopment Framework (2007), http://jade.tilab.com

17. FIPA, FIPA ACL Message Structure Specification F.f.I.P. Agents, Editor (2002)

18. Berne, E.: Games people play: The psychology of human relationships. Penguin, London (1964)

19. Steiner, C.: A Compilation of Core Concepts. Transactional Analysis Journal 33(2) (2003)

20. Velasquez, J.D., Maes, P.: Cathexis: a computational model of emotions. In: Proceedings of the first international conference on Autonomous agents @ 0-89791-877-0, pp. 518–519. ACM, Marina del Rey (1997)

21. Novellino, M.: The Don Juan Syndrome: The Script of the Great Losing Lover. Transactional Analysis Journal 36(1), 35–46 (2006)

22. Gaxiola-Pacheco, C.: Fuzzy Personality Model based on Transactional Analysis and VSM for Socially Intelligent Agents and Robots, Universidad Autónoma de Baja California, Facultad de Ciencias Químicas e Ingeniería: Tijuana (2007)

23. Rodriguez-Diaz, A.: Personality and Behaviour Modelling Based on Cathexis Flux. In: AISB, Joint Symposium on Virtual Social Agents Social Presence Cues for Virtual Humanoids Empathic Interaction with Synthetic Characters Mind Minding Agents. 2005, The Society for the Study of Artificial Intelligence and the Simulation of Behaviour. University of Hertfordshire, Hatfield, UK. pp. 130-136 (2005)

24. Singh, P.: An Architecture for Commonsense Thinking (2003)

25. Singh, P., et al.: Open Mind Common Sense: Knowledge Acquisition from the General Public. In: Conference on Cooperative Information Systems (2002)

26. Jung, C.G.: Psychological Types. In: Collected Works of C.G. Princeton University Press, Jung (1971)

27. Beer, S.: Diagnosing the System for Organizations. John Wiley, London (1985)

Modeling and Simulation by Petri Networks of a Fault Tolerant Agent Node

Arnulfo Alanis Garza, Oscar Castillo, and José Mario García Valdez

Division of Graduate Studies and Research, Calzada Tecnologico,
S/N, Tijuana, Mexico
{alanis,ocastillo,mario}@tectijuana.mx

Abstract. Intelligent Agents have originated a lot discussion about what they are, and how they are a different from general programs. We describe in this paper a new paradigm for intelligent agents, This paradigm helped us deal with failures in an independent and efficient way. We proposed these types of agents to treat the system in a hierarchical way. The Agent Node is also described. A new method to visualize fault tolerant system (FTS) is proposed in this paper with the incorporation of intelligent agents, which as they grow and specialize create the Multi Agent System (MAS).The communications diagrams of the each of the agents is described in diagrams of transaction of states.

1 Introduction

At the moment, the approach of using agents for real applications, has worked with mobile agents, which work at the level of the client-server architecture. However, in systems where the requirements are higher, as in the field of the architecture of embedded industrial systems, the idea is to innovate in this area by working with the paradigm of intelligent agents. Also, it is a good idea in embedded fault tolerant systems, where it is a new and good strategy for the detection and solution of errors.

A rational agent is one that does the right thing. Obviously, this is better than doing the wrong thing, but what does it mean? As a first approximation, we will say that the right action is the one that will cause the agent to be most successful. That leaves us with the problem of deciding how and when to evaluate the agent's success

By an agent-based system, we mean one in which the key abstraction used is that of an agent. In principle, an agent-based system might be conceptualized in terms of agents, but implemented without any software structures corresponding to agents at all. We can again draw a parallel with object-oriented software, where it is entirely possible to design a system in terms of objects, but to implement it without the use of an object-oriented software environment. But this would at best be unusual, and at worst, counterproductive. A similar situation exists with agent technology; we therefore expect an agent-based system to be both designed and implemented in terms of agents. A number of software tools exist that allow a user to implement software systems as agents, and as societies of cooperating agents[2].

2 Agents

For some, the term "agent" means only "*autonomous, intelligent*" agents. An example of this type of thinking can be found in Franklin and Graesser's paper "Is it an Agent,

O. Castillo et al. (Eds.): Soft Computing for Hybrid Intel. Systems, SCI 154, pp. 251–267, 2008.
springerlink.com

or just a Program?: A Taxonomy for Autonomous Agents". Another example of this view is Lenny Foner's excellent article "Agents and Appropriation". (Then there is the other side of the coin: Sverker Janson's list of "Intelligent Software Agents" includes anything called an "agent".)

The Franklin and Graesser paper is a good paper because it 1) surveys various agents, 2) presents a reasoned taxonomy based on features, and 3) avoids assigning any meaning to the word "intelligent". However, it proposes a " formal mathematically " definition: *"An autonomous agent is a system situated within and a part of an environment that senses that environment and acts on it, over time, in pursuit of its own agenda and so as to effect what it senses in the future."* This is, of course, not a definition any mathematician would recognize as being formal. The idea of "senses in the future" is just too open for interpretation to be an objective, much less formal, definition. Moreover, it equates being an agent with this quality of "autonomous" [Only autonomous agents were defined - other kind of agents may exist. - Private Communication from Stan Franklin, June, 1996.]

For Foner, an agent is necessarily "intelligent" and "autonomy" is just one crucial characteristic. His definition of autonomy has a bit more of operational semantics: *"This requires aspects of periodic action, spontaneous execution, and initiative, in that the agent must be able to take preemptive or independent actions that will eventually benefit the user."*

There are three major problems with the attempts to define "agents" as "intelligent". First, as alluded above, the meaning of the adjectives "intelligent" and "autonomous", so far, are subjective labels. Foner´s definition suggests that there might be a test for autonomy, but saying that some action is "preemptive" or "independent" does not get us far. This definition of intelligence, as do all, depends upon the opinion of an intelligent observer after interacting with the candidate agent.

Furthermore, the example agent, Julia, does not exhibit much initiative. The fact that Julia maps a maze without direction from users with whom she interacts does not distinguish Julia from almost any other software that performs a background task while answering queries from users and performing other tasks when directed, such as message forwarding. In fact, Julia never interrupts to volunteer information except to deliver a message as directed: she speaks only when spoken to. Julia's claim of intelligence is much more of the Eliza sort: Julia strikes users as a person. And indeed, the implementation and documentation suggests that Julia is intended to pass a Turing test just above the level of Eliza.

Second, these subjective labels are applicable only to an Epiphenomenon [23] rather than a design objective. Except to pass a Turing test, no one sets out to build an "intelligent agent" as that is a poor target for software. One sets out to build an agent that accomplishes a task in hopes that the task is so difficult or it is so well-accomplished that the agent might be considered intelligent or somehow self-directed. This begs the question of why the agent is one, and not some other kind of software [3],[18].

Third, various definitions of intelligence exist, but the main deficiency of such a label is that it does not sufficiently distinguish the resulting software from other technologies that may also claim intelligence as an attribute. One can take any definition of intelligent software that covers the work in Artificial Intelligence and find that it does not serve to distinguish "agents" as a kind of software. The point is that if it is claimed that to be an agent is to be intelligent, then we have still begged the question of what is an "agent" apart from all of the other intelligent software that has been developed [1],[4],[5].

3 Petri Nets

A Petri net is a graphical and mathematical modeling tool. It consists of places, transitions, and arcs that connect them. Input arcs connect places with transitions, while output arcs start at a transition and end at a place. There are other types of arcs, e.g. inhibitor arcs. Places can contain tokens; the current state of the modeled system (the marking) is given by the number (and type if the tokens are distinguishable) of tokens in each place. Transitions are active components. They model activities which can occur (the transition fires), thus changing the state of the system (the marking of the Petri net). Transitions are only allowed to fire if they are enabled, which means that all the preconditions for the activity must be fulfilled (there are enough tokens available in the input places). When the transition fires, it removes tokens from its input places and adds some at all of its output places. The number of tokens removed / added depends on the cardinality of each arc. The interactive firing of transitions in subsequent markings is called token game.

Petri nets are a promising tool for describing and studying systems that are characterized as being concurrent, asynchronous, distributed, parallel, non-deterministic, and/or stochastic. As a graphical tool, Petri nets can be used as a visual-communication aid similar to flow charts, block diagrams, and networks. In addition, tokens are used in these nets to simulate the dynamic and concurrent activities of systems. As a mathematical tool, it is possible to set up state equations, algebraic equations, and other mathematical models governing the behavior of systems.

To study the performance and dependability issues of systems it is necessary to include a timing concept into the model. There are several possibilities to do this for a Petri net; however, the most common way is to associate a firing delay with each transition. This delay specifies the time that the transition has to be enabled, before it can actually fire. If the delay is a random distribution function, the resulting net class is called stochastic Petri net. Different types of transitions can be distinguished depending on their associated delay, for instance immediate transitions (no delay), exponential transitions (delay is an exponential distribution), and deterministic transitions (delay is fixed) [15].

4 FIPA (The Foundation of Intelligence Physical Agents)

FIPA is specifications represent a collection of standards, which are intended to promote the interoperation of heterogeneous agents and the services that they can represent.

The life cycle [6] of specifications details what stages a specification can attain while it is part of the FIPA standards process. Each specification is assigned a specification identifier [7] as it enters the FIPA specification life cycle. The specifications themselves can be found in the Repository [8].

The Foundation of Intelligent Physical Agents (FIPA) is now an official IEEE Standards Committee.

4.1 FIPA Is Specifications Life Cycle

FIPA's specifications are classified according to their position in the specification life cycle. The intent of the specification life cycle is to chart the progress of a given specification from its inception through its ultimate resolution, figure 1.

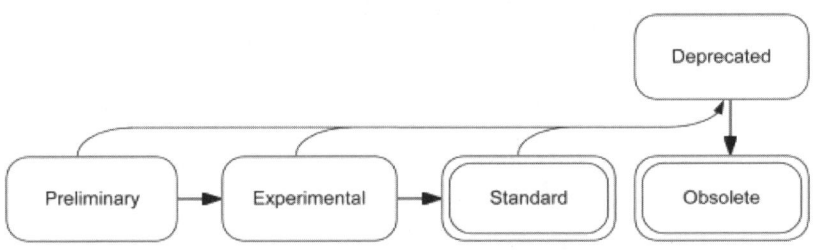

Fig. 1. FIPA's specifications life cycle

4.2 FIPA ACL Message

Over time, failure has come to be defined in terms of specified service delivered by a system. This avoids circular definitions involving essentially synonymous terms such as defect, etc. This distinction appears to have been first proposed by Melliar-Smith [17]. A system is said to have a failure if the service it delivers to the user deviates from compliance with the system specification for a specified period of time. While it may be difficult to arrive at an unambiguous specification of the service to be delivered by any system, the concept of an agreed-to specification is the most reasonable of the options for defining satisfactory service and the absence of satisfactory service, failure.

The definition of failure as the deviation of the service delivered by a system from the system specification essentially eliminates "specification" faults or errors. While this approach may appear to be avoiding the problem by defining it away, it is important to have some reference for the definition of failure, and the specification is a logical choice. The specification can be considered as a boundary to the system's region of concern, discussed later. It is important to recognize that every system has an explicit specification, which is written, and an implicit specification that the system should at least behave as well as a reasonable person could expect based on experience with similar systems and with the world in general. Clearly, it is important to make as much of the specification as explicit as possible.

It has become the practice to define faults in terms of failure(s). The concept closest to the common understanding of the word fault is one that defines a fault as the adjudged cause of a failure. This fits with a common application of the verb form of the word fault, which involves determining cause or affixing blame. However, this requires an understanding of how failures are caused. An alternate view of faults is to consider them failures in other systems that interact with the system under consideration—either a subsystem internal to the system under consideration, a component of the system under consideration, or an external system that interacts with the system under consideration (the environment). In the first instance, the link between faults and failures is cause; in the second case it is level of abstraction or location.

The advantages of defining faults as failures of component/interacting systems are: (1) one can consider faults without the need to establish a direct connection with a failure, so we can discuss faults that do not cause failures, i.e., the system is naturally fault tolerant, (2) the definition of a fault is the same as the definition of a failure with only the boundary of the relevant system or subsystem being different. This means that we can consider an obvious internal defect to be a fault without having to establish a causal relationship between the defect and a failure at the system boundary.

In light of the proceeding discussion, a fault will be defined as the failure of (1) a component of the system, (2) a subsystem of the system, or (3) another system which has interacted or is interacting with the considered system. Every fault is a failure from some point of view. A fault can lead to other faults, or to a failure, or neither.

A system with faults may continue to provide its service, that is, not fail. Such a system is said to be fault tolerant. Thus, an important motivation for differentiating between faults and failures is the need to describe the fault tolerance of a system. An observer inspecting the internals of the system would say that the faulty component had failed, because the observer's viewpoint is now at a lower level of detail [16].

The following terms are used to define the ontology and the abstract syntax of the FIPA ACL message structure:

Frame. This is the mandatory name of this entity that must be used to represent each instance of this class.

Ontology. This is the name of the ontology, whose domain of discourse includes their parameters described in the table.

Parameter. This identifies each component within the frame. The type of the parameter is defined relative to a particular encoding. Encoding specifications for ACL messages are given in their respective specifications.

Description. This is a natural language description of the semantics of each parameter. Notes are included to clarify typical usage.

Reserved Values. This is a list of FIPA-defined constants associated with each parameter. This list is typically defined in the specification referenced.

All of the FIPA is message parameters share the frame and ontology shown in *Table 1*.

Table 1. FIPA ACL Message Frame and Ontology

Frame	fipa-acl-message
Ontology	fipa-acl

5   The KQML Language

Communication takes place on several levels. The content of the message is only a part of the communication. Being able to locate and engage the attention of someone you want to communicate with is a part of the process. Packaging your message in a way which makes your purpose in communicating clear is another.

When using KQML, a software agent transmits content messages, composed in a language of its own choice, wrapped inside of a KQML message. The content message can be expressed in any representation language and written in either ASCII strings or one of many binary notations (e.g. network independent XDR representations).All KQML implementations ignore the content portion of the message except to the extent that they need to recognize where it begin sand ends.

The syntax of KQML is based on a balanced parenthesis list. The initial element of the list is the performative and the remaining elements are the performative's arguments as keyword/value pairs. Because the language is relatively simple, the actual syntax is not significant and can be changed if necessary in the future. The syntax reveals the roots of the initial implementations, which were done in Common Lisp, but has turned out to be quite flexible.

KQML is expected to be supported by a software substrate which makes it possible for agents to locate one another in a distributed environment. Most current implementations come with custom environments of this type; these are commonly based on helper programs called routers or facilitators. These environments are not a specified part of KQML. They are not standardized and most of the current KQML environments will evolve to use some of the emerging commercial frameworks, such as OMG's CORBA or Microsoft's OLE2, as they become more widely used.

The KQML language supports these implementations by allowing the KQML messages to carry information which is useful to them, such as the names and addresses of the sending and receiving agents, a unique message identifier, and notations by any intervening agents. There are also optional features of the KQML language which contain descriptions of the content: its language, the ontology it assumes, and some type of more general description, such as a descriptor naming a topic within the ontology. These optional features make it possible for the supporting environments to analyze, route and deliver messages based on their content, even though the content itself is inaccessible

KQML and Intelligent Information Integration We could address many of the difficulties of communication between intelligent agents described in the Introduction by giving them a common language. In linguistic terms, this means that they would share a common syntax, semantics and pragmatics.

Getting information processes, especially AI processes, to share a common syntax is a major problem. There is no universally accepted language in which to represent information and queries. Languages such as KIF [10], extended SQL, and LOOM [17] have their supporters, but there is also a strong position that it is too early to standardize on any representation language [14]. As a result, it is currently necessary to say that two agents can communicate with each other if they have a common representation language or use languages that are inter-translatable.

Assuming a common or translatable language, it is still necessary for communicating agents to share a framework of knowledge in order to interpret message they exchange. This is not really a shared semantics, but a shared ontology. There is not likely to be one shared ontology, but many. Shared ontologies are under development in many important application domains such as planning and scheduling, biology and medicine. Pragmatics among computer processes includes 1) knowing who to talk with and how to find them and 2) knowing how to initiate and maintain an exchange. KQML is concerned primarily with pragmatics (and secondarily with semantics). It is a language and a set of protocols that support computer programs in identifying, connecting with and exchanging information with other programs [12].

5.1 Typed-Message Agents

Apart from intelligent/autonomous agents, servers, and mobile agents, a fourth common type of agent is the type using a KQML or a similar agent protocol such as the one being developed for SRI's Open Agent Architecture. These KQML agents may also be considered intelligent though not often mobile. An excellent description of this kind of software agent is Michael Genesereth's and Steven Ketchpel's paper "Software Agents". This paper takes more of a systems engineering approach to the definition of agents, which has the advantage that it more objectively distinguishes agents from other types of software. In this paper, software agents communicate using a shared outer language, inner (content) language, and ontology. This approach is also called the "weak" notion of agenthood by Wooldridge and Jennings. Within the engineering community, this view is especially appropriate to the use of agents as an integration technology, as in the paper by Cranefield and Purvis and the Stanford Agent-Based Engineering work.

We follow Genesereth's approach, but differ somewhat from the definition of this paper in light of our experience with Next-Link agents and comparison with other KQML-like agents, our *Typed-Message Agents* are defined in terms of communities of agents. (We may also call these "ACL Agents" after Genesereth.) The community must *exchange messages in order to accomplish a task*. They must use a shared message protocol, such as KQML, in which the some of the *message semantics are typed and independent of the application*. And semantics of the message protocol necessitate that the transport protocol not be only client/server but rather a peer-to-peer protocol. An individual software module is not an agent at all if it can communicate with the other candidate agents *with only a client/server protocol without degradation* of the collective task performance [19].

6   Agent Communication Protocols

There are a variety of interprocess information exchange protocols. In the simplest case, one agent acts as a client and sends a query to another agent acting as a server and then waits for a reply, as is shown between agents A and B in Figure 1. The server's reply might consist of a single answer or a collection or set of answers. In another common case, shown between agents A and C, the server's reply is not the complete answer but a handle which allows the client to ask for the components of the reply, one at a time. A common example of this exchange occurs when a client queries a relational database or a reasoner which produces a sequence of instantiations in

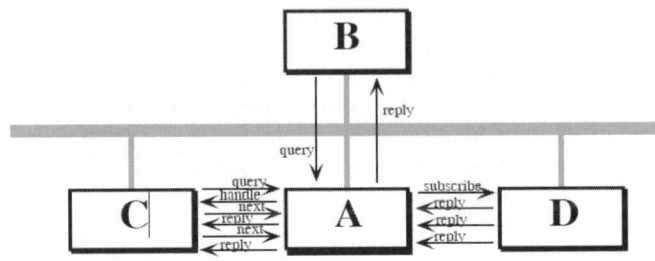

Fig. 2. Several basic communication protocols are support in KQLM

response. Although this exchange requires that the server maintain some internal state, the individual transactions are as before - involving a synchronous communication between the agents. A somewhat different case occurs when the client subscribes to a server's output and an indefinite number of asynchronous replies arrive at irregular intervals, as between agents A and D in Figure 2. The client does not know when each reply message will be arriving and may be busy performing some other task when they do.

There are other variations of these protocols. Messages might not be addressed to specific hosts, but broadcast to a number of them. The replies, arriving synchronously or asynchronously have to be collated and, optionally, associated with the query that they are replying to [13], [9].

7 Proposed Method

Lets suppose that we have a Distributed System (mainly applied to the industrial control), which is made up of a set of Nodes, where each one of them can be constituted by several Devices [21].

On these Nodes a set of ordered Tasks, is executed all of them to have the functionality of the system. In order to identify this Distributed System the following definitions are set out:

Definition 1: is N ={N}, the set of the Nodes of the system, being "n" is the number of units that integrate it.

Definition 2: is [Di, z], the set of devices that contains Node i. Where "z" can take value 1, if it is wanted to see the Node like only device, or greater than 1 if it is desired to be visible to some of the elements that integrate it.

Definition 3: is T = {Tj}, the set of tasks that are executed in the system, being "t" the number of tasks that integrate the system.

Definition 4: A System Distributed like tuple is defined: SD = (N, T) Once characterized what a Distributed System could be denominated Basic (without no characteristic of Tolerance to Failures), one is going away to come to the incorporation on itself from the paradigm of Intelligent Agents with the purpose of equipping it with a layer with Tolerance to Failures. The Fault tolerant Agents will define themselves now that worked in the SD.

Definition 5: An Agent is ANi when is itself a Node, whose mission is the related one to the tolerance to failures at level of the Node N..

Definition 6: {ANi is AN=} a set of Node Agents.

Definition 7: An application of N to AN of form of a is α each node N, of the system associates an Agent to him ANi Node, that is to say: α: N→ANi N → ANi

Definition 8: A System is SMATF Fault tolerant Multi-Agent, formed by tripla of AN, AT, AS. That is to say With it a Distributed System Fault tolerant SDTF is defined as: SMATF = <N, AT, AS>

Definition 9: A Distributed System Fault tolerant SDTF like tuple is defined SDTF =<SD, SMATF>, [22].

7.1 Agent Task ATj

Once having defined the Node Agent, the Agent Task be defined that will be responsible for taking all actions aimed at ensuring that the tasks in the system have an error-free behavior.

In defining the work and behavior of Agent Task, it is necessary to contemplate their allocation which is conditional on those nodes that can be successful.

There will be a copy of this agent in each node of the SMTF where they can execute the task and will have the characteristics of a Mobile Agent, thereby, be able to work throughout the system.

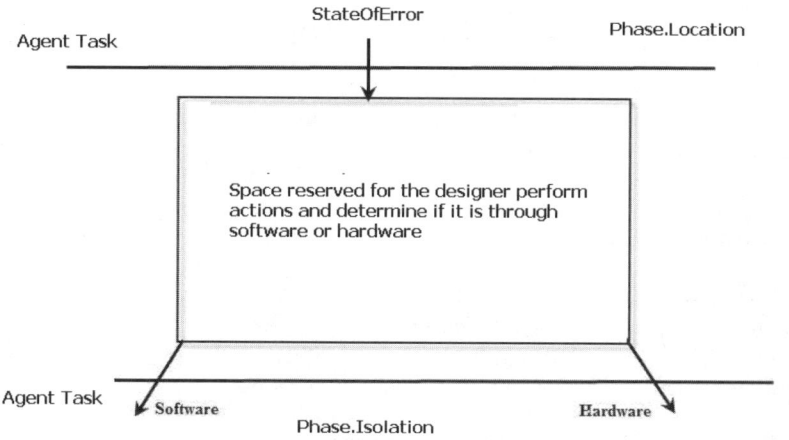

Fig. 3. Location mistake by either hardware or software specified by the designer

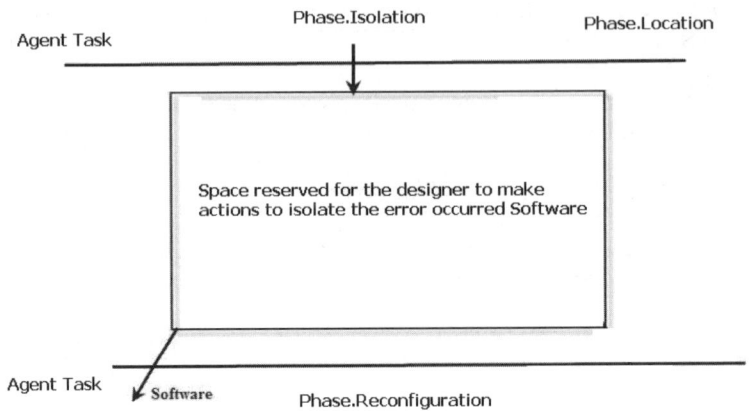

Fig. 4. Isolation of the error by Software determined by the designer

Internally each Agent Task provides a number of variables, defined as standard, which govern its operation:

Definition 10: Be ATj.Phase, is a variable which will be five events: Detection, Location, Privacy, Reconfiguration and Recovery.
Where:

ATj. Phase.Detection This is the phase of troubleshooting if the Agent task is not yet with all entries in error, then operation of the task is correct. This is the phase which initially placed all tasks.

ATj. Phase.Location, this stage is entered after the detection of an error and it intends to locate the error.

Note 1: The designer undertake the actions necessary to determine the error. And determine which is correct one, fig.3.

ATj. Phase.Isolation, after entering this phase, once the localization phase is done, and will try to isolate the task that the previous phase mark as potentially as wrong.

Note 2: If you are at this stage and there is a software error, the designers carry out the necessary actions and determine what is indicated, see fig 4.

ATj. Phase.Reconfigured, at this stage all actions necessary are performed for the reconfigured of the task, this is shown in fig 5.

ATj. Phase.Recovery, in this last phase the actions needed to proceed with the recovery of the task were carried out, this is shown in fig 6.

Note 3: If you are at this stage and an error occurs, the designer can perform the necessary actions and determine what is indicated, and once that determination is done, the result will be obtained.

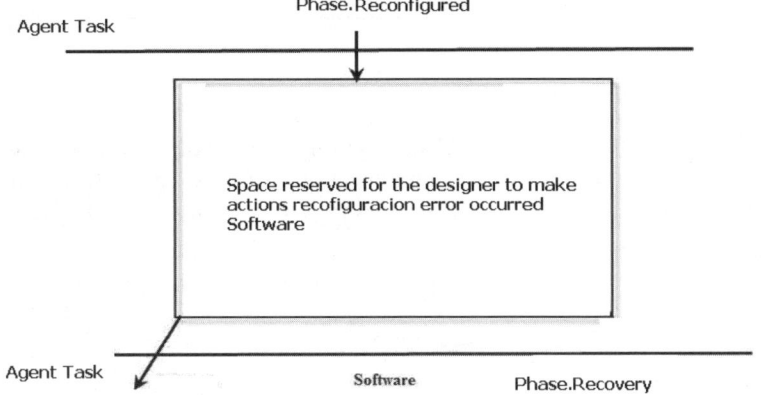

Fig. 5. Reconfiguration of the error by Software determined by the designer

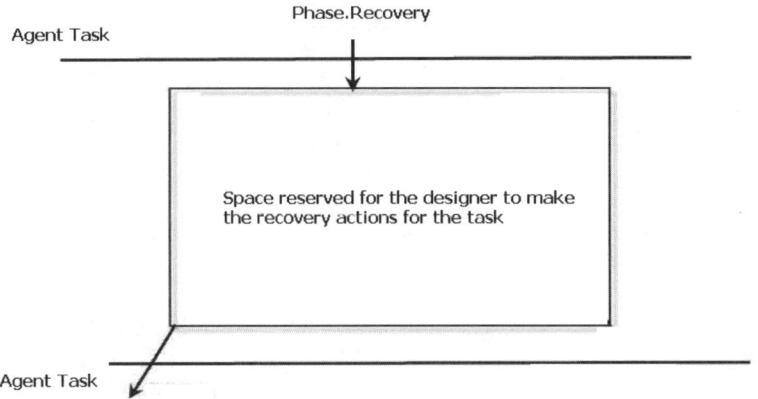

Fig. 6. Recovery actions by the task of the designer

All these mechanisms for detecting errors generate information about the status of the Task that is collected directly by the ATj, at the stage of detection.

The tasks covered in the SMATF as follows:

Once an error occurs will determine whether this is a hardware or software. This can be determined, a) to continue running normally, b) the task to recover c) that Tj is abortion, and therefore will be removed from the system or c) that Tj be stopped because there was a mistake, so that when an error detection mechanism (either hardware or software) Tj of the show, finds an error, and this can come to know where and what to do with the tasks assigned to a node in particularly, in SMAFT.

Having said all this, know we define the AS (Agent System), which contain all the information system and both agents.

7.2 Petri Nets of the Agent Task

The Petri network of the Agent Task is shown in fig 7. The output communication variables are of red color, and the input communication variables are of green color. These interrelate with the agent to whom it must send or receive information to interact.

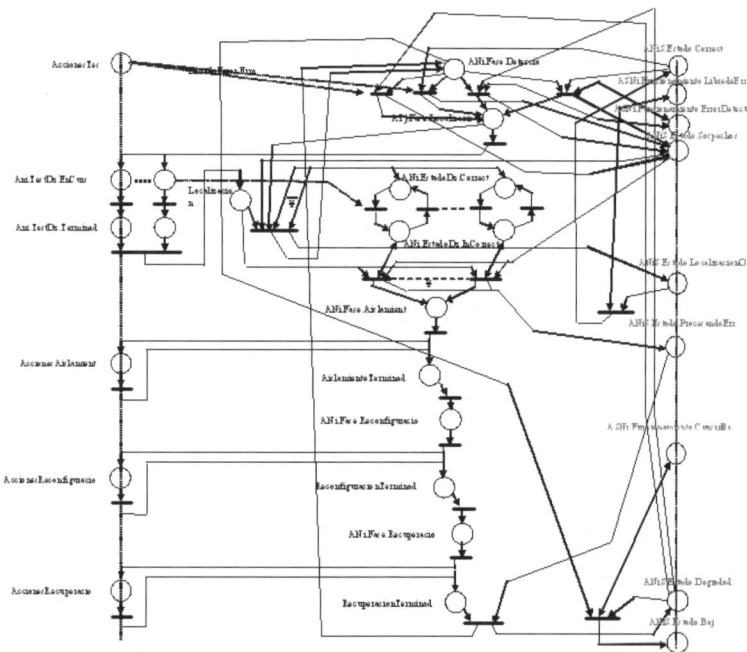

Fig. 7. Petri Network of the Fault Tolerant Agent Node

8 State Transition Diagrams

State transition diagrams were around long before object modeling. They give an explicit, even a formal definition of behavior. A big disadvantage for them is that they mean that you have to define all the possible states of a system. Whilst this is all right for small systems, it soon breaks down in larger systems as there is an exponential growth in the number of states. This state explosion problem leads to state transition diagrams becoming far too complex for much practical use. To combat this state explosion problem, object-oriented methods define separate state-transition diagrams for each class. This pretty much eliminates the explosion problem since each class is simple enough to have a comprehensible state transition diagram. (It does, however,

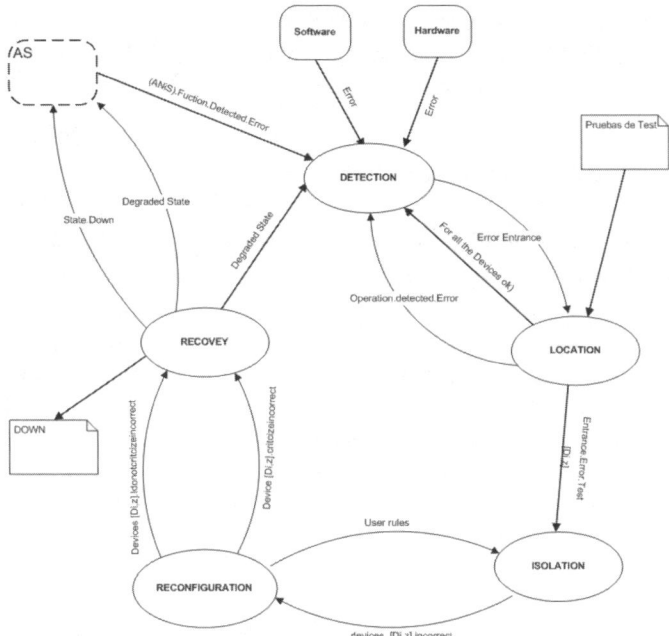

Fig. 8. Several basic communication protocols are supported in KQML

raise a problem in that it is difficult to visualize the behavior of the whole system from a number of diagrams of individual classes - which leads people to interaction and activity modeling) [22].

State transition diagrams have been used right from the beginning in object-oriented modeling. The basic idea is to define a machine that has a number of states (hence the term finite state machine). The machine receives events from the outside world, and each event can cause the machine to transition from one state to another. For an example, take a look at figure 8. Here the machine is a bottle in a bottling plant. It begins in the empty state. In that state it can receive squirt events. If the squirt event causes the bottle to become full, then it transitions to the full state, otherwise it stays in the empty state (indicated by the transition back to its own state). When in the full state the cap event will cause it to transition to the sealed state. The diagram indicates that a full bottle does not receive squirt events, and that an empty bottle does not receive cap events. Thus you can get a good sense of what events should occur, and what effect they can have on the object.

9   Communications Diagram

Now we show the diagram of transition of states of the Agent Task. In figure 8 we show the diagram of transition of states, one is, Agent Task (AT), its operation and interchange of messages, as well as the variables that take part in the passage of their internal communication, in addition to connect with one of the agents of the SMA, the agent system (AS), which as well in its internal states of communication and also, handle its communication with the other agent of the SMATF, the Agent Task [22].

10   Communications Diagram

In SMATF it offers an option of growth system level but single in phase in real time because, and although a state of optimal work can be supposed, could be presented/displayed disadvantages in the execution of a process or task, FIPA does not offer a recovery to short time which could mean lost in time, money or solutions, in SMATF it offers a reaction at the moment to come up lost and to look for the way to solve a problem in the execution of a process. Fig 9 shows is communications diagram based on FIPA-ACL, and table 2 shows the definition in the communication model [11], [22].

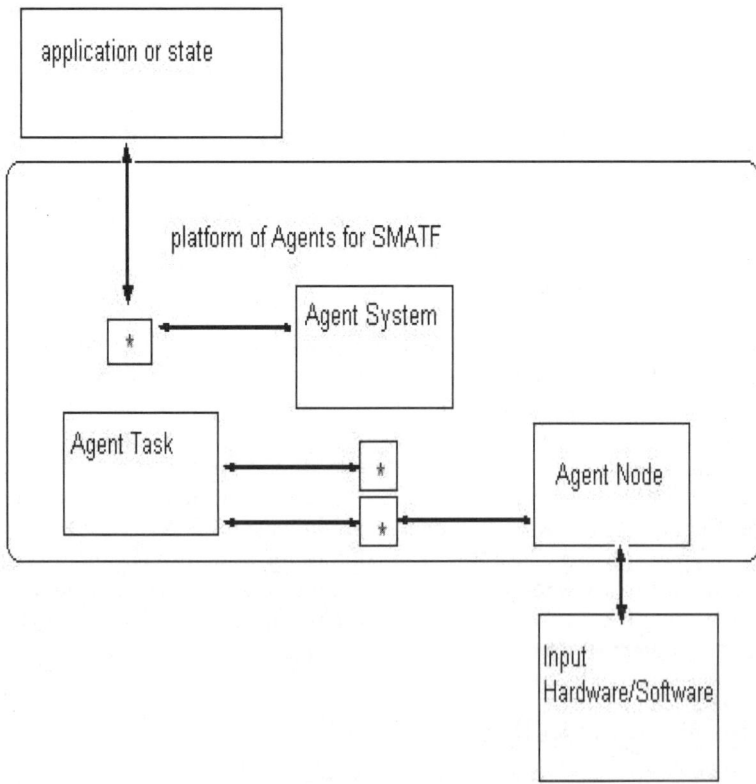

Fig. 9. Communications diagram based on FIPA-ACL in SMATF

Table 2. Definitions in the communication SMATF model

Components	Function
Agent System	the function of this agent is in the components at level of administration of agents has to the Agent System whose mission is the related one to the tolerance to failures system level (what tasks must be executed in the system and on what nodes).
Agent Task	the Agent Task verifies by means of a request (request) at task level if it does not seem that this integral one sends it to recovery.
Agent Node	Receives the message by means of entrances and reviews the integrity of the message at node level if it does not seem that this integral one sends it to
recovery Communication channel or Conditions	Is the three agents (System, Task, Node) since they have a shared channel of communication and this becomes by means of conditions (they if-then).
Phases of Recovery (*)	This has the function to correct the messages that can come in "a suspicious" state only has the task of fixing the message and this consists in: Detection, Location, Isolation, Reconfiguration and Recovery

11 Conclusions

We described in this paper our approach for building multi-agents system for achieving fault tolerant control system in industry. The use of the paradigm of intelligent agents has enabled the profile generation of each of the possible failures in an embedded industrial system. In our approach, each of the intelligent agents is able to deal with a failure and stabilize. It is observed the models and forms to make the communication between the agents' efficient using tools of efficient handling. The system in an independent way, and that the system has a behavior that is transparent for the use application as well as for the user.

References

[1] Griswold, S.: Unleashing Agents The first wave of products incorporation software agent technology has hit the market. See what's afoot,Internet World 7, 5 (1996)

[2] Bratko, I.: Prolog for Programming Artificial Intelligence. Addison-Wesley, Reading (1986)

[3] Or Etzioni, N.: Lesh, and R. Segal Building for Softbots UNIX (preliminary report). Tech. Report 93-09-01. Univ. of Washington, Seattle (1993)

[4] Rich, E., Knight, K.: Artificial intelligence, 2nd edn, pp. 476–478. McGraw-Hill, New York

[5] Durfee, E.H., Lesser, V.R., Corkill, D.D.: Trends in cooperative distributed problem solving. IEEE Transactions on Knowledge and Data Engineering KDE-1, 1, 63–83 (1989)

[6] http://www.fipa.org/specifications/lifecycle.html

[7] http://www.fipa.org/specifications/identifiers.html

[8] http://www.fipa.org/specifications/index.html

[9] Weizenbaum, J.: ELIZA – a computer program for the study of natural language communication between man and machine. Communications of the Association for Computing Machinery 9, 36–45 (1965)

[10] FIPA Abstract Architecture Specification. Foundation for Intelligent Physical Agents (2000), http://www.fipa.org/specs/fipa00001/

[11] FIPA Interaction Protocol Library Specification. Foundation for Intelligent Physical Agents (2000), http://www.fipa.org/specs/fipa00025/

[12] External Interfaces Working Group ARPA Knowledge Sharing Initiative. Specification of the KQML agent-communication language Working, http://www.cs.umbc.edu/kqml/.

[13] Y. Labrou, T. Finin.: A semantics approach for KQML a general purpose communication language for software agents. In: Third International Conference on Information and Knowledge Management, (November 1994)

[14] Finin, T., McKay, D., Fritzson, R., McEntire, R.: KQML: an information and knowledge exchange protocol. In: International Conference on Building and Sharing of Very Large-Scale Knowledge Bases. Knowledge Building and Knowledge Sharing, Ohmsha and IOS Press, Amsterdam (1994)

[15] Desel, J., Esparza, J.: Free Choice Petri Nets. Cambridge University Press, Cambridge (1995)

[16] http://hissa.nist.gov/chissa/SEI_Framework/framework_1.html

[17] Dasgupta, P., Narasimhan, N., Moser, L.E., Melliar-Smith, P.M.: MAgNET: Mobile Agents for Networked Electronic Trading. IEEE Transactions on Knowledge and Data Engineering 11(4), 509–525 (1999)

[18] Jennings, N., Wooldridge, M.: Intelligent agents: Theory and practice. The Knowledge Engineering Review 10, 2 (1995)

[19] Reddy, R.: To Dream the Possible Dream. 1996 Turing Award Lecture in Communications of the ACM 39, 5 (1996)

[20] Booch, G.: Object-Oriented Analysis and Design with Applications, 2nd edn. Addison-Wesley, Menlo Park (1993)

[21] Garza, A.A., Serrano, J.J., Carot, R.O., García Valdez, J.M.: Fault Tolerant Multi-Agent Systems: its communication and co-operation, Engineering Letters, 15:1, EL_15_1_17 (2007)
[22] Garza, A., Serrano, J.J., Carot, R.O., Val-dez, J.M.G.: Modeling and Simulation by Petri Networks of a Fault Tolerant Agnet Node. Analysis and Design of Intelligent Systems Using Soft Computing Techinques, p. 707. Springer, Heidelberg (2007)
[23] Definition of Epiphenomenon,
     http://en.wikipedia.org/wiki/Epiphenomenon

Fuzzy Agents

Eugenio Dante Suarez[1], Antonio Rodríguez-Díaz[2], and Manuel Castañón-Puga[2]

[1] Trinity University, Department of Business Administration, One Trinity Place,
San Antonio, TX, USA, 78212
esuarez@trinity.edu
[2] Baja California Autonomous University, Chemistry and Engineering Faculty,
Calzada Tecnológico 14418, Mesa de Otay, Tijuana, Baja California, México, 22390
{ardiaz,puga}@uabc.mx

Abstract. The common historically reductionist past of evolutionary biology and traditional social sciences such as economics has led way to a nascent holistic perspective of nonlinear science that is capable of describing multiple levels of reality. We propose a novel language for describing human behavior and social phenomena, set within a general theory of collective behavior and structure formation, with a resulting architecture that can be broadly applied. This work represents the blue print for a Multi-Agent Systems (MAS) design language in which agency is granted in a quantitative, rather than the traditional qualitative way. The relevant agents in the proposed system are intermediate in the sense that they are both influenced by an upper level with its own degree of agency, while at the same time they are determined by relatively independent subcomponents that must be 'subdued' into acceptable behavior. Any observed action is considered to be the result of the interplay of multiple distinguishable actors. We put forward this language as a basis for the construction of large-scale simulations and for the description of complex social phenomena.

Keywords: Multi-Agent Systems, Social Simulation, Distributed Agency, Levels of Reality.

1 Introduction

Multi-Agent Systems (MAS) have been increasingly used to model a variety of social phenomena. From ants (Wagner and Bruckstein 2001) to countries forming coalitions in preparation to a cold war (Axelrod 2005), developments in MAS are increasingly allowing breakthroughs in the understanding of systems with multiple interacting agents. The analysis of such systems was previously beyond the reach of science, as the intrinsic nonlinear nature of the phenomena did not permit appropriate analysis by conventional means.

Systems in which multiple, heterogeneous, interacting entities adaptively influence the behavior of each other quickly became intractable for traditional mathematical techniques. To continue the advancement of science and technology, most disciplines generally opted for the simplification of interacting agents, commonly describing them as homogeneous, independent and altogether exogenous. As such, the paradigm of Newtonian physics—where atoms interact in predictable patterns to compose aggregate objects—was followed by the classical economic paradigm of Adam Smith (Smith 1904), in which selfish agents acting independently are guided by an invisible

O. Castillo et al. (Eds.): Soft Computing for Hybrid Intel. Systems, SCI 154, pp. 269–293, 2008.
springerlink.com   © Springer-Verlag Berlin Heidelberg 2008

hand to maximize the greater good (Beinhocker 2006). Similarly, Darwin's (Darwin 1909) revolutionary theory of evolution describes independent agents that survive based upon their individually defined fitness. These three paradigms delineated the boundaries of most research done for more than two centuries in all physical, biological and social realms—even though they represent branches of the same tree; a tree rooted in a linear description of the world around us.

The advent of the computer experiment has turned this tree into a blossoming orchard, in the sense that it does not necessarily contradict the previous way of doing science but rather expands from its foundations. The previously pervasive common paradigm was adopted out of necessity, as any other set up is inaccessible without the aid of a computer. Computer simulations have opened the door to a whole new area of research, where the modeling possibilities are seemingly endless. What is now hindering new advancements in these growing interdisciplinary fields? Is it computing capacity? We believe that the pieces are in place for a novel understanding of the world. What the field needs is a fresh influx of social and evolutionary biology modelers that understand the newfound capabilities brought about by the MAS paradigm and can see their own disciplines through this new light. What this nascent wave of nonlinear modeling needs is a common language to describe social interaction, and it is to this task that we devote this paper. This work represents the blue print for a MAS design language in which agency is granted in a quantitative, rather than the traditional qualitative way.

To describe a more general theory of social interaction, we propose a model in which we drop customary assumptions made in traditional disciplines about what is considered a decision-making unit. The relevant agents in the proposed system are intermediate in the sense that they are both influenced by an upper level with its own degree of agency, while at the same time they are determined by relatively independent subcomponents that must be 'subdued' into acceptable behavior. Any observed action is considered to be the result of the interplay of multiple distinguishable actors. The proposed architecture may be used to build large-scale simulations, as well as models that focus on the interaction between levels of interest.

This language represents a decomposition of intent, based on the idea that an agent's behavior can be seen as an emergent property of a collection of intertwined aims and constraints. We consider a disentangled agent that is formed by multiple and relatively independent components. Part of the resulting agent's task is to present alternatives, or 'fields of action' to its components. Correspondingly, the composed agent is itself constrained by a field of action that the superstructure to which it belongs presents and can thus be ascribed agency and modeled as an agent. To arrive at this view, we redefine what a unit of decision is by unscrambling behavioral influences to the point of not being able to clearly delineate what the individual is, who is part of a group and who is not, or where a realm of influence ends; the boundary between an individual self and its social coordinates is dissolved. The proposed intermediate agent can be thought of as a person, a family, a social class, a political party, a country at war, a species as a whole, or a simple member of a species trying to survive. The archetype of the agents we attempt to describe can be summarized as a group of colluded oligopolists, such as the oil-producing countries of OPEC. As a whole, they share the common interest of jointly behaving like a monopoly and restricting their production, but they cannot avoid having an incentive to deviate and produce above their quota.

The proposed model redefines agents in two ways. First, there are no obvious atomic agents, for all actors represent the emerging force resulting from the organization of (perhaps competing) subsets. The subcomponents in turn form an internal system that is actively reorganized, and shall be referred to as the 'lower level' of a structure. On the other hand, agents are to be described within a group to which they belong, which will be defined as the 'upper level' of the hierarchical representation, and will constrain its subcomponents' behavior. Thus, agents considered in this framework are intermediate; only possessors of some degree of agency that is granted by the higher-level agent to which they belong, and extracted out of the lower-level agents that compose it.

As we discuss in the following sections, we are not advocating for the modeling of all levels of interaction at once; that would obviously be unfeasible as well as impractical. Instead, the modeler can focus at any level of reality of interest. What we are proposing is a modeling language in which the researcher can focus at one level of agency, but without a need for building the description based upon agents that are assumed to be atomic. Instead, atomic agents are only modeled as such because there is no perceived need to model their internal subcomponents, not because it is a defining characteristic. In other words, we propose a view of the world in which all levels display agents that are composed of *lower-level* components that are constrained by the realm of action the *upper-level* agent allows. The composing elements of any 'lower level' could always conceivably consider themselves as relatively independent and ascribed an objective function; in the other direction, the level's agents are also determined by the 'upper level' to which they belong, whether they recognize it or not. We do not assume agency at any of these levels, but we also do not assume that it does not exist.

Our framework stands in contrast to traditional approaches for understanding agency, such as MAS, evolutionary biology or economics. For example, standard economic representations have focused on instantaneous individuals or current profit-maximizing firms as the smallest unit or the ultimate irreducible atom of the paradigm, but we propose further attention to the possibility that such units may actually be agglomerates, and thus the products of internal networks that deserve attention. We can think of a company's organization or a strategic military coalition, and no agency is allowed for identifiable players forming the organizations and alliances, nor emphasis is given to the study of blurry abstract borders dividing the participating members. The possibility of understanding some of the internal complexities that give rise to their positions is essentially ignored. One of the main purposes of this research is then to break apart the threads that make up an agent, whether a person or company, both in an intertemporal and a static sense, in order to recognize the emergent properties that compose it.

Microeconomics has usually relied on simplistic definitions of what represents an indivisible actor in the two main levels of interaction: Individuals or households for consumption decisions, and firms for production ones. Generally speaking, firms have a straightforward behavioral directive: to maximize their profits. Correspondingly, individuals attempt to maximize their utility function. On the other hand, macroeconomic models usually rely on linear aggregation or recursive processes. Economics has thus most commonly modeled behavior through a closed objective function that the agent attempts to maximize, subject to the constraints imposed by the exogenous environment. As the research of Herbert Simon (Simon 1982; Simon 1996) points out, however, many economic structures do not organize themselves in a market scheme, but rather optimize their underlying internal relationships with intricate structures of command. Thus, the company is in a constant reorganization process to create a structure that maximizes profits (its objective function or an aggregate measure of 'fitness'), and then turns around to allocate the spoils among its participating members.

Following this introductory section, Section 2 of the paper describes the workings of a general model of disentangled agents, both those subagents of a lower level, as well as those meta-agents of upper levels. Section 3 proposes a new view of the world, as seen through this proposed paradigm, where agency cannot be pinpointed to any given agent, but is rather distributed across the system. Section 4 summarizes this understanding and further investigates an appropriate description of such fuzzy agents, of how they are created and how they interact. Section 5 provides examples of how this methodology would be used to model human behavior. The concluding section 6 proposes the methodology of distributed agency as an interdisciplinary language that can serve as a modeling taxonomy or as the building block for the construction of large-scale simulations.

2 Fuzzy Agents

Throughout this work, we describe different instances of abstract concepts of distributed agency, attractors and other aspects that influence behavior, and to draw analogies it is imperative that we use welcoming terminology. For example, the word to describe an agglomeration of agents is most commonly 'group', but that might bring to mind a set of humans, and the concept must at all times be kept at an abstract level. It should also not be 'set' because it may not define who are members from who are not: individuals or groups of individuals may be part or partially belong to one or many, in many different coordinates and according to drastically different definitions. The word 'network' could be a better candidate, as it brings to mind a system interconnected with varying intensities. All in all, the problem in finding the right terminology is analogous to the general quandaries facing fuzzy systems, where all is not on or off, black or white, but rather some tone of gray in between.

A useful word that comes to mind is 'bubble', since one can picture them to be joined at the border, one inside of the other or in the process of swallowing it, merging, splitting, or having an intersection. Throughout this work, we use several of these words to convey the sense of a fuzzy agent that is at the same time an agglomeration without clearly defined boundaries. Nonetheless, we will normally feel most comfortable using the word agent—with the understanding that it is a fuzzy agent—because most of the entities we are referring to are commonly thought of as qualifying under a vague definition of agent. Most importantly, what we need is a word that reminds the reader of the novel aspect of the term, and for this reason we officially refer to these entities as "agons," which comes from the same Greek etymological root as "agency." When the agons we are referring to represent subcomponents of our main agents of study, we join the concept with the Greek prefix for lower "kata," and refer to these as "katagons." Similarly, when we refer to upper-level or meta-agents, we use the Greek prefix for above "ana," and refer to these as "anagons."

The reason for using the neologism "agon" to describe these agent-like entities is that most other words might mislead the reader, by invoking preconceived ideas of what the concept stands for, what is being referred to, and what can or cannot qualify to fit the concept. Most commonly, the stereotypical agent people would conceive of is a human, and commonly ascribe human qualities to other entities, such as a greedy corporation, a welcoming culture, or a loving family. We will argue in this work that humans enjoy a limited amount of agency, both because we cannot be understood outside of a context and because we are not in complete control of all of the subsystems that compose us (Cohen 2005). Individuals have social, physical, emotional, and religious selves. They also play different roles depending on circumstances. A subject position incorporates both a conceptual repertoire and a location for persons within the structure of rights for those that use that repertoire (Davies and Harré 1990). Once having taken up a particular position as one's own, a person inevitably sees the world from the vantage point of that position and in terms of the particular images, metaphors, storylines and concepts which are made relevant within the particular discursive practice in which they are positioned.

Similarly, other entities such as an army, a corporation or a family, may be just as well modeled as an agent. For example, if we think of a couple as a higher level for two loving adults, the implication of the proposed paradigm is that the couple, as an entity *per se*, will have purposes which can be thought of as relatively detached from both participants forming the couple, for their individual utility functions are intertwined in the complexities of sharing, and can be returned unrecognizable from their origins. Moreover, if the couple is sanctified by a social, political, or religious institution such as marriage, then we can more clearly see that at some point both parties could want divorce, but it may be best for them to stay together because of social pressures, and it is in this sense that we may consider the couple as independent of its forming parts, and having what we could model as desires of its own.

Moreover, any of the proposed levels will have a historical context that will further restrain the most immediate composing parts. If we want to understand the evolution of organisms that may roam a hypothetical earth-like planet, we must conceive that institutions such as monogamy will be just as important as the people who inhabit them. The upper level may or may not have conscious aspects, as it could be created to fulfill a goal (as a police force), or simply emerge from the evolutionary benefits it offers (as the benefits of mutualism). Either way, we consider the composing agents as having an intrinsic nature that is constrained, transformed, and perhaps consciously channeled by the aggregate to which they belong.

Reductionist linear science has concentrated on the study of entities that are clearly delineated, where one could separate what belongs to an agent's nature against the backdrop of what does not. The relevant agent is taken to be exogenous, and therefore disconnected from the system to which it belongs. At their core, these traditional disciplines are based on a selfish and unitary agent, or atom of description. Implicitly or explicitly, these paradigms claim that all aggregate complexity can be traced back to the lower level of the system: the strategies and actions of the selfish agent. In other words, these represent research agendas that purposely de-emphasize the existence of any level other than that of the individual.

On the other hand, the idea of emergence reflects the fact that different and irreducible levels of interaction will naturally arise in complex systems such as the ones studied by social disciplines, and thus the agent, as we define it in this work, is a combination of levels of interaction. It is through this lens that we would like to consider humans, who will partly be independent creatures, possessors of free will, but who are also partly created by an array of upper levels that 'suggest' agreeable utility functions. This conception stems directly from the concept of complexity, in which wholes are more than the sum of their parts. If we believe in that proposition, then we should expect to find a world full of emergent phenomena, with distinctive levels of interaction that have agency of their own. The proposed language redefines agents in two ways. First, there are no obvious atomic agents, for all actors represent the emerging force resulting from the organization of relatively independent subsets. Second, agents are not created in a vacuum, but are rather the result of what an upper level spawns. To avoid confusion, and with fear of creating a neologism, we propose the term *agon* to represent these intermediate-level agent-like actors.

Figure 1, below defines the *agon* concept in more detail. At the extreme of the X-axis we find a liver, which represents a stereotypical object without relevant subcomponents, but also lacking independence. Even though it is composed of cells, we can think of these as perfectly coordinated under the objective function of the liver. Unless it is not functioning properly, the 'objective function' of the liver is in complete coordination with its upper level *anagon*, the body. As an upper level *anagon*, such as the body, coordinates its subcomponents, its lower level seemingly disappear. In such cases, the lower level components enjoy no agency, just as the soldiers of an army would if they were perfectly trained to follow all orders. On the other extreme, at the end of the Y-axis, we find a group full of unconstrained agents, such as the one conceived of in traditional economic science, where there is no earth to take care of, but only selfish individuals maximizing their own utility functions.

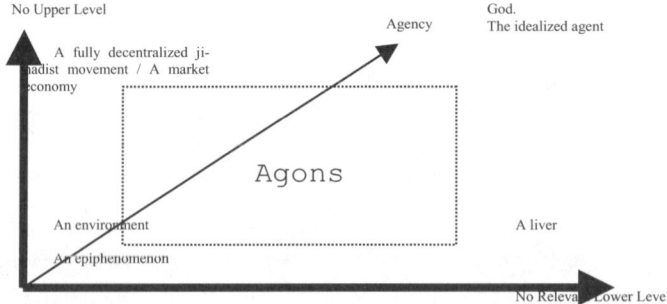

Fig. 1. The concept of a standard agent in MAS and traditional social sciences refers to an indivisible and independent entity. We propose a fuzzy agent, or agon, that is never fully in control of its subcomponents, nor ever fully independent from its context. The concept of agon refers to an entity that enjoys some degree of agency. Agency is thus granted in a quantitative, rather than the traditional qualitative way. As the agon becomes more independent and more internally cohesive, it becomes more of an agent.

In traditional economics, society does not exist, and so the *homo economicus*, or hypothetical consumer is defined in isolation. In contrast, *agons* are beset within the limits imposed by the upper level. For example, we can conceive of a company (formed by groups of investors, managers and employees) as set within the boundaries of an industry, unable to define itself in ways that go against established social norms and applicable laws. Upper levels can therefore represent coordination, identification with others, identities, institutions, implicit laws, religion, a credit bureau, or any human relationship. An environment, however, is not by itself an upper-level *agon*, for the environment may be independent of the creatures that inhabit it, and, unlike the traditional concept of an agent, it does not proactively react to changes in the system. The idealized agent appears in the upper right hand corner of the quadrant. God, as most commonly conceived, is omnipotent and therefore in complete control of all its subcomponents, while at the same time there is nothing above, controlling Him. Aside from these extreme cases, all other agent-like forces in the proposed language, or *agons*, enjoy only limited agency.

Observed behavior as described this language by definition optimizes a *behavioral function*, which is the result of the interplay of all *agons* maximizing their constrained objective functions. The proposed behavioral function tautologically describes how the system unfolds, as any *agon* makes decisions in a setting that implicitly considers the effects of its actions on future selves, offspring, family, other members of its species, and all sorts of lower- and upper-level layers of which it is composed and to which it belongs; in contrast to a more abstract utility function that pertains to its individual, most immediate self. Most importantly, the proposed language will allow for a parsimonious computational description of the world. The simplifications of the past were generally adopted for purposes of tractability, and understandably so, since models grow exponentially more difficult to understand when we consider the interactions of agents defined in different dimensions. With the advent of complexity theory, however, we can now imagine the possibility of tackling these problems directly, with the use of fuzzy MAS, neural networks, cellular automata, numerical approaches, and a whole lot of computer power.

3 A World Divided into Levels

The idea of "emergence" reflects a whole that is more than the sum of its parts(Abbot 2005). Significant research has been devoted to the study of this concept, without coming to a well-accepted consensus. A thorough discussion of the idea is beyond the scope of this paper, but as a working definition, it should suffice to say that the existence of emergent phenomena constrains us to the study of systems that are separated into distinct ontological levels. In other words, a reductionist approach is insufficient for understanding complex entities. Wholes cannot be explained solely by its individual components, but rather the interactions, topology and design of those components matter in the understanding of the whole. Thus, one must analyze an anthill as an organism (Johnson 1999) and a mind as more than a collection of neurons.

Complex social behavior must be studied taking into account the multiple ontological levels that affect it. In particular, an accurate description of human behavior must take into account both the lower-level emotional subcomponents as well as the upper-level context of social influences. Again, the proposition of this paper is that we must consider all traditionally defined agents and other agent-like agons as intermediate, in the sense that they are not independent of the context they inhabit and at the same time are not fully in control of their subcomponents, which may in turn enjoy some degree of agency of their own. Although the analysis could be kept at an abstract level, we mostly discuss the modeling of a human agent as a point of reference.

Humans are traditionally modeled as having a utility function, which is an abstract representation of what makes us tick; our happiness. The idea is that we can conceive of people as having such an objective function and living to maximize it, given the constraints that we face. Economists call for caution in the modeling of humans with this methodology, and stress, for example, that at all times the researcher must recognize that the only observable facts are the preferences by consumers when they make a choice. It is therefore imperative that we do not ascribe mathematical properties to the utility function that are not necessarily present in the revealed preferences of the individuals (VonNeumann and Morgenstern 1944). Aside from that point of caution, the utility function is commonly considered to be exogenous and static (Cohen 2005). Now, stop at this point for a moment and ask the question: Who am I? Or, in other words: Whose utility am I maximizing? The answer to this question has been implicitly avoided in traditional MAS and economic modeling. Figure 2, proposes a general model in which the utility function is the result of an internal system with potentially conflicting desires. It also considers the temporal self as one who is not fully independent, but rather defined within the larger context of a long-term plan. As such, one may want to eat a cake, but is worried about gaining weight. Similarly, one may set lofty new year's resolutions, but finds oneself with no desire to fulfill them later on (McClennen 1998).

A number of scientific disciplines have conceived of humans that have a multiplicity of internal components, such as in the modularity approach of Evolutionary Psychology (Buss 1995; Durrant and Ellis 2003; Tooby and Cosmides 2005; Buss 2008), where each human subsystem was evolved to fulfill a particular need. This internal process is modeled as linear and static, thus making it implicitly irrelevant. In actuality, the process itself may be highly complex, rendering the study of an individual's conflicting desires incomplete and thus inaccurate. This is because such a study both does not account for external pressures and lacks description of the hidden negotiations and transformations that give rise to externally expressed wants, and the corresponding expression of those wants as a utility function. Following Marvin Minsky's research (Minsky 1986; Minsky 2006) we take the position that people may not be the unbreakable wholes, but are rather a collection of many selves entrenched in intense negotiations.

For example, consider the fact that many folk buy a disproportionate amount of lottery tickets, and altogether engage in many risky activities that do not fit the standard description of a risk averse individual. The people in question drastically enjoy buying a $2 lottery ticket with an expected payoff of less than fifty cents, but would never in a lifetime consider a fair bet for a million dollars. A standard representation of this behavior can be embodied in a utility function of income that has decreasing returns to scale in most regions, but highly increasing returns to scale for larger income levels. This set up would ensure that the individual is not prone to take even a fair bet

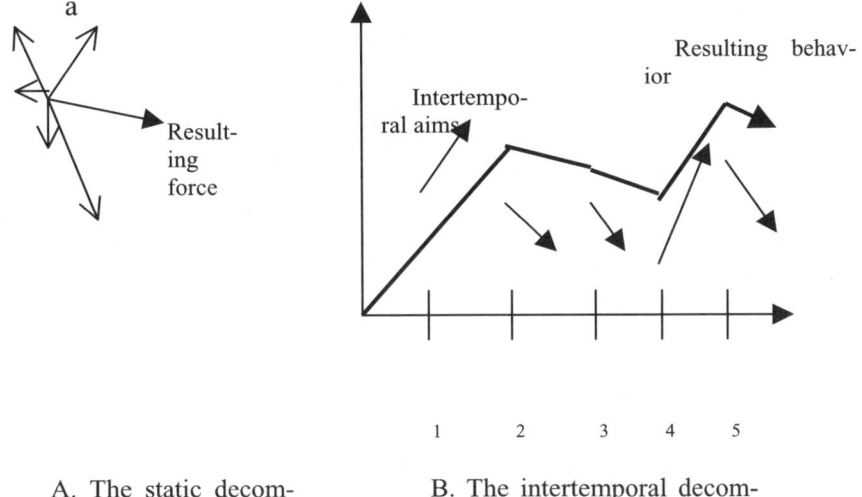

A. The static decom-   B. The intertemporal decomposition of desire   position of utility

Fig. 2. We relax standard assumptions about the utility function and allow for one that can vary though time. This generalized conception of utility is the product of a complex internal process representing the wants of the disconnected subcomponents of a person. At any given moment, a person has internally competing aims that give rise to their external desires. Also, at each time, the person can be thought of as being an independent individual that is constrained by the longer-term plans of the same person.

that will lose or earn him ten thousand dollars, but would be willing to do it if it implied possibly getting a few million (Kahneman and Tversky 1979). Figure 3 depicts such a tradeoff.

The necessary calibration of the particular utility function necessary to obtain these results is not negligible. In other words, it is not easy to conceive of a utility function that would be optimized by the behavior at hand. Moreover, such a person would have an odd view of life, seemingly reaching satisfaction at some level of income (where the marginal utility flattens out), and then suddenly finding himself in a frenzy for more money, as increases in income gradually increase his utility by larger and larger amounts. Most importantly, it is not only the implausibility of the standard utility's interpretation of lottery consumption what makes it inappropriate, it is the fact that it tells us nothing about the internal process that may give rise to such incongruous behavior.

In contrast, our interpretation is based on the way that the network of disconnected selves is organized. The idea is that all selves that are not positively affected by an eventual winning ticket will be so close to indifference that will not veto buying the low-priced tickets, while serious lobbying efforts are devoted by the selves with the millionaire dreams. In the proposed setup, we conceive of internal actors, or katagons whose opinions are heard before taking any decision. Each one of these possesses an objective function that is nonlinear, with a threshold that makes them not care much about petty purchases. As a whole, most humans act this way, taking a long time to

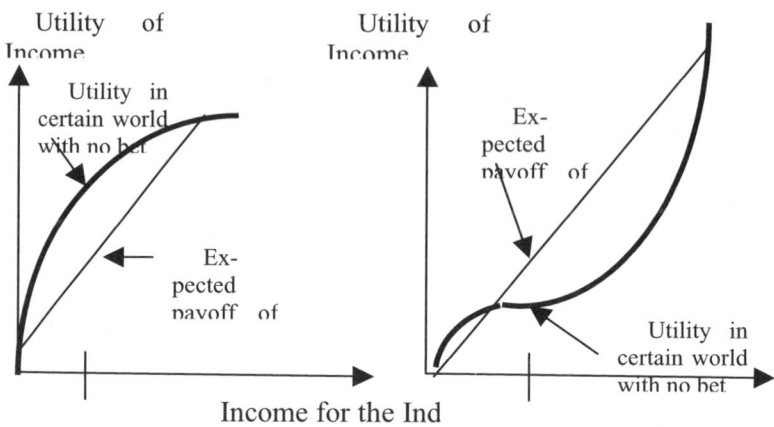

Income for the Ind

A. For normal amounts the preference is for a certain amount of income over the expected

B. For the possibility of an extraordinary amount of income, the individual prefers the expected payoff of the lottery bet

Fig. 3. Here, we depict the standard conception of a utility function that could account for the way in which individuals incur in highly unfair, but small 'lottery' bets, but would not accept large fair bets. The utility function that one could conceive of to account for such behavior is marginally decreasing for 'normal' amounts of income, but marginally increasing for extraordinary large amounts. Because of this particular construction, the utility of the expected payoff for flipping a coin for a thousand dollars is significantly beneath the utility of not having the bet. The expected utility of buying a lottery ticket, however, is above the utility of the certain (perhaps low) income one has without it.

decide upon a big purchase, but not caring much about whether or not to buy bubble gum. Once in a while, the lottery-crazed sub agent comes along and proposes to the person's internal parliament whether to buy in a sweepstakes. All of the other agents then make a comparison of the world with or without the two dollars the ticket costs, and see no general difference in their possible well-being. Therefore, no one objects to the plan. The original proponent of the bet however, will be very happy with the purchase, and will start planning about how he will gladly spend the first prize winnings, as well as all the praises he will get from peers, just for existing. This agent will be happy with a ticket in hand, and no potential agent will have anything to say about it. Moreover, the happiness that he creates from himself can be so great that may have surplus, and conceivably affect the individual as a whole. Quite contrary to this situation is the one in which a considerable sum is at play. Under such circumstances, the potentially affected sub-selves (katagons) come out in force to reject the demand for a bet.

If a model with endogenous utility functions is to be conceived, the way in which the internal decision-making process takes place must be analyzed (Bowles 1998). Only when we understand this internal design will we be able to understand how people make their choices, how they think of themselves, and how they join to form groups. Moreover, the processes that we would like to address in this essay go beyond the way in which the human mind is structured. From now on, we conceive of any possible agent—whether an animal, a family, a university, or a beehive—as an emergent phenomenon possessing a relevant internal system, and we refer to it as a lower level agent (katagon).

In many cases, the lower level may seem irrelevant because it may not be readily apparent. As an upper-level agent coordinates its subcomponents, its lower level may seemingly disappear, leaving us once again with unitary subjects. In such cases, the lower-level components enjoy no agency, just as the soldiers of an army would if they were perfectly trained to follow all orders. In other words, our livers might have a utility function to maximize (making them happiest when they correctly process the body's substances), but that objective function has been perfectly coordinated with upper-level objectives. In this sense, it is not relevant to think of the way in which human cells work when thinking about that person's behavior. However, taking up the case of perfectly trained soldiers, we envision cases in which the lower levels may become exceedingly relevant as they have the ability to reorganize, and may even actively reorganize depending on circumstances. Therefore, the agents we are considering are not time or situation invariant. What makes the attention to the lower levels interesting is not only the possibility but the frequency of the tendency to reorganize.

Cohesion refers to the optimal design of the hierarchical agent. Cohesion does not imply that all lower-level actors are acting in coordination with respect to each other, but rather that an upper level agon exists for whom the actions, design, and general topology of the lower-level agents maximize their objective function. What this implies is that an upper level agon may want some of its lower-level agents to compete, and in this sense perfect competition could be ideal for the upper level that an economy represents. Complete cohesion thus exists when—given a set of constrains, perhaps of the nature of the subcomponents—there are no reorganization possibilities that will better serve an upper-level agon, and it is therefore a relative concept. As agons are defined inside one another, the proposed insight is that a subset of the components of an agglomerate will attempt to maximize what we can define as their objective function, in a process that can be considered selfish or myopic from the point of view of the whole, but optimal for the acting subset.

Aside from an understanding that we may have to consider the subcomponents of any agent, we propose further attention to the context within which the agent is imbedded. Agons are to be described within a group to which they belong, which will be defined as the 'upper level' agents, or anagons, of the hierarchical representation, and will constrain its subcomponents' behavior. In other words, the bee itself is incomplete if not defined as a part of an acting beehive, just as a neuron cannot be fully described outside its complex brain network. In this sense, a beehive reacts to changes in the environment, with the individual bees acting more like neurons rather than decision-making agents in their own right. Similarly, a company will be designed according to somewhat strict guidelines, spanning from the way in which the different departments are organized, to the accounting rules it will be judged by, to the regulations it will abide by, all the way to the competition it must be made prepared to face. So long as all these restrictions are reflective of an economic or survival process that increases the probability of success, we shall refer to all of these constraints encircling the company, or any other agent, as upper-level agons (anagons). Moreover, a social level can be modeled as an agent, in the sense that it can be ascribed an objective function and—given an appropriate time frame—be expected to maximize it by reorganizing its subcomponents.

It is important to stress that an environment is not by itself an anagon, for the environment may be independent of the creatures that inhabit it. An upper level is the superstructure that creates the individual, and without which the nature of the individual would have to be redefined. In this last sense, the environment is an upper-level agon to the degree that it is interconnected with the lower-level agents it breeds. Under this definition, both levels at play coexist, are complicit, and represent indispensable components of the recursive process that gives them existence. Any actor is therefore constructed and defined by the rules of the game that its predecessors instituted, and it is also established in a reality that is constrained by what the recursive system that perpetuates future generations.

Anagons may be created by a conscious decision or by an evolutionary force. In the case of the company, there exist organizational schemes that align the incentives of the administrators with those of the stockholders. Therefore, a conscious attempt to mitigate agency problems will create the optimal structure for the upper level: the corporation. On the other hand, anagons may emerge simply as a product of continued interaction and a corresponding increase in cooperation, as was the case for World War I soldiers stuck in opposite trenches who did not shoot to kill, but rather attempted to wound each other lightly, therefore providing the enemy with a much-desired ticket home. This last example represents an anagon in the sense that the observed behavior cannot be explained in isolation.

The anagon actively arranges its subcomponents so that they find individual maximizations that are in accordance with its overall objective function. We thus find relatively sovereign entities inside the agent that need to be controlled, and who do not necessarily maximize the objective function they would in more isolated circumstances. The analysis distinguishes between a standard objective function describing a simplified, unitary, and clearly-defined agent, and a function that actually describes how the agent reacts to any given situation, once we realize that the true actor belongs to an upper level, and that it has only limited control over itself. We refer to this broader objective as a behavioral function, which implicitly explains what the agent does under any set of circumstances, in a decision-making process that may most often consider the effect of her actions on future selves, offspring, family, peers, and all sorts of lower- and upper-level layers of which she is composed and to which she belongs.

Figure 4, below, describes the sub-composition of a human agent, modeled as having a utility function of the self U^s. This U^s can be thought of as a traditional Von Morgenstern utility function (Morgenstern 1976), but unlike in traditional MAS approaches, it represents the emergent result of a combination of internal aims, traumas, emotions, fears and other wants that can be though of as internal agents with their own agenda, and thus potentially modeled as having their own utility function U^i. Furthermore, the individual's utility function is not to be understood in isolation, but rather as itself having positioned itself within the realm of possibilities, or realm of action allowed by the family he or she belongs to. The family in turn can be modeled as having its own agenda or utility function, U^f. And yet the family is in turn not whatever it wants to be, but positioning itself in accordance with what the community (potentially modeled as having its own utility function U^c) 'wants' it to be.

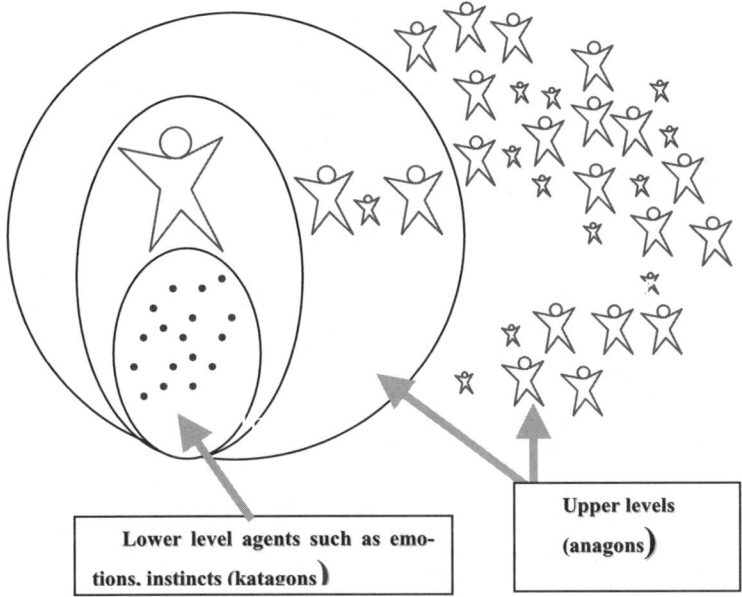

Fig. 4. The utility function of a person cannot be defined in isolation. A human represents an intermediate agent that is controlling its potentially independent subagents (emotions), and to some degree controlled by its social context. Each level of agency can be modeled as having its own utility function. In this sense, a family can be modeled as an agent of its own regard, and thus possessing a utility function. The utility function of the family is maximized when the family members act in agreement with its objectives. Moreover, the utility of the family is defined within the realm of possibilities allowed for by the social agent to which it belongs.

The proposed model therefore subdivides observed behavior into many different levels of agency and interaction. Along the lines of traditional MAS approaches and utility maximization, the actor chooses its best alternative given the set of possibilities it encounters at each one of these levels. The main difference of our approach is that the phase space will include transformations made by an upper-level, agent-like agon to which it belongs. In this fashion, we can think of a rowdy married man as having a utility function of his own, but also belonging to an upper-level relationship that restricts his raucous nature. Moreover, the agent is composed of lower-level subcomponents that may posses agendas of their own. It is the agent's responsibility to present its subcomponents with phase spaces with individual optimal solutions that are agreeable to the upper-level encompassing agent. In other words, the subcomponent agents will optimize the phase spaces in which they find themselves, while the upper-level agon must consider the manipulations of these realms of possibilities that will render the desired aggregate behavior. In this sense, if we consider a firm to be an agent, then this level is composed of the subdivisions that form the company, which in turn are run by a group of people. The company also finds itself in an upper-level anagon that includes all possible regulations for the industry, which is in turn encapsulated in a given society.

4 Distributed Agency

As we have argued, reductionist linear science has concentrated on the study of entities that are clearly delineated, where one could separate what belongs to an agent's nature against the backdrop of what does not. The relevant agent is taken to be exogenous, and therefore disconnected from the system to which it belongs. At their core, the traditional scientific disciplines we have discussed are based on a selfish and unitary agent, or atom of description. Implicitly or explicitly, these paradigms claim that all aggregate complexity can be traced back to the lower level of the system: the strategies and actions of the selfish agent. In other words, these represent research agendas that purposely de-emphasize the existence of any level other than that of the individual. Similarly, MAS have commonly been applied to systems in which only one level of agency is allowed, and where agents are modeled as exogenous entities.

On the other hand, the idea of emergence reflects the fact that different and irreducible levels of interaction will naturally arise in complex systems such as the ones studied by social disciplines, and thus the agent, as we define it in this work, is a combination of levels of interaction. It is through this lens that we would like to consider humans, who will partly be independent creatures, possessors of free will, but who are also partly created by an array of upper-level agons that 'suggest' agreeable utility functions. This conception stems directly from the basic formulation of complexity, in which wholes are more than the sum of their parts. This conception of reality forces us to not analyze agents in isolation but rather in context. If we believe in that proposition, then we should expect to find a world where emergent phenomena are ubiquitous, with distinctive levels of interaction that have agency of their own.

Along this train of thought, we have argued that the traditional concept of an agent is outdated and must therefore be refined. The problem with the concept of an autarkic and unitary agent is twofold: on the one hand, it does not consider the fact that agents belong to an upper-level system that gives them life and meaning, and on the other hand, it does not consider the relevance of lower-level subsystems that compose the agent, but that may possess some degree of agency in their own right. Furthermore, agency may exist in many different dimensions, and each dimension may not overlap in terms of how it defines its agents. We thus propose redefining agency by creating a concept that can include these necessary modifications. In order to stress the novelty of the concept, we believe it is necessary to use a new word to describe the concept, and we have proposed—as a preliminary term—the word *agon* to describe these fuzzy agents.

What exactly is an agon? An agon shares many of the attributes of a traditional agent, but is more general in the sense that it represents a building block for a model where agency is distributed across many actors. Agons are defined in many dimensions, and only possess some amount of agency. Agents are here therefore not defined in an either-or fashion, but rather in a fuzzy, quantitative way. An agon must possess an objective function that grants valence to a number of different possibilities the agon is choosing from. Valence can be defined as the capacity of one person or thing to react with or affect another in some special way, as by attraction or the facilitation of a function or activity. Following traditional MAS jargon, we can think of this substance as some sort of sugar (Epstein and Axtell 1996) or cathexis flux (Castañón-Puga, Rodriguez-Diaz et al. 2007).

In other words, a traditional economic agent is thought to have a utility function that ranks different alternatives. The agent maximizes its objective function, given the reality of its constraints. Similarly, an agon represents an entity that is modeled as having an objective function that it attempts to maximize. Unlike the traditional economic agent and its utility function, the agon has an objective function that cannot be understood in isolation, and it is the emergent result of the objective functions of the lower-level agons that compose it. In this sense, an agon represents a building block of a system of distributed agency, which stands in direct contrast to the traditional agents of economic theory and MAS that represent complete and independent beings, by definition disconnected from their environment and other agents.

The analysis of distributed agency distinguishes between a standard objective function describing a simplified, unitary, and clearly-defined agent, and a function that actually describes how the agent reacts to any given situation, once we realize that the true actor belongs to an upper level, and that it has only limited control over itself. We refer to this broader objective as a behavioral function. It implicitly explains what the agent does under any set of circumstances, in a decision-making process that may most often consider the effect of her actions on future selves, offspring, family, peers, and all sorts of lower- and upper-level layers of which it is composed and to which it belongs.

Our proposed methodology is therefore intrinsically descriptive. It does not test any particular hypothesis. Rather, we propose that any observed social behavior can be expressed as the result of an array of intertwined aims and constrains, hereby modeled as agons. The behavioral function tautologically represents observed behavior, and we propose that any such behavioral function can be built by an appropriate combination of agons, with their corresponding objective function. Following Ossorio (Ossorio 1978), the idea is that once the researcher is able to describe our world in detail, then the researcher does not have much left to explain.

What we propose is a benchmark position in which all observed behavior is the result of the optimization of a myriad of agons' objective functions, so long as we understand the active agons involved in the observed behavior, and always taking into account that the relevant agons may be abstractly defined. In this sense, when we arbitrarily zoom in and analyze a relatively well-defined agent, we may classify its behavior as suboptimal in relationship to its own abstract objective function, but only because we would be artificially studying it in isolation, or without regard for the struggles of its internal nature. The upper level may force lower level members into behaviors that are only considered optimal for the former, and it is in this sense that we can understand the behavior of a kamikaze pilot. In other words, the actions of a suicide bomber cannot be accurately modeled as the result of an individual maximizing his objective function, and must instead be modeled as reflections of the intentions of a larger social agent. Similarly, one may say that the actions of a self-conscious heroin addict are irrational, since he is aware that such behavior is detrimental to his well-being. In our methodology, we would state that there must be a lower-level agent—the internal drug addict—that considers 'getting a fix' as optimal.

After the language of distributed agency is implemented, the researcher may wonder how these agons are created. The answers to these deep questions may well represent a research agenda in its own right, and is therefore beyond the scope of this paper. Nonetheless, as an introductory discussion we can consider the electoral process in a simple majority-based democracy, where the winning party must do its best to find an objective that satisfies the desires of a social contraption attempting to win the election, and in order to do so, they must bring the majority of the population to their side. The process is naturally a very complex one: on the one hand, the political party searches for positions that are most agreeable for its core identity, while on the other, it must be careful to choose the issues that bring to its side the necessary amount of people to win the election. Naturally, these two sides are interwoven, as finding an acceptable political position for the base must include a directive not to alienate the general population by doing so.

In addition, the winning party must take into account that new elections will come again later, and that its actions must not disturb the losing party too much, for the disenfranchised could become a liability to the ruling coalition, as the opposition could let go of most of its issues in order to concentrate on just winning. The entity created will find its position in a longer-term upper level of existence, with an objective function that does not get trapped in immediacy. Although it may reach across many different coordinates of concerns, the identity, platform, ideology or political party will most commonly have the incentive to present itself as internally consistent and relatively homogeneous, as its internal cohesion will likely prove key to its success. Groups that have found cooperative structures to abide by will tend to survive longer than groups that are penalizing themselves internally for lack of such coordination.

As the history of the world's game unfolds, evolutionary pressures will solicit larger, more complex and adaptive organisms or networks of coordinated organisms that find themselves better at exploiting the ever changing environments offered. These larger organisms will have a nature, a design that has optimized the possibility for exploitation *at that level*. The upper level agon or 'agglomerate individual' will have to become cohesive, and organize its subcomponents in order to maximize its emergent objective function, since suboptimal internal coalitions can materialize. We can think of living organisms as smoothly coordinated societies, as their subcomponents 'magically' aid each other, quasi-perfectly specialize in tasks, and have well-developed methods of internal communication (such as a central nervous system). Disconnected organisms are therefore a near oxymoron. There must be an evolutionary force connecting otherwise autarkic selves and individuals.

5 Modeling Human Behavior with Distributed Agency

Distributed Agency represents a language in which social behavior can be described, allowing the researcher to state hypothesis and to represent multiple and interactive levels of reality. Within this framework, all observed behavior is said to be the result of a myriad of agent-like entities—broadly described as agons—that maximize their respective objective functions, given constraints. Our approach is therefore intrinsically descriptive (Ossorio 1978).

In economics, humans are said to have a utility function, which basically represents the happiness of the individual, who is normally described as an independent, rational and well-informed agent. In contrast, we propose a model in which humans are made of relatively independent components, and who are also best described as parts of larger wholes (Schwartz 1986). Upper-level meta-agents, or anagons may influence both the constraints and rewards faced by a human, as well as the utility function itself. We cannot think of the traditional utility function as the relatively unrestricted happiness of the agent, in absence of upper-level restrictions. In a prisoner's dilemma, for example, the utility function is represented by the proposed payoffs presented to the prisoners, while the behavioral function we propose is represented by what the individual actually *does*. Moreover, the constraints to which the behavioral function has to abide by are different than those presented by a typical budget constraint, as they literally define the agent, in a sense compelling the individual to 'like' consumption bundles that he would otherwise object. One could only dread to wonder, for example, what a suicide bomber feels as he presses the deadly button, but it must be nonetheless something that he would not have felt in the absence of social pressures. The behavioral function describes at all times how the system unfolds, in a setting where the agents make decisions that may most often consider the effects of her actions on future selves, offspring, family, other members of her species, and all sorts of lower and upper level layers of which she is composed and to which she belongs; in contrast to a more abstract utility function that pertains to her individual, most immediate self.

The prisoner's dilemma model predicts lack of cooperation between the two participants. What such a model is implicitly doing is defining the relevant agons, which in this case represent selfish, rational, myopic, exogenous and autarkic individuals. When in experiment after experiment (Smith 1982; Gurerk, Irlenbusch et al. 2006) we observe cooperation, the descriptive language of distributed agency defines an upper-level agon (anagon) that has enough power to influence the behavior of these individuals into a cooperative result. The anagon is represented as possessor of a given quantity of reward currency to coerce the player to behave in unison. Anagons thus are modeled as owners of a given amount of 'sugars' that the lower-level agents, or katagons, crave (Epstein and Axtell 1996). Anagons in turn have an objective function, which they maximize by allocating the sugar they have to bring about the desired behavior. In this recursive process, anagons represent themselves subagents to a yet higher agon that is itself reorganizing its own sugar. As an analogy, we can think of a CEO that allocates salaries, bonuses and other incentives to induce the appropriate coordinated work of her employees. The CEO may have incentives of her own, but mostly she should have the objective of maximizing shareholder wealth.

Empirical evidence has consistently shown that humans cooperate when facing a prisoner's dilemma situation. One of the pioneers in experimental economics is Nobel Laureate Vernon Smith, who was forced to pay significantly more money to his irrationally cooperating students facing this situation, than he would have if they acted as detached individuals. Given this dilemma and these circumstances, not cooperating becomes, rationally, the best choice, if only we make reference to the anagon representing the coalition. An individual who does not choose its maximizing strategy in a prisoner's dilemma is obviously choosing based on something it deems better, whether by a pre-commitment or by an evolutionary influence. We must look for these fuzzy upper-level agents, or agons that are explicitly defining behavior; binding the individualistic nature of the prisoner to a strategy that does not make sense when analyzed in isolation.

The better-than rational results of Smith's experiments may then be understood by appropriately characterizing the acting agent: in this case, the people who decided to cooperate against their incentive to defect were not acting as individuals, but as representations of the anagon of evolved and cooperating persons. In this way, the anagon asserted its agency. Individuals are born into structures that are usually in many ways set for them, with incentives for cooperation and trade that are inescapable for the individual, as their nature intrinsically possesses an appetite for the resources that only the superstructure can provide. Other examples of this process include the behavior of working ants, unmonitored employees committed to the success of a company that is not their own, or soldiers acting in the name of their religion.

The payoff structure that a player faces in the prisoner's dilemma refers to its abstract utility function, which could only be understood in isolation. The payoff must not be the sole source of information for the decision maker. The participants in experimental games such as the one discussed above are making decisions as representatives of potential intertemporal selves, or of the human race, and not as isolated individuals. One of the most important reasons that humans have been so successful on this earth is because they learned to take advantage of mutual cooperation, of the possibility of social synergies for finding better scales of environmental exploitation.

In terms of rationality as it is most commonly defined (Marschak 1946), one can think of the actions of benevolent cooperators as irrational, since they do not follow their individual best interest. However, as these cooperators generally end up prevailing, one may wonder if an agent in more realistic circumstances may be aware of such outcomes—as well as of the likely retaliations of deceived players—and be therefore acting in a perfectly rational way, once she acknowledges the better information of peers' characteristics, retaliation functions and gains of cohesion, and in that sense develops a less myopic mind. In comparing intrapersonal choice and personal decisions, it is interesting to note that an individual who acts in a resolute manner (that is, forcing through a pre-commitment a future self to act in a manner which is then not optimal for her) is deemed by some as irrational; a statement which is in direct contrast with calling myopic someone who does not recognize the opportunities of intrapersonal rearrangements to achieve preferred outcomes, when having a longer horizon to base their decisions.

Moreover, aside from an unclear definition of what is to be considered 'I' or 'we', we must of course see the other side of the coin: at some point we leave the 'we' territory to enter into the world of 'them'. Another incentive for individuals in a group to merge their incentives lies in the fact that discontented individuals have the tendency to be nasty. The payoffs of a more realistic version of the prisoner's dilemma are naturally determined by the history of the game—most importantly, whether the participants are anonymous or known participants, and the previous experiences of the participants. Non-cooperative actions can always set a precedent for something that will later affect an agent directly related to the defector, as cooperation and lack of it are normally persistent. An angry participant may thus set drastic reactionary payoffs, possibly to instill fear in possible rivals of future similarly afflicted selves, as any malicious action played against a peer can become institutionalized and potentially harmful for the meta-temporal self.

In summary, anagons have abstract objective functions with varied ontological outlets. As such, humans live in a vast net of positive and negative incentives that trap us with or without us being aware of them. We may think of ethics as given to us by God, but we may also model them as the reflection of the objective function of the social agent. Individuals could become unfriendly, criminals, traitors, and generally undesirable, but the society will create mechanisms to discourage such outcomes, here modeled by an 'expensive' region in the realm of action of the agent's utility function, or directly as sugar with negative valance (poison) that the anagon allocates.

By the same token, honorability, courage, honesty, friendliness, and all sorts of likeable characteristics will be praised and encouraged, thus modeled as regions with more sugar. People may encounter temptations they avoid because of previous experiences in which the then current self became uncontrollable and detrimental to the welfare of the more general self. Such an agent does not act in isolation, as the anagons it belongs to impose their muscle in her decisions, through forces that can be translated into the same dimensions as the ones in which the utility is defined. Aside from other incentives for joint effort (such as specialization, risk sharing, and the avoidance of violence), individuals mostly make decisions based on incomplete information, and often without a clear objective to maximize. Therefore, the current individual may often find herself in a position where she may evade searching for drastic, egoistic outcomes, and instead settle for strategies that have proven useful at maintaining the long-run individual alive, healthy and wealthy.

The payoff structure of the standard prisoner's dilemma is thus relatively hard to conceive of for a realistic human, since most current selves and individual agents will naturally be afraid of—or compelled by evolution to avoid—strategies that overemphasize rampant immediacy. The benefits of cooperation cannot be understood if we zoom in too close to the individual's decision without understanding her within the context of the superstructure to which she belongs, as the setup of a one-time deal given to an individual with no recourse to an uncertain future is unrealistic, as organisms who think like this would shortly perish without intertemporal cooperation.

More than altruism, what we are referring to is coordination. Any activity one could think of, from walking to talking to tool-making, requires some degree of coordination, and for activities with longer-term horizons intertemporal coordination is indispensable. Like humans, animals living in an environment with seasons must also coordinate thrifty summer selves with needy winter partners. The upper level to which we am referring works as a process of identification with intertemporal selves as well as with other related individuals in a group.

In terms of describing individual behavior, we can also conceive of a person as an anagon, with internal subagents possessing relatively independent agendas, such as it is described by transactional analysis (Berne 1964). We nonetheless conceive of the potentially vast legion of selves as naturally related, and perhaps having a more static and institutional complex function ordering their preferences; that is, we can think of individuals as possessing an internal government body that may be relatively stable. In terms of having different 'selves' inside, most people refer to the one that they are in the very moment they talk about this subject, and claim that the current spokesperson is the only version of themselves that exists; granted, sometimes this person can be angry or on a particularly good mood, but the variation does not grant them in any way as having multiple personalities. The view we propose recognizes that a person plays very diverse roles in life, and that each role can be represented as an independent agent that embodies different 'subject positions'(Davies and Harré 1990).

The katagons, subagents, aims, or desires that compose the more general and unified perspective in which we normally conceive of humans as not entrenched in a lawless competition for attention, but rather as confined within a sort of political system, full of laws, traditions, history, habits, coalitions, and all imaginable aspects of a full-fledged bureaucracy of mind. This internal system may be well established and generally permanent, perhaps reflecting what we think of when we say 'I'. From the point of view of the internal katagon, the person's system represents an anagon that offers a rugged landscape for the subagents to maximize. In this sense, the desire to go on vacation is always there, but in order to be actualized, it must form a coalition based on the current political circumstances, as well as wait for the opportunity when the protocols allow for a vote on the proposed bill. This could be modeled as a neural network in which receptor neurons receive the person's partitions wishes. The receptors can be in turn anthropomorphized as a system of extremely capable political analysts who calculate all the effects of any particular decision, determining whether or not it would be approved by the particular political system in place.

6   A Language to Build Large-Scale Simulations

We have proposed the basis of a language for describing social behavior in which agents are replaced by a more general fuzzy or distributed concept that grants agency in a relative, rather than the traditional absolute fashion. In this generalized multi-agent language, all agents are intermediate, in the sense that they are potentially part of another, higher-level agent, and potentially formed by lower-level agents. Agents are also fuzzy in the sense that they are defined in more than one dimension, such as a person that belongs to a political party but does not share its position on a wedge issue. To stress the novelty of the concept we are putting forward, we distinguish the fact that the agents we are defining may be abstract, and thus use the working concept of "agons" to describe these intermediate building blocks. Distributed agency, as a generalized language of description, can be used to construct large-scale simulations, because the proposed agons are intrinsically designed to integrate to multiple levels of interaction.

Most importantly, we establish a benchmark position in which all observed behavior is the result of interacting agons described in the ontological process of maximizing their objective functions. The discussion so far has left us with an important absolute principle: that all possible definitions of optimality in a hierarchically decomposed world are relative. The relativity is a direct result of the hierarchical nature of the system, where each agent binds the subagents that compose it, and is bounded by the super-agent to which it belongs. Therefore, defining optimality implies defining an agon, for it is only by placing it within its appropriate context that we can describe how one behavior is better than another.

Observed behavior is then classified as optimal only in the sense that we are modeling it under the assumption of entities that can control their environment, given their constraints. The behavioral function represents the description of actual behavior, and it is the emergent result of the interconnection of the minimum number of agons that the researcher needs to describe reality at an appropriate level of detail. In building a simulation, we must abide by Occam's razor and put forth the simplest possible description, albeit one that captures the desired level of realism and complexity. As part of this process, the researcher must decide which actors are going to be granted agency in the simulation. In this way, when studying cultural shifts over time, the researcher—perhaps a sociologist—may opt for only granting agency at the level of social groups and ethnicities, while an economist may prefer to focus at the household level. This notwithstanding, if built on the proposed principles of distributed agency, either model may still be able to potentially interact with each other, as they would both be expressed in a similar and interconnecting language.

In recent years, there has been ample discussion of the building of a parallel world inside the computer, such as Second Life® or the one depicted in the classic sci-fi movie "The Matrix®" (Wachowski, Wachowski et al. 2001). Naturally, the computing capabilities necessary to build a complete recreation of our world are impossible (Hofstadter 1979), so the researcher must choose what aspects of reality to focus on in the process of building a simulation. Complex phenomena refers to systems that cannot be described in a reductionist fashion, as the algorithm that describes observed results is itself the shortest solution, and thus cannot be adequately simplified into a set of dynamic equations. Nonetheless, the researcher may not be particularly interested in detail at all levels of description, and must therefore proactively choose what levels to study in a nonlinear fashion, and which levels can be ignored and correspondingly studied linearly (that is, as a direct aggregation of atomic components). Each resulting level, however, is 'scalable' in the sense that if it is essentially ignored in one model, it may be naturally resurrected in another.

Following the discussion, an example of a large-scale simulation may clarify the power of distributed agency. Suppose that we are interested in understanding the reason why Latin American countries are relatively poor, and in the process of creating an agent-based simulation to analyze the problem, we find ourselves lost as to how to define our agents. Should we have a macro view in which we make the Latin American bloc an agent that deals with its position in world politics? Or should we consider each one of the participating countries as separate entities? Should we rather adopt a micro model that starts from individual people? One of the main ideas of this work is to bring attention to the way in which a more general model would deal with the connection between the different dimensions of analysis. In the twentieth century, a push was made for developing macroeconomic models with strong microeconomic foundations, but the need for tractability left the discipline with no choice but to resort to linear models of interaction, as well as unrealistic assumptions about rationality. The era of agent-based modeling, however, can forgo linearity and perfect foresight, and take a leap towards a more general and realistic model that can describe the emergent strategic interactions at each relevant level.

Within that description, supposed that we are interested in understanding the political effects of granting Mexican immigrants living in the United States the right to vote in Mexican elections. If we think of Mexican culture, political parties, and other sociological actors as abstract agents or agons, should we then classify the affected group as a subset of Latin America or as an independent entity? Neither one necessarily, since not all aspects of Mexico can be defined inside the concept of Latin America. In particular, the Mexican business cycle is now much more related to Canada and the U.S., its partners in the North American Free Trade Agreement (NAFTA).

The example of Mexican immigrants in the U.S. sheds light on the fact that no agent we describe under the proposed paradigm may necessarily have clear borders delineating what belongs to it from what does not. Consequently, any agent considered will be fuzzy in the sense that it will be described according to many different

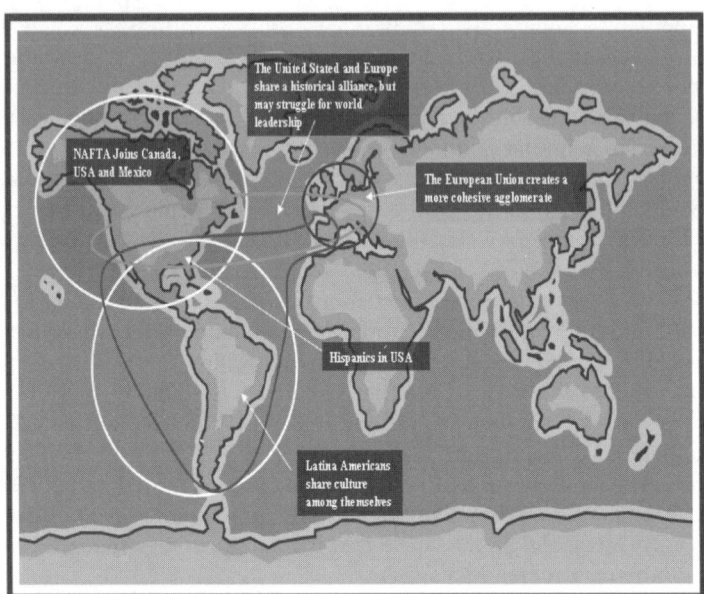

Fig. 5. A large-scale simulation describing the Hispanic population of the United States must take into account aspects of agency that take place in multiple dimensions. Any one individual of this group may share agency and be part of an agon in some respects and not in others. If a researcher wanted to understand how the group may stand in terms of the debate on illegal immigration, the analysis must include representations of actors that affect behavior in distinctly separate dimensions, such as national ties, culture, commerce and economic treaties.

criteria, that it will have relatively blurry borders, and that it may or may not have overlap with other agents, depending on the criteria we consider for defining it. When we think of people at a micro level, we think of individuals who are to some degree the one they are in the very moment we look at them, but also to some degree a reflection of a longer time horizon of the self; to some degree the relationships they are involved in; as well as to some degree a piece of their family, of their community, and of their nation.

Moreover, simulations of this nature will allow social scientists to describe abstract, but most relevant aspects of social interaction, such as norms, traditions, institutions and individual freedom. It is here, we claim, that wealth is created (Ball 1991; Beinhocker 2006). By contrast, society does not exist in traditional economics, as it merely represents the linear agglomeration of independent actors. This is in direct contradiction with the language of distributed agency, where societies can more accurately described as anagons, that is, upper-level agents that attempt to maximize an abstract objective function. This notwithstanding, the concept of an agon is one that is quantitative, rather than qualitative, in the sense that one can have entities that are more or less of an agon, that is, have more or less agency, be more of an agent or more of a simple aggregate. In this sense, we argue that one can understand the reason why North America is significantly wealthier than Latin America, as the former represents more of a cohesive agon, while the latter represents more of a direct aggregation of uncoordinated individuals and other social actors.

To understand the agency imbedded in the American social agent, let us consider its history. The American constitution represents one of the greatest experiments in the history of human evolution, and the scope of its results are written every day in the history of the most powerful social object the world has ever seen. The Founding Fathers of America had a great practical vision for how to organize a society with such great potential, as it had a fresh start with a relatively equal society (of its free men) as well as vast resources, both natural and human, as they inherited the rich cultural and scientific tradition of European nations. With the grand ideas of the French revolution, backed by the condensed knowledge of the Western world, the Founding Fathers applied all their scientific and pragmatic wisdom into a constitution that separates powers very effectively, with an executive branch that remains relatively external to the government, and a commander in chief with significant control over bellicose decisions; a legislative power designed to 'listen' more intently to the needs of its heterogeneous constituency, and is somewhat comparable to a living organism's nervous system; and a judicial power that ensures the permanency of rules that should remain relatively constant, providing fertile ground for projects that need longer time horizons to develop fully, and makes sure that rules are applied equally to all individuals, thereby maintaining the hard fought equality of the citizens that will eventually make the country strongest, with reliable institutions, networks, generalized habits and the corresponding social norms.

7 Conclusions

One of the most important aspects that the separation of powers is that it allows for the social organism to better discern the many different aspects of change. We would expect a social object to be 'fearful' of change, as it is hard to digest all the implications of an organizational scheme that has never been tested, and that affects citizens through so many different and regularly unquantifiable dimensions. The separation of powers represents a mechanism by which individuals can discern and disentangle the all-encompassing effects of social change, knowing that their government will follow rules of engagement that will attempt to protect their need for security and the exploitation of known benign structures and norms, while accepting the possibility of change and further progress through the exploration of other dimensions of the social conundrum. Aside from the original design, the fact that the American system has been bipartisan has made for a government that is most ready for immediate action, a trait most important in the art of inter-group conflict (i.e. war), since the ruling coalition of a multi-partisan political system could partition, particularly in times of political stress.

In contrast, social systems such as the one Latin American countries have present a political ruling class for which it is convenient to have a significant part of the electorate left in the dark, and this can be seen in the vague—and mostly devoid of controversial issues—electoral campaigns that their candidates run. The most important example is that no significant attack on generalized poverty is proposed, and this is the direct reflection of the fact that the system is implicitly colluded with the

economic ruling classes, in an agreement that precludes an evolution that could bring
about the birth of a more equitable state. Would such a state be preferable? The improvement could be measured in any reasonable dimension, and definitively provide
visible monetary results after a generation. The objective function of a well-oiled social anagon will take into account the nature and corresponding complexities of all the
levels of agents it encompasses, quantifying such abstract notions as social cohesive-
ness, identification, and play allowed for the individual's utility functions. Such a
social welfare function would then provide a scale in which we can evaluate alternative social structures. Most directly, the proposed function could map social structures
to corresponding future changes in per capita GDP. Such a scale is unavailable until
now in traditional economic theory, and it would be most useful in the understanding
of real-world economic development.

References

Abbot, R.: Emergence Explained: Getting epiphenomena to do real work. In: Working Paper.
    Lake Arrowhead Conference on human complex systems (2005)
Axelrod, R.: Agent-Based Modeling as Bridge Between Disciplines. In: Judd, K.L., Tesfatsion,
    L. (eds.) Handbook of Computational Economics, North-Holland, vol. 2 (2005)
Ball, J.: The creation of wealth. London, London Business School Centre for Economic Fore-
    casting (1991)
Beinhocker, E.D.: The Origin of Wealth: Evolution, Complexity, and the Radical Remaking of
    Economics. Harvard Business School Press, Boston (2006)
Berne, E.: Games people play: The psychology of human relationships. Penguin, London
    (1964)
Bowles, S.: Endogenous preferences: The cultural consequences of markets and other economic
    institutions. Journal of Economic Literature 36, 75–111 (1998)
Buss, D.: Evolutionary Psychology: The New Science of the Mind. Allyn & Bacon, Austin
    (2008)
Buss, D.M.: Evolutionary psychology: A new paradigm for psychological science. Psychologi-
    cal Inquiry 6(1), 1–30 (1995)
Castañón-Puga, M., Rodriguez-Diaz, A., et al.: Social Systems Simulation Person Modeling as
    Systemic Constructivist Approach. Working Paper. Uni-versidad Autonoma de Baja Cali-
    fornia (2007)
Cohen, J.D.: The Vulcanization of the Human Brain A Neural Perspective on Interactions Be-
    tween Cognition and Emotion. Journal of Economic Perspectives 19(4), 3–24 (2005)
Darwin, C.R.: The Origin of Species. Collier & Son, New York (1909)
Davies, B., Harré, R.: Positioning: The Discursive Production of Selves. Journal for the Theory
    of Social Behaviour 20(1), 43–63 (1990)
Durrant, R., Ellis, B.J.: Evolutionary Psychology. In: Gallagher, M., Nelson, R.J. (eds.) Com-
    prehensive Handbook of Psychology. Biological Psychology, vol. 3, Wiley & Sons, New
    York (2003)
Epstein, J.M., Axtell, R.: Growing artificial societies social science from the bottom up. Brook-
    ings Institution Press, Washington (1996)
Gurerk, O., Irlenbusch, B., et al.: The Competitive Advantage of Sanctioning Institutions.
    SCIENCE 312 (2006)
Hofstadter, D.R.: Gödel, Escher, Bach an eternal golden braid. Basic Books, New York (1979)
Johnson, G.: Mindless Creatures Acting 'Mindfully', New York Times. New York (1999)

Kahneman, D., Tversky, A.: Prospect theory: An analysis of decisions under risk. Econometrica 47, 263–291 (1979)

Marschak, J.: Neumann's and Morgenstern's New Approach to Static Economics. The Journal of Political Economy 54(2), 97–115 (1946)

McClennen, E.F.: Rationality and Rules. Modelling Rationality, Morality, and Evolution. P. A. Danielson, Vancouver Studies in Cognitive Science. 7 (1998)

Minsky, M.: The Society of Mind. Simon and Schuster, New York (1986)

Minsky, M.: The Emotion Machine: Commonsense Thinking, Artificial Intelligence, and the Future of the Human Mind. Simon & Schuster (2006)

Morgenstern, O.: The Collaboration Between Oskar Morgenstern and John von Neumann on the Theory of Games. Journal of Economic Literature 14(3), 805–816 (1976)

Ossorio, P.: What Actually Happens: The Representation of Real-World Phenomena. University of South Carolina Press (1978)

Schwartz, B.: The battle for human nature: Science, morality, and modern life, New York, Norton (1986)

Simon, H.A.: Models of bounded rationality. MIT Press, Cambridge (1982)

Simon, H.A.: The sciences of the artificial, 3rd edn. MIT Press, Cambridge (1996)

Smith, A.: An Inquiry into the Nature and Causes of the Wealth of Nations. Methuen and Co. Ltd., London (1904)

Smith, V.: Microeconomic Systems as an Experimental Science. The American Economic Review (1982)

Tooby, J., Cosmides, L.: Conceptual Foundations of Evolutionary Psychology. In: The Handbook of Evolutionary Psychology, pp. 5–67. Wiley, Hoboken (2005)

VonNeumann, J., Morgenstern, O.: Theory of games and economic behavior. Princeton. University Press, Princeton (1944)

Wachowski, A., Wachowski, L., et al.: The Matrix. Burbank, CA, Warner Bros. Pictures: Distributed by Warner Home Video (2001)

Wagner, I.A., Bruckstein, A.M.: From Ants to A(ge)nts: A Special Issue on Ant-Robotics. Annals of Mathematics and Artificial Intelligence 31(1-4), 1–5 (2001)

Part IV

Hardware Implementations

Design and Simulation of the Fuzzification Stage through the Xilinx System Generator

Yazmín Maldonado[1,a], Oscar Montiel[1], Roberto Sepúlveda[1], and Oscar Castillo[2]

[1] Centro de Investigación y Desarrollo de Tecnología Digital (CITEDI)
del IPN. Av. del Parque No.1310, Mesa de Otay, 22510, Tijuana, BC, México
maldonado@citedi.mx, {o.montiel,r.sepulveda}@ieee.org
[2] Division of Graduate Studies and Research, Tijuana Institute of Technology,
Calzada Tecnológico S/N, Tijuana, México
ocastillo@hafsamx.org
[a] M.S. Student of CITEDI

Abstract. In the last years, several algorithms to implement the fuzzification stage for Very Large Scale of Integration (VLSI) Integrated Circuits (IC) using a Hardware Description Language (HDL) have been developed. In this work it is presented a proposal based in the arithmetic calculation of the slopes in triangular and trapezoidal membership functions to obtain a fuzzified value. We used an arithmetic calculation algorithm to implement trapezoidal and triangular membership functions. This proposal is different to others that at present time are currently used. We discuss the advantages and disadvantages of this implementation. A methodology to test and validate this stage through the Xilinx System Generator is described.

1 Introduction

Theory of fuzzy sets proposes to obtain solutions to problems with vague information, formalizing human reasoning mathematically to get conclusions from facts observed vague or subjective [1].

The use of fuzzy logic systems is getting more common because they can tolerate inaccurate information, they can be used to model nonlinear functions of arbitrary complexity, they make possible to build a system based on the experts' knowledge using a natural language. The main drawbacks associated with the realization of microelectronics fuzzy logic systems, come from the high cost and development time associated with the design and manufacture of an IC. A solution to this problem is to use FPGA's (Field Programmable Gate Array) to implement specific processing architectures with the advantage of the reusability of large amount of existing code that was developed using a Hardware Description Language, this provide an excellent relationship "cost-performance" [2,3].

An FPGA is a semiconductor device that contains in its interior components such as gates, multiplexers, etc. It is interconnected with each other, according to a given design. These devices use the VHDL programming language, which is an acronym that represents the combination of VHSIC (Very High Speed Integrated Circuit) and HDL (Hardware Description Language) [2].

A fuzzy logic system consists mainly of three stages: fuzzification, inference, and defuzzification [4,5].

O. Castillo et al. (Eds.): Soft Computing for Hybrid Intel. Systems, SCI 154, pp. 297–305, 2008.
springerlink.com

Fuzzification comprises the process of transforming crisp values into grades of membership for linguistic terms of fuzzy sets. The membership function is used to associate a grade to each linguistic term.

The inference process is the brain of the fuzzy logic system, here are proposed rules of the form IF-THEN that describe the behavior of a system.

Defuzzification is the process of producing a quantifiable result in fuzzy logic.

In this work it is presented a proposal based in the arithmetic calculation of the slopes in triangular and trapezoidal membership functions to obtain a fuzzified value. We used an arithmetic calculation algorithm to implement trapezoidal and triangular membership functions. This proposal is different to others that at present time are currently used.

This paper is organized as follows, section 2 describes diverse methods to implement the fuzzification stage; section 3 explains our proposal to implement the fuzzification stage; in section 4 is the description of the experiments realized. Finally, section 5 presents the conclusions.

2   VHDL Methods to Implement the Fuzzification Stage

In literature, there are two main methods to design the fuzzification stage using VHDL codification, they are:

1) Storage Memory.
2) Arithmetic Calculation.

The Storage Memory method uses tags values instead of names to identify the linguistic variable names [6,7]. The method consists in storing into a memory the tags values (L_a, L_b, and L_c) and their corresponding membership function degrees (μ_a, μ_b, μ_c). To obtain a fuzzified value, it will be necessary to search into the look up table the corresponding binary word for the crisp input x [4], Fig. 1 illustrates this method.

The arithmetic method consists in performing a progressive calculation of the degree of membership of the antecedents. This method is restricted to use only normalized

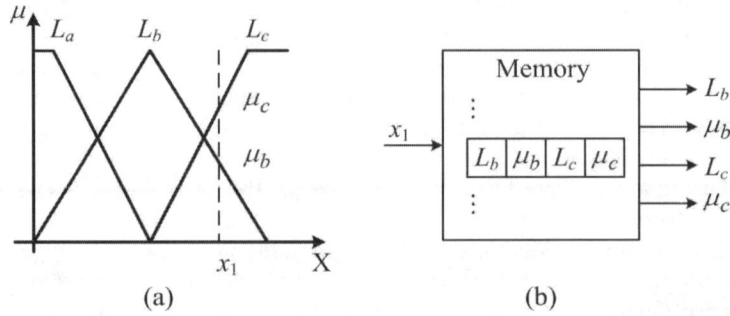

Fig. 1. The input x_l showed in (a) has two membership function values to it. In (b) is shown how the tags and fuzzified values are stored into a memory for a determined input, more specifically for the input x_l.

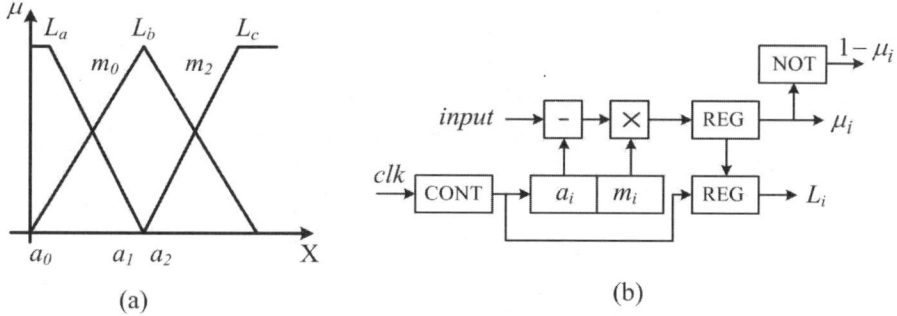

(a)   (b)

Fig. 2. Arithmetic calculation method. In (a) is showed the intercept points (a_i's) and slopes (m_i's) of each straight line. In (b) the a_i's and m_i's are used to calculate the degree of membership and tags.

triangular MFs [6-9]. The method uses two memory locations for each straight line, where their slopes values and interception points are saved. An arithmetic circuit for each input shown in Fig. 2(a), solves the corresponding straight-line equation as can be seen in Fig. 2(b).

3 New Algorithm to Implement the Fuzzification Stage

This proposal is based on the arithmetic calculation method; however, there exist a significant difference to those we found reported using the same algorithm. This difference increases the flexibility to implement an application since the slope value is calculated dynamically at demanding time, although we have to expect an increase of the execution time that is not important for many applications considering the high speed of calculation that a VLSI can reach.

This algorithm can be divided in the next three steps:

1. Calculate the value of the slope.
2. Calculate the degree of membership.
3. Assign to the outputs the degree of membership as well as the respective tag of the linguistic variable.

Figure 3 shows a flow diagram to explain the abovementioned steps.

One more advantage that we can obtain implementing this proposal is that it is possible to implement triangular and trapezoidal MFs using formulas (1), and (2) [10,11].

$$trapezoid\ (x;a;b;c;d) = \begin{cases} 0 & x \leq a \\ \dfrac{x-a}{b-a} & a \leq x \leq b \\ 1 & b \leq x \leq c \\ \dfrac{d-x}{d-c} & c \leq x \leq d \\ 0 & d \leq x \end{cases} \tag{1}$$

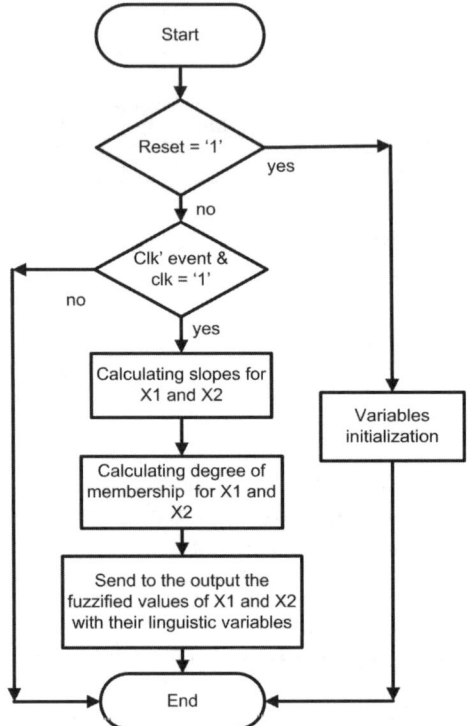

Fig. 3. Flow diagram of the fuzzification stage

$$triangle \ (x;a;b;c) = \begin{cases} 0 & x \le a \\ \dfrac{x-a}{b-a} & a \le x \le b \\ \dfrac{c-x}{c-b} & b \le x \le c \\ 0 & c \le x \end{cases} \qquad (2)$$

4   Experiment Design and Validation

4.1   Experiment Set Up

To test the proposed algorithm a comparative experiment was designed using Matlab/Simulink and Xilinx System Generator [12,13] which is an integrated designed envorinmet (IDE) tool for Matlab/Simulink, in [14,15] are some application examples.

To achieve the comparison between methods, we follow the next steps:

1. We considered a two input system with five MFs for each input, two trapezoi-dal MFs and three triangular MFs. Figures 4 and 5 show the MF of these inputs.
2. A fuzzifing VHDL entity was codified using the proposed algorithm of section 3. The obtained schematic diagram and design entity is shown in Fig. 6. We used 8 bits, but the entity can be easily scalable to any common size. Hence, for this case the universe of discourse is in the range 0 to 255.
3. The fuzzification stage coded in VHL was exported to the graphical simulation platform Simulink, through the use of a "black box" provided by Xilinx Sys-tem Generator, see Fig. 7.
4. The simulation model shown in Fig. 8 was implemented. There are three inputs and nine outputs. The Inputs x1 and x2 are going to be fuzzified. Attached to each input x1 and x2, there are three block sequentially connected blocks. Go-ing from left to right, in the first block is given the numeric value that we want to fuzzify. Next is a conversion function to the range 0 to 255, and the last block is a Xilinx System Generator function known as "gateway in" func-tion which is used to convert numeric values from the Simulink to the format needed by the Xilinx "black box", i.e., fixed point. The reset input is used to initialize the system. In the model, there are two classes of outputs, one class

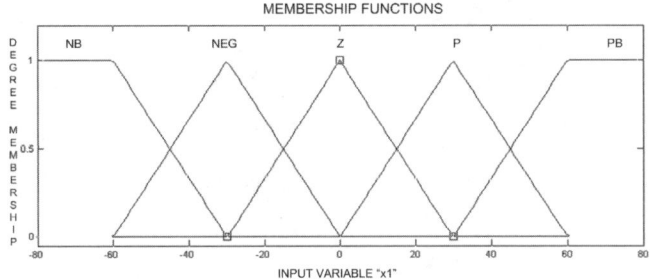

Fig. 4. Membership functions for input x1

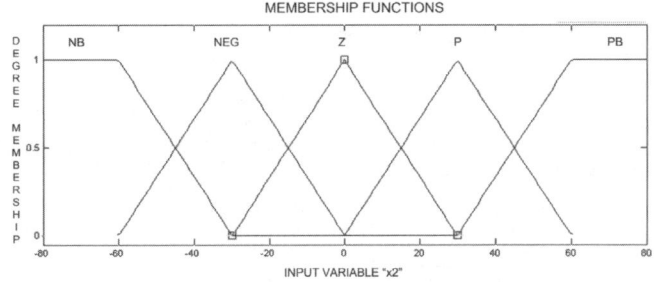

Fig. 5. Membership functions for input x2

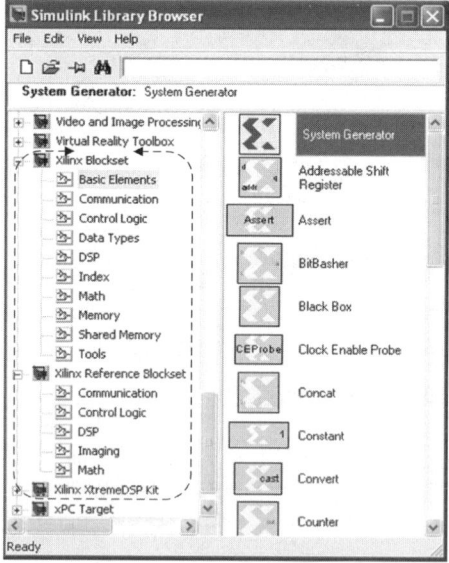

```
                                              entity fuzzification is
  x1(8:1)    degree_in1a(8:1)                    Port (

             degree_in1b(8:1)
                                                    clk : in  STD_LOGIC;
             degree_in2a(8:1)                       reset : in  STD_LOGIC;
  x2(8:1)                                           x1 : in  STD_LOGIC_VECTOR (8 downto 1);
             degree_in2b(8:1)                       x2 : in  STD_LOGIC_VECTOR (8 downto 1);
                                                    degree_in1a : out  STD_LOGIC_VECTOR (8 downto 1);
             ling_v_in1a(3:1)                       degree_in1b : out  STD_LOGIC_VECTOR (8 downto 1);
                                                    degree_in2a : out  STD_LOGIC_VECTOR (8 downto 1);
  clk        ling_v_in1b(3:1)                       degree_in2b : out  STD_LOGIC_VECTOR (8 downto 1);
                                                    ling__in1a : out  STD_LOGIC_VECTOR (3 downto 1);
             ling_v_in2a(3:1)                       ling__in1b : out  STD_LOGIC_VECTOR (3 downto 1);
                                                    ling__in2a : out  STD_LOGIC_VECTOR (3 downto 1);
             ling_v_in2b(3:1)                       ling__in2b : out  STD_LOGIC_VECTOR (3 downto 1);
                                                    send_data : out  STD_LOGIC
  reset                                             );
             send_data
                                              end fuzzification;
        (a)                                                  (b)
```

Fig. 6. a) Schematic diagram of the Fuzzification stage. **b)** Entity of Fuzzification stage in VHDL.

Fig. 7. Xilinx Blockset Library after its installation. We used the Matlab/Simulink version 7.2, and the the Xilinx ISE software version 8.2i.

is to provide the degree of membership, and the other class is to indicate which linguistic variable was activated. The first output class is labeled as "degree_in1a", degree_in1b", etc. Attached to these outputs are three blocks, going from right to left, the first block is a Xilinx System Generator function known as "gateway out". The goal of this block is to perform a numeric conversion from eight bits fixed point, which is the output format of the fuzzification entity embedded in the "black box", to double precision needed in Simulink. The next block converts from 0 to 255 to values in the universe of discourse of the membership functions. The last Simulink block is a display used to visualize the results.

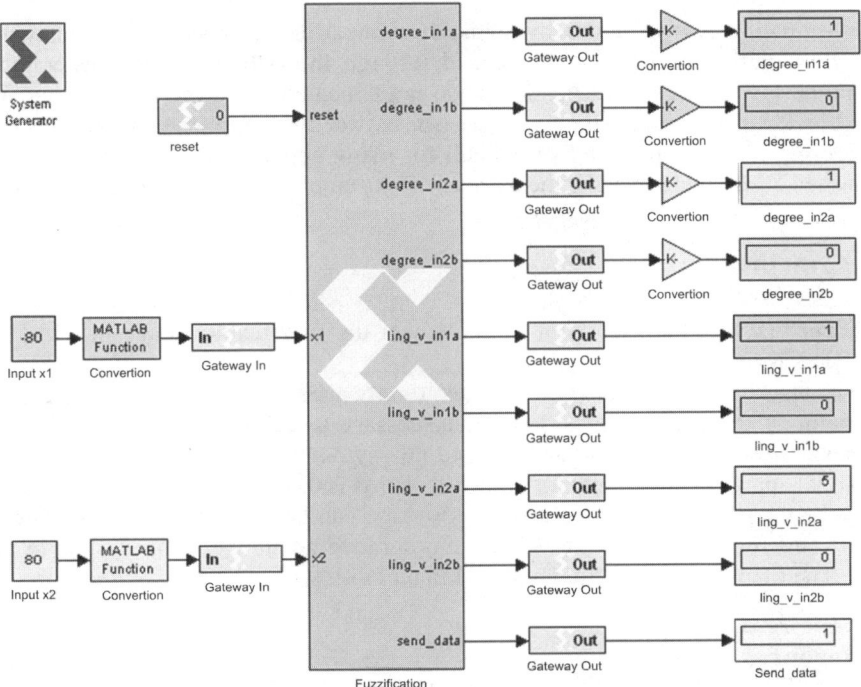

Fig. 8. Fuzzification stage coded in VHDL embedded in a "black box" of the Xilinx System Generator. The simulation was achieved using Simulink.

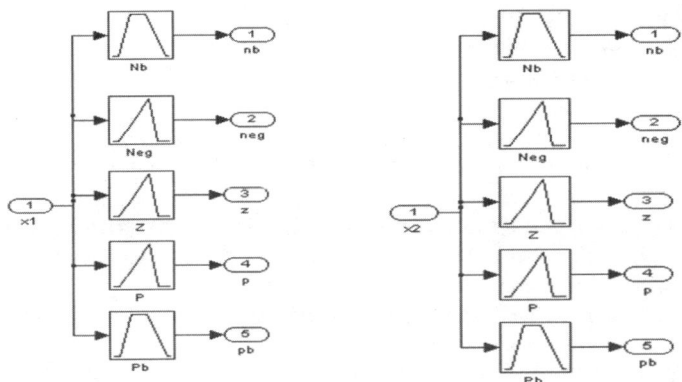

Fig. 9. Subsystem of the two inputs of fuzzification stage in Matlab

5. The same inputs given in figures 4 and 5 were programmed with the Fuzzy Logic Toolbox for Simulink. Fig. 9 shows the two subsystems, one for each input. Note that the idea is to visualize inputs and outputs values, hence we used the appropriated Simulink blocks to achieve this task.

4.2 Experiment Validation

The validation was achieved by statically comparing the results obtained with the VHDL fuzzification stage implemented through the Xilinx System Generator and Simulink [12,13], and fuzzification stage implemented using the Fuzzy Logic Toolbox block for Simulink. The maximal error between the implementations was 0.001. This error can be considered very small for many applications. The error can be diminished easily by increasing the number of bits employed in the fuzzification stage.

5 Conclusions

A novel fuzzification algorithm to implement the fuzzification stage in VLSI using VHDL was presented.

The algorithm has the next advantages: It only needs the values of each characteristic point of the straight lines, being unnecessary to calculate previously the slopes values. Allows the use of symmetric and non-symmetric membership functions. The maximal approximation error for eight bits words is 0.001.

Once the code has been validated, this stage can be combined with the other two stages (inference and defuzzification), in order to simulate the whole system in Simulink. The final code can be implemented in a FPGA.

References

1. Zadeh, L.A.: Fuzzy Sets. Information and Control 8, 338–353 (1965)
2. Web page of FPGA's, http://www.xilinx.com
3. Jose, O., Yazmin, M., Oscar, M., Roberto, S.: Logica Difusa en FPGA, agosto, pp. 23–24 (2007); difu100ci@
4. George, J.K., Yuan, B.: Fuzzy Sets and Fuzzy Logic Theory and Applications. Prentice Hall, Englewood Cliffs (1995)
5. Tsoukalas, L.H., Ohrig, R.E.: Fuzzy and Neura Approaches in Engineering. Wiley-Interscience, Chichester (1997)
6. Sánchez Solano, S., Cabrera, A., Jiménez, C.J., Brox, P., Baturone, I., Barriga, A.: Implementación sobre FPGA's de Sistemas Difusos Programables (2001), http://www.imse.cnm.es
7. Lago, E., Jiménez, C.J., Lopez, D.R. Solano, S. Barriga, A.: XFVHDL: A tool for Síntesis of Fuzzy logic Copntrollers, Design Automation and Test in Europe, pp. 102-107 (1998), http://www.imse.cnm.es
8. Miguel, A.: Melgarejo, Desarrollo de un Sistema de Inferencia Difusa sobre FPGA, Universidad Distrital Francisco José De Caldas (2003)
9. Philip, T.V., Asad, M.M., Jim, B.V.: VHDL Implementation for a Fuzzy Logic Controller. World Automation Congress, WAC 2006, 1–8 (2006)

10. Sanchez Solano, S., Barriga, A., Brox, P., Baturone, I.: Síntesis de Sistemas Difusos a partir de VHDL. In: Proc. XII Espanol Conference of Tecnologic and fuzzy logic (ESTYLF 2004), pp. 107-112 (2004), http://www.imse.cnm.es
11. Cirstea, M.N., Khor, J.G., McCormick, M.: Neural and fuzzy logic control of drives and power system, Newnes (2002)
12. Web page of Matlab-Simulink (2007), http://www.mathworks.com
13. Web page Xilinx System generator manufacturer, http://www.xilinx.com
14. Yazmin, M., Angel, O.J., Oscar, M., Roberto, S.: Implementacion de Maquinas Difusas en FPGA. In: Congreso Internacional de Ingenieria Electronica ELECTRO 2007, pp. 97–102 (2007)
15. Serra, M., Navas, O., Escrig, J., Bonamusa, M., Marti, P., Carrabina, J.: Metodologia de prototipado rapido desde Matlab: herramientas visuals para flujo de datos, (2004), http://www.uvic.cat

High Performance Parallel Programming of a GA Using Multi-core Technology

Rogelio Serrano[1,a], Juan Tapia[1], Oscar Montiel[1], Roberto Sepúlveda[1], and Patricia Melin[2]

[1] Centro de Investigación y Desarrollo de Tecnología Digital (CITEDI)
del IPN. Av. del Parque No. 1310, Mesa de Otay, 22510, Tijuana, BC, México
{serrano,jjtapia}@citedi.mx, {o.montiel,r.sepulveda}@ieee.org
[2] Division of Graduate Studies and Research, Calzada Tecnológico S/N, Tijuana,
México, Tijuana Institute of Technology, México
pmelin@tectijuana.mx
[a] M.S. Student of CITEDI

Abstract. Multi-core computers give the opportunity to solve high-performance applications more efficiently by using parallel computing. In this way, it is possible to achieve the same results in less time compared to the non-parallel version. Since computers continue to grow on the number of cores, we need to make our parallel applications scalable. This paper shows how a Genetic Algorithm (GA) in a non-parallel version takes long time to solve an optimization problem; in comparison, using multi-core parallel computing the processing time can be reduced significantly as the number of cores grows. The tests were made on a quad-core computer; a comparison of the speeding up in relation to the number of cores is shown.

1 Introduction

Genetic Algorithms as well as other Soft Computing techniques take lots of time to solve a problem, and before the development of multi-core computers, the most practical way to speed this system up was by applying distributed computation using a very expensive multi-processor computer or a cluster [1]. However, these relatively new architectures allow us to take the same techniques used in distributed computation and apply them on a single multi-core computer, so that we can take advantages of all its processing capability to solve a high performance application. Some of the advantages of multi-core architectures are shown at [2,3]. Parallel Genetic algorithms (PGAs) can provide considerable gains in terms of performance and scalability. PGAs can easily be implemented on networks of heterogeneous computers or on parallel mainframes. Several interesting applications of parallel computing are presented by Dongarra et al. [4], the white paper [5] presents a description on multi-core processor architecture. An introduction to Multi-Objective Evolutionary Algorithm can be found in [6].

The Distributed Computing Toolbox extends the Matlab technical computing environment to solve computationally and data-intensive problems using a multi-processor computing environment. The toolbox provides high-level constructs, such as parallel for loops and MPI- based functions, as well as low-level constructs for job and task management.

O. Castillo et al. (Eds.): Soft Computing for Hybrid Intel. Systems, SCI 154, pp. 307–314, 2008.
springerlink.com   © Springer-Verlag Berlin Heidelberg 2008

The Genetic Algorithm presented in this paper was implemented using the Distributed Computing Toolbox of MatLab 7.4a. which uses the Message Passing Interface model for distributed computation [7].

The goal of this work is to achieve an analysis of multi-core technology applied to the problem of GAs to solve the optimization of a path of a 2 Degree Of Freedom (DOF) arm in presence of an obstacle.

This paper is organized as follows, in Section 2 it is explained the problem to be solved as well as the characteristics of the GA; in Section 3 shows the experiments to evaluate the multi-core parallelization, and finally in Section 4 are the conclusions.

2 Problem Statement

The parallel algorithm consists in simulate a mechanical arm, which will be moved from its initial position **PA** to the final position **PB** on a plane surface as shown in Fig 1. In the movement it is considered the following conditions:

- There is an obstacle within the working area with a circular shape.
- The application must find a four-point trajectory to get from **PA** to **PB** without colliding with the obstacle.
- The distance between the ends of all points of the trajectory must be minimal.

The Genetic Algorithm has an initial chromosome composed of six pairs of genes; each pair correspond to the angles that defines the position of each articulation for the mechanical arm. The first and last pair of genes corresponds to the initial and final position, **PA** and **PB** respectively; the other four pairs correspond to the four points who will define the trajectory of the arm.

Each part of the arm is 4 units long. The first articulation can only be moved in the 90 degrees that define the first quadrant; the second articulation can be moved freely in his 360 degrees.

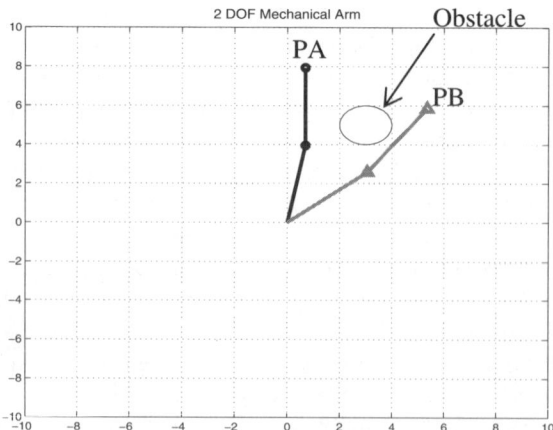

Fig. 1. The 2 DOF mechanical arm is avoiding the obstacle by going from point PA to PB

The Genetic Algorithm has the following characteristics:

- **Population size.** Defines the number of individuals that will compose each generation of the population.
    - *Population_Size =10*
- **Maximum of Generations.** Determinates how many generations will the genetic algorithm generate for solving the problem.
    - *Max_Generations=30*
- **Selection Probability.** When using the Roulette Wheel selection, each individual has a probability of been selected defined in part by their fitness.
    - *Selection_Prob=0.6*
- **Crossover Probability.** The selected individuals have a probability of mating, acting as parents to generate two new individuals that will represent them in the next generation. The crossing point can be random or can follow a certain pattern.
    - *Crossover_Prob=0.8*
- **Mutation Probability.** Represents the probability that an arbitrary bit in the individual sequence will be changed from its original stat.
    - *Mutation_Prob=0.4*
- **Chromosome Size.** It is the number of genes in each individual. For this application, each individual is defined by six pairs of values, representing the two angles of the mechanical arm. Each angle represents a gene; therefore each chromosome has 12 genes.

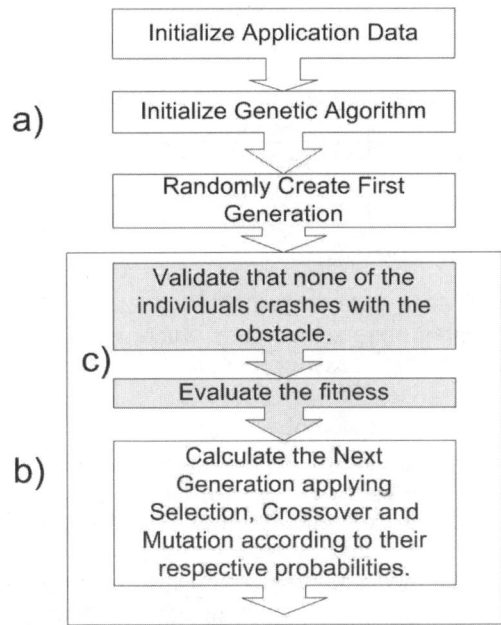

Fig. 2. Genetic Algorithm Diagram. (a) Initial configuration.- This block is done just one time. (b) Evaluation of each Generation.- Operations done for each generation. (c) Validation and Fitness Evaluation.- High performance module to be parallelized.

- **Obstacle.** The obstacle has a circle shape with radium of one, and its center is located at the coordinates (x,y)=(3,5).

This Genetic Algorithm is applied as shown in Fig 2. Other methods for solving a Parallel GA can be seen at [8,9,10].

3   Experiments to Evaluate Multi-core Parallelization

The Genetic Algorithm will be tested in a Multi-Core computer with following characteristics: CPU Intel Core 2 Quad Q6600, x86_64 2.4 GHz, Bus 1066 MHz, 8MB of L2 cache, Memory 2 GBytes DDR2 of main memory,   Local Disk 300 GBytes SATA, Operating System GNU/Linux Fedora 8, Kernel version 2.6.23.8-63.fc8. All the experiments were achieved in the MatLab Version 7.4a (2007).

Since we have written the application in MatLab, we can use its Implicit Parallel Mode [11] to get a free speedup. The Speedup is calculated by

$$S_p = \frac{T_1}{T_N},$$ \hfill (1)

where T_1 is the execution time in one core and T_N is the execution time in N cores.

The results of the parallelization are shown in Table 1, it can be seen that the explicit parallel mode presents a higher speedup compared to implicit parallel mode.

Table 1. Implicit vs Explicit Parallel Speedup

Cores	Implicit parallel mode		Explicit parallel mode	
	Time (s)	Speedup	Time (s)	Speedup
2	120.21	1.10	74.32	1.78
3	111.54	1.18	48.46	2.73
4	108.13	1.22	35.95	3.68

The MatLab Parallel Mode follows the MPI model for SIMD [12] (Single Instruction Multiple Data) programming architecture, where we have N initial threads who communicates through Message Passing. For this application we will use a Master-Slave Single Population model for solving a Parallel Genetic Algorithm [13]. This model defines that the Master node solves all the methodic modules of the Genetic Algorithm (Fig. 2(a) and (b)); and the Validation and Fitness modules will be solved in conjunction with the slaves nodes as shown on Fig. 2(c).

In MatLab's Parallel Mode we can find a pair of functions to communicate between two process, `labSend` and `labReceive`; but since we need to make our application scalable so that the same program runs from an N-core computer, we will use the function `labBroadcast`, which can be used both to send and receive a message.

After a successful parallelization of the Genetic Algorithm, using techniques for multi-threading programming [14,15], the speedup becomes higher as shown on

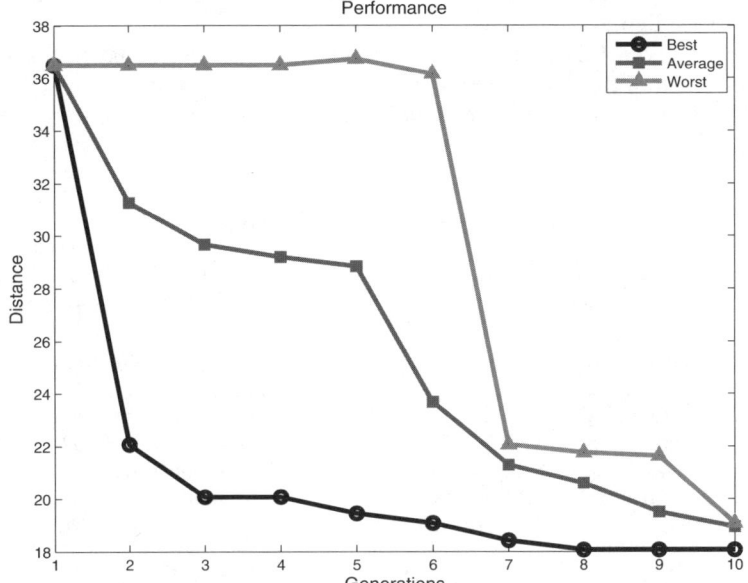

Fig. 3. Application performance

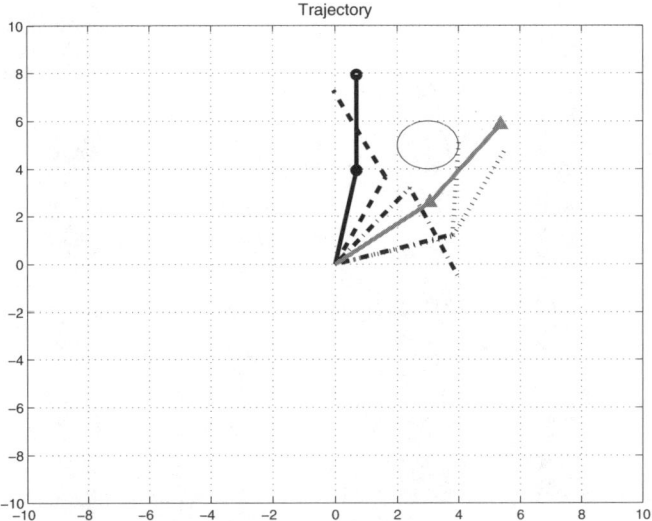

Fig. 4. Different trajectories followed by the 2 DOF arm robot

Table 1, now we can see that the explicit parallel mode presents a higher speedup compared to implicit parallel mode.

The performance of the Genetic Algorithm is shown in Fig 3, the best, average and worst case. The trajectory followed by the robot arm is shown in Fig. 4.

In Table 2 we can see the first and last configuration created by the genetic algorithm.

Table 2. Application Results

GENERATION 1: Best Chromosome: 1, with a Distance = **36.5**	GENERATION 10: Best Chromosome: 6, with a Distance = **18.1**
Robot Configurations:	Robot Configurations:
Initial Position : q1= 80, q2= 10	Initial Position : q1= 80, q2= 10
Configuration 2: q1= 74, q2= 91	Configuration 2: q1= 66, q2= 49
Configuration 3: q1= 12, q2= 92	Configuration 3: q1= 53, q2= 240
Configuration 4: q1= 57, q2= 100	Configuration 4: q1= 18, q2= 69
Configuration 5: q1= 50, q2= 96	Configuration 5: q1= 17, q2= 48
Final Position : q1= 40, q2= 15	Final Position : q1= 40, q2= 15

In Fig. 5 we can se the time taken by the Genetic Algorithm in each execution mode. Fig. 6 shows the final speedup performed by the Genetic Algorithms.

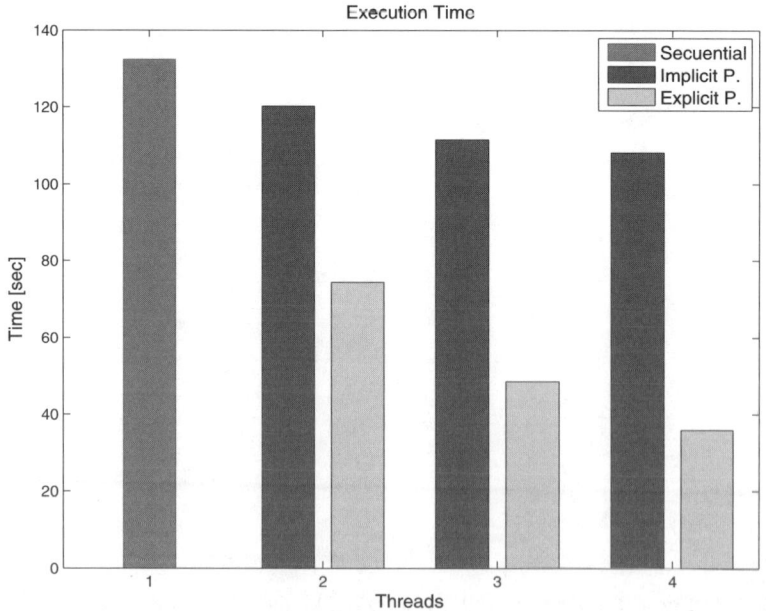

Fig. 5. Genetic Algorithm execution time

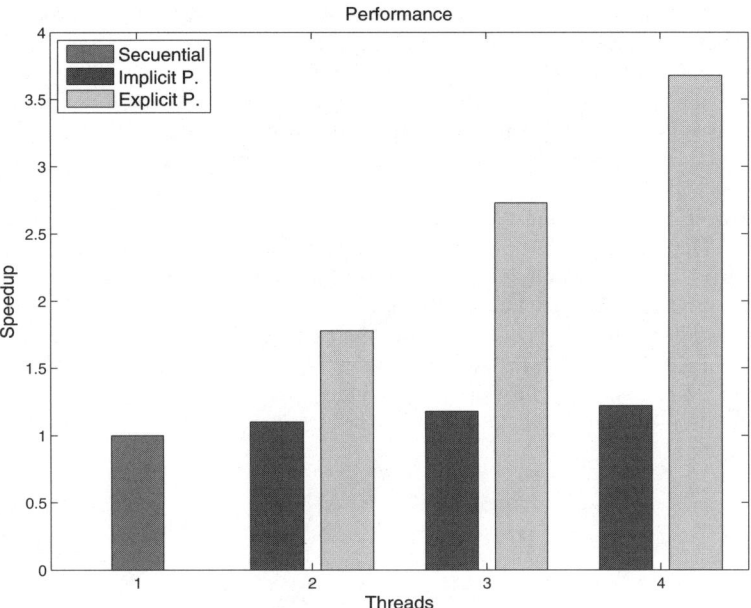

Fig. 6. Parallel Genetic Algorithm Speedup

4   Conclusions

Multi-core computers can help us solve high performance applications in a more efficient way by using parallel computation. On the other hand, Genetic Algorithm can be parallelized to speedup its execution; and if we use Explicit Parallelization we can achieve much better speedup than when using Implicit Parallelization. Therefore the study of parallelization techniques is required to implement high performance applications in a multi-core computer.

References

[1] Alba, E., Luna, F., Nebro, A.J.: Advances in Parallel Heterogeneous Genetic Algorithms for Continuous Optimization. International Journal of Applied Mathematics and Computer Science 14, 317–333 (2004)
[2] Domeika, M., Kane, L.: Optimization Techniques for Intel Multi-Core Processors, http://softwarecommunity.intel.com/articles/eng/2674.htm
[3] Chai, L., Gao, Q., Panda, D.K.: Understanding the Impact of Multi-Core Architecture in Cluster Computing: A Case Study with Intel Dual-Core System. In: The 7th IEEE International Symposium on Cluster Computing and the Grid (CCGrid 2007) (2007)
[4] Dongarra, J., et al.: Sourcebook of Parallel Computing. Morgan Kaufmann Publishers, San Francisco (2003)
[5] Burger, T.W.: Intel Multi-Core Processors: Quick Reference Guide, http://cache-www.intel.com/cd/00/00/20/57/205707_205707.pdf

[6] Coello, C.A., Lamont, G.B., Van Veldhuizen, D.A.: Evolutionary Algorithms for Solving Multi-Objective Problem. Springer, Heidelberg (2004)

[7] Snir, M., et al.: MPI: The complete Reference. MIT Press, Cambridge (1996)

[8] Sahab, M.G., Toropov, V.V., Ashour, A.F.: A Hybrid Genetic Algorithm For Structural Optimization Problems. Asian journal of civil Engineering, 121–143 (2004)

[9] Haupt, R.L., Haupt, S.E.: Practical Genetic Algorithms. Wiley-Interscience, Chichester (2004)

[10] Cantu-Paz, E.: Efficient and Accurate Parallel Genetic Algorithms. Kluwer Academic Publisher, Dordrecht (2001)

[11] Distributed Computing Toolbox User's Guide, Mathworks (2007)

[12] Edelman, A.: Applied Parallel Computing (2004)

[13] Nowostakski, M., Poli, R.: Parallel Genetic Algorithm Taxonomy

[14] Hwang, K., Xu, Z.: Scalable Parallel Computing. McGraw-Hill, New York (1998)

[15] Akhter, S., Roberts, J.: Multi-Core Programming. In: Increasing Performance through Software Multi-Threading. Intel Press (2006)

Scalability Potential of Multi-core Architecture in a Neuro-Fuzzy System

Martha Cárdenas[1,a], Juan Tapia[1], Roberto Sepúlveda[1], Oscar Montiel[1], and Patricia Melín[2]

[1] Centro de Investigación y Desarrollo de Tecnología Digital (CITEDI)
del IPN. Av. del Parque No. 1310, Mesa de Otay, 22510, Tijuana, BC, México
{cardenas,jjtapia}@citedi.mx, {o.montiel,r.sepulveda}@ieee.org
[2] Division of Graduate Studies and Research, Calzada Tecnológico S/N, Tijuana,
México Tijuana Institute of Technology, México
pmelin@tectijuana.mx
[a] M.S. Student of CITEDI

Abstract. Parallelism in hardware and software is necessary to solve aplications that require high processing. Parallel computers provide great amounts of computing power, the multi-core technology will be designed to increase performance and minimize heat. This paper shows the performance improvements with multi-core architecture and parallel programming applied in a Multiple Adaptive Neuro-Fuzzy Inference System, obtaining with this a significantly reduction of the processing time. In addition, it shows the comparison between the non-parallel and parallel implementation and the results obtained.

1 Introduction

Some applications develop large dimensions because the amount of data involved and their processing. In the last years, fuzzy logic has been used in a wide number of applications [1]. An important area of application is in the fuzzy logic controllers, however, to select the appropriate parameters, like a number, type, parameters, of the membership functions and rules for the performance desired is difficult in most cases. Adaptive Neuro-fuzzy Systems (ANFIS) facilitate the learning and adaptation [2], ANFIS systems combine the theory of artificial neural networks and fuzzy systems. The artificial neural networks provide effective learning methods whereas fuzzy theory allows working with uncertain data in an effective manner. Part of the problem with ANFIS is the Neural Networks, whose disadvantage is that the learning process can be long processing time.

This paper introduces the multi-core concept, and shows the performance and advantages results in an implementation of large time processing with parallel computing applied in the multi-core technology.

This paper is organized as follows, Section 2 describes the differences between single-core, multiprocessors, and multi-core architectures; it explains the models of parallel programming. Section 3 explains the ANFIS implementation to solve a problem of control position and motion of a robotic arm of three Degree Of Freedom (DOF), using non-parallel and parallel training; Section 5 presents the experimental results. Finally, section 5 presents the conclusions.

O. Castillo et al. (Eds.): Soft Computing for Hybrid Intel. Systems, SCI 154, pp. 315–323, 2008.
springerlink.com © Springer-Verlag Berlin Heidelberg 2008

2 Parallelism

In recent years, computing has undergone a change in technology. Architecture, etc., although this has not been able to evolve in as far as performance concerns, and that because do not fully exploit new technologies and architectures. Today, the computation demand higher performance, which can provide a conventional computer mono-processor in which programs followed the standard model of straight-line instruction execution proposed by the Von Neumann architecture.

The most commonly used metric in measuring computing performance is CPU clock frequency. Over the past 40 years, CPU clock speed has tended to follow Moore's law (the number of transistors available to semiconductor manufacturers would double approximately every 18 to 24 months), unfortunately impose limits in design of microprocessors. Also, rising the rate at which signals move through the processor, create a larger increase in heat and more energy consumption.

Currently the hardware advances faster than the software, the processor technologies now offer parallel personal computers; this opens the door to the implementation of parallel computing. The main goal of parallel computing is to optimize the performance and processing speed.

2.1 Multi-core

In order to achieve parallel execution in software, hardware must provide a platform that supports the simultaneous execution of multiple threads. Software threads of execution are running in parallel, it means that the active threads are running simultaneously on different hardware resources, or processing elements. Multiple threads may make progress simultaneously [3]. Now it is important to understand that the parallelism occurs at the hardware level too.

Multi-core processors technology is the implementation of two or more "execution cores" within a single processor. These cores are essentially two individual processors on a single chip. Depending on design, these processors may or may not share a large on-chip cache [3], the operating system perceives each of its execution cores as a discrete logical processor with all the associated execution resources [4].

These features allow us to see a single computer with a multi-core architecture, like a high-performance computer. The fast communication between the cores in the processor makes these great systems speedup compared with other multiprocessors systems, because of the intra-node speed and communication (intra-cores). Some advantages and experiments for evaluating multi-core architectures are shown at [5,6].

Figure 1, shows graphically the comparison of a) single-core, b) Multi-processor, and c) Multi-core architectures. In c) Multi-core, it has its own execution units, cache memory, etc., it allows the truly parallel computing and therefore increases the speedup of large applications.

The improvement measure or speedup takes as reference, the time of execution of a program in a mono-core system regarding the time of execution of the same program in a multiprocessor or multi-core system that is represented as follows:

$$speedup = \frac{t_1}{t_j}, \qquad (1)$$

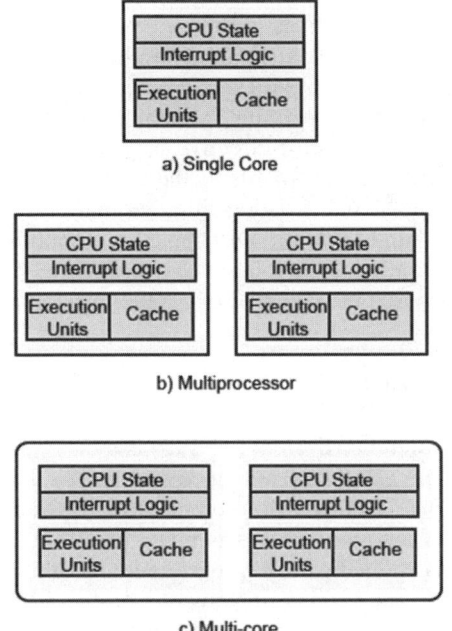

Fig. 1. Comparisons of a) single-core, b) Multi-processor, c) Multi-core architectures

where t_1 is the time it takes to run the program in a mono-processor system and t_j is the time it takes to run the same program in a system with j execution units, see Fig. 1.

2.2 Parallel Programming

There are many models of parallel programming, the two main choices and the most common are:

- Shared-memory programming model, all data accessed to a global memory, is accessible from all parallel processors; each processor can fetch and store data to any location in the memory independently.
- Message-passing model, data are associated with particular processors, so communication through messaging is required to access a remote data location.

Although these two programming models are inspired by the corresponding parallel computer architectures, their use is not restricted. It is possible to implement the shared-memory model on a distributed-memory computer, either through hardware (distributed shared memory). Symmetrically, message passing can be made to work with reasonable efficiency on a shared-memory system [7]. Programming techniques to effective parallelism and optimization techniques are shown in [8,9].

Furthermore, the parallelism can be implemented in two ways, implicit parallelism, that some compilers do automatically. These are responsible to generate the parallel code for the parts of the program that are parallel, on the other hand, the explicit parallelism is implemented using parallel languages, and the responsible of the parallelism is the programmer, that defines the threads to work, the code of each thread, the communication, etc., this last parallelism get higher performance.

The Message Passing Interface (MPI), is an important and increasingly popular standardized and portable message passing system, that brings us closer to the potential development of practical and cost-effective large-scale parallel applications. The major goal of MPI, as with most standards, is the degree of portability across the different machines [10]. The Matlab existing libraries that correspond to the functions MPI are denominated MatlabMPI, the Distributed Computing Toolbox works with MPI, and the MatLab Parallel Mode follows the MPI model [11,12], so with explicit parallelism it can be increased the potential of parallel programming.

3 ANFIS Implementation

Adaptive Neuro-Fuzzy Inference System (ANFIS) is a class of adaptive networks that are functionally equivalent to a fuzzy inference system [2]. The advantage of ANFIS is the Neural Network like facility to represent a fuzzy system, however one of its disadvantages is that the learning process is relatively slow, i.e. many train epochs. This can be solved if it is applied parallel programming models.

The problem to solve is the parallel implementation of a MANFIS model to control position and movement of a robotic arm. The programming took place in Matlab 7.4 R2007a, like the original implementation Non-parallel. The parallel implementation carries the same architecture that the original network, this is in order to have a comparison and show the advantages of parallelism.

The objective is obtain a model to solve the inverse kinematics model, using a direct kinematics model , to solve the problem of control position and motion for a robotic arm of 3 DOF as it is shown in Fig. 2 in 3 dimensions [13].

First there were obtained the coordinates in space based on the angles chosen, this was carried out by the equation of direct kinematics, this gives a data table, to be used for training the proposed model.

Taking the table of data training, it was perform the inverse modelling, i.e. we used the values of the coordinates in space to get the right angles of each articulation of the robot when it is placed in a desired position.

The proposed model learns the behaviour of the data of the table, comparing the desired output, t which is in the table of data training, with the output calculated, to get the difference error that exists through the use of optimization methods. The goal is that the error tends to zero as time progresses.

The system consists of 3 ANFIS with a set of independent rules, corresponding to the angles of the three degrees of freedom of the arm, theta1, theta2, theta3. The architecture is a MANFIS model of 3 input and 3 output as shown in Fig. 3, where x,y and z are the corresponding inputs of the coordinates in the space and $\theta_1, \theta_2, \theta_3$ the corresponding outputs to the angles of each articulation of the robot.

The number of Membership Functions for $\theta_1, \theta_2, \theta_3$ is 7, 6 and 5 for ANFIS1, ANFIS2 and ANFIS3 respectively. There were built training tables for each degree of freedom, i.e. (x,y,z,θ_1), (x,y,z,θ_2), (x,y,z,θ_3) for each ANFIS.

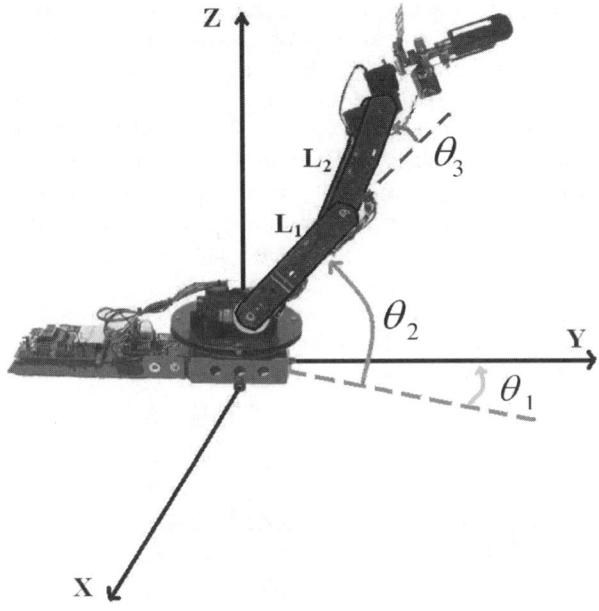

Fig. 2. Robotic arm used in this problem

Table 1. Characteristics of the computers for the model training

Non-Parallel Training	Parallel Training
Intel Pentium IV 2.4 Ghz	Intel Core 2 Quad core 2.4 Ghz
RAM 512 MB	RAM 2 Gb
40 Gb HD	320 Gb HD

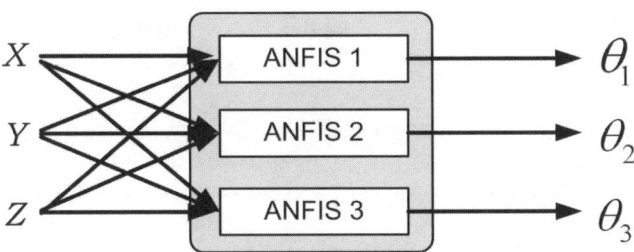

Fig. 3. MANFIS architecture

As shown in Table 1, the parallel training was carried out on a computer with parallel architecture, a Multi-Core computer Quad-Core, among them were distributed the 3 ANFIS system.

The parallel training was implemented using the tool Interactive Parallel Mode or Pmode of Matlab, it is part of the toolbox of distributed computing. This tool allows us to work with up to 4 laboratories or instances simultaneously of Matlab, and lets you work interactively with a parallel job running simultaneously on several labs. The labs receive commands entered in the Parallel Command Window process them, and send the command output back to the Parallel Command Window. Variables can be transfered between the MATLAB client and the labs [14]. Besides of Matlab functions, Pmode provides functions, makes parallel sections of code, the message passing between different laboratories, i.e. we can send and receive data from one laboratory to others, copy variables, etc.

One of the most important advantages of working with the multi-core architecture and the Pmode tool, is that each laboratory of Matlab runs in a specific core. These features allow us to view this as a distributed architecture.

4 Experimental Results

This section presents the performance numeric results of the training experiments in Non-parallel and parallel mode. First it is presented the time consumed to train the MANFIS in a mono-core system, next the results obtained in the training of the same MANFIS, but in a multi-core system. The differences in time are very significantly, as can be seen on Tables 2, 3 and 4.

Table 2. Training the MANFIS with 150 epoch and 8000 data

Angles	Hours of training	Error
θ_1	240 hrs	0.06486
θ_2	168 hrs	0.01162
θ_3	84 hrs	0.01508

Table 3. Train times with 300 epoch and 4096 data

Angles	Hours of train	Error
θ_1	27.2217 hrs	0.00816
θ_2	12.01379 hrs	0.0036
θ_3	3.7852 hrs	0.0064

Table 4. Train Times with 150 epoch and 8000 data

Angles	Hours of train	Error
θ_1	26.10744 hrs	0.00816
θ_2	12.01379 hrs	0.0036
θ_3	3.8066 hrs	0.0064

Non-Parallel Original Mode

Table 2, shows the time of training in hours, of non-parallel training for the MANFIS model proposed, with 150 epoch and 8000 data. The next Tables 3 and 4, show the time of training in hours of parallel training with 300 epoch and 4096 data, and with 150 epoch and 8000 data.

Parallel Mode

In Fig 4, we can see the first 3 cores that correspond to the 3 ANFIS, they work at 100%, while the fourth core remains coordinating the others, this model is known as master-slave.

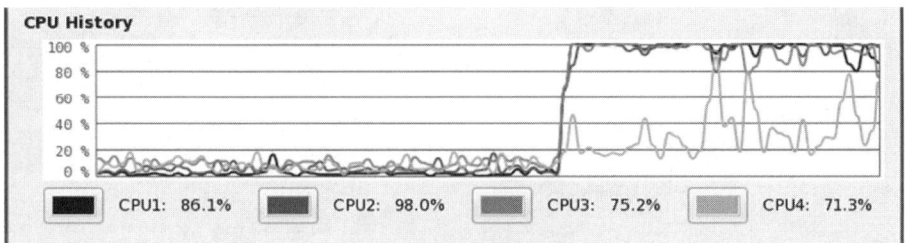

Fig. 4. CPU performance

Figure 5, shows a performance comparison between a parallel and a non-parallel training.

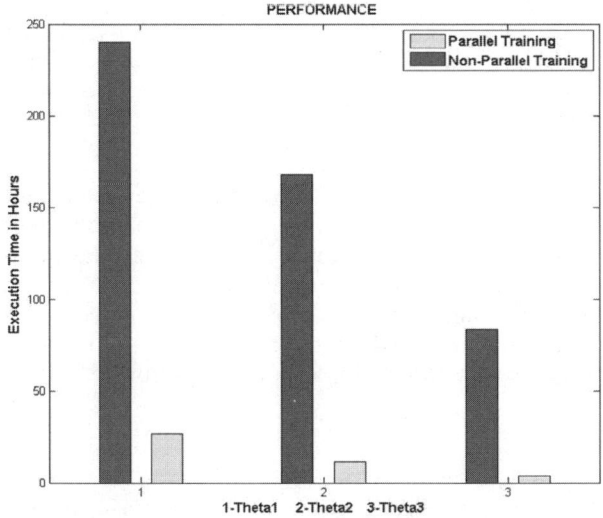

Fig. 5. Performance between non-parallel training and parallel training

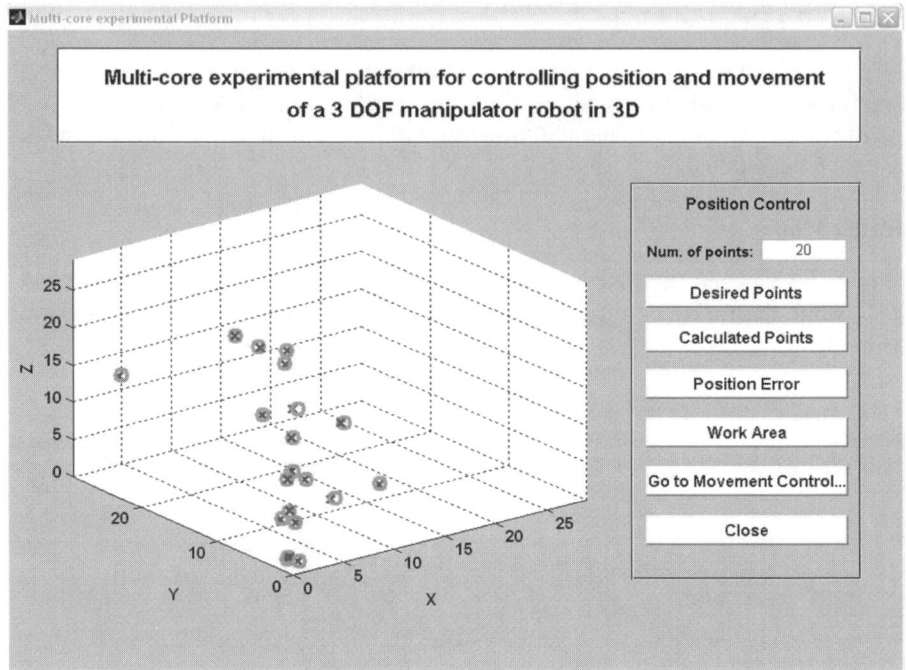

Fig. 6. Multi-core experimental platform for controlling position and movement of a 3 DOF robotic arm

The resultant trained ANFIS was tested in a multi-core experimental platform for controlling position and movement of a manipulator robot, and the results are shown in the Fig. 6. The 'O' light grey points are the Desired Points generated random, and the 'x' dark grey points are the calculated points for the train MANFIS system.

5   Conclusions

In this paper, improvements of using multi-core architectures and parallel programming in an application with processing large time were shown, the results of performance take a full advantage of this technology to minimize in a large-scale the processing time.

Multi-core processing continues with a significant impact on software evolution resulting in potential performance increases providing more processing at a lower price with lower power consumption and heat problems. Multi-core technology will be predominant in the next years, this encourages writing software in a parallel form.

References

[1] Melin, P., Castillo, O.: Hybrid Intelligent Systems for Pattern Recognition Using Soft Computing. In: An Evolutionary Approach for Neural Networks and Fuzzy Systems. Springer, Heidelberg (2005)

 [2] Jang, J.-S.R., Sun, C.-T., Mizutani, E.: Neuro-Fuzzy and Sof Computing. In: A computational Approach to learning and Machine Intelligence. Prentice-Hall, Englewood Cliffs (1997)
 [3] Akhter, S., Roberts, J.: Multi-Core Programming. In: Increasing Performance Through Software Multi-threading. Intel Press (2006)
 [4] Burger, T.W.: Intel Multi-Core Processors: Quick Reference Guide, http://cache-www.intel.com/cd/00/00/20/57/205707_205707.pdf
 [5] Chai, L., Gao, Q., Panda, D.K.: Understanding the Impact of Multi-Core Architecture in Cluster Computing: A Case Study with Intel Dual-Core System. In: The 7th IEEE International Symposium on Cluster Computing and the Grid (CCGrid 2007)
 [6] Chai, L., Hartono, A., Panda, D.K.: Designing High Performance and Scalable MPI Intranode Communication Support for Clusters. In: The IEEE International Conference on Cluster Computing (Cluster 2006) (September 2006)
 [7] Dongarra, J., et al.: Sourcebook of Parallel Computing. Morgan Kaufmann Publishers, San Francisco (2003)
 [8] Tian, T. Shih, C.-P.: Software Techniques for Shared-Cache Multi-Core Systems, http://softwarecommunity.intel.com/articles/eng/2760.htm
 [9] Domeika, M., Kane, L.: Optimization Techniques for Intel Multi-Core Processors, http://softwarecommunity.intel.com/articles/eng/2674.htm
[10] Snir, M., Otto, S., Huss-Lenderman, S., Walker, A., Dongarra, J.: MPI: The complete Reference. MIT Press, Cambridge (1996)
[11] Hwang, K., Xu, Z.: Scalable Parallel Computing. McGraw-Hill, New York (1998)
[12] Edelman, A.: Applied Parallel Computing (2004)
[13] Saldivar, P.M.: Control de un Brazo Mecánico Mediante Técnicas de Computación Suave. M.S. thesis, CITEDI-IPN (2007)
[14] Distributed Computing Toolbox 3 User's Guide, Mathworks (2007)

Methodology to Test and Validate a VHDL Inference Engine through the Xilinx System Generator

José Á. Olivas[1,a], Roberto Sepúlveda[1], Oscar Montiel[1], and Oscar Castillo[2]

[1] Centro de Investigación y Desarrollo de Tecnología Digital (CITEDI)
del IPN. Av. del Parque No.1310, Mesa de Otay, 22510, Tijuana, BC, México
olivas@citedi.mx, {o.montiel,r.sepulveda}@ieee.org
[2] Division of Graduate Studies and Research, Calzada Tecnológico S/N, Tijuana, México
ocastillo@hafsamx.org
[a] M.S. Student of CITEDI

Abstract. There exists an increasing interest in the field of digital intelligent systems, being one of the current research target the computational efficiency. This work presents the implementation of a fuzzy inference system faced to achieve high performance computations since the highly flexibility that a specific tailored FPGA implementation can offer to parallelize processes. A methodology to simulate and validate the inference engine developed in VHDL is given. Improvements over an exciting inference engine are proposed. The resulting code can be implemented in specific application hardware.

1 Introduction

Fuzzy logic is a branch of artificial intelligence that is based on the concept of perception, which allows handling vague information or of difficult specification. Fuzziness concept is born when one left to think that all phenomena founded in everyday world may qualify under crisp sets, when it is admitted the necessity to mathematically express vague concepts that can not be represented in an adequate way using crisp values [1].

The fuzzy systems are increasingly used because they tolerate inaccurate information, can be used to model nonlinear functions of arbitrary complexity to build a system based on the knowledge of experts using a natural language [2].

There are many choices to implement a fuzzy inference system [3,4,5], this work presents the implementation of a fuzzy inference system faced to achieve high performance computations since the highly flexibility that a specific tailored FPGA implementation can offer to parallelize processes. Any hardware implementation of an electronic system requires a complex methodology to test and validate every stage in the design process to guarantee its correct functionality. In particular, the use of VHDL to develop a product is an appealing tool, because it can be used in the different stages involved in taking a conceptual idea by means of its formal hardware description to the final product [6].

In this work we are presenting a slightly different architecture of an inference engine that the one proposed in [7,8,9]. Although the difference is substantially small it gives the user some improvements in flexibility to make modifications.

O. Castillo et al. (Eds.): Soft Computing for Hybrid Intel. Systems, SCI 154, pp. 325–331, 2008.
springerlink.com

A methodology to simulate and validate the inference engine developed in VHDL is given. In order to achieve this goal, it was used the Xilinx System Generator (XSG) [10], which is an integrated design environment (IDE) for FPGAs level system, that uses Simulink as a development environment and is present in the form of blocks [11]. It has an integrated design flow, to move straight to the configuration file (*. bit) necessary for the programming of FPGA. XSG can automatically generate VHDL code and an ISE project of the model being developed [12]. It is possible to make hierarchical synthesis VHDL, expansion and mapping hardware, in addition to generating UCF files, files simulation, test vectors and test bench files, among other things. XSG was created principally to work with complex DSPs (Digital Signal Processing) applications [13], but it has other applications as is the topic of this paper.

This paper is organized as follows Section 2 explains the High performance Inference engine architecture, it is explained trough an easy example how the fuzzification stage works; in Section 3 it is presented the Simulation and test of the inference engine using the XSG, Section 4 shows the evaluation of the operation of this stage with the remaining stages of the fuzzy system created in Matlab/Simulink; finally Section 5 discusses some conclusions of this work.

2 High Performance Inference Engine Architecture

The inference engine described in this section allows the realization of fuzzy inference systems known as SISC (Singleton Input Singleton Consequent) that are very common in control applications where inputs are crisp values usually collected from sensors and can be represented with fuzzy singletons. In this kind of realization, each rule has its own consequent defined by its firing strength.

The SISC realization has two main advantages that help to reduce the hardware cost by mean of reducing its complexity. Hence, the system considers processing only the active rules, and limiting to two the degree of overlapping between membership functions.

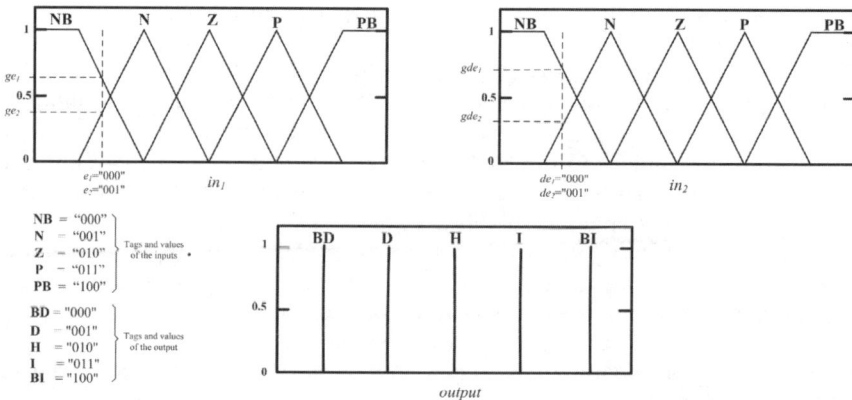

Fig. 1. Fuzzy inputs and outputs. Each fuzzy input has membership named as Negative Big (NB), Negative (N), Zero (Z), Positive (P), Positive Big (PB). The output has five singletons called Big Decrease (BD), Decrease (D), Hold (H), Increase (I), Big Increase (BI).

The inputs and output names were used in a motor speed fuzzy controller application. Each input has five MFs (NB, N, Z, P, PB) and their respective labels have associated binary values; for example, the label NB (Negative Big) is associated with the binary value "000", etc. The output has five singleton values labeled as (BD, D, H, I, BI) with their corresponding associated binary values, i.e., the label BD (Big Decrease) is associated with the binary value "000", etc.

The architecture used in this work is illustrated in Fig. 2. First, it is the fuzzification stage, it has two inputs, each input produces two fuzzified values, i.e., in input 1 (in_1) the dashed line indicates a crisp value that is being fuzzified, so it will produce the labels e_1 and e_2 with their corresponding degree of membership ge_1 and ge_2. The label value for e_1 is "000", and "001" for e_2. The input 2 (in_2) is handle in similar way.

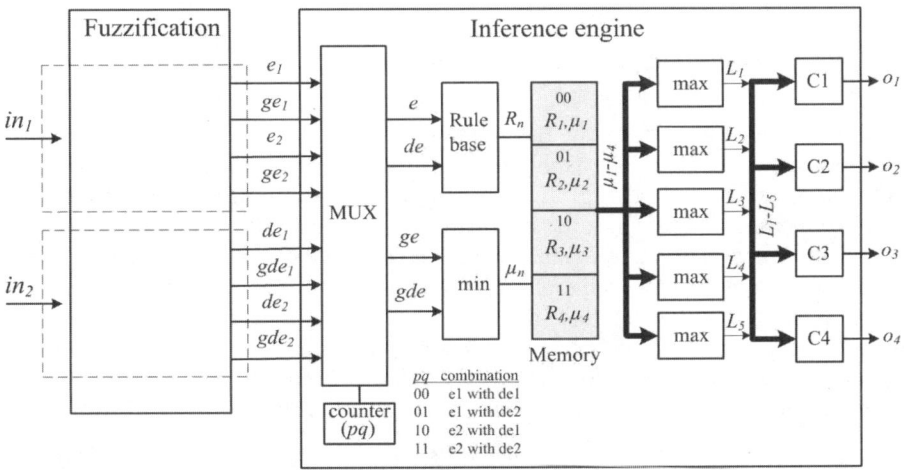

Fig. 2. Fuzzification and Inference engine stages. The system has two crisp inputs named in_1 and in_2. The fuzzification stage gives four output values for each input, they are the input of the Inference engine, which has four outputs (o_1-o_4). Each output has two values, the label and the firing strength of a set of rules.

The values e_1, e_2, ge_1 and ge_2 from the fuzzification stage are the inputs of the inference engine. They are connected to the MUX block which function is to address each combination of e_i and de_i to joint the labels values in order to obtain a rule combination to determine which output fuzzy set is involved. For example if the counter value pq in Fig. 2 is "00", and considering the dashed line of in_1 in Fig. 1, we will have the rule combination "000000", for pq="01" the rule combination is "000001", for pq="10" the combination is "001000", and for pq="11" is "001001". A rule that uses this codification is written as follows:

If e_1 is "000" and de_1 is "000" then BI

In Fig. 3 is illustrated a code piece that handles the rules. Using the above rule, the antecedent (ante) is form by the concatenation of the values e_1 and e_2, so we have the "ante" value of "000000".

```
case ante is
    when "000000" => c_n <= BI;          when "010011" => c_n <= H;
    when "000001" => c_n <= BI;          when "010100" => c_n <= D;
    when "000010" => c_n <= BI;          when "011000" => c_n <= D;
    when "000011" => c_n <= BI;          when "011001" => c_n <= H;
    when "000100" => c_n <= BI;          when "011010" => c_n <= D;
    when "001000" => c_n <= BI;          when "011011" => c_n <= D;
    when "001001" => c_n <= I;           when "011100" => c_n <= BD;
    when "001010" => c_n <= I;           when "100000" => c_n <= BD;
    when "001011" => c_n <= H;           when "100001" => c_n <= BD;
    when "001100" => c_n <= I;           when "100010" => c_n <= BD;
    when "010000" => c_n <= I;           when "100011" => c_n <= BD;
    when "010001" => c_n <= H;           when "100100" => c_n <= BD;
    when "010010" => c_n <= H;           when others => c_n <= "111";
                                     end case;
```

Fig. 3. VHDL code used to determine the labels of the active rules

For this rule, we also calculate the "min" of both fuzzified values, since we are using the max-min method. The label and the firing strength are saved into a memory position, tagged as "00" in Fig. 2. The whole process is repeated for all the active rules; the maximal is four since the method that we are using.

Once the rules have been evaluated and saved their results in memory, the next step is to calculate the max value of all memory positions tagged with the same label. Finally, at the outputs we can have the four possible consequents tags and theirs firing strength, this is indicated in Fig. 2 by four output values o_1-o_4. For example for the output o_1 can be given by the content of C1 and the resulting firing strength after applying the "max" operation to the corresponding set of rules.

3   Test of the Inference Engine

Every Simulink model that uses any block from the Xilinx Blockset must contain the block called "System Generator" since in this block is the relevant information of the card where the final application will be implemented, as well as one of the System Generator block. It is possible to specify how code generation and simulation should be handled.

In order to provide bit-accurate simulation of hardware, System Generator blocks operate on Boolean and arbitrary precision fixed-point values.

In contrast, Simulink uses double precision floating point as the fundamental scalar signal type.

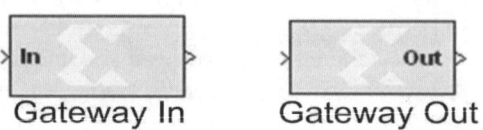

Fig. 4. Gateway Blocks of the XSG

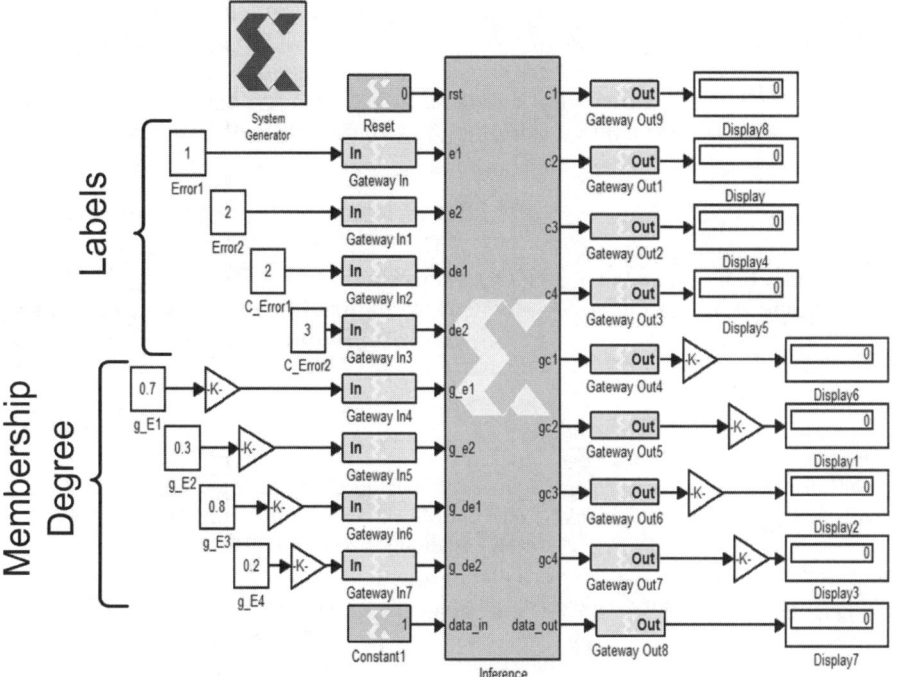

Fig. 5. Simulink model to test the Inference Engine

The connection between Xilinx blocks and non-Xilinx blocks is made through input and output "gateway blocks", they are shown in Fig. 4. The "Gateway In" converts a double precision signal into a Xilinx signal, and the "Gateway Out" converts a Xilinx signal into double precision. Simulink continuous time signals must be sampled by the Gateway In block [10].

To test the inference engine we simulated the VHDL developed code in Matlab/Simulink and the XSG. The XSG allows integrating the developed VHDL code into the Simulink through the use of a Black Box. The created Simulink model and the VHDL design entity must be kept in the same directory. The VHDL code is imported using a XSG Black Box which automatically opens a wizard to add the file to be imported. At this time, the file automatically generates a Matlab function (M-file) which is associated with the Black Box. Fig. 5 shows the simulation model of the inference engine.

4   Testing the Fuzzy System

Once the simulation and evaluation of the VHDL inference engine has been done, the next step is to test it with the fuzzification and defuzzification stages programmed using the appropriated Matlab function from the Fuzzy Matlab Toolbox. The idea is to provide the inference engine with labels and membership degrees values generated using a known sequence to obtain at the output, after the defuzzification stage, known values. The Simulink model is shown in Fig. 6.

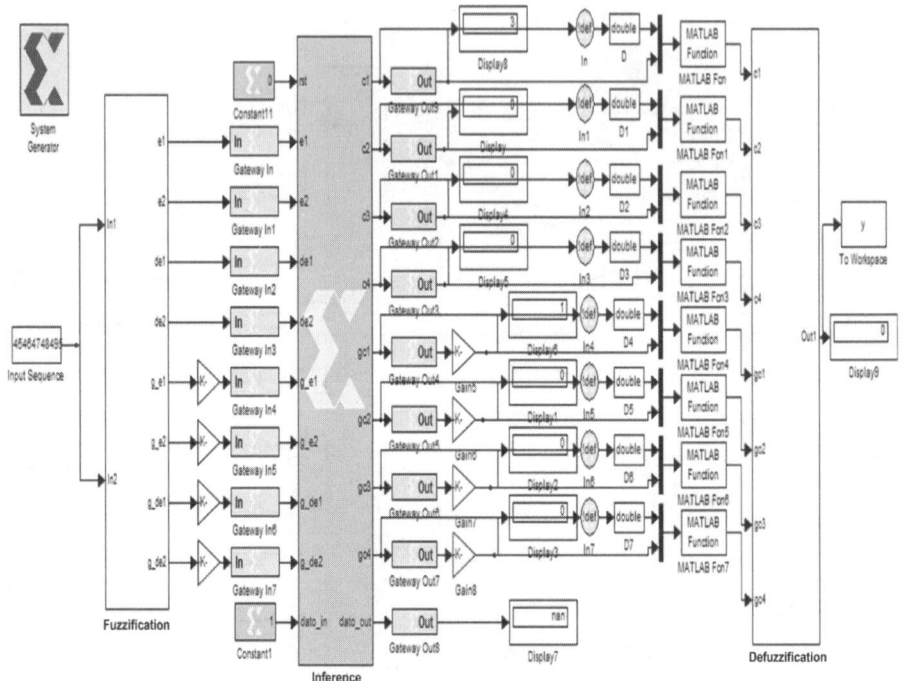

Fig. 6. Simulink model of the Fuzzy System

5 Conclusions

Modifications made to the original model to implement an inference engine, using a hardware description language, give a more practical and flexible model since the process of updating any change in the rule base can be made easily. Modeling and simulating a hardware design to obtain a final product is not easy. The methodology proposed in this paper is an alternative to quickly achieve both goals, since the developed code to describe the hardware, i.e. the system model, can be used with any modification to simulate the system and make the system implementation in the final target, for example in an FPGA.

References

1. Tsoukalas, L.H., Uhring, R.E.: Fuzzy and Neural Approaches in Engineering. Wiley-Interscience, Chichester (1997)
2. Yazmín, M., Ángel, O.J., Oscar, M. Roberto, S.: Implementación de Maquinas Difusas en FPGA´s, XXIX Congreso Internacional De Ingeniería Electrónica (Electro 2007), Chihuahua, México, (17-19 de octubre 2007)
3. Barriga, A., Marban, M.A., Solano, S., Brox, P., Cabrera, A.: Modelado de Alto Nivel e Implementación Sobre FPGA's de Sistemas Difusos. III jornada de computación reconfigurable y aplicaciones (JRCA 2003), 359–366 Madrid, September 10-12 (2003)

4. Vuong, P.T., Madni, A.M., Vuong, J.B.: VHDL Implementation For a Fuzzy Logic Controller. World Automation Congress, July 24-26, pp. 1–8 (2006)
5. Sánchez Solano, S., Barriga, A., Brox, P., Baturone, I.: Síntesis de Sistemas Difusos a Partir de VHDL, XII Congreso Español de Tecnologías y Logica Fuzzy (ESTYLF 2004), pp. 107-112, Jaén, Septembre 15-17 (2004)
6. Galan, D., Jiménez, C.J., Barriga, A., Sanchez Solano, S.: VHDL Package for Description of Fuzzy Logic Controllers. In: European Design Automation Conference (EURO-VHDL 1995), Brihton-Great Britain, September 18-22, pp. 528–533 (1995)
7. Miguel, A., Melgarejo, R.: Desarrollo de un sistema de inferencia difusa sobre FPGA, universidad distrital francisco josé de caldas, centro de investigación y desarrollo cientifico, 21 (2003)
8. Cirstea, M.N., Dinu, A., Khor, J.G., McCormick, M.: Neural and Fuzzy Logic Control of Drives and Power Systems, Newnes (2002)
9. Jimenez, C.J., Barriga, A., Sánchez Solano, S.: Digital Implementation of SISC Fuzzy Controllers. In: III Internacional Conference on Fuzzy Logic, Neural Nets and Soft Computing (IIZUKA 1994), pp. 651–665, IIzuka-Japan, (August 1-7, 1994)
10. Manual of Xilinx System Generator, http://www.xilinx.com
11. Moctezuma Eugenio, J.C., Sanchez Galvez, A., Ata Perez, A.: Implementacion Hardware de Funciones de Transferencia para Redes Neuronales Artificiales, Iberchip (2006)
12. Moisés Serra, Oscar Navas, Jordi Escrig, Marti Bonamusa, Pere Marti, Jordi Carabina, Metodología de prototipado Rápido desde Matlab: Herramientas Visuales para Flujo de Datos, http://www.uvic.cat
13. Moctezuma,J.C., Huitzil, E. C. T.: Estudio Sobre La Implementación de Redes Neuronales Artificiales Usando Xilinx System Generator, Facultad de Ciencias de la Computación, Benemérita Universidad Autónoma de Puebla, México (2006)

Modeling and Simulation of the Defuzzification Stage Using Xilinx System Generator and Simulink

Gabriel Lizárraga[1,a], Roberto Sepúlveda[1], Oscar Montiel[1], and Oscar Castillo[2]

[1] Centro de Investigación y Desarrollo de Tecnología Digital (CITEDI)
del IPN. Av. del Parque No.1310, Mesa de Otay, 22510, Tijuana, BC, México
lizarraga@citedi.mx, {o.montiel,r.sepulveda}@ieee.org
[2] Division of Graduate Studies and Research, Calzada Tecnológico S/N, Tijuana, México
ocastillo@hafsamx.org
[a] M.S. Student of CITEDI

Abstract. Nowadays, there is an increasing interest in using FPGA devices to design digital controller, and a growing interest in control systems based on fuzzy logic where the Defuzzification stage is of primordial importance. In this work we are presenting the design, modeling and simulation of a fixed point defuzzification VHDL method. The modeling and simulation of this stage is realized in Simulink through the Xilinx System Generator, and a second inference system was implemented with Matlab code. Comparative analysis of both systems and result are shown.

1 Introduction

Since the development of digital technology, there has been a trend in manufacturing products sharing the common goals of being smaller in integration, highly efficient in power consumption, faster regarding processing speed, and others. Several technologies have emerged and dead in the last 40 years following the Moore's Law crawling to be ever smaller in size. Alternative but parallel ways to develop high performance applications is to use VLSI programmable devices with the idea of developing the applications in software. In this field, the use of devices such as Field Programmable Gate Array (FPGA) is a very good option because this technology offers appealing characteristics for designers, some of them are the high scale of integration, low power consumption, the existence of high level languages to develop and simulate the application code, and reprogrammability.

Nowadays, there is an increasing interest in using FPGA devices to design digital controllers, and a growing interest in control systems based on fuzzy logic, since they allow compensating inaccuracies in the data from the instrumentation systems, such as noise. This work is about digital fuzzy controllers, it is focused in the Defuzzification stage which is of primordial importance in this kind of controllers.

There are several works around this topic; however it is well known that designing of functional VHDL modules involves a several step process where it is common to write VHDL simulation test benches to make exhaustive simulations, in [1] is given simulation code to achieve this task.

O. Castillo et al. (Eds.): Soft Computing for Hybrid Intel. Systems, SCI 154, pp. 333–343, 2008.
springerlink.com   © Springer-Verlag Berlin Heidelberg 2008

In this case the Defuzzification stage was simulated using the Software Xilinx System Generator (XSG) [2], which is a software tool that creates and verify hardware designs for Xilinx FPGA's[3,4].

One of the contributions of this work is that we are giving an alternative way to test the final VHDL module avoiding writing a VHDL test bench.

In [1] was proposed a deffuzzified architecture that handles fixed point with arithmetic for real values. In contrast, we are proposing the use of a modified high performance fixed point architecture for positive numbers, and make at the final stage the conversion to real numbers. Moreover, comparative numerical analysis is achieved.

This paper is organized as follows, Section 2 presents in a general context the defuzzification method using VHDL, it is explained trough an easy example how the Defuzzification stage works; in Section 3 it is presented the experimental set up, as well as the software tools used; Section 4 discusses the experiments and results with a VHDL Simulink model of the Defuzzification stage using Xilinx System Generator and shows the evaluation of the operation of this stage with the remaining stages of the fuzzy system created in Matlab/Simulink, with three cases of study. Section 5 presents the conclusions of this work.

2   General Contexts

Fuzzy logic is a mathematical method to obtain approximate reasoning, emulating the human brain mechanism from facts described using natural language with uncertainty. A Fuzzy Inference System (FIS) is based on fuzzy logic and consists of three stages which are called: Fuzzification, Inference and Defuzzification.

The Fuzzification stage transforms the crisp values to fuzzy values [5,6,7]. The Inference Engine is the core of the fuzzy logic system, here are proposed rules of the form IF-THEN that describe the behavior of a system [8,9]. The Defuzzification stage, involves extracting a crisp numerical value from a fuzzy set [1, 10, 11].

The explanation of this work is made using a two input and one output FIS. Figure 1 shows the general scheme of the FIS with the three stages, Fuzzification, Inference and Defuzzification.

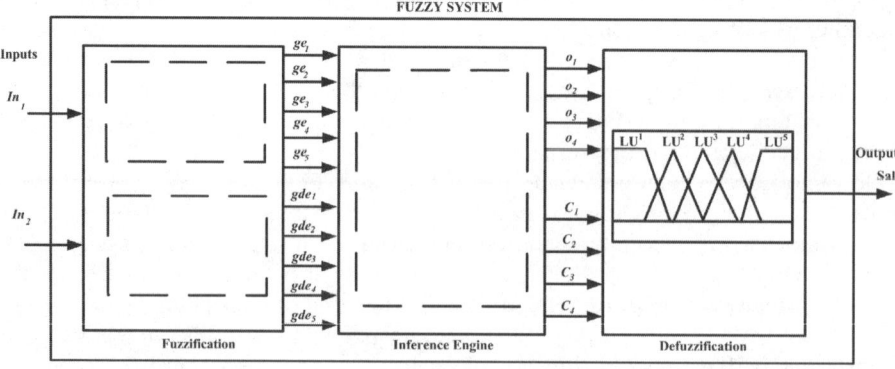

Fig. 1. Fuzzy Inference System

The fuzzy controller has two inputs and one output, the first ones are connected to the Fuzzification stage which produces a fuzzy output for each evaluation. A fuzzy outputs has two values, the membership grade and the linguistic value, that are represented in this work as *ge* values for the input *In₁*, and *gde* for the input *In₂* arranged in order according how the linguistic values are ordered in their universe of discourse; in this way, for a five membership function we have ge_1, ge_2,..., ge_5 and gde_1, gde_2,..., gde_5 respectively, that are the Inference Engine inputs. The output values of the Inference Engine named as o_1, o_2, o_3, o_4 and C_1, C_2, C_3, C_4 are the four possible grades of activation with their respective consequent values. The Defuzzication stage using the *o's* values and *C's* tags produces a crisp value using the Height defuzzification method.

Considering that the FIS is going to be used in a dc speed control application, we can set some parameters in order to make numeric calculation to compare the VHDL implementation against the code developed in Matlab. In this controller, there are two inputs called error and cerror (change of error). Each input has five membership functions, where two are trapezoidal and three are of triangular type. The output has five membership functions, two trapezoidal type and three triangular type. The universe of discourse for inputs and outputs is in the range [-80, +80].

2.1 Defuzzification Method for Fuzzy System Using VHDL Code

Defuzzification stage is in many practical applications an essential step, especially where the fuzzy inference system is going to be use as a controller, where it is necessary to have a crisp output value instead of having a fuzzy set.

There are several methods to achieve this task, and their selection usually depends on the application and processing capacity available. The method used in this work is known as Height [10] that calculates a weighted average value which is a good option for a FPGA implementation. A remarkable characteristic is that its performance depends on how symmetrical the MFs are [12, 13, 14].

Based on the diagrams of Figure 2, the Defuzzification process using the Height method can be expressed by

$$f(y) = \frac{\sum_{m=1}^{n} C^m \cdot o^m}{\sum_{m=1}^{n} o^m} \tag{1}$$

where

f(y) is the crisp output value.
C^m is the peak value for the linguistic value LU^m.
o^m is the height of the linguistic value LU^m.

This method was implemented developed in VHDL using the simplified block diagram shown in Figure 3, where C^m is the peak value for the linguistic value LU^m, o^m is the height of the linguistic value LU^m. These two inputs are connected to a block multiplier which corresponds to the part of the numerator in (1); only the addition of the o^m will produce the denominator. Both, numerator and denominator are connected to the block divider. The result of that division is *f(y)* [15].

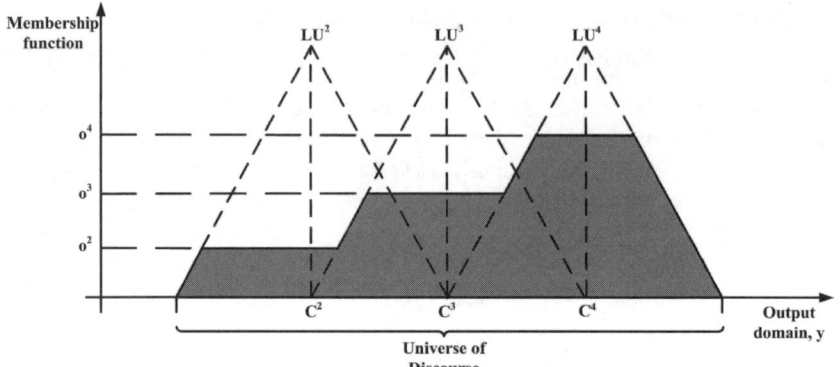

Fig. 2. Distribution possibility of an output condition

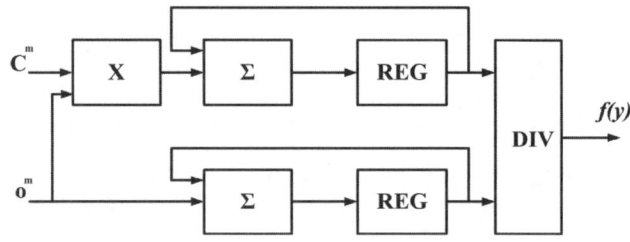

Fig. 3. Block diagram of Height defuzzification method

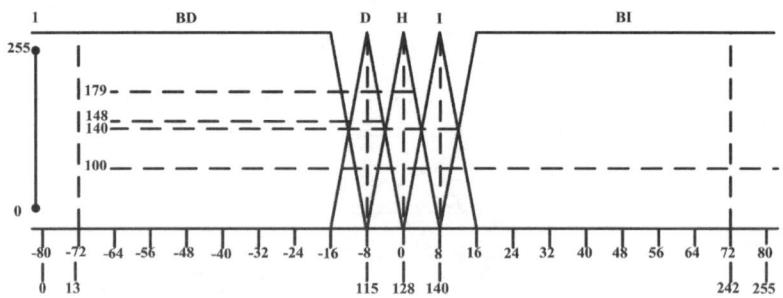

Fig. 4. Membership functions of output variable

Note in Fig. 4 that the useful scale is in the real domain. Because this implementation works with positive integer numbers, it was necessary to adapt the original scale [-80,80], to a positive scale [0,160]. For 8 bit representation, finally we used the scale [0,255], this is illustrated in Fig. 4.

The design entity of the Height Defuzzification method is shown in Figure 5. Such entity was programmed in VHDL to be implemented in a FPGA, but it can be used to simulate the Defuzzification stage without the necessity of designing and implementing any test bench.

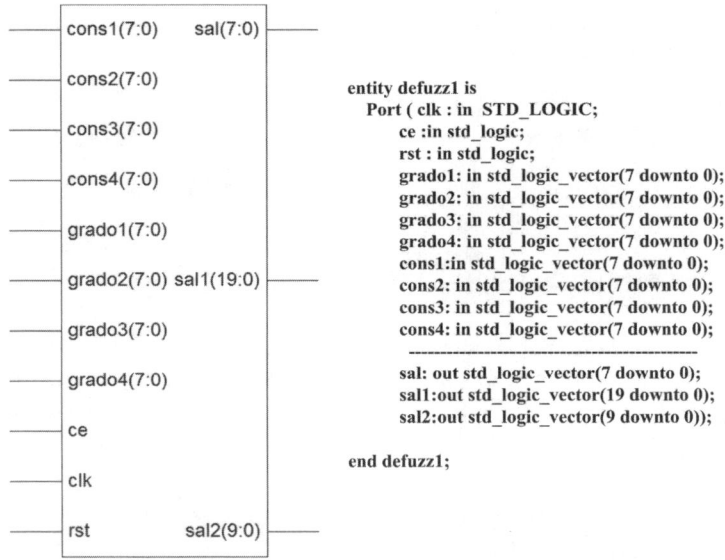

Fig. 5. Entity and RTL scheme of the Height Defuzzification stage

This entity has 11 inputs and three outputs. The first eight inputs correspond to the four possible activation grades with their respective consequent. The remaining three input signals are the clock enable, clock, and reset.

The defuzzification output $f(y)$ is given as an 8 bits word in "sal(7:0)". The other two outputs, "sal1(9:0)" and "sal2(9:0)" correspond to numerator and denominator results, and they are used for debugging purposes, so they can be removed at the final implementation.

3 Experimental Set-Up

We used three main different software tools; they are:

1. Simulink from Mathwork which is a very attractive high-level design and simulation tool because it provides a flexible design and simulation platform that allows to test and correct designs at high level.
2. Xilinx Integrated Software Environment (Xilinx ISE) is a Hardware Description Language (HDL) design software suite that allows taking designs throw several steps in the ISE design flow finishing with final verified modules that can be implemented in a hardware target such a Field Programmable Gate Array (FPGA). Top level designs can be created using VHDL (Very High Speed Integrated Circuits VHSIC and HDL), Verilog, or Abel.
3. Xilinx System Generator is a DSP design tool that enables the use of the Simulink for FPGA design. This tool allows generating VHDL code from the System Generator Simulink modules; and vice versa, VHDL modules can be included in the Simulink design platform by placing the VHDL code in a System Generator "Black box". The last characteristic is the one we used to test the designed VHDL module of the inference engine.

To make this set-up works, it is very important to have the adequate versions of each software tool. In this case we have the next setting:

1. Matlab/Simulink version is: 7.1 (R14).
2. Xilinx ISE Project Navigator: Release version 8.2.03i, application version 1.34.
3. Xilinx System Generator: v8.2.

4   Experiments and Results

Two experimental Simulink models were created to achieve a comparative test. The difference between them is in the Defuzzification stage since we are interested in testing it. In the first system, the Defuzzification stage was coded in VHDL and the other two fuzzy stages using models from Fuzzy Logic Toolbox. In the second one, the

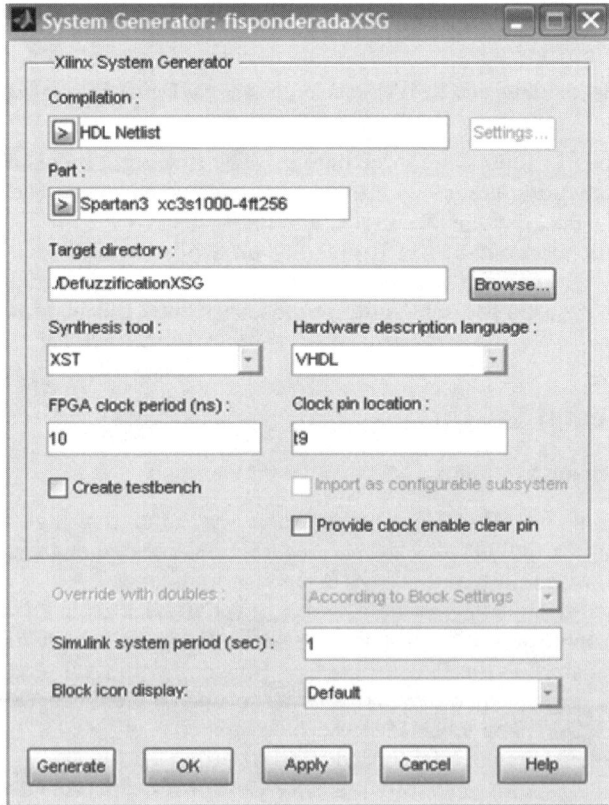

Fig. 6. Configuration of the Xilinx System Generator to work with the Spartan 3, starter board from Digilent

whole system was designed with the Fuzzy Logic Toolbox. Several comparative experiments were made with both models.

4.1 VHDL Simulink Model Using Xilinx System Generator

In Fig. 6 is presented the necessary configuration for the FPGA Spartan 3. These experiments were made considering that the final implementation will be achieved in the development starter board Spartan 3 from Digilent.

In Fig. 7 is the whole Simulink model where the three main blocks are the Fuzzification, Inference Engine, and Defuzzification stages, the two first stages were programmed using the Fuzzy Logic Toolbox, and the Defuzification stage using VHDL codification, and it was simulated with Xilinx System Generator. The system was test with several values, in this figure are shown two of them, -45 for the "error" and 5 for "cerror".

4.2 Simulink Model with the Fuzzy Logic Toolbox

In Fig. 8 it is shown the system where the three stages are an implementation of the "max-min" method using Matlab codification. We used this system to compare results with the system shown in Fig. 7, where the Defuzzification stage was created using VHDL.

The values at the inputs of this system are the same values used in the previous system. This allows a comparison between the two systems.

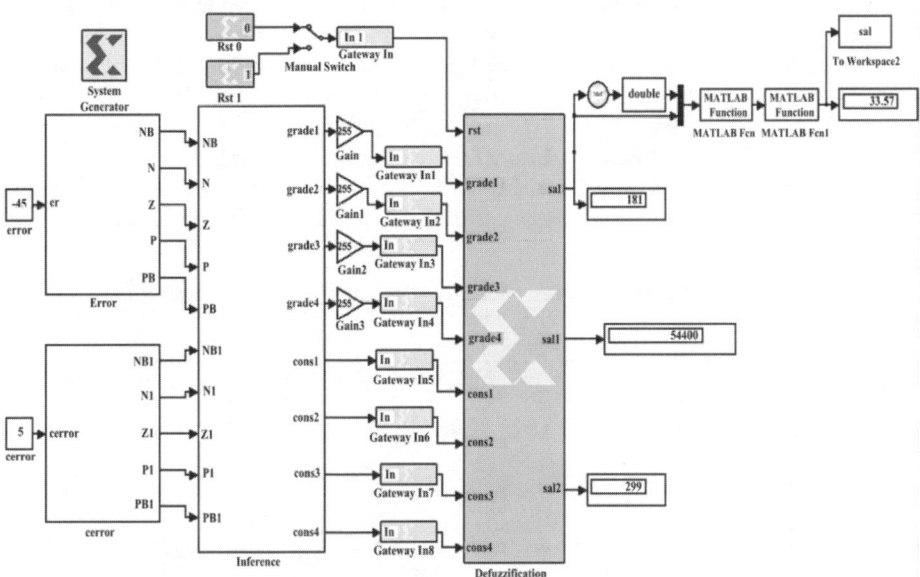

Fig. 7. System 1. Simulation of the FIS, the Height Defuzzification stage is in VHDL contained in a XSG black box.

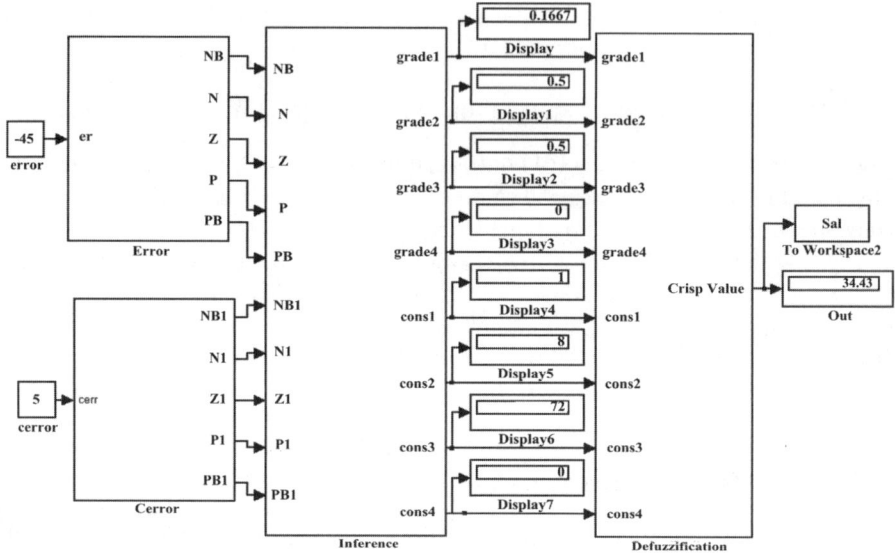

Fig. 8. System 2. Simulation of the FIS. It is an implementation of the "max-min" method in Matlab.

4.3 Comparison of Results

In Table 1, a comparison between the performances of the two systems is shown.

From Table 1, we chose three cases of study to present numeric analysis of the error in order to justify the differences. For all the cases, we used (1) to calculate the Defuzzified output; where first, we present the numeric results of System 1, an then for System 2.

Fig. 4 shows that for D (LU^2 in Fig. 2) the center has a value of 115, in H (LU^3 in Fig. 2) the center is in 128, and for I (LU^4 in Fig. 2) the center is in 140. Other values not shown are BD (LU^1) with a center value in 13, and BI (LU^5) in 242. In Fig. 4, we

Table 1. Percentage difference between systems. The inputs are error and change of error (cerror).

Inputs			Output of System 1 (VHDL)	Output of System 2	Difference of two systems (%)
Experiment #	error	cerror			
1	-45	5	33.57	34.29	-2.14
2	-28	-10	6.58	7.27	-10.48
3	-10	-8	2.82	2.66	5.67
4	15	6	-4.07	-4	1.71
5	30	15	-7.84	-8	2.04

can see that the universe of discourse is [-80,80]; because this design was made for positive numbers, it was necessary to make a domain transformation to handle only positive numbers in eight bits representation, so we obtain the transformed positive universe of discourse [0,160] ; for example the number -80 correspond to 00 hex, -72 is 0D hex, 0 is 80 hex, 72 is F2 hex.

One important thing to note is that, to revert to the original values we will not obtain the same values because there are truncation and roundoff errors. For example, when we first transform -72 to 8 in the positive decimal scale, the number 8 is represented in eight bits to obtain the value 0D hex. In this step we have introduced a truncation error since 0D hex corresponds to 12.75 in the original scale instead of 13. As we are using the integer digital representation, we are obligated to use the 0D hex value; hence we have also a roundoff error.

Case of Study 1

The inputs are: error = -45 and cerror= 5.

System 1: Simulink with XSG calculates the next digital output:

$$out = \frac{(128*43)+(140*128)+(242*128)}{(43+128+128)} = 181 \tag{2}$$

The value 181 decimal (0B5 hex) is **33.57** in the original universe of discourse [-80,80].

System 2: Matlab/Simulink.

$$out = \frac{(0*0.1667)+(8*0.5)+(72*0.5)}{(0.1667+0.5+0.5)} = 34.2847 \cong 34.29 \tag{3}$$

Now, it is clear that 33.57-34.29=-0.72, hence the percentage difference between systems is 2.14%

Case of Study 2

error = -28 and cerror= -10, using

System 1: Simulink with XSG calculates the next digital output

$$out = \frac{(128*17)+(140*170)}{(17+170)} = 138 \tag{4}$$

The value 138 decimal (8A hex) is **6.58** in the original universe of discourse [-80,80].

It is very important to note that the floating point answer of (4) is 138.9091 that is closer to 139 (8B hex), which is 7.21 in the original scale, so the difference is 7.21-6.58=.63. In this case the roundoff error produces the biggest difference possible (±1 bit) for this implementation.

System 2: Matlab/Simulink.

$$out = \frac{(0*0.06667) + (8*0.6667)}{(0.06667 + 0.6667)} = \frac{5.3336}{0.73337} = 7.273 \tag{5}$$

In the same way as in the previous case, we have 6.58-7.23=-0.65, hence the percentage difference between systems is 10.48%.

Case of Study 3

error = -10 and cerror= -8, using

System 1: Simulink with XSG calculates the next digital output:

$$out = \frac{(128*170) + (140*85)}{(170+85)} = 132 \tag{6}$$

The value 132 decimal (84 hex) is **2.8235** in the original universe of discourse [-80,80].

System 2: Matlab/Simulink.

$$out = \frac{(0*6667) + (8*0.3333)}{(0.6667 + 0.3333)} = 2.667 \tag{7}$$

Similar to the other two study cases, the difference between System 1 and System 2 is 2.8235-2.667=0.1565, hence the percentage difference between systems is 5.67%.

5 Conclusions

We inspected the control surfaces of System 1 and System 2 and observed that in general terms the behavior for both systems is very similar. Some numerical differences are given in Table 1, they are attributable to the numerical 8 bits implementation of System 1, being the most important the roundoff and truncation errors that are spread in all the stages suffering multiplicative effects. We conclude that the maximal expected error is ±1 bit of resolution. Depending on the application as well as the users constrains, this error can be acceptable. If it is needed more resolution it is indispensable to used more bits to reduce the magnitude of errors.

In general terms, considering the chosen resolution, both systems behave very similar, so we can conclude that the developed VHDL code to implement the Defuzzification stage will work fine in the final implementation in an FPGA.

References

[1] Cirstea, M.N., Dinu, A.: Neural and Fuzzy Logic Control of Drives and Power Systems, Newnes (2002)

[2] Manual of Xilinx System Generator, http://www.xilinx.com

[3] Lee, C.-C.: Fuzzy Logic Toolbox. For Use with Matlab, User Guide, Version (2000)

[4] Moctezuma, J.C., Torres, C.: Estudio sobre la implementación de redes neuronales artificiales usando Xilinx System Generator. XII Taller Hiberchip, IWS (2006)

[5] Roger Jang, J., Tsai Sun, C., Mizutani, E.: Neuro-Fuzzy and Soft Computing. In: A computational Approach to Learning and Machine Intelligence. Prentice-Hall, Englewood Cliffs (1997)

[6] Zadeh, L.A.: A Fuzzy-Set-Theoretic Interpretation of Linguistic Hedge. Journal of Cybernetics 2, 434 (1972)

[7] Bojadziev, G., Bojadziev, M.: Fuzzy Sets, Fuzzy Logic, Applications. World Scientific, Singapore (1995)

[8] Ross, T.J.: Fuzzy logic with engineering applications, 2nd edn. John Wiley and Sons, Chichester (2004)

[9] Zhang, R., Phillis, A., Kouikoglou, V.S.: Fuzzy Control of Queuing Systems. Springer, Heidelberg (2005)

[10] Driankov, D., Hellendoorn, H.: An Introduction to Fuzzy Control, 2nd edn. Springer, Heidelberg (1996)

[11] Mamdani, E.H.: Applications of fuzzy algorithms for control of simple dynamic plant. In: Proc. IEE, vol. 121(12) (1974)

[12] Poorani, S., Urmila Priya, T.V.S.: FPGA Based Fuzzy Logic Controller for Electric Vehicle. Journal of the Institution of Engineers 45(5) (2005)

[13] Brown, S., Vranesic, Z.: Fundamentals of Digital Logic with VHDL Design. McGraw-Hill, New York

[14] Sivanandam, S.N., Sumathi, S.: Introduction to Fuzzy Logic Using Matlab. Springer, Heidelberg (2006)

[15] Sanchez-Solano, S., Cabrera, A., Jimenez, C., Jimenez, P., Castillo, I.B., Barros, A.: Implementación sobre FPGA de Sistemas Difusos Programables. In: IBERCHIP 2003. Workshop IBERCHIP La Habana, Cuba (2003)

Part V

Modeling, Simulation and Optimization

A New Evolutionary Method Combining Particle Swarm Optimization and Genetic Algorithms Using Fuzzy Logic

Fevrier Valdez[1], Patricia Melin[2], and Oscar Castillo[2]

[1] PhD Student of Universidad Autónoma de Baja California,
[2] Tijuana Institute of Technology, Tijuana BC. México

Abstract. We describe in this paper a new hybrid approach for mathematical function optimization combining Particle Swarm Optimization (PSO) and Genetic Algorithms (GAs) using Fuzzy Logic to integrate the results. The new evolutionary method combines the advantages of PSO and GA to give us an improved PSO+GA hybrid method. Fuzzy Logic is used to combine the results of the PSO and GA in the best way possible. The new hybrid PSO+GA approach is compared with the PSO and GA methods with a set of benchmark mathematical functions. The new hybrid PSO+GA method is shown to be superior than the individual evolutionary methods.

1 Introduction

We describe in this paper a new evolutionary method combining PSO and GA, to give us an improved PSO+GA hybrid method. We apply the hybrid method to mathematical function optimization to validate the new approach. Also in this paper the application of a Genetic Algorithm (GA) [12] and Particle Swarm Optimization (PSO) [5] for the optimization of mathematical functions is considered. In this case, we are using the Rastrigin's, Rosenbrock's, Ackley's, Sphere's and Griewank's functions [4][13] to compare the optimization results between a GA, PSO and PSO+GA.

The paper is organized as follows: in part 2 a description about the Genetic Algorithms for optimization problems is given, in part 3 the Particle Swarm Optimization is presented, in part 4 we can appreciate the proposed PSO+GA method and the fuzzy system, in part 5 we can appreciate the mathematical functions that were used for this research, in part 6 the simulations results are described, in part 7 we can appreciate a comparison between GA, PSO and PSO+GA, in part 8 we can see the conclusions reached after the study of the proposed evolutionary computing methods.

2 Genetic Algorithm for Optimization

John Holland, from the University of Michigan initiated his work on genetic algorithms at the beginning of the 1960s. His first achievement was the publication of *Adaptation in Natural and Artificial System* [1] in 1975.

He had two goals in mind: to improve the understanding of natural adaptation process, and to design artificial systems having properties similar to natural systems [7].

O. Castillo et al. (Eds.): Soft Computing for Hybrid Intel. Systems, SCI 154, pp. 347–361, 2008.
springerlink.com

The basic idea is as follows: the genetic pool of a given population potentially contains the solution, or a better solution, to a given adaptive problem. This solution is not "active" because the genetic combination on which it relies is split between several subjects. Only the association of different genomes can lead to the solution.

Holland's method is especially effective because it not only considers the role of mutation, but it also uses genetic recombination, (crossover) [7]. The crossover of partial solutions greatly improves the capability of the algorithm to approach, and eventually find, the optimal solution.

The essence of the GA in both theoretical and practical domains has been well demonstrated [12]. The concept of applying a GA to solve engineering problems is feasible and sound. However, despite the distinct advantages of a GA for solving complicated, constrained and multiobjective functions where other techniques may have failed, the full power of the GA in application is yet to be exploited [12] [4].

To bring out the best use of the GA, we should explore further the study of genetic characteristics so that we can fully understand that the GA is not merely a unique technique for solving engineering problems, but that it also fulfils its potential for tackling scientific deadlocks that, in the past, were considered impossible to solve. In figure 1 we show the reproduction cycle of the Genetic Algorithm.

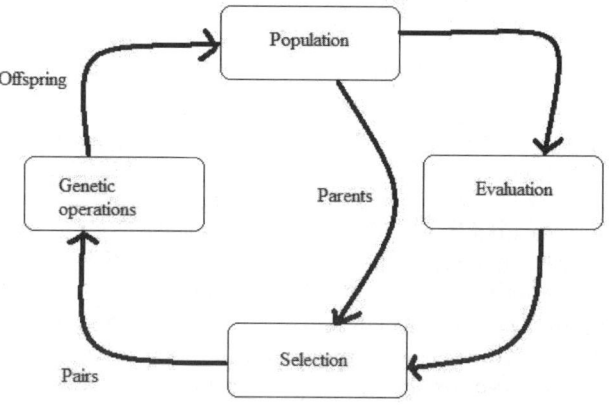

Fig. 1. The Reproduction cycle

The Simple Genetic Algorithm can be expressed in pseudo code with the following cycle:

1. Generate the initial population of individuals aleatorily P(0).
2. While (number _ generations <= maximum _ numbers _ generations)
     Do:
        {
        Evaluation;

```
        Selection;
        Reproduction;
        Generation ++;
    }
3. Show results
    4. End of the generation
```

3 Particle Swarm Optimization

Particle swarm optimization (PSO) is a population based stochastic optimization technique developed by Eberhart and Kennedy in 1995, inspired by social behavior of bird flocking or fish schooling [2].

PSO shares many similarities with evolutionary computation techniques such as Genetic Algorithms (GA) [5]. The system is initialized with a population of random solutions and searches for optima by updating generations. However, unlike GA, the PSO has no evolution operators such as crossover and mutation. In PSO, the potential solutions, called particles, fly through the problem space by following the current optimum particles [1].

Each particle keeps track of its coordinates in the problem space, which are associated with the best solution (fitness) it has achieved so far (The fitness value is also stored). This value is called *pbest*. Another "best" value that is tracked by the particle swarm optimizer is the best value, obtained so far by any particle in the neighbors of the particle. This location is called *lbest*. When a particle takes all the population as its topological neighbors, the best value is a global best and is called *gbest*.

The particle swarm optimization concept consists of, at each time step, changing the velocity of (accelerating) each particle toward its *pbest* and *lbest* locations (local version of PSO). Acceleration is weighted by a random term, with separate random numbers being generated for acceleration toward *pbest* and *lbest* locations.

In the past several years, PSO has been successfully applied in many research and application areas. It is demonstrated that PSO gets better results in a faster, cheaper way compared with other methods [2].

Another reason that PSO is attractive is that there are few parameters to adjust. One version, with slight variations, works well in a wide variety of applications. Particle swarm optimization has been used for approaches that can be used across a wide range of applications, as well as for specific applications focused on a specific requirement.

The pseudo code of the PSO is illustrated in figure 2.

```
Choose the particle with the best fitness value of all the particles as the gBest
    For each particle
        Calculate particle velocity
        Update particle position
    End
While maximum iterations or minimum error criteria is not attained
```

```
    For each particle
    Initialize particle
End
Do
    For each particle
        Calculate fitness value
        If the fitness value is better than the best fitness value
(pBest) in history
            set current value as the new pBest
    End
```

Fig. 2. Pseudocode of PSO

4 PSO+GA Method

The general approach of the proposed method PSO+GA can be seen in figure 3. The method can be described as follows:

1. It receives a mathematical function to be optimized
2. It evaluates the role of both GA and PSO.
3. A main fuzzy system is responsible for receiving values resulting from step 2.
4. The main fuzzy system decides which method to take (GA or PSO)
5. After, another fuzzy system receives the Error and DError as inputs to evaluates if is necessary change the parameters in GA or PSO.

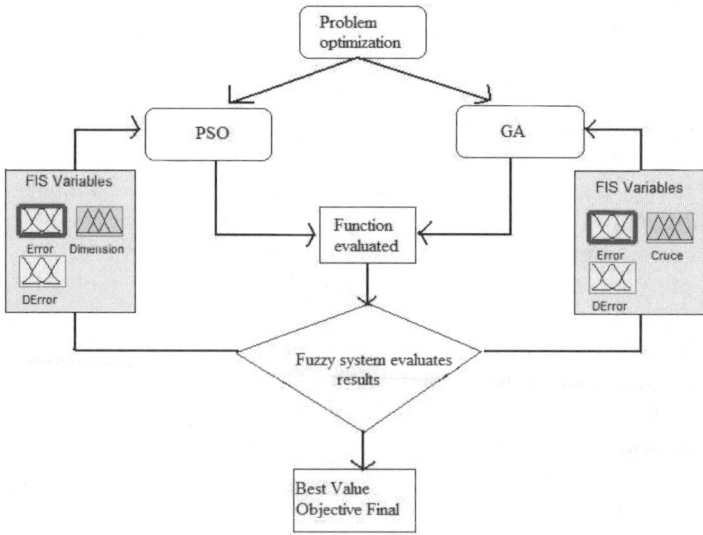

Fig. 3. The PSO+GA scheme

6. There are 3 fuzzy systems. One is for decision making (is called main fuzzy), the second one is for change parameters the GA (is called fuzzyga) in this case change the value of crossover and the third fuzzy system is used for change parameters the PSO(is called fuzzypso) in this case change the value of dimension.

7. The main fuzzy system decides in the final step the optimum value for the function introduced in step 1.

8. Repeat the above steps until the termination criterion of the algorithm is met.

The basic idea of the PSO+GA scheme is to combine the advantage of the individual methods using a fuzzy system for decision making and the others two fuzzy systems to improve the parameters of the GA and PSO when is necessary.

4.1 Fuzzy System

As can be seen in the proposed hybrid PSO+GA method, it is the internal fuzzy system structure, which has the primary function of receiving as inputs (Error and DError) the results of the outputs GA and PSO. The fuzzy system is responsible for integrating and decides which is the best results being generated at run time of the PSO+GA. It is also responsible for selecting and sending the problem to the "fuzzypso" fuzzy system when is activated the PSO or to the "fuzzyga" fuzzy system when is activated GA. Also activating or temporarily stopping depending on the results being generated. Figure 4 shows the membership functions of the main fuzzy system that is implemented in this method. The fuzzy system is of Mamdani type and the defuzzification method is the centroid. The membership functions are triangular in the inputs and outputs as is shown in the figure 4. The fuzzy system has 9 rules, for example one rule is if error is P and DError is P then best value is P (view figure 5). Figure 6 shows the fuzzy system rules viewer. Figure 7 shows the surface corresponding for this fuzzy system. The other two fuzzy system are as the fuzzy system main.

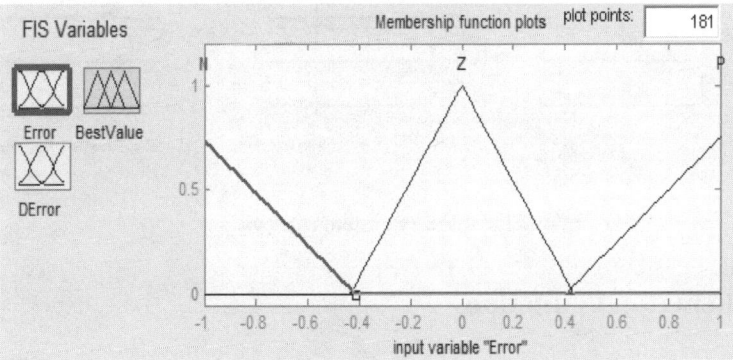

Fig. 4. Fuzzy system membership functions

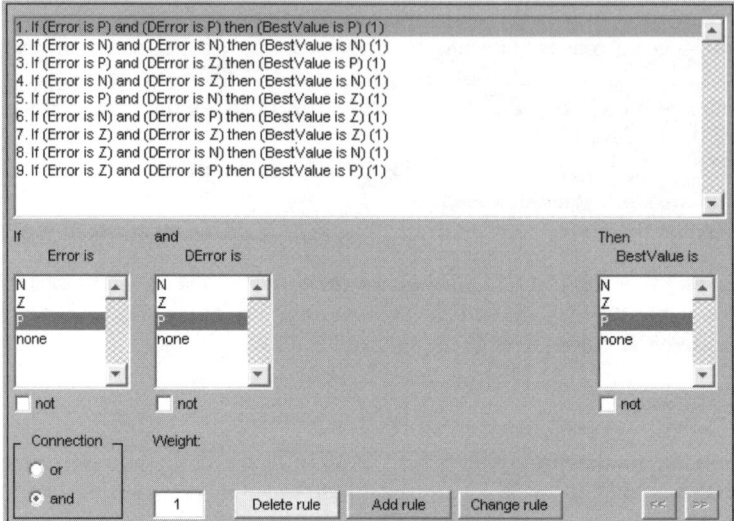

Fig. 5. Fuzzy system rules

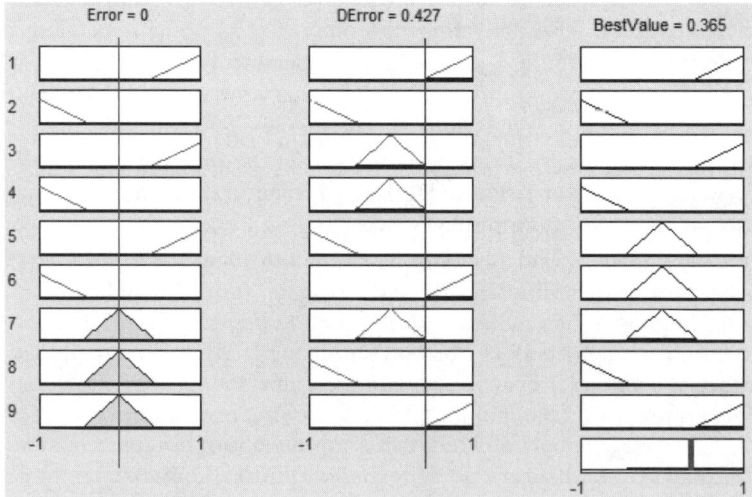

Fig. 6. Fuzzy system rules viewer

5   Mathematical Functions

In the field of evolutionary computation, it is common to compare different algorithms using a large test set, especially when the test set involves function optimization. However, the effectiveness of an algorithm against another algorithm cannot be

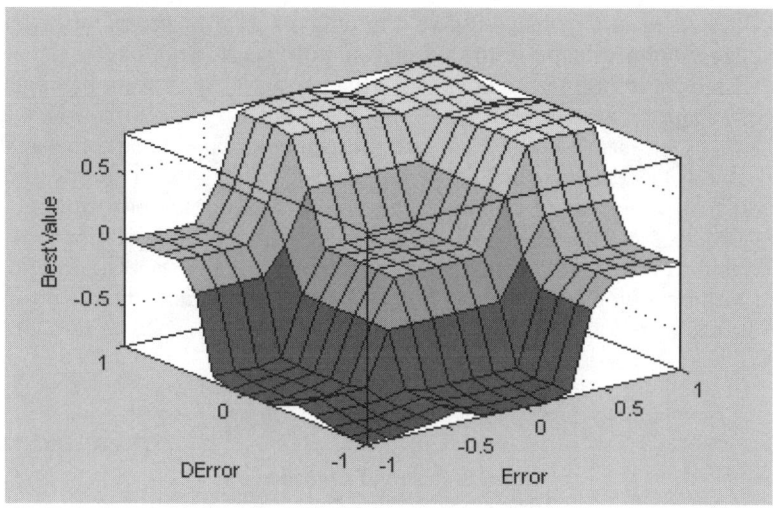

Fig. 7. Surface of fuzzy system

measured by the number of problems that it solves better. If we compare 3 searching algorithms with all possible functions, the performance of any 3 algorithms will be, on average, the same. As a result, attempting to design a perfect test set where all the functions are present in order to determine whether an algorithm is better than any other for every function is impossible. The reason is because, when an algorithm is evaluated, we must look for the kind of problems where its performance is good, in order to characterize the type of problems for which the algorithm is suitable. In this way, we have made a previous study of the functions to be optimized for constructing a test set with five benchmark functions. This allows us to obtain conclusions of the performance of the algorithm depending on the type of function. The mathematical functions analyzed in this paper are the Rastrigin's function, Rosenbrock's function, Ackley's function, Sphere's function and Griewank's function [7] [15]. All the functions were evaluated considering 2 variables.

5.1 Rastrigin's Function

The Rastrigin's function is given by the following equation :

$$f(x) = 10n + \sum_{i=1}^{n} (x^2 - 10\cos(2\pi x_i)) \tag{1}$$

Where the global minima is: $x^* = (0, ..., 0), f(x^*) = 0$

In figure 8 the Rastrigin's function is shown, in which it can be appreciated that there are several local minima.

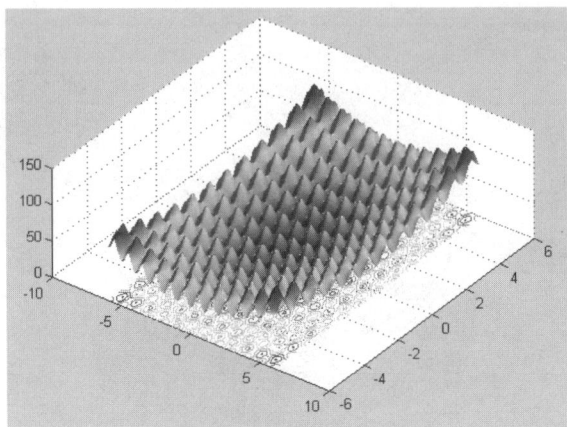

Fig. 8. Rastrigin's function

5.2 Rosenbrock's Function

The Rosenbrock's function is given by the following   equation:

$$f(x) = \sum_{i=1}^{n-1} [100(x_i^2 - x_i^2 + 1)^2 + (x_i - 1)^2] \tag{2}$$

Where the global minima is: $x^* = (1, ..., 1), f(x^*) = 0.$

In figure 9 the Rosenbrock's function is shown, in which it can be appreciated that there are several local minima.

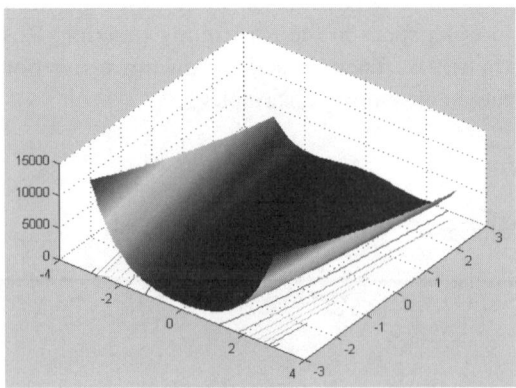

Fig. 9. Rosenbrock's function

5.3 Ackley's Function

The Ackley's function is given by the following equation:

$$f(x) = 20 + e - 20e^{-1/5}\sqrt{1/n\sum_{i=1}^{n} x_i^2} - e^{1/n}\sum_{i=1}^{n} \cos(2\pi x_1) \tag{3}$$

Where the global minima is: $x^* = (0, ..., 0), f(x^*) = 0$.

In figure 10 the Ackley's function is shown, in which it can be appreciated that there are several local minima.

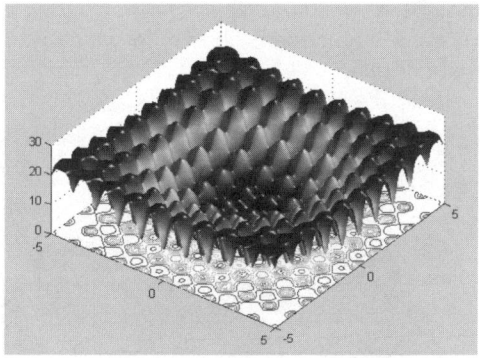

Fig. 10. Ackley's function

5.4 Sphere's Function

The Sphere's function is given by the following   equation:

$$f(x) = \sum_{i=1}^{n} x_i^2 \tag{4}$$

Where the global minima is: $x^* = (0, ..., 0), f(x^*) = 0$.

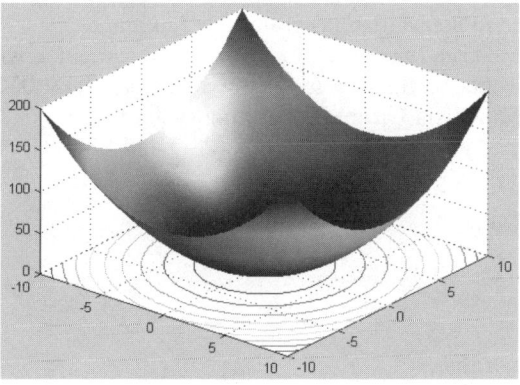

Fig. 11. Spheres's function

In figure 11 the Sphere's function is shown, in which it can be appreciated that there are only one global minima.

5.5 Griewank's Function

The Griewank's function is given by the following   equation:

$$f(x) = \sum_{i=1}^{n} \frac{x_i^2}{4000} - \prod_{i=1}^{n} \cos(x_i / \sqrt{i}) + 1 \tag{5}$$

Where the global minima is: $x^* = (0, ..., 0), f(x^*) = 0$.

In figure 12 the Griewank's function is shown, in which it can be appreciated that there are only one global minima.

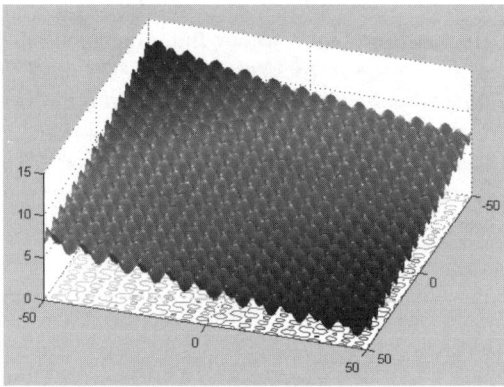

Fig. 12. Griewank's function

6 Experimental Results

Several tests of the PSO, GA and PSO+GA algorithms were made with and implementation done in the Matlab programming language.

All the implementations were developed using a computer with processor AMD turion X2 of 64 bits that works to a frequency of clock of 1800MHz, 2 GB of RAM Memory and Windows Vista Ultimate operating system.

The results obtained after applying the GA, PSO and PSO+GA to the mathematical functions are shown in tables 1, 2, 3, 4 and 5:

The parameters of Tables 1, 2 and 3:

POP= Population size
CROS= % crossover
MUT = % mutation
BEST= Best Fitness Value
MEAN= Mean of 50 tests
DIM=Dimensions

6.1 Simulation Results with the Genetic Algorithm (GA)

From Table 1 it can be appreciated that after executing the GA 50 times, for each of the tested functions, we can see the better results and their corresponding parameters that were able to achieve the global minimum with the method. In figure 13 it can be appreciated the experimental results of table 1. In figure 13 it can be appreciated that the genetic algorithm was not able to find the global minimum for the Ackley's function because the closest obtained value was 2.98.

Table 1. Experimental results with GA

MATHEMATICAL FUNCTION	POP	%CROS	%MUT	BEST	MEAN
Rastrigin	100	80	2	7.36E-07	2.15E-03
Rosenbrock	150	50	1	2.33E-07	1.02E-05
Ackley	100	80	2	2.981	2.980
Sphere	20	80	1	3.49-07	1.62E-04
Griewank	80	90	6	1.84E-07	2.552E-05

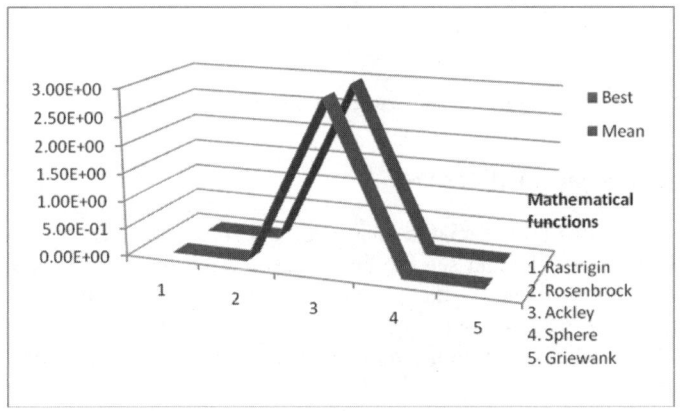

Fig. 13. Experimental results with GA

6.2 Simulation Results with Particle Swarm Optimization (PSO)

From Table 2 it can be appreciated that after executing the PSO 50 times, for each of the tested functions, we can see the better results and their corresponding parameters that were able to achieve the global minimum with the method. In figure 13 it can be appreciated the experimental results of table 2. In figure 14 it can be appreciated that the particle swarm optimization was not able to find the global minimum for Ackley's function because the closest obtained value was 2.98.

Table 2. Experimental results with GA

MATHEMATICAL FUNCTION	POP	DIM	BEST	MEAN
Rastrigin	20	10	2.48E-05	5.47
Rosenbrock	40	10	2.46E-03	1.97
Ackley	30	1	2.98	2.98
Sphere	20	10	4.88E-11	8.26E-11
Griewank	40	20	9.77E-11	2.56E-02

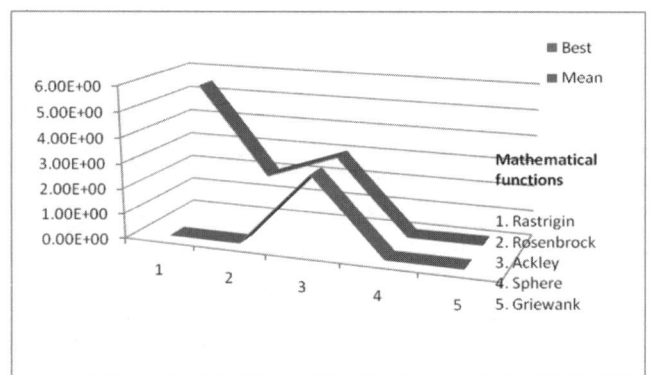

Fig. 14. Experimental results with PSO

6.3   Simulation Results with PSO+GA

From Table 4 it can be appreciated that after executing the GA 50 times, for each of the tested functions, we can see the better results and their corresponding parameters that were able to achieve the global minimum with the method. We can see that the crossover in the GA and dimension in the PSO are two variables parameters, because are changing every time the fuzzy system decides modify the two parameters.

Table 4. Experimental results with PSO+GA

MATH	POP	%CR OS	% MUT	DIM	POP	BEST	MEAN
		GA		PSO		RESULTS	
Rastrigin	100	Variable	5	variable	100	7.03E-06	1.88E-04
Rosenbrock	100	Va-riable	2	variable	100	3.23E-07	3.41E-04
Ackley	100	Va-riable	3	variable	100	1.76E-04	1.84E-03
Sphere	100	Va-riable	3	variable	10	2.80E-09	5.91E-07
Grienwak	160	Va-riable	2	variable	80	6.24E-09	9.03E-07

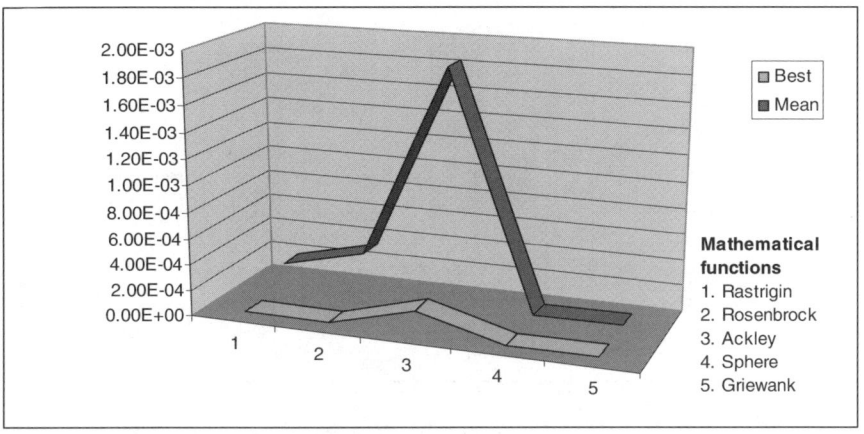

Fig. 15. Experimental results with PSO+GA

In figure 15 it can be appreciated the experimental results for table IV. In figure 15 it can be appreciated that the PSO+GA was able to find the global minimum for all test functions in this paper because the objective value was reached, and in all cases was 0.

7 Comparison Results between GA, PSO and PSO+GA

In Table 5 the comparison of the results obtained between the GA, PSO and PSO+GA methods for the optimization of the 5 proposed mathematical functions is shown. Table 5 shows the results of figure 16, it can be appreciated that the proposed PSO+GA method was better than GA and PSO, because with this method all test functions were optimized. In some cases the GA was better but in table 5 and figure 16 it can be seen that the better mean values were obtained with the PSO+GA, only in the Sphere function was better the PSO than the other two methods. Also in the Rosenbrock's function was better the GA than the other two methods.

Table 5. Experimental results with PSO+GA

Mathematical Functions	GA	PSO	PSO+GA	Objective Value
Rastrigin	2.15E-03	5.47	1.88E-04	0
Rosenbrock	1.02E-05	1.97	3.41E-04	0
Ackley	2.980	2.98	1.84E-03	0
Sphere	1.62E-04	8.26E-11	5.91E-07	0
Griewank	2.552E-05	2.56E-02	9.03E-07	0

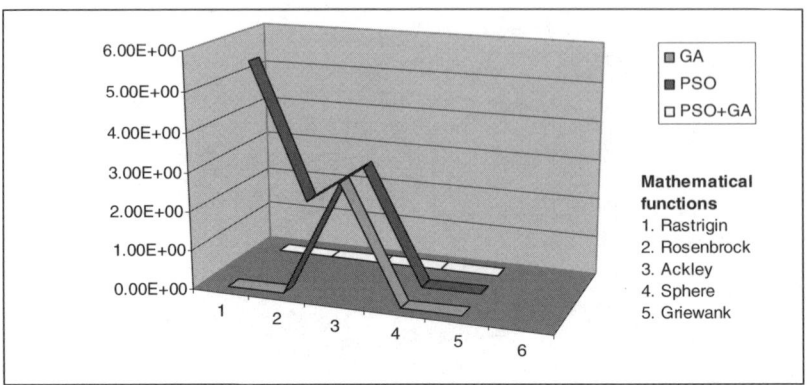

Fig. 16. Comparison results between the proposed methods

8 Conclusions

The analysis of the simulation results of the 3 evolutionary methods considered in this paper, in this case the Genetic Algorithm (GA), the Particle Swarm Optimization (PSO), and PSO+GA lead us to the conclusion that for the optimization of these 5 mathematical functions, in all cases one can say that the 3 proposed methods work correctly and they can be applied for this type of problems. But we can appreciate that for the Ackley's function the GA and PSO were not able of reach the global minimum.

After studying the 3 methods of evolutionary computing (GA, PSO and PSO+GA), we reach the conclusion that for the optimization of these 5 mathematical functions, GA and PSO evolved in a similar form, achieving both methods the optimization of 4 of the 5 proposed functions, with values very similar and near the objectives. Also it is possible to observe that even if the GA as the PSO did not achieve the optimization of the Ackley's function, this may have happened because they were trapped in local minima. However we can appreciate that the proposed hybrid method in this paper (PSO+GA) was able of optimize all test functions. Also, in general PSO+GA had the better average optimization values. The advantage the use this method is that it incorporates a fuzzy system to improve the optimization results.

Figure 16 shows the comparison of the results obtained for these 5 test functions and it can be appreciated that the values that were taken from the tables above mentioned, the GA and the PSO obtained very good results and was very little the difference between of them. But the PSO+GA was able of optimize the Ackley's function while that other two methods were not able to reach the global minimum for this function. Table 5 shows the values corresponding to figure 16. The advantage to use PSO is that there are few parameters used for the implementation. The genetic algorithm uses more parameters for its implementation. The PSO+GA is a more complex method but is more reliable because is a hybrid method that combines the PSO and GA, also uses a fuzzy system for decision and integration of the final results and other two fuzzy systems for changes the parameters of method, therefore in this research was better that the other two methods.

Table 5 shows the final results of figure 16, it can be appreciated in some cases, the GA was better than the PSO, for example, for the Rastrigin's function, Rosenbrock's function and Griewank's function. In other cases, the PSO was better than the GA, for example, for the Sphere's function. But as above is mentioned GA+PSO was better because with this method all functions were optimized with a smaller error.

Acknowledgments

We would like to express our gratitude to the CONACYT, Universidad Autónoma de Baja California and Tijuana Institute of Technology for the facilities and resources granted for the development of this research.

References

[1] Angeline, P.J.: Evolutionary Optimization Versus Particle Swarm Optimization: Philosophy and Performance Differences. In: Porto, V.W., Waagen, D. (eds.) EP 1998. LNCS, vol. 1447, pp. 601–610. Springer, Heidelberg (1998)

[2] Angeline, P.J.: Using Selection to Improve Particle Swarm Optimization. In: Proceedings 1998 IEEE World Congress on Computational Intelligence, Anchorage, Alaska, pp. 84–89. IEEE, Los Alamitos (1998)

[3] Back, T., Fogel, D.B., Michalewicz, Z. (eds.): Handbook of Evolutionary Computation. Oxford University Press, Oxford (1997)

[4] Castillo, O., Valdez, F., Melin, P.: Hierarchical Genetic Algorithms for topology optimization in fuzzy control systems. International Journal of General Systems 36(5), 575–591 (2007)

[5] Fogel, D.B.: An introduction to simulated evolutionary optimization. IEEE transactions on neural networks 5(1) (January 1994)

[6] Eberhart, R.C., Kennedy, J.: A new optimizer using particle swarm theory. In: Proceedings of the Sixth International Symposium on Micromachine and Human Science, Nagoya, Japan, pp. 39–43 (1995)

[7] Emmeche, C.: Garden in the Machine. In: The Emerging Science of Artificial Life, p. 114. Princeton University Press, Princeton (1994)

[8] Germundsson, R.: Mathematical Version 4. Mathematical J. 7, 497–524 (2000)

[9] Goldberg, D.: Genetic Algorithms. Addison Wesley, Reading (1988)

[10] Holland, J.H.: Adaptation in natural and artificial system. University of Michigan Press, Ann Arbor (1975)

[11] Kennedy, J., Eberhart, R.C.: Particle swarm optimization. In: Proceedings of IEEE International Conference on Neural Networks, Piscataway, NJ, pp. 1942–1948 (1995)

[12] Man, K.F., Tang, K.S., Kwong, S.: Genetic Algorithms: Concepts and Designs. Springer, Heidelberg (1999)

[13] Montiel, O., Castillo, O., Melin, P., Rodriguez, A., Sepulveda, R.: Human evolutionary model: A new approach to optimization. Inf. Sci. 177(10), 2075–2098 (2007)

[14] Valdez, F., Melin, P., Castillo, O.: Evolutionay Computing for the Optimi-zation of Mathematical Functions. Analysis and Design of intelligent Systems Using Soft Computing Techniques. Advances in Soft Computing 41 (June 2007)

[15] Valdez, F., Melin, P.: Parallel Evolutionary Computing using a cluster for Mathematical Function Optimization, San Diego,CA, USA, pp. 598–602 (June 2007)

A Hybrid Learning Algorithm for Interval Type-2 Fuzzy Neural Networks: The Case of Time Series Prediction

Juan R. Castro, Oscar Castillo, Patricia Melin, and Antonio Rodríguez-Díaz

Baja California Autonomous University, UABC. Tijuana, Mexico
jrcastror@uabc.mx, ardiaz@uabc.mx.
Division of Graduate Studies, Tijuana Institute of Technology, Tijuana, Mexico
ocastillo@tectijuana.mx, pmelin@tectijuana.mx

Abstract. In this work, a class of Interval Type-2 Fuzzy Neural Networks (IT2FNN) is proposed, which is functionally equivalent to interval type-2 fuzzy inference systems. The computational process envisioned for fuzzy neural systems is as follows: it starts with the development of an "Interval Type-2 Fuzzy Neuron", which is based on biological neural morphologies, followed by the learning mechanisms. We describe how to decompose the parameter set such that the hybrid learning rule of adaptive networks can be applied to the IT2FNN architecture for the Takagi-Sugeno-Kang reasoning.

Keywords: Interval type-2 Fuzzy Neural Networks, Interval Type-2 Fuzzy Neuron, Hybrid Learning Algorithm, Interval Type-2 Fuzzy Systems.

1 Introduction

Intelligent hybrid systems combining fuzzy logic (FL), neural networks (NN), genetic algorithms (GA), and expert systems (ES) are proving their effectiveness in a wide variety of real-world problems [1], [22]-[25]. Every intelligent technique has particular computational properties (e.g. ability to learn and explanation of decisions) that make them suitable for special kinds of problems and not for others. For example, while neural networks are good at recognizing patterns, they are not good at explaining how they reach their decisions. Fuzzy systems [2], [20], [11]-[13], which can reason with imprecise information and uncertainty, are good at explaining their decisions but they cannot automatically acquire the rules they use to make those decisions. These limitations have been a central driving force behind the creation of intelligent hybrid systems, where two or more techniques are combined in a manner that overcomes the limitations of individual techniques. Intelligent hybrid systems are also important when considering the varied nature of application domains. Many complex domains have many different component problems, each of which may require different types of processing. If there is a complex application, which has two distinct sub-problems (e.g. a signal processing task and a serial reasoning task), then a neural network and an expert system respectively can be used for solving these separate tasks. The use of intelligent hybrid systems is growing rapidly with successful applications in many areas including process control, engineering design, financial trading, credit evaluation, medical diagnosis, and cognitive simulation [1],[21],[30],[31].

O. Castillo et al. (Eds.): Soft Computing for Hybrid Intel. Systems, SCI 154, pp. 363–386, 2008.
springerlink.com

While interval type-2 fuzzy logic (IT2FL) provides an inference mechanism under cognitive uncertainty, computational neural networks offer exciting advantages, such as learning, adaptation, fault-tolerance, parallelism and generalization. To enable a system to deal with cognitive uncertainties in a manner more like humans do, one may incorporate the concept of interval type-2 fuzzy logic into neural networks.

The idea of considering type-1 fuzzy neurons (T1FN) was introduced by Hirota and Pedrycz [3] and Pedrycz and Rocha [4]. While the models of those neurons involve *max* and *min* operators and triangular norms, the neuron presented in this paper utilizes a particular extension of fuzzy sets.

The computational process envisioned for fuzzy neural systems is as follows: it starts with the development of an "interval type-2 fuzzy neuron (IT2FN) " based on the understanding of biological neural morphologies, followed by the learning mechanisms. This leads to the following three steps in an interval type-2 fuzzy neural computational process:

- Development of interval type-2 fuzzy neural models motivated by biological neurons.
- Models of synaptic connections, which incorporate fuzziness into neural networks.
- Development of learning algorithms (that is, the method of adjusting the synaptic weights)

An interval type-2 fuzzy neural network (IT2FNN) is a neural network with interval type-2 fuzzy signals and/or interval type-2 fuzzy weights, gaussian, generalized bell and sigmoid transfer function, and all operations defined by Zadeh's [5]-[10] extension principle.

2   Interval Type-2 Fuzzy Logic Systems

A general interval type-2 fuzzy logic system (IT2FLS) is depicted in Figure 1. An IT2FLS is very similar to a type-1 fuzzy logic systems (FLS) [1,2], the major structural difference being that the defuzzifier block of type-1 FLS is replaced by the output processing block in an interval type-2 FLS, which consists of type-reduction followed by defuzzification.

Consider an interval type-2 Takagi-Sugeno-Kang (TSK) FLS having n inputs $x_1 \in X_1,..., x_i \in X_i,..., x_n \in X_n$ and m outputs $y_1 \in Y_1, ..., y_j \in Y_j,...,$ $y_m \in Y_m$. An interval type-2 TSK FLS is also described by fuzzy IF-THEN rules that represent input-output relations of a system. In general, a first-order interval type-2 TSK [1], [15], [16], [30], [31] models with a rule base of M rules, each having n antecedents, the kth rule can be expressed as follows:

$$R^k : IF \ x_1 \ is \ \tilde{A}_1^k \ and \ ...and \ x_i \ is \ \tilde{A}_i^k \ and \ ...and \ x_n \ is \ \tilde{A}_n^k$$
$$THEN \ y_j \ is \ \tilde{y}_k^j = C_{k,1}^j x_1 + ... + C_{k,i}^j x_i + ... + C_{k,n}^j x_n + C_{k,0}^j$$

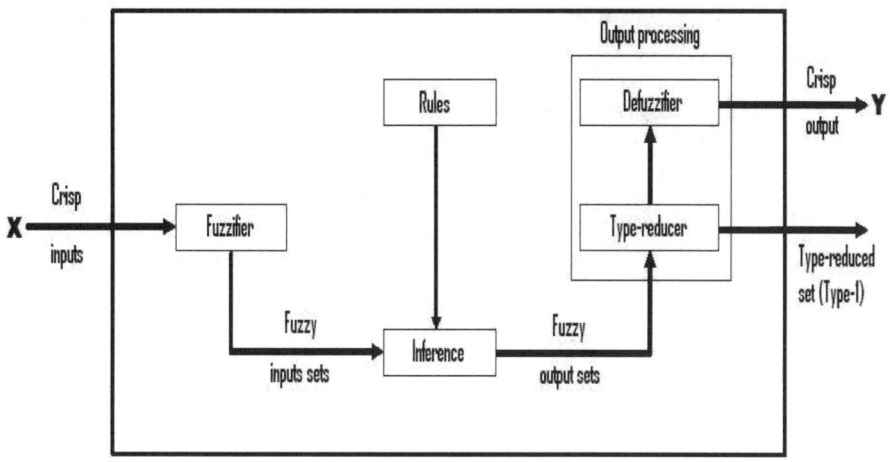

Fig. 1. Type-2 Fuzzy Logic Systems

where $k = 1,\ldots,M$, $C_{k,i}^{j}$, $(i=0,1,\ldots, n; j=1,\ldots, m)$ are consequent interval type-1 fuzzy sets; $\tilde{y}_{k}^{j}$. The jth output of the kth rule is also an interval type-1 fuzzy set (since it is a linear combination of interval type-1 fuzzy sets); and $\tilde{A}_{i}^{k}$ $(i=1,\ldots,n)$ are interval type-2 antecedent fuzzy sets. These rules take into account simultaneously uncertainty about antecedent membership functions and consequent parameter values.

In an interval type-2 TSK FLS with *meet* under product or minimum t-norm, the firing set of the kth rule is $F^{k}(x)$, which is an interval type-1 set defined as:

$$F^{k}(x) = \sqcap_{i=1}^{n} \mu_{\tilde{A}_{i}^{k}}(x_{i}) = \left[\underline{f}^{k}(x), \overline{f}^{k}(x)\right] \quad (1)$$

where

$$\underline{f}^{k}(x) = \overset{n}{\underset{i=1}{*}}\left[\mu_{\tilde{A}_{i}^{k}}(x_{i})\right] \text{ and } \overline{f}^{k}(x) = \overset{n}{\underset{i=1}{*}}\left[\overline{\mu}_{\tilde{A}_{i}^{k}}(x_{i})\right] \quad (2)$$

The consequent of rule R^{k}, $\tilde{y}_{k}^{j} = [{}_{l}y_{k}^{j}, {}_{r}y_{k}^{j}]$, is also an interval; $\tilde{y}_{k}^{j} = C_{k,1}^{j}x_{1} + \ldots + C_{k,i}^{j}x_{i} + \ldots + C_{k,n}^{j}x_{n} + C_{k,0}^{j}$ set; and $C_{k,i}^{j} \in \left[c_{k,i}^{j} - s_{k,i}^{j}, c_{k,i}^{j} + s_{k,i}^{j}\right]$ where $c_{k,i}^{j}$ denotes the center (mean) of $C_{k,i}^{j}$, $s_{k,i}^{j}$ denotes the spread of $C_{k,i}^{j}$ and ${}_{l}y_{k}^{j}, {}_{r}y_{k}^{j}$ are defined by:

$$\begin{aligned} {}_{l}y_{k}^{j} &= \sum_{i=1}^{n} c_{k,i}^{j}x_{i} + c_{k,0}^{j} - \sum_{i=1}^{n} s_{k,i}^{j}\left|x_{i}\right| - s_{k,0}^{j} \\ {}_{r}y_{k}^{j} &= \sum_{i=1}^{n} c_{k,i}^{j}x_{i} + c_{k,0}^{j} + \sum_{i=1}^{n} s_{k,i}^{j}\left|x_{i}\right| + s_{k,0}^{j} \end{aligned} \quad (3)$$

The jth output of a first-order interval type-2 TSK FLS is obtained by applying the extension principle, where now both $f^k(x)$ and $y_k^j(x)$ are replaced by interval type-1 fuzzy sets. Hence, $\tilde{Y}_{TSK,2}^j(x)$ is an interval type-1 set. To compute $\tilde{Y}_{TSK,2}^j(x)$, we therefore only need to compute its two end-points $_ly^j$ and $_ry^j$ as follows:

$$\tilde{Y}_{TSK2}^j(x) = [_ly^j,_ry^j] = \int \cdots \int_{y_1^j \in [_ly_1^j,_ry_1^j]} \int \cdots \int_{y_M^j \in [_ly_M^j,_ry_M^j]\; f^1 \in [f_l^1,f_r^1]\; f^M \in [f_l^M,f_r^M]} 1 \Bigg/ \frac{\sum\limits_{k=1}^{M} f^k y_k^j}{\sum\limits_{k=1}^{M} f^k} \tag{4}$$

In an interval type-2 TSK FLS, $\tilde{Y}_{TSK,2}^j(x)$ is an interval type-1 fuzzy set. $\tilde{Y}_{TSK,2}^j(x)$ is defuzzified using the average of $_ly^j$ and $_ry^j$; hence, the defuzzified output of any interval type-2 TSK FLS is

$$y_{TSK,2}^j(x) = \frac{_ly^j + _ry^j}{2} \tag{5}$$

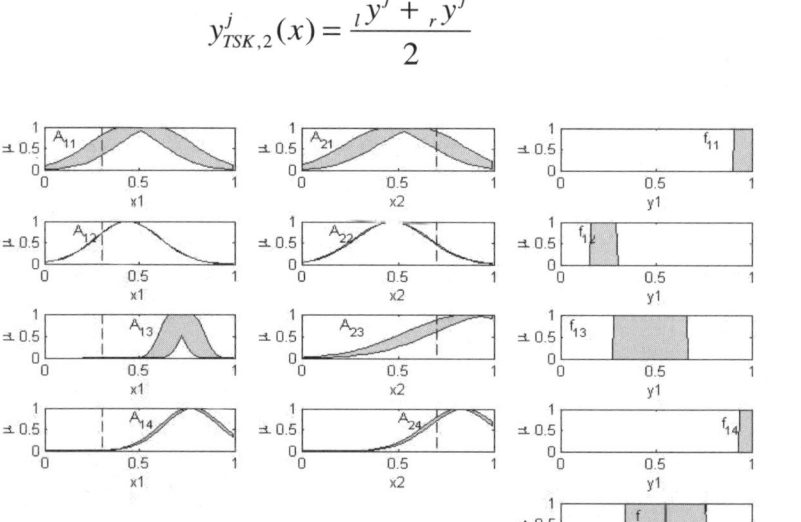

Fig. 2. Interval Type-2 TSK FLS

For example, Figure 2 shows an interval type-2 TSK FLS [14], [26] with two inputs (x_1, x_2), one output (y_1) and four rules ($k=1,\ldots,4$); each input fuzzified by four membership functions $(A_{1,k}, A_{2,k})$ and each output is fuzzified by four interval linear membership functions $f_{1,k} \in (y_l^k, y_r^k)$.

3   Interval Type-2 Fuzzy Neural Networks

One way to build interval type-2 fuzzy neural networks (IT2FNN) is to fuzzify a conventional neural network. Each part of a neural network (the activation function, the weights, and the inputs and outputs) can be fuzzified. A fuzzy neuron is basically similar to an artificial neuron, except that it has the ability to process fuzzy information.

An IT2FNN with TSK reasoning and processing elements called interval type-2 fuzzy neurons (IT2FN) for defining antecedents and interval type-1 fuzzy neurons (IT1FN) for defining the consequents of rules R^k is proposed.

Figure 3 shows an IT2FN with crisp input signals (x), crisp synaptic weights (w, b) and type-1 fuzzy outputs (μ_1, μ_1, $\underline{\mu}, \overline{\mu}$). This kind of neuron is build from two conventional neurons with transference functions $\mu(net)$, gaussian, generalized bell and logistic for fuzzifier the inputs. Each neuron equation is defined as follows:

$$
\begin{aligned}
net_1 &= w_{1,1}x + b_1 \quad ; \mu_1 = \mu(net_1) \\
net_2 &= w_{2,1}x + b_1 \quad ; \mu_2 = \mu(net2)
\end{aligned}
\tag{6}
$$

$$
\begin{aligned}
\underline{\mu}(x) &= \mu(net_1) \cdot \mu(net_2) \\
\overline{\mu}(x) &= \mu(net_1) + \mu(net_1) - \underline{\mu}(x)
\end{aligned}
\tag{7}
$$

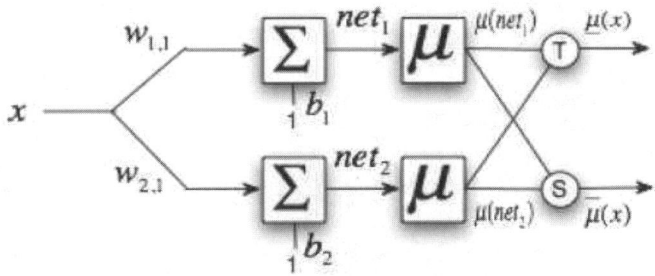

Fig. 3. Interval type-2 fuzzy neuron (IT2FN)

Each IT2FN adapts an interval type-2 fuzzy set [2], $\tilde{A}$, expressed in terms of output $\underline{\mu}(x)$, of type-1 fuzzy neuron with T-norm and $\overline{\mu}(x)$ of type-1 fuzzy neuron with S-norm. An interval type-2 fuzzy set is denoted as:

$$
\tilde{A} = \int_{x \in X} \left[\int_{\mu(x) \in [\underline{\mu}(x), \overline{\mu}(x)]} 1 / \mu \right] \Big/ x
\tag{8}
$$

Figure 4 shows an interval type-1 fuzzy neuron build from two conventional adaptive linear neurons (ADALINE) [29] for adapting consequents $\tilde{y}_k^j \in [\,_l y_k^j ,\,_r y_k^j\,]$ from rules R^k , for the jth output defined by:

$$_l y_k^j = \sum_{i=1}^{n} c_{k,i}^j x_i + c_{k,0}^j - \sum_{i=1}^{n} s_{k,i}^j \left| x_i \right| - s_{k,0}^j$$

$$_r y_k^j = \sum_{i=1}^{n} c_{k,i}^j x_i + c_{k,0}^j + \sum_{i=1}^{n} s_{k,i}^j \left| x_i \right| + s_{k,0}^j$$

(9)

Thus, consequents can be adapted with a network of adaptive linear networks (MADALINE) [29]. This network weights are the parameters of consequents $c_{k,i}^j$ and $s_{k,i}^j$ for the kth rule. The outputs represent interval linear membership functions of the rule's consequents, as shown in Figure 5.

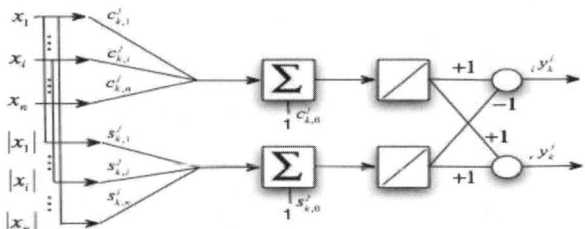

Fig. 4. Interval type-1 fuzzy neuron

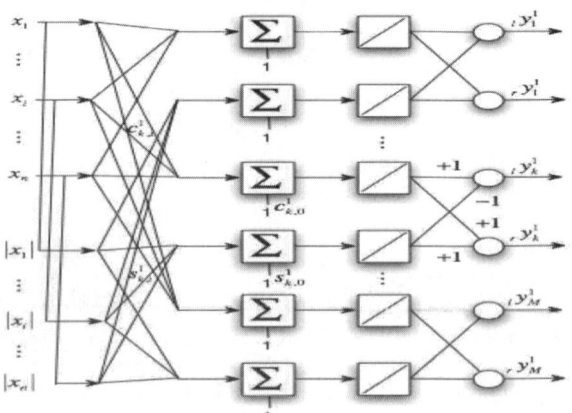

Fig. 5. Interval type-1 fuzzy neural network

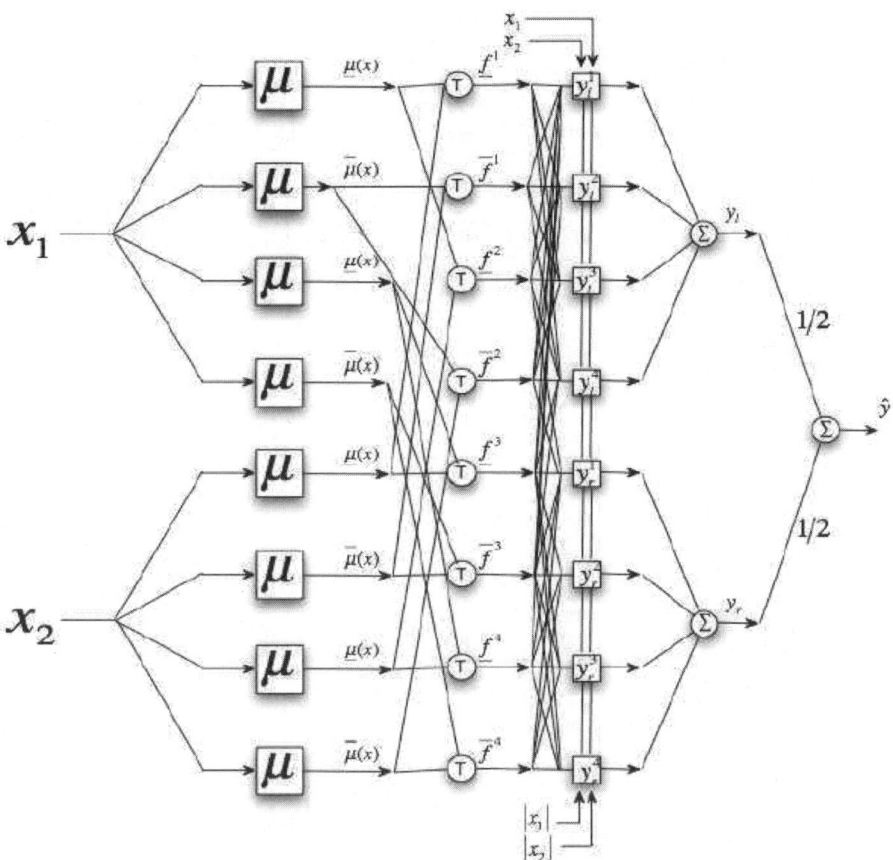

Fig. 6. Interval type-2 fuzzy neural network

In this paper, three IT2FNN architectures are proposed. The main difference between them is the way inputs are fuzzified and how type-reduction of the IT2FLS TSK is made. The learning rule used is back-propagation with descendent gradient, with the alternative of using recursive LSE algorithm to upgrade the consequent rules.

The first architecture (Figure 6) is an adaptive five layer NN. Nodes of the first layer fuzzify the inputs; each node is a *lower-upper* membership function. Layer two has non-adaptive nodes that represent *lower-upper* firing set rules of the antecedents. In layer three each node evaluates consequent rules. In layer four the left-right points are evaluated using Karnik-Mendel type-reduction algorithm [2]. Layer five defuzzifies the system's output. For simplicity, we assume the IT2FNN under consideration has **n** inputs and one output. The forward-propagation procedure is described as follows:

Layer 0. Inputs

$$x = (x_1, \ldots, x_i, \ldots, x_n)^t$$

Layer 1. Interval type-2 member functions, $\tilde{\mu}_{k,i}(x_i) = \{\underline{\mu}_{k,i}(x_i), \overline{\mu}_{k,i}(x_i)\}$

For example:

$$\mu_{k,i}(x_i) = \{\underline{\mu}_{k,i}(x_i), \overline{\mu}_{k,i}(x_i)\} = igaussmtype2(x_i, [\sigma_{k,i}, {}^1m_{k,i}, {}^2m_{k,i}])$$

where k=1,2,…,M ; i= 1,2,…,n

$${}^1\mu_{k,i}\left(x_i, \left[\sigma_{k,i}, {}^1m_{k,i}\right]\right) = e^{-\frac{1}{2}\left(\frac{x_i - {}^1m_{k,i}}{\sigma_{k,i}}\right)^2}$$

$${}^2\mu_{k,i}\left(x_i, \left[\sigma_{k,i}, {}^2m_{k,i}\right]\right) = e^{-\frac{1}{2}\left(\frac{x_i - {}^2m_{k,i}}{\sigma_{k,i}}\right)^2}$$

$$\overline{\mu}_{k,i}(x_i) = \begin{cases} {}^1\mu(x_i, [\sigma_{k,i}, {}^1m_{k,i}]) & x_i < {}^1m_{k,i} \\ 1 & {}^1m_{k,i} \le x_i \ge {}^2m_{k,i} \\ {}^2\mu(x_i, [\sigma_{k,i}, {}^2m_{k,i}]) & x_i > {}^2m_{k,i} \end{cases}$$

$$\underline{\mu}_{k,i}(x_i) = \begin{cases} {}^2\mu(x_i, [\sigma_{k,i}, {}^2m_{k,i}]) & x_i \le \dfrac{{}^1m_{k,i} + {}^2m_{k,i}}{2} \\ {}^1\mu(x_i, [\sigma_{k,i}, {}^1m_{k,i}]) & x_i > \dfrac{{}^1m_{k,i} + {}^2m_{k,i}}{2} \end{cases}$$

Layer 2. Rules

$$\underline{f}^k = \mathop{*}_{i=1}^{n}\left(\underline{\mu}_{k,i}\right) \qquad ; \overline{f}^k = \mathop{*}_{i=1}^{n}\left(\overline{\mu}_{k,i}\right)$$

Layer 3. Consequent left-right firing points

$$y_l^k = \sum_{i=1}^{n} c_{k,i} x_i + c_{k,0} - \sum_{i=1}^{n} s_{k,i} |x_i| - s_{k,0}$$

$$y_r^k = \sum_{i=1}^{n} c_{k,i} x_i + c_{k,0} + \sum_{i=1}^{n} s_{k,i} |x_i| + s_{k,0}$$

Layer 4. Left-right points (Type-reduction using KM Algorithm)

$$\hat{y}_l = \hat{y}_l(\overline{f}^1, ..., \overline{f}^L, \underline{f}^{L+1}, ..., \underline{f}^M, y_l^1, ..., y_l^M) = \frac{\sum_{k=1}^{M} f_l^k \cdot y_l^k}{\sum_{k=1}^{M} f_l^k}$$

$$= \frac{\sum_{k=1}^{L} \overline{f}^k \cdot y_l^k + \sum_{k=L+1}^{M} \underline{f}^k \cdot y_l^k}{\sum_{k=1}^{L} \overline{f}^k + \sum_{k=L+1}^{M} \underline{f}^k}$$

$$\hat{y}_r = \hat{y}_r(\underline{f}^1, ..., \underline{f}^R, \overline{f}^{R+1}, ..., \overline{f}^M, y_r^1, ..., y_r^M) = \frac{\sum_{k=1}^{M} f_r^k \cdot y_r^k}{\sum_{k=1}^{M} f_r^k}$$

$$= \frac{\sum_{k=1}^{R} \underline{f}^k \cdot y_r^k + \sum_{k=R+1}^{M} \overline{f}^k \cdot y_r^k}{\sum_{k=1}^{R} \underline{f}^k + \sum_{k=R+1}^{M} \overline{f}^k}$$

Layer 5. Defuzzification

$$\hat{y} = \frac{\hat{y}_l + \hat{y}_r}{2}$$

The IT2FNN uses back-propagation (steepest descent) method [27] for learning how to determine premise parameters (to find the parameters related to interval membership functions) and consequent parameters. The learning procedure has two parts: In the first part the input patterns are propagated, the consequent parameters and the premise parameters are assumed to be fixed for the current cycle through the training set. In the second part the patterns are propagated again, and at this moment, back-propagation is used to modify the premise parameters, and consequent parameters. These two parts are considered to be an epoch.

Given an input-output training pair $\{(x_p : t_p)\} \; \forall \; p = 1, ..., q$, in order to get the design of the IT2FNN, the error function (E) must be minimized.

$$e_p = t_p - \hat{y}_p \tag{10}$$

$$E_p = \tfrac{1}{2} e_p^2 = \tfrac{1}{2}\left(t_p - \hat{y}_p\right)^2 \tag{11}$$

$$E = \sum_{p=1}^{q} E_p \qquad (12)$$

Accordingly, the update formulas for the generic antecedents ($\xi_{k,i}$) and consequents ($c_{k,i}, s_{k,i}$), are given by equations (13) to (15) respectively

$$^{new}c_{k,i} = {}^{old}c_{k,i} - \eta \cdot \frac{\partial E_p}{\partial c_{k,i}} \qquad (13)$$

$$^{new}s_{k,i} = {}^{old}s_{k,i} - \eta \cdot \frac{\partial E_p}{\partial s_{k,i}} \qquad (14)$$

$$^{new}\xi_{k,i} = {}^{old}\xi_{k,i} - \eta \cdot \frac{\partial E_p}{\partial \xi_{k,i}} \qquad (15)$$

where η is a learning rate.

Equations from (16) to (19) update the parameters of the consequents of the rules

$$^{new}c_{k,i} = {}^{old}c_{k,i} + \eta \cdot \tfrac{1}{2} \cdot e_p \left[\frac{f_r^k}{\sum\limits_{k=1}^{M} f_r^k} + \frac{f_l^k}{\sum\limits_{k=1}^{M} f_l^k} \right] \cdot x_i \qquad (16)$$

$$^{new}c_{k,0} = {}^{old}c_{k,0} + \eta \cdot \tfrac{1}{2} \cdot e_p \left[\frac{f_r^k}{\sum\limits_{k=1}^{M} f_r^k} + \frac{f_l^k}{\sum\limits_{k=1}^{M} f_l^k} \right] \qquad (17)$$

$$^{new}s_{k,i} = {}^{old}s_{k,i} + \eta \cdot \tfrac{1}{2} \cdot e_p \left[\frac{f_r^k}{\sum\limits_{k=1}^{M} f_r^k} - \frac{f_l^k}{\sum\limits_{k=1}^{M} f_l^k} \right] \cdot |x_i| \qquad (18)$$

$$^{new}s_{k,0} = {}^{old}s_{k,0} + \eta \cdot \tfrac{1}{2} \cdot e_p \left[\frac{f_r^k}{\sum\limits_{k=1}^{M} f_r^k} - \frac{f_l^k}{\sum\limits_{k=1}^{M} f_l^k} \right] \qquad (19)$$

Equations from (20) to (23) update the parameters of the antecedents of the rules

$$\xi_{k,i}^{new} = \xi_{k,i}^{old} + \eta \cdot \frac{1}{2} \cdot e_p \cdot \frac{y_l^k - \hat{y}_l}{D_l} \cdot \underset{l=1,l\neq i}{\overset{n}{*}} \left(\overline{\mu}_{k,l}\right) \cdot \frac{\partial \overline{\mu}_{k,i}}{\partial \xi_{k,i}} \quad ; k \leq L \tag{20}$$

$$\xi_{k,i}^{new} = \xi_{k,i}^{old} + \eta \cdot \frac{1}{2} \cdot e_p \cdot \frac{y_l^k - \hat{y}_l}{D_l} \cdot \underset{l=1,l\neq i}{\overset{n}{*}} \left(\underline{\mu}_{k,l}\right) \cdot \frac{\partial \underline{\mu}_{k,i}}{\partial \xi_{k,i}} \quad ; k > L \tag{21}$$

$$\xi_{k,i}^{new} = \xi_{k,i}^{old} + \eta \cdot \frac{1}{2} \cdot e_p \cdot \frac{y_r^k - \hat{y}_r}{D_r} \cdot \underset{l=1,l\neq i}{\overset{n}{*}} \left(\underline{\mu}_{k,l}\right) \cdot \frac{\partial \underline{\mu}_{k,i}}{\partial \xi_{k,i}} \quad ; k \leq R \tag{22}$$

$$\xi_{k,i}^{new} = \xi_{k,i}^{old} + \eta \cdot \frac{1}{2} \cdot e_p \cdot \frac{y_r^k - \hat{y}_r}{D_r} \cdot \underset{l=1,l\neq i}{\overset{n}{*}} \left(\overline{\mu}_{k,l}\right) \cdot \frac{\partial \overline{\mu}_{k,i}}{\partial \xi_{k,i}} \quad ; k > R \tag{23}$$

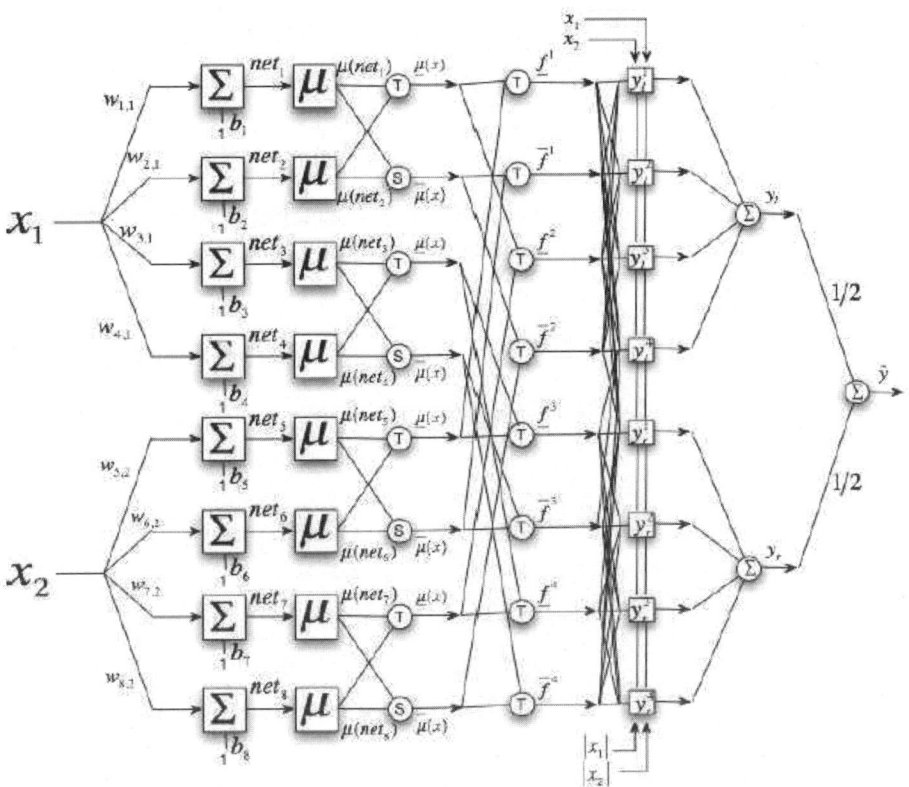

Fig. 7. Interval type-2 fuzzy neural network

The second architecture (Figure 7) is an adaptive six layer NN. The nodes of the first layer fuzzify the inputs, each node is a type-1 fuzzy neuron. Nodes from the second layer are T-norm type-1 fuzzy neurons whose outputs are the lower membership function and S-norm type-1 fuzzy neurons whose outputs are the upper membership functions. Layer three contains non-adaptive nodes that represent *lower-upper* firing set of the antecedents of the rules. In layer four each node evaluates the consequents rules. In layer five left-right points are evaluated using the Karnik-Mendel type-reduction algorithm [2]. Layer six defuzzifies the output of the system. The forward-propagation procedure is described as follows:

Layer 0. Inputs

$$x = (x_1, \ldots, x_i, \ldots, x_n)^t$$

Layer 1. Every node ℓ in this layer is a square (Figure 7) with a node function.

for k=1 to M
      for i=1 to n

$$^1net_{k,i} = {}^1w_{k,i}x_i + {}^1b_{k,i} \quad ; {}^1\mu_{k,i}(x_i) = \mu({}^1net_{k,i})$$

$$^2net_{k,i} = {}^2w_{k,i}x_i + {}^2b_{k,i} \quad ; {}^2\mu_{k,i}(x_i) = \mu({}^2net_{k,i})$$

      end
end

where μ is the transfer function (which can be gaussian, gbell or logistic). For example gaussian with uncertain mean (igaussmsttype2) is defined as:

$$^1\mu_{k,i}(x_i) = e^{-{}^1net_{k,i}^2} = e^{-\frac{1}{2}\left(\frac{x_i - {}^1m_{k,i}}{\sigma_{k,i}}\right)^2}$$

$$^2\mu_{k,i}(x_i) = e^{-{}^2net_{k,i}^2} = e^{-\frac{1}{2}\left(\frac{x_i - {}^2m_{k,i}}{\sigma_{k,i}}\right)^2}$$

and the transfer function gbell with uncertain mean (igbellmsttype2) is defined as:

$$^1\mu_{k,i}(x_i) = \frac{1}{1 + \left({}^1net_{k,i}^2\right)^{b_{k,i}}} = \frac{1}{1 + \left[\left(\frac{x_i - {}^1m_{k,i}}{a_{k,i}}\right)^2\right]^{b_{k,i}}}$$

$$^2\mu_{k,i}(x_i) = \frac{1}{1 + \left({}^2net_{k,i}^2\right)^{b_{k,i}}} = \frac{1}{1 + \left[\left(\frac{x_i - {}^2m_{k,i}}{a_{k,i}}\right)^2\right]^{b_{k,i}}}$$

k=1,2,…,M ; i= 1,2,…,n

Layer 2. Every node ℓ in this layer is a circle labeled with T-norm and S-norm alternated.

$$\underline{\mu}_{k,i}(x_i) = {}^1\mu_{k,i}(x_i) \cdot {}^2\mu_{k,i}(x_1)$$

$$\overline{\mu}_{k,i}(x_i) = {}^1\mu_{k,i}(x_i) + {}^2\mu_{k,i}(x_i) - \underline{\mu}_{k,i}(x_i)$$

k=1,2,...,M ; i= 1,2,...,n

Layer 3. Rules

$$\underline{f}^k = \overset{n}{\underset{i=1}{\circledast}}\left(\underline{\mu}_{k,i}\right) \qquad ; \overline{f}^k = \overset{n}{\underset{i=1}{\circledast}}\left(\overline{\mu}_{k,i}\right)$$

Layer 4. Consequent left-right firing points

$$y_l^k = \sum_{i=1}^{n} c_{k,i} x_i + c_{k,0} - \sum_{i=1}^{n} s_{k,i}|x_i| - s_{k,0}$$

$$y_r^k = \sum_{i=1}^{n} c_{k,i} x_i + c_{k,0} + \sum_{i=1}^{n} s_{k,i}|x_i| + s_{k,0}$$

Layer 5. Left and right firing points (Type-reduction)

$$\hat{y}_l = \hat{y}_l(\overline{f}^1,....,\overline{f}^L, \underline{f}^{L+1},....,\underline{f}^M, y_l^1,...,y_l^M) = \frac{\sum_{k=1}^{M} f_l^k \cdot y_l^k}{\sum_{k=1}^{M} f_l^k}$$

$$= \frac{\sum_{k=1}^{L} \overline{f}^k \cdot y_l^k + \sum_{k=L+1}^{M} \underline{f}^k \cdot y_l^k}{\sum_{k=1}^{L} \overline{f}^k + \sum_{k=L+1}^{M} \underline{f}^k}$$

$$\hat{y}_r = \hat{y}_r(\underline{f}^1,....,\underline{f}^R, \overline{f}^{R+1},....,\overline{f}^M, y_r^1,...,y_r^M) = \frac{\sum_{k=1}^{M} f_r^k \cdot y_r^k}{\sum_{k=1}^{M} f_r^k}$$

$$= \frac{\sum_{k=1}^{R} \underline{f}^k \cdot y_r^k + \sum_{k=R+1}^{M} \overline{f}^k \cdot y_r^k}{\sum_{k=1}^{R} \underline{f}^k + \sum_{k=R+1}^{M} \overline{f}^k}$$

Layer 6. Defuzzification

$$\hat{y} = \frac{\hat{y}_l + \hat{y}_r}{2}$$

Equations from (16) to (19) update the parameters of the consequents and (20) to (23) update the parameters of the antecedents of the rules.

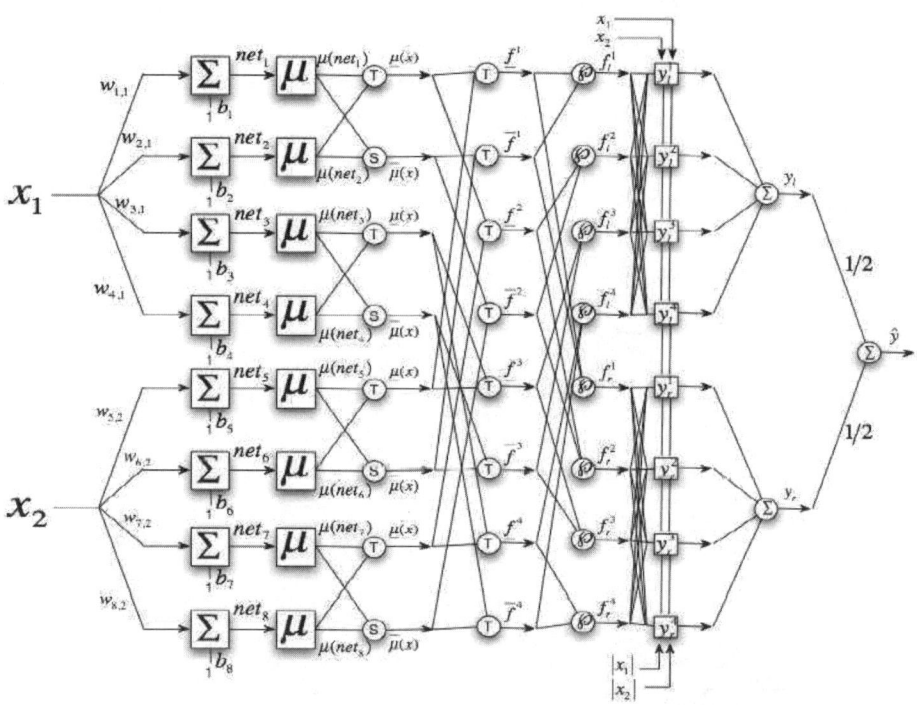

Fig. 8. Interval type-2 fuzzy neural network

The third IT2FNN architecture implements a TSK IT2FIS and has a seven layered architecture as shown in Figure 8. The first and second hidden layers are for fuzzifying input variables, and T-norm operators are deployed in the third hidden layer to compute the rule antecedent part. The fourth hidden layer evaluates firing left-most and right-most points, $f^k(x) \in [\underline{f}^k(x), \overline{f}^k(x)]$, the rule strengths followed by the fifth hidden layer where the consequent for each rule are determined by $\tilde{y}_k^j \in [_l y_k^j, _r y_k^j]$ (see Figures 5 and 8). The sixth layer computes its two end-points, $_l y^j$ and $_r y^j$, and in the seventh layer, we defuzzify the jth output using the average of $_l y^j$ and $_r y^j$. The forward-propagation procedure is described as follows:

Layer 0. Inputs

$$x = (x_1, \ldots, x_i, \ldots, x_n)^t$$

Layer 1. Every node ℓ in this layer is a square (Figure 6) with a node function.
for k=1 to M

      for i=1 to n

$$^1net_{k,i} = {}^1w_{k,i}x_i + {}^1b_{k,i} \quad ; {}^1\mu_{k,i}(x_i) = \mu({}^1net_{k,i})$$

$$^2net_{k,i} = {}^2w_{k,i}x_i + {}^2b_{k,i} \quad ; {}^2\mu_{k,i}(x_i) = \mu({}^2net_{k,i})$$

      end

end

where μ is this transfer function (which can be gaussian, gbell or logistic). For example gaussian with uncertain mean:

$$^1\mu_{k,i}(x_i) = e^{-{}^1net_{k,i}^2} = e^{-\frac{1}{2}\left(\frac{x_i - {}^1m_{k,i}}{\sigma_{k,i}}\right)^2}$$

$$^2\mu_{k,i}(x_i) = e^{-{}^2net_{k,i}^2} = e^{-\frac{1}{2}\left(\frac{x_i - {}^2m_{k,i}}{\sigma_{k,i}}\right)^2}$$

or gbell transfer function:

$$^1\mu_{k,i}(x_i) = \frac{1}{1+\left({}^1net_{k,i}^2\right)^{b_{k,i}}} = \frac{1}{1+\left[\left(\frac{x_i - {}^1m_{k,i}}{a_{k,i}}\right)^2\right]^{b_{k,i}}}$$

$$^2\mu_{k,i}(x_i) = \frac{1}{1+\left({}^2net_{k,i}^2\right)^{b_{k,i}}} = \frac{1}{1+\left[\left(\frac{x_i - {}^2m_{k,i}}{a_{k,i}}\right)^2\right]^{b_{k,i}}}$$

k=1,2,…,M ; i= 1,2,…,n

Layer 2. Every node ℓ in this layer is a circle labeled with T-norm and S-norm alternated.

$$\underline{\mu}_{k,i}(x_i) = {}^1\mu_{k,i}(x_i) \cdot {}^2\mu_{k,i}(x_i)$$

$$\overline{\mu}_{k,i}(x_i) = {}^1\mu_{k,i}(x_i) + {}^2\mu_{k,i}(x_i) - \underline{\mu}_{k,i}(x_i)$$

k=1,2,…,M ; i= 1,2,…,n

Layer 3. Every node ℓ in this layer is a circle labeled T, which multiplies the incoming signals and sends the product out. Each output node represents the lower ($\underline{f}^k$) and upper ($\overline{f}^k$) firing strength of a rule.

$$\underline{f}^k = \underset{i=1}{\overset{n}{*}}\left(\underline{\mu}_{k,i}\right) \qquad ; \overline{f}^k = \underset{i=1}{\overset{n}{*}}\left(\overline{\mu}_{k,i}\right)$$

Layer 4. Every node ℓ in this layer is a circle labeled $\wp$ which evaluates the left-most and right-most firing points denoted by:

$$f_l^k = \frac{\overline{\omega}_l^k \overline{f}^k + \underline{\omega}_l^k \underline{f}^k}{\overline{\omega}_l^k + \underline{\omega}_l^k} \qquad ; f_r^k = \frac{\underline{\omega}_r^k \underline{f}^k + \overline{\omega}_r^k \overline{f}^k}{\underline{\omega}_r^k + \overline{\omega}_r^k}$$

where ω are adjustable weights.

Layer 5. Every node ℓ in this layer is a square labeled y_l and y_r, which computes y_l^k, y_r^k.

$$y_l^k = \sum_{i=1}^{n} c_{k,i} x_i + c_{k,0} - \sum_{i=1}^{n} s_{k,i}|x_i| - s_{k,0}$$

$$y_r^k = \sum_{i=1}^{n} c_{k,i} x_i + c_{k,0} + \sum_{i=1}^{n} s_{k,i}|x_i| + s_{k,0}$$

Layer 6. The two nodes in this layer are circles labeled with "Σ" that evaluates the two end-points, y_l and y_r

$$\hat{y}_l = \hat{y}_l(\overline{f}^1,....,\overline{f}^L, \underline{f}^{L+1},....,\underline{f}^M, y_l^1,...,y_l^M) = \frac{\sum_{k=1}^{M} f_l^k \cdot y_l^k}{\sum_{k=1}^{M} f_l^k}$$

$$\hat{y}_r = \hat{y}_r(\underline{f}^1,....,\underline{f}^R, \overline{f}^{R+1},....,\overline{f}^M, y_r^1,...,y_r^M) = \frac{\sum_{k=1}^{M} f_r^k \cdot y_r^k}{\sum_{k=1}^{M} f_r^k}$$

Layer 7. The single node in this layer is a circle labeled "Σ" that computes the output.

$$\hat{y} = \frac{\hat{y}_l + \hat{y}_r}{2}$$

Equations from (16) to (19) update the parameters of the consequents of the rules and (24) to (27) update the left-right most firing set points for type-reduction

$$^{new}\underline{\omega}_l^k = {}^{old}\underline{\omega}_l^k + \eta \cdot \tfrac{1}{2} \cdot e_p \cdot \frac{y_l^k - \hat{y}_l}{\sum\limits_{j=1}^{M} f_l^j} \cdot \frac{\underline{f}^k - f_l^k}{\overline{\omega}_l^k + \underline{\omega}_l^k} \tag{24}$$

$$^{new}\overline{\omega}_l^k = {}^{old}\overline{\omega}_l^k + \eta \cdot \tfrac{1}{2} \cdot e_p \cdot \frac{y_l^k - \hat{y}_l}{\sum\limits_{j=1}^{M} f_l^j} \cdot \frac{\overline{f}^k - f_l^k}{\overline{\omega}_l^k + \underline{\omega}_l^k} \tag{25}$$

$$^{new}\underline{\omega}_r^k = {}^{old}\underline{\omega}_r^k + \eta \cdot \tfrac{1}{2} \cdot e_p \cdot \frac{y_r^k - \hat{y}_r}{\sum\limits_{j=1}^{M} f_r^j} \cdot \frac{\underline{f}^k - f_r^k}{\underline{\omega}_r^k + \overline{\omega}_r^k} \tag{26}$$

$$^{new}\overline{\omega}_r^k = {}^{old}\overline{\omega}_r^k + \eta \cdot \tfrac{1}{2} \cdot e_p \cdot \frac{y_r^k - \hat{y}_r}{\sum\limits_{j=1}^{M} f_r^j} \cdot \frac{\overline{f}^k - f_r^k}{\underline{\omega}_r^k + \overline{\omega}_r^k} \tag{27}$$

$$\frac{\partial f_l^k}{\partial \xi_{k,i}} = \frac{\overline{\omega}_l^k \cdot \underset{l=1,l\neq i}{\overset{n}{*}} \left(\overline{\mu}_{k,l}\right) \cdot \frac{\partial \overline{\mu}_{k,i}}{\partial \xi_{k,i}} + \underline{\omega}_l^k \cdot \underset{l=1,l\neq i}{\overset{n}{*}} \left(\underline{\mu}_{k,l}\right) \cdot \frac{\partial \underline{\mu}_{k,i}}{\partial \xi_{k,i}}}{\overline{\omega}_l^k + \underline{\omega}_l^k} \tag{28}$$

$$\frac{\partial f_r^k}{\partial \xi_{k,i}} = \frac{\underline{\omega}_r^k \cdot \underset{l=1,l\neq i}{\overset{n}{*}} \left(\underline{\mu}_{k,l}\right) \cdot \frac{\partial \underline{\mu}_{k,i}}{\partial \xi_{k,i}} + \overline{\omega}_r^k \cdot \underset{l=1,l\neq i}{\overset{n}{*}} \left(\overline{\mu}_{k,l}\right) \cdot \frac{\partial \overline{\mu}_{k,i}}{\partial \xi_{k,i}}}{\underline{\omega}_r^k + \overline{\omega}_r^k} \tag{29}$$

$$^{new}\xi_{k,i} = {}^{old}\xi_{k,i} + \eta \cdot \tfrac{1}{2} \cdot e_p \left[\frac{y_l^k - \hat{y}_l}{\sum\limits_{j=1}^{M} f_l^j} \cdot \frac{\partial f_l^k}{\partial \xi_{k,i}} + \frac{y_r^k - \hat{y}_r}{\sum\limits_{j=1}^{M} f_r^j} \cdot \frac{\partial f_r^k}{\partial \xi_{k,i}} \right] \tag{30}$$

4 Simulation Results

We present results form simulations of three different IT2FNN architectures for the TSK model. Tables 1 and 2 show the results of the hybrid models for the on-line identification in a control system and prediction of Mackey-Glass chaotic series respectively. The first column in tables show indexes of the hybrid model, first index represent the architecture and second index the transference function type where 1 is for Gaussian symmetric (igaussmtype2) with uncertain mean and 2 for asymmetric transference function (igausssttype2) with uncertain mean and standard deviation.

4.1 On-Line Identification in Control Systems

In this paper, we compare our IT2FNN with a simulation example given in [1], [18], where a 1-20-10-1 back-propagation MLP is employed to identify a nonlinear component in a control system. The plant under consideration is governed by the following difference equation:

$$y(k+1) = 0.3y(k) + 0.6 * y(k-1) + f(u(k)) \tag{31}$$

where y(k) and u(k) are the output and input, respectively, at time step k. The unknown function $f(\cdot)$ has the form

$$f(u) = 0.6\sin(\pi u) + 0.3\sin(3\pi u) + 0.1\sin(5\pi u) \tag{32}$$

In order to identify the plant, a series-parallel model governed by the difference equation

$$\hat{y}(k+1) = 0.3\hat{y}(k) + 0.6 * \hat{y}(k-1) + F(u(k)) \tag{33}$$

was used, where $F(\cdot)$ is the function implemented by the IT2FNN and its parameters are updated at each time step.

The number of membership functions assigned to each input of the IT2FNN was arbitrarily set to 5, so the rule number is 5. After 50 epochs (Table I), we obtained a RMSE = 0.0055 (training). The desired and predicted values for both training data are essentially the same in Figure 7.

Table 1. Training RMSE After 50 epochs

Hybrid Model	IT2MF	RMSE
11	igaussmtype2	0.0061
12	igausssttype2	0.0064
21	igaussmtype2	0.0063
22	igausssttype2	0.0057
31	igaussmtype2	0.0056
32	igausssttype2	0.0055

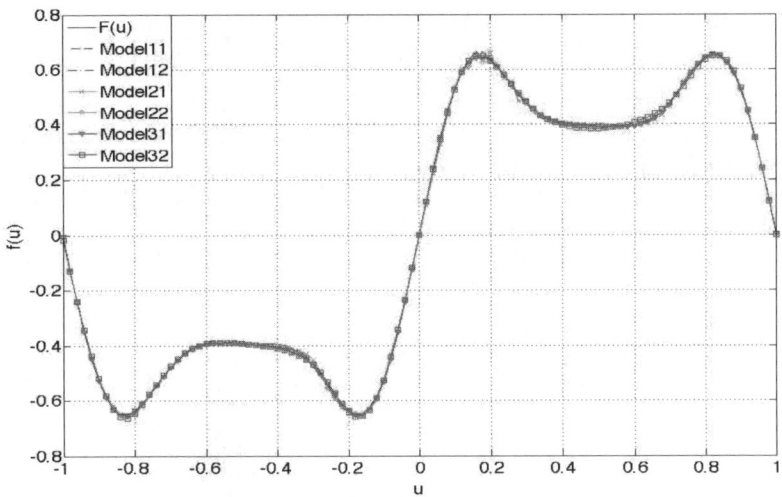

Fig. 7. Off-line learning with five rules

In table 1 and figure 7 it is shown that proposed architectures clearly identify control function and the best architecture is the third one (Models 31 and 32) with asymmetric transference functions.

4.2 Predicting Chaotic Time Series

The time series used for the simulation was generated by the chaotic Mackey-Glass differential delay equation [17] defined below:

$$\dot{x}(t) = \frac{0.2x(t - \tau)}{1 + x^{10}(t - \tau)} - 0.1x(t) \qquad (34)$$

The prediction of future values of these time series is a benchmark problem, which has been considered by a number of connectionist researchers [18].

The goal is to use known values of the time series up to the point $x=t$ to predict the value at some point in the future $x=t+P$. The standard method for this type of prediction is to create a mapping from D points of the time series spaced Δ units apart, that is $(x(t-(D-1)\Delta),...,x(t-\Delta),x(t))$, to a predicted future value $x(t+P)$. To allow comparison with earlier work [18], the values D=4 and $\Delta =P=6$ were used. All other simulation setting in this example were purposely arranged to be as close as possible to those reported in [19].

To obtain the time series value at each integer point, we applied the fourth-order Runge-Kutta method to find the numerical solution to the equation. The time step used in the method is 0.1, initial condition $x(0)$=1.2, τ=17, and $x(t)$ is thus derived for $0<=t<=2000$. From the Mackey-Glass time series $x(t)$, we extracted 1000 input-output data pairs of the following format:

$$[x(t-24),x(t-18),x(t-12),x(t-6);x(t)]$$

where t=124 to 1123. The first 500 pairs (training data set) were used for training the IT2FNN while the remaining 500 pairs (checking data set) were used for validating the identified model. The number of membership functions assigned to each input of the IT2FNN was arbitrarily set to 4, so the number of rules is 4.

After 50 epochs (Table 2), we obtained a RMSE = 0.0135 (training) and RMSE = 0.0155 (checking). The desired and predicted values for both training data and checking data are essentially the same in Figure 7. Rules and surface after 50 epochs are shown in Figures 8, 9 and 10.

It can be seen in table 2 that architecture 3 identifies best Mackey-Glass series and figure 10 shows that architecture 3 (models 31 and 32) converges faster with asymmetric transference functions.

Table 2. Training RMSE After 50 epochs

Hybrid Model	IT2MF	RMSE
11	igaussmtype2	0.0257
12	igausssttype2	0.0250
21	igaussmtype2	0.0196
22	igausssttype2	0.0221
31	igaussmtype2	0.0153
32	igausssttype2	0.0135

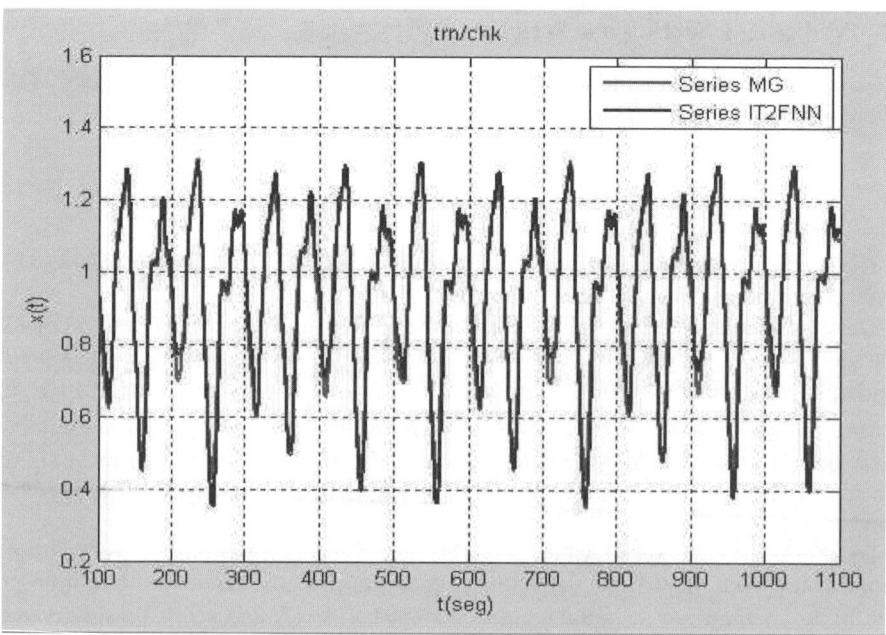

Fig. 7. Mackey-Glass time series from t=124 to 1123 and six-step ahead prediction

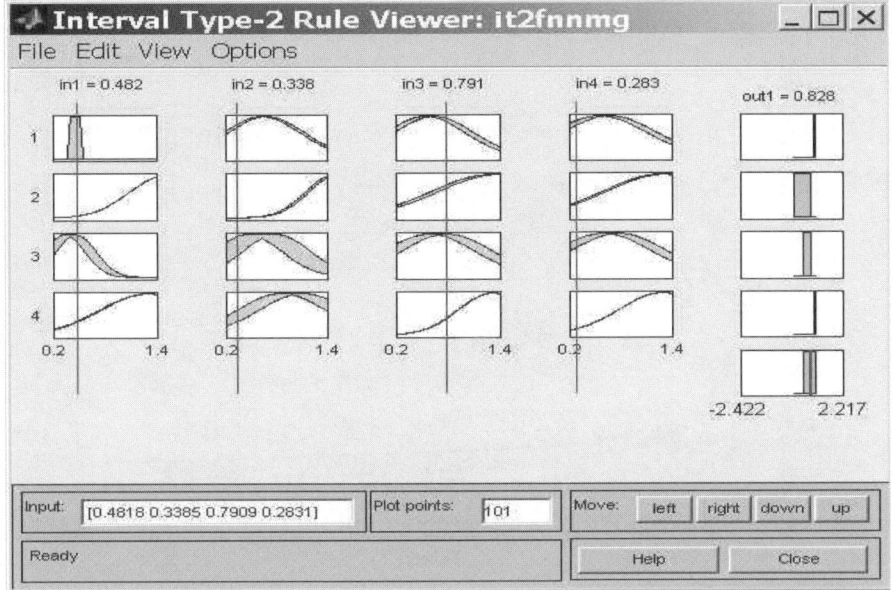

Fig. 8. Rules and Membership functions after learning

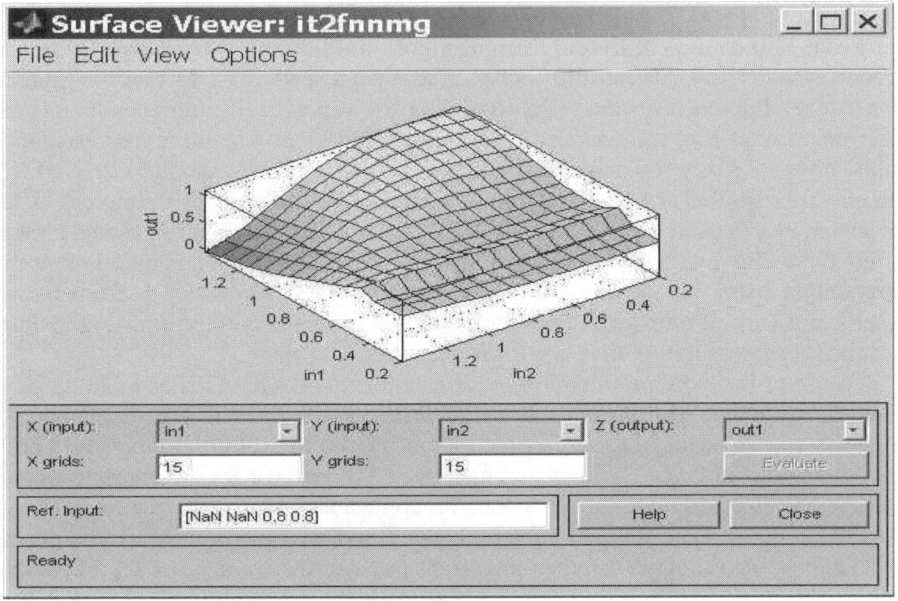

Fig. 9. Final non-linear surface

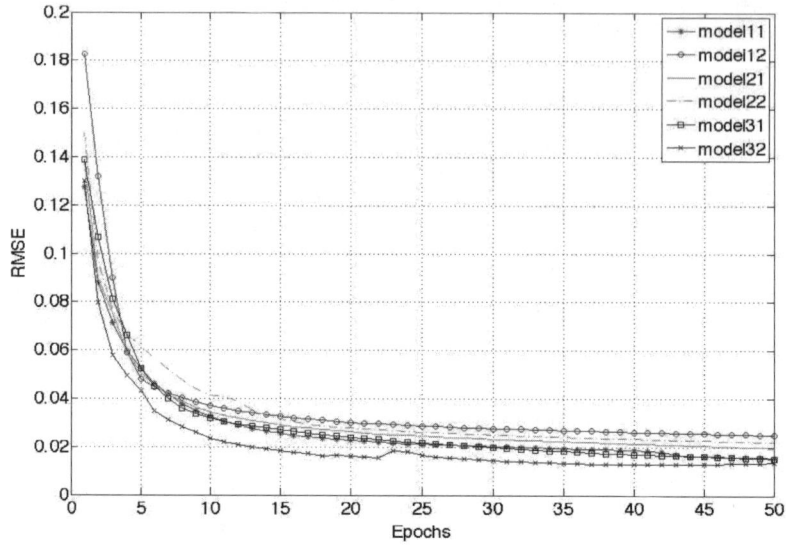

Fig. 10. RMSE curves for the Hybrid Model IT2FNN

5 Conclusions

Before the introduction of hybrid learning algorithms into the interval type-2 fuzzy inference systems, the relationships between the neural network (NN) and the interval type-2 fuzzy inference system (IT2FIS) can be viewed as two extreme endpoints on a spectrum of modeling approaches. At one end, IT2FIS has meaningful representations (interval type-2 fuzzy If-Then rules and interval type-2 fuzzy reasoning) derived from human expertise, but it has no adaptive capability (learning from examples) to take advantage of a desired input-output data set. At the other end, NN represents a totally different paradigm with learning capability that adapts its parameters based on desired input-output pairs, but neither can accommodate *a priori* knowledge from humans experts, nor can we transform a network configuration and connection weights into a meaningful representation to account for structured knowledge.

Because of the extreme flexibility of adaptive networks the IT2FNN can be generalized in a number of different ways. For instance, the interval type-2 membership functions can be changed to any of the parameterized MFs. Furthermore; we can replace the nodes in layer 3 with a parameterized T-norm and let the learning rule decide the best T-norm operator for a specific application. Moreover, the realization of rules with OR'ed antecedents, linguistic hedges, and multiple outputs can be put into the IT2FNN accordingly. Another important issue in the training of the IT2FNN is how to preserve some intuitive features that make the resulting interval type-2 fuzzy rules easy to interpret.

Throughout this paper, we have assumed that the structure of IT2FNN is fixed and that the parameter identification is solved through the hybrid learning rule. However, to make the whole approach more complete, the structure identification (which is concerned with the selection of an appropriate input-space partition style, the number of membership functions on each input, and so on) is equally important to the successful application of IT2FNN, especially for modeling problems with a large of inputs. Effective partitioning of the input space can decrease the number of rules and thus increase the speed in both the learning and application phases.

The time series results show that intelligent hybrid methods can be derived as a generalization of the autoregressive non-linear models in the context of time series. This derivation allows a practical specification for a general class of prognosis and identification time series models, where a set of input-output variables are part of the dynamics of the time series knowledge base. This helps the application of the methodology to a series of diverse dynamics, with a very small number of causal variables to explain behavior.

Acknowledgments

The authors would like to thank CONACYT and DGEST for the financial support given to this research project. The student (Juan R. Castro) is supported by a scholarship from UABC-CONACYT.

References

1. Jang, J.-S.R., Sun, C.-T.: Neuro-fuzzy and soft computing: a computational approach to learning and machine intelligence. Prentice-Hall, Inc., Upper Saddle River (1996)
2. Mendel, J.: Uncertain Rule-Based Fuzzy Logic Systems: Introduction and New Directions. Prentice-Hall, NJ (2001)
3. Pedrycz, W.: Neurocomputations in relational systems. IEEE Trans. on Pattern Anal. Mach. Intell. PAMI 13(3), 289–297 (1991)
4. Pedrycz, W., Rocha, A.F.: Fuzzy-set based model of neurons and knowledge-based networks. IEEE Transactions on Fuzzy Systems 1, 254–266 (1993)
5. Zadeh, L.A.: Fuzzy sets. Information and Control 8, 338–353 (1965)
6. Zadeh, L.A.: Outline of a new approach to the analysis of complex systems and decision processes. IEEE Transactions on Systems, Man, and Cybernetics 3(1), 28–44 (1973)
7. Zadeh, L.A.: The concept of a linguistic variable and its application to approximate reasoning, Parts 1, 2, and 3. Information Sciences, 8, 9, 199–249, 301–357, 43–80 (1975)
8. Zadeh, L.A.: Fuzzy Logic. Computer 1(4), 83–93 (1988)
9. Zadeh, L.A.: Knowledge representation in fuzzy logic. IEEE Transactions on Knowledge and Data Engineering 1, 89–100 (1989)
10. Zadeh, L.A.: Fuzzy logic = computing with words. IEEE Transactions on Fuzzy Systems 2, 103–111 (1996)
11. Karnik, N.N., Mendel, J.M.: An Introduction to Type-2 Fuzzy Logic Systems, Univ. of Southern Calif., Los Angeles, CA., (June 1998b)
12. Liang, Q., Mendel, J.: Interval type-2 fuzzy logic systems: Theory and design. IEEE Transactions Fuzzy Systems 8, 535–550 (2000)
13. Mizumoto, M., Tanaka, K.: Some Properties of Fuzzy Sets of Type-2. Information and Control 31, 312–340 (1976)
14. Castro, J.R.: Hybrid Intelligent Architecture Development for Time Series Forecasting. Masters Degree Thesis. Tijuana Institute of Technology (December 2005)

15. Sugeno, M., Kang, G.T.: Structure identification of fuzzy model. Fuzzy Sets and Systems. In: Sugeno, M. (ed.) Industrial applications of fuzzy control, vol. 28, pp. 15–33. Elsevier Science Pub. Co, Amsterdam (1985)
16. Takagi, T., Sugeno, M.: Fuzzy identification of systems and its applications to modeling and control. IEEE Transactions on Systems, Man, and Cybernetics 15, 116–132 (1985)
17. Mackey, M.C., Glass, L.: Oscillation and chaos in physiological control systems. Science 197, 287–289 (1977)
18. Lapedes, S., Farber, R.: Nonlinear signal processing using neural networks: prediction and system modeling. Technical Report LA-UR-87-2662, Los Alamos National Laboratory, Los Alamos, New Mexico 87545 (1987)
19. Crowder, R.S.: Predicting the Mackey-Glass time series with cascade-correlation learning. In: Touretzky, D., Hinton, G., Sejnowski, T. (eds.) Proc. of the 1990 Connectionist Models Summer School, Carnegic Mellon University, pp. 117–123 (1990)
20. Abraham, A.: Intelligent Systems: Architectures and Perspectives. In: Abraham, A., Jain, L., Kacprzyk, J. (eds.) Recent Advances in Intelligent Paradigms and Applications. Studies in Fuzziness and Soft Computing, ch. 1, pp. 1–35. Springer, Germany (2002)
21. Abraham, A., Khan, M.R.: Neuro-Fuzzy Paradigms for Intelligent Energy Management. In: Abraham, A., Jain, L., Jan van der Zwaag, B. (eds.) Innovations in Intelligent Systems: Design, Management and Applications. Studies in Fuzziness and Soft Computing, ch. 12, pp. 285–314. Springer, Germany (2003)
22. Nauck, D., Kruse, R.: NEFCLASS: A Neuro-Fuzzy Approach for the Classification of Data. In: George, K., et al. (eds.) Proceedings of ACM Symposium on Applied Computing, pp. 461–465. ACM Press, Nashville (1995)
23. Nauck, D., Kruse, R.: A Neuro-Fuzzy Method to Learn Fuzzy Classification Rules from Data. Fuzzy Sets and Systems 89, 277–288 (1997)
24. Nauck D, D., Kruse, R.: Neuro-Fuzzy Systems for Function Approximation. Fuzzy Sets and Systems 101, 261–271 (1999)
25. Yager, R.R., Filev, D.P.: Adaptive Defuzzication for Fuzzy System Modeling. In: Proceedings of the Workshop of the North American Fuzzy Information Processing Society, pp. 135-142 (1992)
26. Castro, J.R., Castillo, O., Melin, P.: Building Fuzzy Inference Systems with the Interval Type-2 Fuzzy Logic Toolbox. In: Proceedings of IFSA 2007, vol. 41, Part I, pp. 53–62 (2007)
27. Werbos, P.: Beyond regression: New tools for prediction and analysis in the behavioral sciences. PhD thesis, Harvard University (1974)
28. Narendra, K.S., Parthsarathy, K.: Identification and control of dynamical systems using neural networks. IEEE Transactions on Neural Networks 1(1), 4–27 (1990)
29. Hagan, M.T., Demuth, H.B., Beale, M.H.: Neural Network Design. PWS Publishing, Boston (1996)
30. Sepúlveda, R., Castillo, O., Melin, P., Díaz, A.R., Montiel, O.: Experimental study of intelligent controllers under uncertainty using type-1 and type-2 fuzzy logic. Inf. Sci. 177(10), 2023–2048 (2007)
31. Sepulveda, R., Castillo, O., Melin, P., Montiel, O.: An Efficient Computational Method to Implement Type-2 Fuzzy Logic in Control Applications. In: Melin, P., et al. (eds.) Analysis and Design of Intelligent Systems using Soft Computing Techniques, 1st edn. Studies in Fuzziness and Soft Computing, vol. 5, pp. 45–52. Springer, Germany (2007)

Optimization of Artificial Neural Network Architectures for Time Series Prediction Using Parallel Genetic Algorithms

Salvador González Mendivil, Oscar Castillo, and Patricia Melin

Tijuana Institute of Technology

Abstract. This paper considers the application of parallel genetic algorithms to the optimization of modular neural network architectures for time series prediction. We have a cluster configuration of 16 computers and the application is executed using the Matlab Distributed Computing Engine included in MATLAB r2006b. The Linux Fedora Core VI Operating System was installed and configured for the cluster execution due to its high performance, scalability and because it presents innumerable benefits that facilitate the implementation of distributed computing applications. The first part of this paper presents the theoretical framework with basic concepts like times series, artificial neural networks, genetic algorithms, and parallel genetic algorithms. The second part of this paper presents the procedure for configuring the cluster of computers, requirements, experiences and main problems that were encountered. Also, the development of the project is presented explaining as it was initially proposed and the adjustments that were required. The third part of this paper presents the obtained results for the time series prediction using tables, graphics and describing each one of them. Finally the conclusions and future works are presented.

1 Introduction

Forecasting refers to a process by which the future behavior of a dynamical system is estimated based on our understanding and characterization of the system. If the dynamical system is not stable, the initial conditions become one of the most important parameters of the time series response, i.e. small differences in the start position can lead to a completely different time evolution. This is what is called sensitive dependence on initial conditions, and is associated with chaotic behavior [1, 2] for the dynamical system.

The financial markets are well known for wide variations in prices over short and long terms. These fluctuations are due to a large number of deals produced by agents that act independently from each other. However, even in the middle of the apparently chaotic world, there are opportunities for making good predictions [3, 4]. Traditionally, brokers have relied on technical analysis, based mainly on looking at trends, moving averages, and certain graphical patterns, for performing predictions and subsequently making deals. Most of these linear approaches, such as the well-known Box-Jenkins method, have disadvantages [5].

More recently, soft computing [6] methodologies, such as neural networks, fuzzy logic, and genetic algorithms, have been applied to the problem of forecasting complex

O. Castillo et al. (Eds.): Soft Computing for Hybrid Intel. Systems, SCI 154, pp. 387–399, 2008.
springerlink.com   © Springer-Verlag Berlin Heidelberg 2008

time series. These methods have shown clear advantages over the traditional statistical ones [7]. The main advantage of soft computing methodologies is that, we do not need to specify the structure of a model a-priory, which is clearly needed in the classical regression analysis [8]. Also, soft computing models are non-linear in nature and they can approximate more easily complex dynamical systems, than simple linear statistical models. Of course, there are also disadvantages in using soft computing models instead of statistical ones. In classical regression models, we can use the information given by the parameters to understand the process, i. e. the coefficients for the model can represent the elasticity of price for a certain good in the market. However, if the main objective is to forecast as closely as possible the time series, then the use of soft computing methodologies for prediction is clearly justified.

2 Monolithic Neural Network Models

A neural network model takes an input vector X and produces an output vector Y. The relationship between X and Y is determined by the network architecture, there are many forms of network architecture (inspired by the neural architecture of the brain). The neural network generally consists of at least three layers: one input layer, one output layer, and one or more hidden layers. Figure 1 illustrates a neural network with n neurons in the input layer, one hidden layer with m neurons, and one output layer with one neuron.

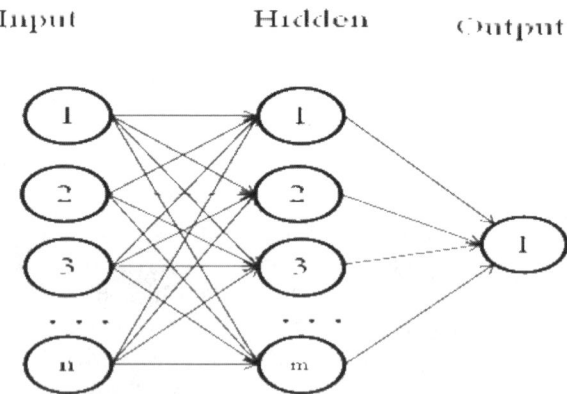

Fig. 1. Single Hidden Layer feedforward network

3 Modular Neural Networks

There exists a lot of neural network architectures in the literature that work well when the number of inputs is relatively small, but when the complexity of the problem grows or the number of inputs increases, their performance decreases very quickly. For this reason, there has also been research work in compensating in some way the problems in learning of a single neural network over high dimensional spaces.

A. Ensemble

B. Ensemble(Subtask)

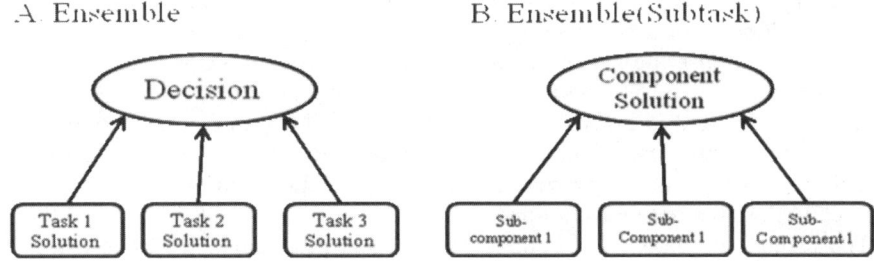

Fig. 2. Ensembles for one task and subtask

A. Modular

B. Modular(Subtask)

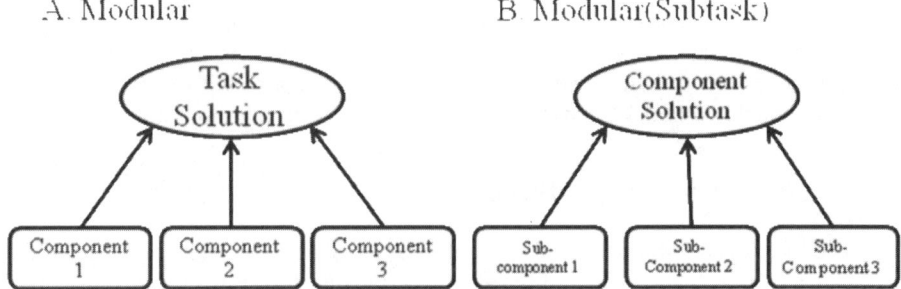

Fig. 3. Modular approach for task and subtask

In the work of Sharkey [9], the use of multiple neural systems (Multi-Nets) is described. It is claimed that multi-nets have better performance over even solve problems that monolithic neural networks are not able to solve. It is also claimed that multi-nets or *modular systems* have also the advantage of being easier to understand or modify, if necessary.

In the literature there is also mention of the terms "ensemble" and "modular" for this type of neural networks. The term "ensemble" is used when a redundant set of neural networks is utilized, as described in Hansen and Salomon [10], in this case, each of the neural networks is redundant because it is providing a solution for the same task, as it is shown in Figure 2. On the other and, in the modular approach, one task or problem is decompose in subtasks, and the complete solution requires the contribution of all the modules, as it is shown in Figure 3.

4   Genetic Algorithms for Optimization

John Holland, from the University of Michigan began his work on genetic algorithms at the beginning of the 60's. Hi first achievement was the publication of *Adaptation in Natural and Artificial System* [12] in 1975.

Holland had two goals in mind_ to improve the understanding of natural adaptation process, and to design artificial systems having properties similar to natural systems [13].

The basic idea is as follows: the genetic pool of a given population potentially contains the solutions, or a better solution, to a given adaptive problem. This solution is not "active" because the genetic combination on which it relies is split between several subjects. Only the association of different genomes can lead to the solution.

Holland's method is especially effective because it not only considers the role of mutation, but it also uses genetic recombination (crossover) [14]. The crossover of partial solutions greatly improves the capability of the algorithm to approach, and eventually find, the optimal solution.

The essence of the GA in both theoretical and practical domains has been well demonstrated [15]. The concept of applying a GA to solve engineering problems is feasible and sound. However, despite the distinct advantages of a GA for solving complicated, constrained and multi-objective functions where other techniques may have failed, the full power of the GA in applications is yet to be exploited [16].

To bring out the best use of the GA, we should explore further the study of genetic characteristics so that we can fully understand that the GA is not merely a unique technique for solving engineering problems, but that it also fulfils its potential for tackling scientific deadlocks that, in the past, were considered impossible to solve.

5 Parallel Genetic Algorithms

The basic idea behind most parallel programs is to divide a large problem into smaller tasks and to solve the tasks simultaneously using multiple processors. This divide-and-conquer approach can be applied to GAs in many different ways, and the literature contains numerous examples of successful parallel implementations. Some parallelization methods use a single population, while others are better suited to multicomputers with fewer and more powerful processing elements connected by a slower network.

We can recognize four major types of parallel GAs [17]:

1. Single-population master-slave GAs.
2. Multiple-population GAs.
3. Fine-grained GAs.
4. Hierarchical Hybrids.

Master-slave GAs have a single population. One master node executes the GA (selection, crossover, and mutation) and the evaluation of fitness is distributed among

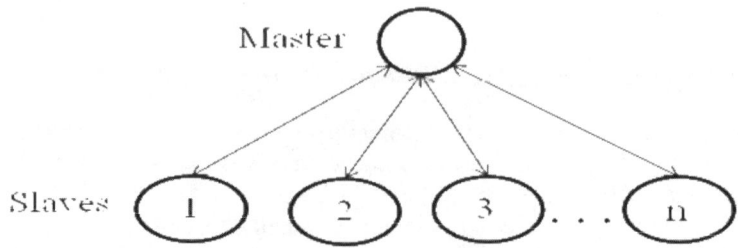

Fig. 4. A schematic of a master-slave parallel GA

several slave processors (see Figure 4). The slaves evaluate the fitness of the individuals that they receive from the master and return the results. Since in this type of parallel GAs selection and crossover consider the entire population, master-slave GAs are also known as "global" parallel GAs.

Probably the easiest way to implement GAs on parallel computers is to distribute the evaluation of fitness among several slave processors while one master executes the GA operations. This paper examines master-slave parallel GAs. These algorithms are important for several reasons:

- They explore the search space in exactly the same manner as serial GAs, and therefore the existing design guidelines for simple GAs are directly applicable,
- They are very easy to implement, which makes them popular with practitioners, and
- In many cases master-slave GAs result in significant improvements in performance.

6 Cluster's Configuration

6.1 Network Configuration

- ➢ Operating System Fedora Core VI
- ➢ Matlab r2006b for UNIX/LINUX
- ➢ Computer's name:
    - ❖ CLUSTER1 - CLUSTER16
- ➢ Static IP directions
    - ❖ For computers:
        - 192.168.1.1 - 192.168.1.16
    - ❖ In the network switch:
        - 192.168.1.254
- ➢ Resolution of names
    - ❖ /etc/hosts

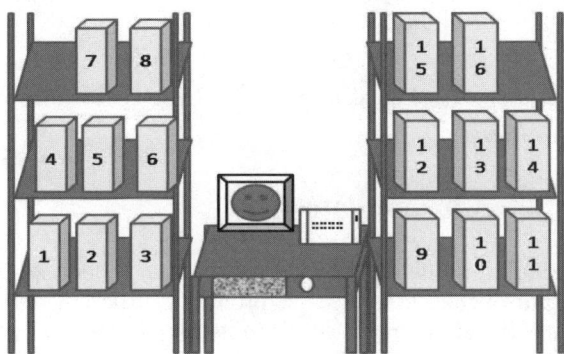

Fig. 5. Graphical representation for the cluster

6.2 Matlab Distributed Computing Toolbox

Distributed Computing Toolbox and MATLAB Distributed Computing Engine enable you to coordinate and execute independent MATLAB operations simultaneously on a cluster of computers, speeding up execution of large MATLAB jobs.

A *job* is some large operation that you need to perform in your MATLAB session. A job is broken down into segments called *tasks*. You decide how best to divide your job into tasks. You could divide your job into identical tasks, but tasks do not have to be identical.

The MATLAB session in which the job and its tasks are defined is called the *client* session. Often, this is on the machine where you program MATLAB. The client uses Distributed Computing Toolbox to perform the definition of jobs and tasks. MATLAB Distributed Computing Engine is the product that performs the execution of your job by evaluating each of its tasks and returning the result to your client session.

The *job manager* is the part of the engine that coordinates the execution of jobs and the evaluation of their tasks. The job manager distributes the tasks for evaluation to the engine's individual MATLAB sessions called *workers*. Use of the MathWorks job manager is optional; the distribution of tasks to workers can also be performed by a third-party scheduler, such as Windows CCS or Platform LSF [18].

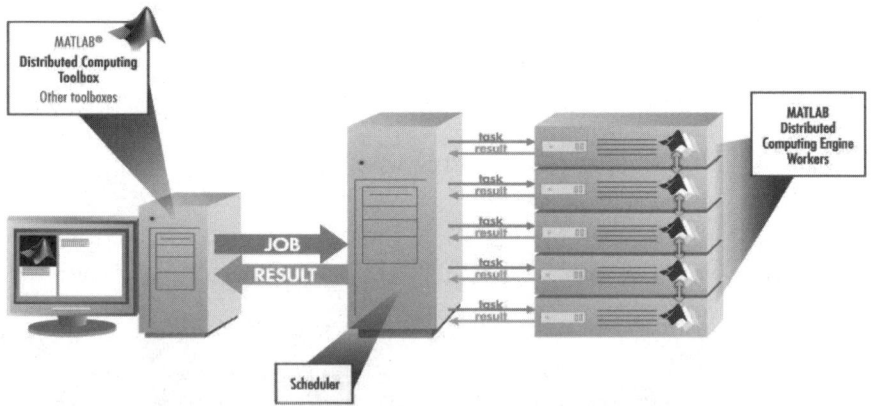

Fig. 6. Developing distributed and parallel applications with the Distributed Computing Toolbox of Matlab

7 Structure of the Chromosome

The Number of modules, hidden layers and neurons per hidden layer were optimized by genetic algorithm. Table 1 shows the proposed structure of the chromosome, Equation 1 calculates the total quantity of neurons for a hidden layer.

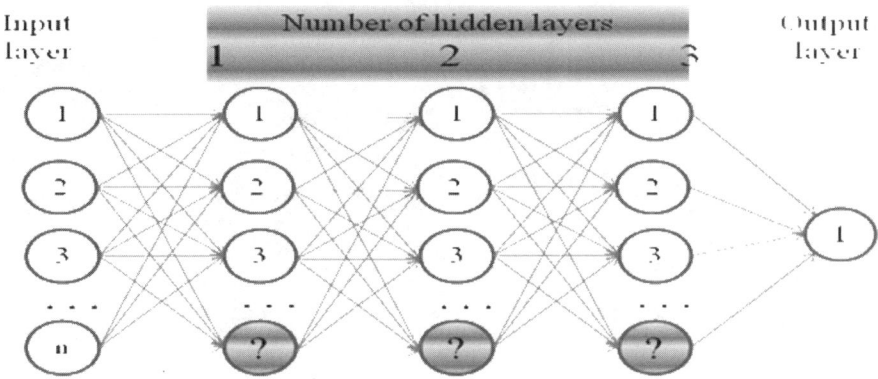

Fig. 7. Variables to optimize using genetic algorithm

Table 1. Structure of the Chromosome

	GA1 (31 Neurons)		GA2 (63 Neurons)	
	MONOLITHIC	**MODULAR**	**MONOLITHIC**	**MODULAR**
MODULES		6		6
LAYERS	3	3	3	3
NEURONS	5	5	6	6
Length(bits)	18	114	21	132

Equation 1:

$$2^n - 1$$

Table 2. Binary-decimal conversion

BINARY	**Equation 1**	**DECIMAL**
11111	$2^5 - 1$	31
111111	$2^6 - 1$	63

Figure 8 is graphical representation for the structure of the chromosome to represent a monolithic neural network called GA1 MONOLITHIC: first 3 bits represent the hidden layers, next 5 bits represent the neurons of first hidden layer, next 5 bits represent the neurons of second hidden layer and, last 5 bits represent the neurons of last hidden layer.

Figure 9 is graphical representation for the structure of the chromosome to represent a modular neural network called GA2 MODULAR: first 6 bits represent the number of modules to be training, next 6 sections of 21 bits represent a structure type GA2 MONOLITHIC: first 3 bits represent the hidden layers, next 6 bits represent the neurons of first hidden layer, next 6 bits represent the neurons of second hidden layer and, last 6 bits represent the neurons of last hidden layer.

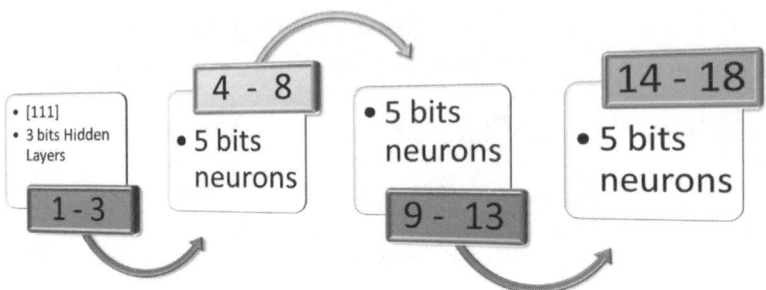

Fig. 8. Graphical representation for minimum structure of the chromosome (GA1 monolithic) with 18 bits

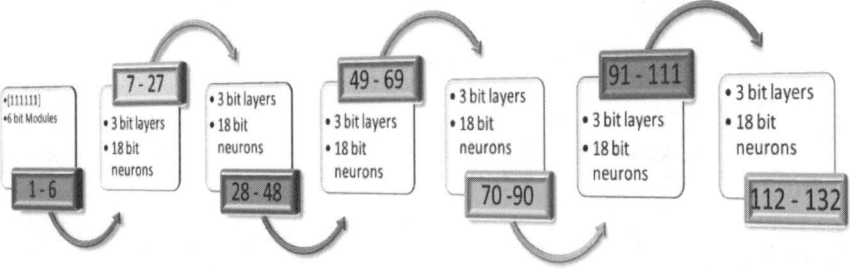

Fig. 9. Graphical representation for maximum structure of the chromosome (GA2 modular) with 132 bits

8 Objective Function

The objective of training the network is to minimize two different parameters: the accuracy of the network (f_1) and the complexity of the network (f_2) which is simply defined by the number of *active connections* or *neurons* in the network. This is calculated based upon the summation of the total number of active connections taking place. The accuracy of the network (f_1) is defined as:

$$f_1 = \frac{1}{N} \sum_{i=1}^{N} (y'_i - y_i)^2$$

Where N is the size of the testing vector, y'_i and y_i are the network output and desired output for the i-th pattern of the test vector respectively [15].

8.1 Selection Process

Since there are two different objective functions, (f_1) and (f_2) of the network optimization process, the fitness value of chromosome z is thus determined:

$$f(z) = \alpha \cdot rank[f_1(z)] + \beta \cdot f_2(z)$$

Where α is accuracy weighting coefficient; β is complexity weighting coefficient; and $rank[f_1(z)] \in Z^+$ is the rank value [15].

α is set as following to ensure $f(z_j) > f(z_i)$.

$$\varepsilon > \beta \cdot M$$

9 Simulation Results

Table 3 show the parameters used in execution of the genetic algorithm in the cluster. Levenberg-Marquardt was used like Training function due previous investigations [19] it obtained the best results.

Table 3. Parameters for execution of genetic algorithm

Number of individuals	40
Generations	10
Training function	Levenberg-Marquardt
Selection	Roulette
Mutation	10%
Crossover	70%
Elitism	10%

9.1 Simulation and Forecasting the Logistic Equation

We will consider the problem forecasting the logistic equation series calculated using

$$x_{t+1} = 4x_t(1 - x_t) \tag{2}$$

When $x_0 = 0.4$

Figure 10 shows the time series for first 200 point calculated with logistic equation. We show in Table 4 the evolution of the genetic algorithm for logistic equation series.

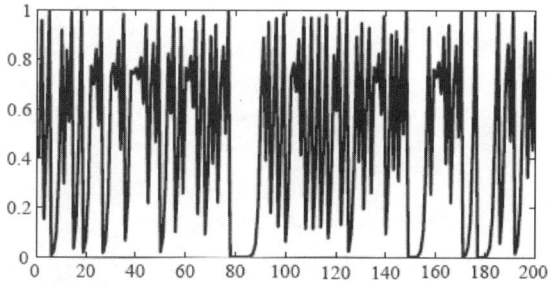

Fig. 10. First 200 values for logistic equation

Table 4. Evolution of the genetic algorithm for logistic equation series

GENERATION	FITNESS VALUE	CHROMOSOME	ARCHITECTURE
1	3.0835	[1 1 1 0 0 0 0 1 1 0 1 0 1 1 0 1 1 1]	[1 21 23]
2	3.0835	[1 1 1 0 0 0 0 1 1 0 1 0 1 1 0 1 1 1]	[1 21 23]
3	3.0835	[1 1 1 0 0 0 0 1 1 0 1 0 1 1 0 1 1 1]	[1 21 23]
4	3.0835	[1 1 1 0 0 0 0 1 1 0 1 0 1 1 0 1 1 1]	[1 21 23]
5	3.0835	[1 1 1 0 0 0 0 1 1 0 1 0 1 1 0 1 1 1]	[1 21 23]
6	3.0835	[1 1 1 0 0 0 0 1 1 0 1 0 1 1 0 1 1 1]	[1 21 23]
7	3.0835	[1 1 1 0 0 0 0 1 1 0 1 0 1 1 0 1 1 1]	[1 21 23]
8	1.3801	[1 1 1 0 0 0 0 1 1 1 1 0 1 1 0 1 1 0]	[1 29 22]
9	1.3801	[1 1 1 0 0 0 0 1 1 1 1 0 1 1 0 1 1 0]	[1 29 22]
10	1.3801	[1 1 1 0 0 0 0 1 1 1 1 0 1 1 0 1 1 0]	[1 29 22]

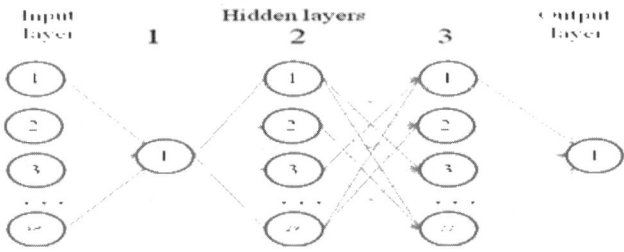

Fig. 11. Optimized neural network architecture for forecasting the logistic equation series

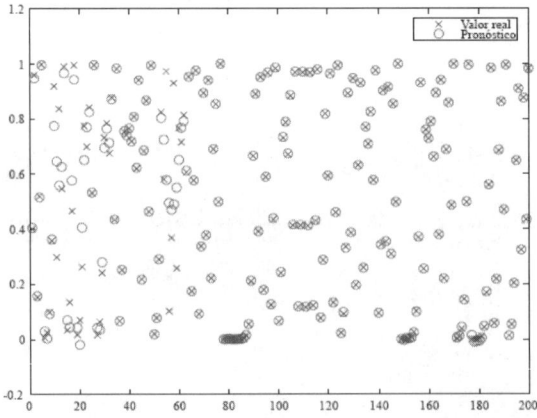

Fig. 12. Monolithic network results for logistic equation series

In Figure 11 we show the optimized neural network architecture for forecasting logistic equation series. We have to mention that the architecture shown in Figure 11 is result of the genetic algorithm execution.

We need to mention that the results shown in Figure 12 are for the best individual that the genetic algorithm was able to find in its execution with a minimum square error of **0.00472857**.

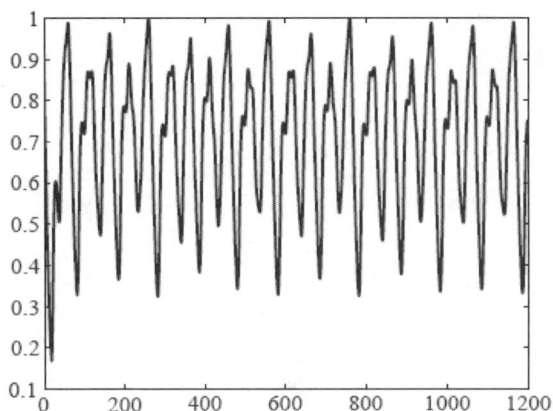

Fig. 13. First 1200 values for Mackey-glass series

Table 5. Evolution of the genetic algorithm for Mackey-Glass series

GENERATION	FITNESS VALUE	CHROMOSOME	ARCHITECTURE
0	27.1113	[0 1 1 0 1 0 1 1 0 1 0 1 0 0 1 1 0 1]	[10 13]
1	25.7873	[0 1 0 1 0 0 1 1 1 1 1 1 0 1 1 0 1 1]	30
2	25.7873	[0 1 0 1 0 0 1 1 1 1 1 1 0 1 1 0 1 1]	30
3	13.4572	[1 1 1 0 0 0 1 1 0 1 1 1 0 1 0 0 0 1]	[3 14 17]
4	13.4572	[1 1 1 0 0 0 1 1 0 1 1 1 0 1 0 0 0 1]	[3 14 17]
5	13.4572	[1 1 1 0 0 0 1 1 0 1 1 1 0 1 0 0 0 1]	[3 14 17]
6	13.4572	[1 1 1 0 0 0 1 1 0 1 1 1 0 1 0 0 0 1]	[3 14 17]
7	10.1092	[1 1 1 0 0 0 1 1 0 0 1 1 0 1 0 0 0 1]	[3 6 17]
8	10.1092	[1 1 1 0 0 0 1 1 0 0 1 1 0 1 0 0 0 1]	[3 6 17]
9	7.1342	[1 1 1 0 0 0 0 1 0 1 1 1 0 1 0 0 0 1]	[1 14 17]
10	7.1342	[1 1 1 0 0 0 0 1 0 1 1 1 0 1 0 0 0 1]	[1 14 17]

9.2 Simulation and Forecasting Mackey-Glass Series

We show in Figure 13 the values of Mackey-Glass series provided by Matlab r2006b
Table 5 shows the evolution of genetic algorithm for Mackey-Glass series.

In Figure 14 we show the optimized neural network architecture for forecasting logistic equation series. We have to mention that the architecture shown in Figure 14 is
result of the genetic algorithm execution.

We need to mention that the results shown in Figure 15 are for the best individual
that the genetic algorithm was able to find in its execution with a minimum square
error of **0.000394663**.

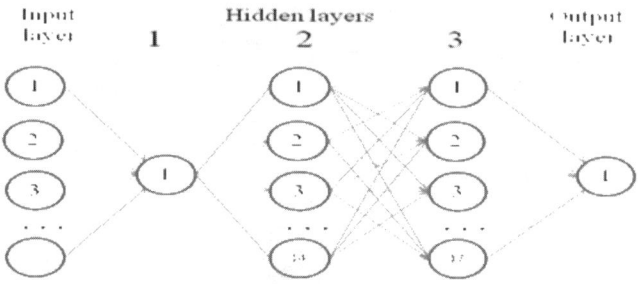

Fig. 14. Optimized neural network architecture for forecasting logistic equation series

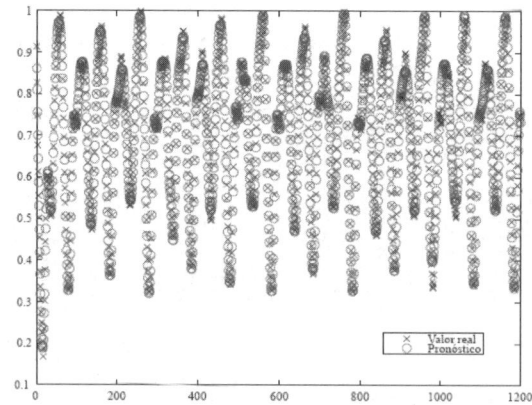

Fig. 15. Monolithic network results for Mackey-Glass series

10   Conclusions

We describe in this paper the use of parallel genetic algorithms for optimization of
neural networks architectures for simulation and forecasting of chaotic time series.
We have considered two chaotic series: logistic equation and Mackey-Glass series.

The genetic algorithm train both monolithic and modular neural networks and find the best architecture for each times series.

References

1. Brock, W.A., Hsieh, D.A., LeBaron, B.: Nonlinear Dynamics, chaos and Instability. MIT Press, Cambridge (1991)
2. Shdmit, A., Bandar, Z.: A Modular Neural Network Architecture with Additional Generalization Abilities for High Dimensional Input Vectors. In: Proceedings of ICANNGA 1997, Norwich, England (1997)
3. Castillo, O., Melin, P.: A new Fuzzy-Genetic Approach for the Simulation and Forecasting of International Trade Non-Linear Dynamics. In: Preccedings of CIFEr 1998, New York, USA, pp. 189–196. IEEE Press, Los Alamitos (1998)
4. Castillo, O., Melin, P.: Automated Mathematical Modelling for Financial Times Series Prediction Combinning Fuzzy Logic and Fractal Theory. In: Soft Computing for Financial Engineering, pp. 93–106. Springer, Germany (1999)
5. Haykin, S.: Adaptive Filter Theory, 3rd edn. Prentice Hall, Englewood Cliffs (1996)
6. Jang, J.-S.R., Sun, C.-T., Misutani, E.: Neuro Fuzzy and Soft Comuting: A Computational Approach to Learning and Machine Intelligence. Prentice Hall, Englewood Cliffs (1997)
7. Maddala, G.S.: Introduction to Econometrics. Prentice Hall, Englewood Cliffs (1996)
8. Castillo, O., Melin, P.: Automated Mathematical Modelling for Financial Time Series Prediction using Fuzzy Logic, Diynamical System Theory and Fractal Theory. In: Proceedigns of CIFEr 1996, pp. 120–126. IEEE Press, New York (1996)
9. Sharkey, A.: Combining Artificial Neural Nets: Ensemble ando Modular Multi-Nets Systems. Springer, London (1999)
10. Hansen, L.K., Salomon, P.: Neural Network Ensembles. IEEE Transactions on Pattern Analysis and Machine Intelligence 12, 993–1001 (1990)
11. Valdez, F., Melin, P., Castillo, O.: Evolutionary Computing for the Opti-mization of Mathematical Functions. In: Melin, P., et al. (eds.) Analysis and Design of Intelligent Systems Using Soft Computing Techniques, pp. 463–472. Springer, Heidelberg (2007)
12. Holland, J.H.: Adaptation in natual and artificial system. University of Michigan Press, Ann Arbor (1975)
13. Goldberg, D.: Genetic Algorithms. Addison Wesley, Reading (1988)
14. Emmeche, C.: Garden in the Machine. In: The Emerging Science of Artificial Life, Princeton University Press, Princeton (1994)
15. Man, K.F., Tang, K.S., Kwong, S.: Genetic Algorithms: Concepts and Designs. Springer, Heidelberg (1999)
16. Back, T., Fogel, D.B., Michalewicz, Z.: Handbook of Evolutionary Computation. Oxford University Press, Oxford (1997)
17. Cantú-Paz, E.: Efficient and Accurate Parallel Genetic Algorithms. Kluwer Academic Publishers, Norwell (2001); ISBN 0-7923-7221-2
18. The MathWorks, Inc. Distributed Computing Toolbox 3: User's Guide (2007)
19. Ramirez, A.M.: Optimización de la Arquitectura de Redes Neuronales aplicada a la Predicción de Series de Tiempo mediante Algoritmos Genéticos. Tijuana, B.C. México s.n (2006)

Optimized Algorithm of Discovering Functional Dependencies with Degrees of Satisfaction Based on Attribute Pre-scanning Operation*

Qiang Wei, Guoqing Chen, and Xiaocang Zhou

School of Economics and Management, Tsinghua University, Beijing 100084, China

Abstract. Functional dependency (FD) is an important type of semantic knowledge reflecting integrity constraints in databases. Traditionally, FDs are proposed by managers or domain experts, which is regarded as a logic-driven method. FD has nowadays attracted an increasing amount of research attention in data mining and many efforts have been made to discover FDs in large-scale databases automatically. In mining FDs, two major problems exist. First, imprecise or noisy data may often exist in massive databases which will lead to missing precise FDs. Second, how to efficiently discover the so-called minimal set of FDs is still a hot issue. In order to tolerate partial truth due to imprecise or incomplete data, or due to a very tiny insignificance of tuple differences in a huge volume of data, the notion of functional dependency with degree of satisfaction, denoted as $(FD)_d$, has been proposed in [32], along with Armstrong-like properties and the concept of minimal set. Moreover, the efficient mining algorithm MFDD has been proposed in [29, 30, 33], by which some inference rules could be used to improve efficiency in mining process and the minimal set of satisfied $(FD)_d$ could be discovered. Based on the MFDD algorithm, this paper will further propose the concept of degree of diversity of attribute, which will be proved consistent to the framework of degree of satisfaction. Morcover, some important properties along with some optimization strategies will be presented. Furthermore, by measuring the degree of diversities of attributes with pre-scanning operation, quite many $(FD)_d$ could be determined satisfied or dissatisfied using the strategies. This process could highly save the computational consumption for further scanning databases in MFDD algorithm, which could effectively improve the efficiency of the whole mining algorithm. Furthermore, the experimental results show the optimization strategies could take significant effects to improve the computational efficiency. Finally, some concluding remarks and future works will be presented.

1 Introduction

Nowadays, more than 99% commercial databases are relational databases (RDB). RDB theories and modeling methods are also the mainstream in research and application fields of databases. One of the reasons that RDBs are so popular is that, the RDB modeling process has been standardized and supported by several ripe theories and methods, e.g., Entity-Relationship (ER) conceptual modeling, ER algebra, Normalization theory, SQL query language, etc [11, 26-27].

* Partly supported by the National Natural Science Foundation of China (79925001/70231010), the Tsinghua Research Center of Contemporary Management and the Bilateral Scientific and Technological Cooperation between China and Flanders.

O. Castillo et al. (Eds.): Soft Computing for Hybrid Intel. Systems, SCI 154, pp. 401–415, 2008.
springerlink.com

Traditionally, in typical RDB modeling, the ER conceptual modeling should be processed first, by which the entities, relationships, attributes, etc., could be extracted based on context information and experts' knowledge. Moreover, the conditions and constraints regarding an objects' static aspects and its dynamic aspects, which are often referred as the business rules, should be considered and integrated into data modeling. A particular business rules can be categorized as data constraints, e.g., domain constraints, key constraints, integrity constraints and data dependencies. Specifically, a kind of data dependency, called functional dependency, play a very important role in RDB modeling [26-27].

Generally, for two collections A and B of attributes, a functional dependency $A{\rightarrow}B$ means that A values uniquely determine B values. An example of $A{\rightarrow}B$ is (*Student#*, *Course#*)$\rightarrow$*Grade*, meaning that the value of grade can be uniquely determined by a given value of *Student#* and a given value of *Course#*. Formally, let $\Re(I_1, I_2, ..., I_m)$ be an n-ary relational scheme on domains $D_1, D_2,..., D_n$ with $Dom(I_i) = D_i$, A and B be subsets of the attribute set $U = \{I_1, I_2, ..., I_m\}$, i.e., $A, B \subseteq U$, and R be a relation of scheme $\Re(U)$, $R \subseteq D_1{\times}D_2{\times}...{\times}D_m$. A functionally determines B (or B is functionally dependent on A), denoted as $A{\rightarrow}B$, if and only if $\forall t, t' \in R$, if $t(A) = t'(A)$ then $t(B) = t'(B)$, where t and t' are tuples of R, and $t(A)$, $t'(A)$, $t(B)$ and $t'(B)$ are values of t and t' for A and B respectively [26-27].

In this way, the original relational scheme, or called relational model, could be obtained, e.g., $\Re(U, F)$, where U is the set of attributes in $\Re$, and F is the set of functional dependencies on U, which is regarded as first normal form (1NF) [29-30]. In RDB, functional dependencies reflect some kind of important business rules, which can be used to avoiding data redundancy and update anomalies. It is important to note that functional dependency possesses several desirable properties, including so-called Armstrong axioms that constitute a FD inference system [26-27].

Concretely, the axiomatic system composed of the axioms (A1, A2, A3) is as follows:

A1: If $B \subseteq A$, then $A{\rightarrow}B$;
A2: If $A{\rightarrow}B$, then $AC{\rightarrow}BC$;
A3: If $A{\rightarrow}B$ and $B{\rightarrow}C$, then $A{\rightarrow}C$.

Suppose that F is a set of FDs on relational model $\Re$. Then the set of all FDs that are derived from F using the inference rules A1, A2 and A3 is denoted as F^A. Based on Armstrong axioms, normalization theory and decomposition operations could be conducted for further RDB model optimization.

In 1NF data model, potential data redundancy and update anomalies may exist inevitably and highly affect the efficiency and effectiveness of RDB modeling, especially in large-scale data modeling problem. With normalization theory, the original relational schemes could be optimized and normalized with decomposition algorithms, so the so-called 1NF could be converted to 2NF, 3NF, BCNF, etc. Accordingly, 3NF is the highest normal form which supports lossless-join and dependency-preservation [26-27], which is called the optimized relational data model in this paper. It could be formalized that, after normalization, the original $\Re(U, F)$ could be decomposed into $\cup\Re_i(U_i, F_i)$, each of which satisfies 3NF, where $U = \cup U_i$ and

$F = \cup F_i$. In so doing, the RDB modeling process is finished, following which the databases could be constructed.

Clearly, functional dependency (FD) play a very important role in RDB modeling. The quality of FDs highly affect the quality of RDB modeling. Traditionally, the way to deriving FDs is based on context and expert knowledge, which is regarded as a logic-driven method. This method is effective and efficient while the problem scale is not so large. However, nowadays, with the IT fusion in all levels of business management, normally, the scales of RDB-enabled modeling problems are usually large and the structures are very complex. It is not easy for managers or domain experts to conclude all the necessary business rules in terms of functional dependencies.

On the other hand, several decades of IT applications had resulted in a large number of databases that were constructed and maintained in which useful and interesting FDs might have already exist in the databases. However, these FDs are not explicitly known or are hidden, and therefore need to be discovered.

Go a step further, nowadays, in the great amount of databases and data warehouses, some previously unknown and potentially useful FDs may exist. This kind of FDs are actually much more important, since they express some innovatively new business rules, which will play more important roles to improve business process and further optimize the business databases. This could also be regarded as one of the parts of data mining and knowledge discovery [14, 20].

For the above three aspects, it will be very interesting to develop some data mining methods to discover hidden FDs in large-scaled databases. Furthermore, if the hidden FDs could be discovered from data mountains, then the optimized RDB could be constructed automatically.

In this paper, a method to discover FDs efficiently in large-scale databases will be further discussed. This method is the key step in the process of constructing and optimizing RDB model automatically. First, given the large-scale databases, the set of all attributes, e.g., U, could be collected easily. Then the original $\Re(U, F)$ could be construct, where $F = \varnothing$. Second, the databases could be scanned and the satisfied FDs could be discovered, then F could be updated as F'. So $\Re'(U, F')$ could be retrieved. Third, based on normalization theory and decomposition operation, $\cup \Re_i(U_i, F_i)$ could be retrieved, each of which satisfies 3NF, where $U = \cup U_i$ and $F' = \cup F_i$. In so doing, given data, the RDB modeling could be done using this automatically data-driven method.

According to the data-driven process, the key step is to discovering FDs efficiently. Since 1990s, an increasing number of efforts have been devoted to mining FDs and related issues [1-7, 15-16, 24, 34-35]. Moreover, some other attempts centered on extended forms of FD, such as functional dependencies with null values [21], partial determination [19], approximate functional dependencies [17-18], fuzzy functional dependencies (FFDs) [8-10, 12-13, 22, 28, 36], functional dependencies with degrees of satisfactions (FDs)$_d$ [29-33], etc.

Apparently, by definition, FDs do not tolerate such noisy or disturbing data, which strictly express the semantics that "Equal A values determine equal B values for all tuples". In discovering functional dependencies, there still exist some open problems. First, in large existent databases, noises often pertain, such as conflicts, nulls, and errors that may result from, for instance, inaccurate data entry, transformation or updates. Even without noisy data, sometimes a partial truth of a FD may still make sense. For instance, "a FD almost holds in a database" or "Equal A values determine equal B values for most of the tuples" expresses a sort of partial knowledge, meaning that the FD satisfies the relational databases of concern to a large extent.

Second, in developing corresponding mining methods, the computational efficiency needs to be further improved. Since FD inference is desirable but still needs to be further investigated. That is, deriving a FD by inference from discovered FDs without scanning the database may help improve the computational efficiency of the mining process. For example, if both $A{\rightarrow}B$ and $B{\rightarrow}C$ satisfy a relational database, and if $A{\rightarrow}C$ could be inferred directly, then the effort in scanning the database for checking whether $A{\rightarrow}C$ holds can be saved.

In 2002, Wei and Chen [32] presented a notion of functional dependency with degree of satisfaction (FD$_d$: $(A{\rightarrow}B)_\alpha$) to reflect the semantic that equal B values correspond to equal A values at a certain degree (α). Moreover, Wei and Chen presented the Armstrong-like inference rules, along with an inference system, based on which the minimal set of (FDs)$_d$ has been proposed [29-30]. Furthermore, a fuzzy relation matrix-based algorithm has been constructed to perform transitivity-type FD inference. Accordingly, the algorithm for mining (FDs)$_d$, called MFDD, has been provided, which can discover the minimal set of (FDs)$_d$ efficiently [33]. Further, two optimized strategies have been incorporated to further improve the efficiency in [31].

In this paper, based on [31, 33], we will further discuss the optimization strategies by attribute pre-scanning operation. The paper is organized as follows. Some preliminaries about the functional dependency with degree of satisfaction will be briefly reviewed in Section 2. Section 3 will discuss how to evaluate the diversity of an attribute based on degree of satisfaction by pre-scanning operation and propose several derivatives, by which more (FDs)$_d$ could be inferred satisfied or dissatisfied without scanning database. In Section 4, the optimized algorithm will be proposed along with the theoretical analysis on corresponding computational efficiency. Data experimental results will be proposed in Section 5 to show how the strategies take significant effects on computational efficiency. Finally, some concluding remarks will be discussed in Section 6.

2 Preliminaries

The concept of functional dependency with degree of satisfaction, e.g., (FD)$_d$, should be introduced first along with some important preliminaries [31, 33].

Definition 1: Let $\Re(I_1, I_2, \ldots, I_m)$ be a relation scheme on domains $D_1, D_2, \ldots, D_m, A$, $B \subseteq U$, and R be a relation of $\Re(U)$, $R \subseteq D_1{\times}D_2{\times}\ldots{\times}D_m$, where tuples $t_i, t_j \in R$ and $t_i \neq t_j$. Then B is called to functionally depend on A for a tuple pair (t_i, t_j), denoted as $_{(t_i,t_j)}(A{\rightarrow}B)$, if $t_i(A) = t_j(A)$ then $t_i(B) = t_j(B)$.

It can easily be seen that the FD for a tuple pair could be represented in terms of degree of satisfaction, $d_{(t_i,t_j)}(A{\rightarrow}B)$, where if $t_i(A) = t_j(A)$ and $t_i(B) \neq t_j(B)$, then $d_{(t_i,t_j)}(A{\rightarrow}B) = 0$; otherwise 1. Subsequently, FD for relation R can be defined in terms of degree of satisfaction.

Definition 2: Let $\mathfrak{R}(U)$ be a relation scheme, A, $B \subseteq U$, and R be a relation of $\mathfrak{R}(U)$ with n tuples. Then the degree that R satisfies $A{\to}B$ is $d_R(A{\to}B)$ or $d(A{\to}B)$ in brief:

$$d(A{\to}B) = d_R(A{\to}B) = \frac{\displaystyle\sum_{\substack{\forall t_i, t_j \in R \\ t_i \neq t_j}} d_{(t_i, t_j)}(A \to B)}{NTP},$$

where NTP represents the number of tuple pairs in R and equals $n(n-1)/2$.

Normally, for a $(FD)_d$ $A{\to}B$, if $d(A{\to}B) = \alpha$, we will denote this $(FD)_d$ as $(A{\to}B)_\alpha$. Given a minimum satisfaction threshold θ, $0 \le \theta \le 1$, if $d(A{\to}B) \ge \theta$, then $A{\to}B$ is called a satisfied functional dependency. Moreover, some properties could be derived. Let R be a relation on $\mathfrak{R}(U)$ and A, B, $C \subseteq U$, we have [31-33]:

A1':   If $B \subseteq A$, then $d(A{\to}B) = 1$.

A2':   If $d(A{\to}B) = \alpha$, then $d(AC{\to}BC) \ge \alpha$, $0 \le \alpha \le 1$.

A3':   If $d(A{\to}B) = \alpha$ and $d(B{\to}C) = \beta$, then $d(A{\to}C) \ge \alpha + \beta - 1$.

A4':   If $d(A{\to}B) = \alpha$, then $d(B{\to}C) \ge 1 - \alpha$.

The first three properties are similar to the three classical Armstrong inference rules, except for A3' in that it guarantees a lower-bound $d(A{\to}B)$ value for a transitive $(FD)_d$ that could be inferred without scanning database. Moreover, A4' is important to guarantee that invalid values less than 0 will not be generated in transitive inference. Based on A1', A2' and A3', the Armstrong-like inference system has be defined, as well as the θ-equivalence and minimal set of $(FDs)_d$.

Theorem 1: (Armstrong-like Axioms) Let R be a relation on $\mathfrak{R}(U)$ and A, B, $C \subseteq U$. Then for any R in $\mathfrak{R}$, the following inference rules, denoted as extended Armstrong-like axioms (A1', A2', A3'), hold:

A1': If $B \subseteq A$, then $d(A{\to}B) = 1$;
A2': If $d(A{\to}B) \ge \alpha$, then $d(AC{\to}BC) \ge \alpha$;
A3': If $d(A{\to}B) \ge \alpha$ and $d(B{\to}C) \ge \beta$, then $d(A{\to}C) \ge \gamma$, where $\alpha + \beta - 1 \le \gamma \le 1$.

Furthermore, Suppose that F is a set of $(FDs)_d$ on $\mathfrak{R}$. Then we denote $F^{A'}$ as the set of all $(FDs)_d$ that are derived from F using A1', A2' and A3'. And F^{A+} could be defined as a set containing the $(FDs)_d$ with upper bound degrees of satisfaction. So we can have the following definitions.

Definition 3: Let F and G be two sets of $(FDs)_d$ on $\mathfrak{R}$. Then F and G are called equivalent if and only if $F^{A+} = G^{A+}$.

In the mining process, however, we are only concerned with those $(FDs)_d$ with satisfied degrees as mining outcomes. For $\theta = 0.7$, both $(A{\to}C)_{0.8}$ and $(A{\to}C)_{0.7}$ will be regarded satisfied. If the set of (FDs)$_d$ could be viewed as a fuzzy set with the corresponding degrees of satisfaction as the grades of membership, then F and G are the same in terms of the 0.7-cuts of F^{A+} and G^{A+}. That is, $(F^{A+})_{0.7} = \{A{\rightarrow}B, B{\rightarrow}C, A{\rightarrow}C, AC{\rightarrow}B, AB{\rightarrow}C, AB{\rightarrow}C\} = (G^{A+})_{0.7}$. In general, the θ-cut of a set of (FDs)$_d$ F is denoted as $(F)_\theta = \{A{\rightarrow}B \mid (A{\rightarrow}B)_\alpha \in F$, and $\alpha \geq \theta\}$. Based on $(F)_\theta$, the notion of θ-equivalence could be defined.

Definition 4: Let F and G be two sets of (FDs)$_d$ and θ be an θ-cut threshold with $\theta \in [0, 1]$. Then F and G are called θ-equivalent if and only if θ-cuts of F^{A+} and G^{A+} are equal, i.e., $(F^{A+})^\theta = (G^{A+})^\theta$.

Note that, for any (FDs)$_d$ F and its F^{A+}, it can be proved that there exists a minimal set MF of (FDs)$_d$ such that MF is a subset of F^{A+} and is θ-equivalent to F^{A+}. It seems desirable and efficient if we could develop an approach to discovering F^{A+} by only scanning databases for obtaining MF and deriving all other (FDs)$_d$ in $(F^{A+} - MF)$ using the extended Armstrong-like axioms [31, 33].

Accordingly, an algorithm called MFDD based on fuzzy relation matrix operation has been proposed, by which the minimal set of satisfied (FDs)$_d$ could be discovered efficiently. For details, please refer to [33].

In the framework of MFDD, all the satisfied (FDs)$_d$ could be classified into 2 groups, Scanned (FDs)$_d$ and Inferred (FDs)$_d$. The group of Scanned (FDs)$_d$ represent the (FDs)$_d$ in MF, which could be obtained only through scanning databases. The group of Inferred (FDs)$_d$ represent the (FDs)$_d$ in $(F^{A+} - MF)$, which could be inferred by Scanned (FDs)$_d$ without scanning databases.

It should be emphasized that, for a database with m attributes and n tuples, the computational complexity of database scanning operation for each (FD)$_d$ generally highly exceeds the computational complexity of inference, because usually m is far less than n, e.g., a database with 10 attributes could have 10,000 tuples. So the orientation of optimization is to infer as many (FDs)$_d$ as possible instead of scanning database.

In [31], the MFDD algorithm has been further optimized on two aspects. First, the algorithm of computing the degree of satisfaction of a certain (FD)$_d$ have been optimized with group operation. Second, the A4' inference rule has been further investigated to a greater extent for efficiency purposes.

In this paper, we will further investigate the framework to find further optimization strategies to improve the computational efficiency by inferring as many (FDs)$_d$ as possible instead of scanning database.

3   Evaluation on Attribute Diversity and Derivatives

According to Definition 1 and 2, it could be found that the degree of (FD)$_d$ $A{\rightarrow}B$ is highly related to the diversity of attribute A and B. Roughly speaking, diversity of an attribute represents the level of degree that the values are different. Generally, the more the diversity of A is, the more the degree of satisfaction of $A{\rightarrow}B$, e.g., $d(A{\rightarrow}B)$, is. The less the diversity of B is, the less of the degree of satisfaction of $A{\rightarrow}B$, e.g., $d(A{\rightarrow}B)$, is. This could be easily conducted according to definition 1 and 2. So, if the attributes could be scanned before mining process, called pre-scanning operation, some attributes could be determine to have more or less diversity. Further, some corresponding (FDs)$_d$ could be inferred satisfied or dissatisfied without scanning databases in mining process afterwards. Theoretically, the cost of pre-scanning operation is to scan m attribute with n tuples, e.g., $O(m \times n^2)$. However, for a (FD)$_d$ $A \to B$, A or B are single attribute, the scanning cost is $O(2 \times n^2)$. Roughly estimating, if $m/2$ (FDs)$_d$ could be inferred after pre-scanning operation, the pre-scanning operation is worth doing.

In this section, we will discuss in detail how to evaluate the diversity in the framework of degree of satisfaction. Secondly, some important derivatives could be further obtained, based on which the optimization strategies to improve algorithmic efficiency could be performed.

Definition 5: Let $\Re(U)$ be a relation scheme, $A \subseteq U$, and R be a relation of $\Re(U)$ with n tuples, then $d_{(t_i, t_j)}(A) = 1$, while $t_i(A) \neq t_j(A)$, otherwise 1. Moreover,

$$d_R(A) = d(A) = \frac{\displaystyle\sum_{\substack{\forall t_i, t_j \in R \\ t_i \neq t_j}} d_{(t_i, t_j)}(A)}{NTP},$$

where NTP represents the number of tuple pairs in R and equals $n(n-1)/2$.

Clearly, $d(A)$ could be regarded as a measure to evaluate the diversity of the values of attribute A. Specifically, if A values of all tuple are identical, then $d(A) = 0$, representing attribute A are not diversified. If A value of each tuple is unique, then $d(A) = 1$, representing attribute A are totally diversified.

Importantly, we can deem $d(A)$ in another way. Let R be a relation of $\Re(U)$ with n tuples, $A \subseteq U$, we can construct a new attribute X, where for any $t_i, t_j \in R$, $t_i(X) \neq t_j(X)$, which means X is totally diversified . Then the following property could be obtained.

Property 1: $d(A \to X) = d(A)$.
Proof: Without loss of generality, all the tuple pairs could be classified into two groups. The A values of each tuple pairs which are equal are classified into Group 1, otherwise into Group 2. Then the following table could be constructed.

It could be found that $d(A \to X) = N_2/(N_1 + N_2) = d(A)$.   □

Tuple pair group	Number of tuple pairs	$t_i(A) \neq t_j(A)$	$t_i(X) \neq t_j(X)$	$d_{(t_i, t_j)}(A)$	$d_{(t_i, t_j)}(A \to X)$
1	N_1	No	Yes	0	0
2	N_2	Yes	Yes	1	1

Moreover, one of the byproduct of Property 1 is that, since X is totally diversified, then $d(X \to Y) = 1$, for any $Y \in U$.

Property 1 is very important on two aspects. First, Property 1 links diversity of attribute to degree of satisfaction. Second, with Property 1, it is proved that the diversity of attribute is consistent to the degree of satisfaction, and so the entire framework of (FD)$_d$, which will be illustrated further in following parts of the paper.

Based on Definition 3 and Property 1, some important derivatives could be obtained. Let R be a relation on $\mathfrak{R}(U)$ and A, $B \subseteq U$, and X is a totally diversified constructed attribute, then we have the following derivatives.

Derivative 1: $d(AB) \geq d(A)$.
Proof: Since $d(AB) = d(AB{\rightarrow}X)$, and $d(A) = d(A{\rightarrow}X)$. Then according to A2' and A3',

$$d(AB{\rightarrow}X) \geq d(AB{\rightarrow}A) + d(A{\rightarrow}X) - 1$$
$$\geq 1 + d(A{\rightarrow}X) - 1 = d(A{\rightarrow}X) = d(A).$$

So $d(AB) \geq d(A)$. $\square$

Derivative 2: $d(AB) \leq d(A) + d(B)$.
Proof: Since X is totally diversified, $d(AB) = d(AB{\rightarrow}X)$, $d(A) = d(A{\rightarrow}X)$ and $d(B) = d(B{\rightarrow}X)$.

According to A1', A2' and A3', we have
$$d(A{\rightarrow}X) \geq d(A{\rightarrow}AB) + d(AB{\rightarrow}X) - 1$$
$$= d(A{\rightarrow}B) + d(AB{\rightarrow}X) - 1.$$

According to A4', we have $d(A{\rightarrow}B) + d(B{\rightarrow}X) \geq 1$, so $d(A{\rightarrow}B) \geq 1 - d(B{\rightarrow}X)$.

Then we have
$$d(A{\rightarrow}X) \geq 1 - d(B{\rightarrow}X) + d(AB{\rightarrow}X) - 1.$$

So $d(A{\rightarrow}X) + d(B{\rightarrow}X) \geq d(AB{\rightarrow}X)$.

According to Derivative 1, $d(A) + d(B) \geq d(AB)$. $\square$

The above two derivatives describe the characteristics on attributes themselves, however, in the process of mining so-called satisfied $(FDs)_d$, we care more on the characteristics on how diversity of attributes will affect the degree of satisfaction of $(FDs)_d$. Based on Definition 3, Property 1 and Armstrong-Like Axioms, let R be a relation on $\mathfrak{R}(U)$ and A, $B \subseteq U$, and X is a totally diversified constructed attribute, then we have the following derivatives.

Derivative 3: $d(A{\rightarrow}B) \geq d(A)$.
Proof: According to A3', $d(A{\rightarrow}B) \geq d(A{\rightarrow}X) + d(X{\rightarrow}B) - 1$

Because $d(X{\rightarrow}B) = 1$, since X is totally diversified, then $d(A{\rightarrow}B) \geq d(A{\rightarrow}X)$,

so $d(A{\rightarrow}B) \geq d(A)$. $\square$

Derivative 4: $d(A{\rightarrow}B) \geq 1 - d(B)$.
Proof: According to A4', $d(A{\rightarrow}B) + d(B{\rightarrow}X) \geq 1$, then $d(A{\rightarrow}B) \geq 1 - d(B{\rightarrow}X)$,

so $d(A{\rightarrow}B) \geq 1 - d(B)$. $\square$

Derivative 5: $d(A{\rightarrow}B) \leq d(A) + 1 - d(B)$.
Proof: According to A3', $d(A{\rightarrow}X) \geq d(A{\rightarrow}B) + d(B{\rightarrow}X) - 1$, then $d(A) \geq d(A{\rightarrow}B) + d(B) - 1$,

so $d(A{\rightarrow}B) \leq d(A) + 1 - d(B)$. $\square$

With Derivative 3, 4 and 5, before mining process with MFDD, if the attributes could be pre-scanned to compute $d(Y)$, for each $Y \in U$, then some $(FDs)_d$ could be inferred satisfied or dissatisfied without scanning database. Then some optimized strategies could be constructed.

Strategy 1: If $d(A) \geq \theta$, then $d(A \rightarrow Y) \geq \theta$, so $A \rightarrow Y$ is satisfied, for $\forall Y \in U$.
Proof: It could be directly inferred based on Derivative 3.   □

Strategy 1 means that, for a $(FD)_d$ $A \rightarrow Y$, if A values are too diversified, then $A \rightarrow Y$ is satisfied, whatever Y is. This is because the degree of satisfaction is major contributed by the diversified A values.

Strategy 2: If $1 - d(A) \geq \theta$, then $d(Y \rightarrow A) \geq \theta$, so $Y \rightarrow A$ is satisfied, for $\forall Y \in U$.
Proof: It could be directly inferred based on Derivative 4.   □

Strategy 2 means that, for a $(FD)_d$ $Y \rightarrow A$, if A values are not so diversified, then $Y \rightarrow A$ is satisfied, whatever Y is. This is because the degree of satisfaction is major contributed by the equivalent A values.

Strategy 3: If $d(B) - d(A) \geq 1 - \theta$, then $d(A \rightarrow B) \leq \theta$, so $A \rightarrow B$ is dissatisfied.
Proof: Since $d(B) - d(A) \geq 1 - \theta$, then $d(B) - 1 - d(A) \geq 1 - \theta - 1$. Further, it could
be derived that $d(A) + 1 - d(B) \leq \theta$.
According to Derivative 5, $d(A \rightarrow B) \leq d(A) + 1 - d(B) \leq \theta$, so $d(A \rightarrow B) \leq \theta$.
So $A \rightarrow B$ is dissatisfied.   □

Strategy 3 means that, for a $(FD)_d$ $A \rightarrow B$, if the B values are much more diversified than A values, then $A \rightarrow B$ could be dissatisfied. This is because B values vary a lot while A values are equivalent.

Strategy 4: If $d(B) - d(A) \geq \theta$, then $d(B \rightarrow Y) \geq \theta$, so $B \rightarrow Y$ is satisfied, for $\forall Y \in U$.
Proof: Since $d(B) - d(A) \geq \theta$, then $d(B) - 1 - d(A) \geq \theta - 1$. Then it could be derived
that $d(A) + 1 - d(B) \leq 1 - \theta$. Because of Derivative 5, $d(A \rightarrow B) \leq d(A) + 1 -$
$d(B) \leq 1 - \theta$, so $d(A \rightarrow B) \leq 1 - \theta$.
According to A4', it could be derived that $d(B \rightarrow Y) \geq 1 - d(A \rightarrow B) \geq 1 - 1 +$
θ. Then $d(B \rightarrow Y) \geq \theta$.
So $B \rightarrow Y$ is satisfied, for $\forall Y \in U$.   □

Strategy 4 means that, for a $(FD)_d$ $A \rightarrow B$, if B values are more diversified than A values to some extent, then $A \rightarrow B$ is not only dissatisfied but less than a rather small value $1 - \theta$, so according to A4', $B \rightarrow Y$ is satisfied, whatever Y is.

The above 4 inference strategies could be used to infer as many $(FDs)_d$ as possible before mining process, which could optimize the MFDD algorithm.

To do so, the relation R should be pre-scanned before mining process to determine the degree of diversity of each attribute in U. Based on the computed $d(Y)$, $Y \in U$, the 4 optimized strategies could be performed.

4 Attribute Pre-scanning Algorithm

According to the discussion previously, if each attribute Y, e.g., $Y \in U$, could be pre-scanned before mining process, $d(Y)$ could be derived, then the 4 inference strategies could be performed to infer satisfied or dissatisfied $(FDs)_d$. The algorithm is as shown in Table 2.

Table 2. Attribute Pre-Scanning Algorithm

// Attribute Pre-Scanning Algorithm
for $A \in U$ { Calculating $d(A)$ }
for $A \in U$ { if $d(A) \geq \theta$ then $d(A{\rightarrow}Y) = \theta$, $Y \in U$; }   //
Strategy 1
for $A \in U$ { if $1 - d(A) \geq \theta$ then $d(Y{\rightarrow}A) = \theta$, $Y \in U$; }   //
Strategy 2
for $A \in U$
{ for $B \in U$
{ if $d(B) - d(A) \geq 1 - \theta$, then $d(A{\rightarrow}B) = 0$;   //
Strategy 3
if $d(B) - d(A) \geq \theta$, then $d(B{\rightarrow}Y) = \theta$, $Y \in U$;   //
Strategy 4
}
}

It has been emphasized that, the pre-scanning algorithm will be performed before the mining process. Therefore, theoretically, if at least $m/2$ $(FDs)_d$ could be inferred satisfied or dissatisfied with the strategies, then the algorithm is worth doing.

How the 4 strategies could take effects on efficiency depends on the concrete attribute values in the given database. However, some qualitative aspects could be induced as follows.

First, since θ is no less than 50% semantically, if $d(A)$ is large ($\geq \theta$) or small ($\leq 1 - \theta$) enough, then Strategy 1 or 2 could take effects. And if there exists an attribute A that $d(A) \geq \theta$ or $d(A) \leq 1 - \theta$, then $A{\rightarrow}Y$ or $Y{\rightarrow}A$ could be inferred satisfied, Y could be any attribute in U. The number of inferred $(FDs)_d$ will be $m - 1$. Generally, the effect of Strategy 1 and 2 will be quite significant. This is also the reason why we put Strategy 1 and 2 together in further experimental analysis.

Second, if θ is small enough, Strategy 4 will take effects, then $m - 1$ $(FDs)_d$ could be inferred satisfied. On the other hand, if θ is large enough, then Strategy 3 will take effect, then one $(FD)_d$ could be inferred dissatisfied.

Finally, it could also be found that the effects of the 4 strategies are independent and could be superpositioned.

According to analysis, intuitively, it seems that it is not difficult that $m/2$ $(FDs)_d$ could be inferred by pre-scanning operation. How the pre-scanning operation will take effect, please refer to Section 5.

5 Experimental Results and Analysis on Scalabilities

A real-world database was used for testing the presented approach. The database contains the real business data of The Insurance Company (TIC) Benchmark provided by Dutch Data Mining Company Sentient Machine Research, which is usually used as benchmarking data for evaluating data mining algorithms. The database is from "The Insurance Company 2000," containing 5,822 tuples with 86 attributes [25]. The experiments were conducted in an environment with a Pentium IV 2.2GHz computer, with 512M RAM and Visual C++ 6.0. Due to the exponential exploration on increase of number of attributes, we project the whole dataset on the first 10 attributes ($m = 10$, $n = 5,822$), which can illustrate the problem well.

The following Table 3 shows the experimental results. For illustration purpose, we focus on the number of scanned $(FDs)_d$ and inferred $(FDs)_d$ in order to evaluate the effect of pre-scanning operation. Moreover, in order to illustrate the effect of the 4 strategies. We test the experiments using 3 methods. Method 1 represents the method

Table 3. Experimental Results with Method 1, Method 2 and Method 3

θ	Method 1		Method 2			Method 3		
	Scanned	Inferred	Scanned	Inferred	Method 2 – Method 1	Scanned	Inferred	Method 3 – Method 1
0.50	19	71	9	81	10	4	86	15
0.60	28	62	9	81	19	12	78	16
0.70	55	35	33	57	22	37	53	18
0.80	70	20	41	49	29	54	36	16
0.90	90	0	81	9	9	48	42	42
0.92	90	0	81	9	9	47	43	43
0.94	90	0	81	9	9	46	44	44
0.96	90	0	90	0	0	53	37	37
0.98	90	0	90	0	0	49	41	41
0.99	90	0	90	0	0	49	41	41
1.00	90	0	90	0	0	45	45	45

in [5, 11] which does not incorporate pre-scanning operation. Method 2 represents the method which incorporates Strategy 1 and 2. Method 3 represents the method which incorporates all the 4 strategies. For method k, $k = 2$ and 3, the Method k – Method 1 represents the number of increased inferred $(FDs)_d$ (also represents the number of decreased scanned $(FDs)_d$) against Method 1.

Clearly, given a threshold θ, the total number of scanned $(FDs)_d$ and inferred $(FDs)_d$ are the same in all the 3 methods, which also show the correctness of the 3 methods from a certain aspect. For comparison, Figure 1 shows the number of scanned $(FDs)_d$ for all the 3 methods.

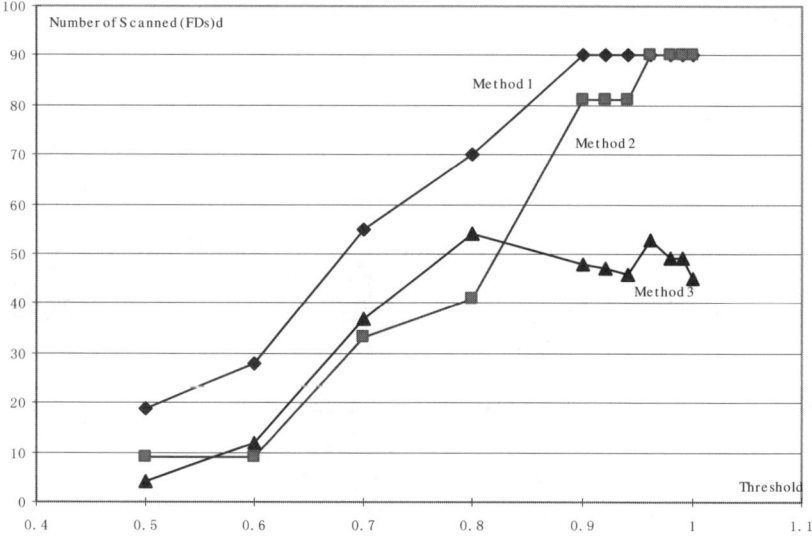

Fig. 1. Number of Scanned $(FDs)_d$ for Methods 1, 2 and 3

On the experimental results, some conclusions could be discovered.

First, in Method 2, Strategy 1 and 2 take significant effects (Method 2 – Method 1 $\geq m/2 = 5$) while $\theta \leq 0.94$, however, with the increase of θ ($\theta \geq 0.96$), Strategy 1 and 2 gradually take no effects since the increased number of inferred $(FDs)_d$ is no more than $m/2$. According to Strategy 1 and 2, while θ is large to some extent, it is not easy to find some $d(A) \geq \theta$ or $d(A) \leq 1 - \theta$.

Second, in Method 3, all the 4 strategies take significant effects in any situation. According to the pre-scanning algorithm, the effects for all the 4 strategies are independent and could be superpositioned. So Method 3 remedies the shortcoming of Method 2 due to Strategy 3 and 4. The bigger θ is, the more possibly $d(B) - d(A) \geq 1 - \theta$ take effects. This characteristic is very important, since, in this situation, B values are much more diversified than A values, then $A \rightarrow B$ is definitely dissatisfied.

So the pre-scanning operation could significantly improve the computational efficiency.

6   Concluding Remarks and Future Works

In this paper, we have further discussed the functional dependency with degree of satisfaction, which could tolerate noisy data and express partial knowledge and play an important role in relational data modeling. Moreover, in order to further improve the performance of the discovering algorithm, this paper has focused on analyzing diversity of attribute. In Section 3, it has been proved that the diversity characteristic of attribute, in form of $d(A)$, will highly influence inference of $(FDs)_d$. Furthermore, based on the definition of degree of diversity and 5 inferred derivatives, 4 important strategies could be derived. Many $(FDs)_d$ could be inferred satisfied or dissatisfied without scanning databases. According to the experimental results, it shows that the pre-scanning operation can significantly improve the computational efficiency.

Future works will focus on two aspects. The first is to further optimize the algorithm and test on more large-scaled real databases. The second is to, based on the optimized $(FD)_d$ mining algorithm, construct the automatically data-driven method of RDB modeling.

References

[1] Andersson, M.: Extracting an Entity Relationship Schema from A Relational Database through Reverse Engineering, http://citeseer.nj.nec.com/

[2] Baudinet, M., Chomicki, J., Wolper, P.: Constraint-generating dependencies. Journal of Computer and System Science 59(1), 94–115 (1999)

[3] Baudinet, M., Chomicki, J., Wolper, P.: Constraint-Generating Dependencies, http://citeseer.nj.nec.com/

[4] Bell, S., Brockhausen, P.: Discovery of Data Dependencies in Relational Databases, University of Dortmund, German, Computer Science Department, LS-8 Report 14 (1995)

[5] Bell, S., Brockhausen, P.: Discovery of data dependencies in relational databases. LS-8 Report 14. University of Dortmund, Germany (1995)

[6] Castellanos, M., Saltor, F.: Extraction of Data Dependencies. European-Japanese conferences on Information Modelling and Knowledge Bases, Budapest, Hungary, May 31 - June 3. pp. 401-421 (1993)

[7] Castellanos, M., Saltor, F.: Extraction of data dependencies. Report LSI-93-2-R. Barcelona: University of Catalonia (1993)

[8] Chen, G.Q.: Fuzzy Logic in Data Modeling: semantics, constraints and database design. Kluwer Academic Publishers, Boston (1998)

[9] Chen, G.Q., Vandenbulcke, J., Kerre, E.E.: A step towards the theory of fuzzy database design. In: The 2nd World Congress of International Fuzzy Systems Association (IFSA 1991), pp. 44–47 (1991)

[10] Chen, G.Q., Kerre, E.E., Vandenbulcke, J.: A computational algorithm for the FFD closure and a complete axiomatization of fuzzy functional dependency (FFD). International Journal of Intelligent Systems 9, 421–439 (1994)

[11] Codd, E.F.: A Relational Model for Large Shared Data Banks. Communications of the ACM 13(6), 377–387 (1970)

[12] Cubero, J.C., et al.: Data Summarization in Relational Databases through Fuzzy Dependencies. Information Sciences 121(3-4), 233–270 (1999)

[13] Cubero, J.C., Medina, J.M., Pons, O., Vila, M.A.: Rules discovery in fuzzy relational databases. In: Conference of the North American Fuzzy Information Processing Society, NAFIPS 1995, pp. 414–419. IEEE Computer Society Press, Maryland (1995)

[14] Fayyad, U., Piatetsky-Shapiro, G., Smyth, P.: From Data Mining to Knowledge Discovery: An Overview. In: Fayyad, U., Piatetsky-Shapiro, G., Smyth, P., Uthurusamy, R. (eds.) Advances in Knowledge Discovery and Data Mining, pp. 1–30. AAAI Press/The MIT Press, Cambridge (1996)

[15] Flach, P.A., Savnik, I.: Database Dependency Discovery: A Machine Learning Approach, AI Communications, ISSN 0921-7126, 2001.

[16] Giannella, C., Wyss, C. M.: Finding Minimal Keys in a Relation Instance, http://citeseer.nj.nec.com/

[17] Huhtala, Y., Karkkainen, J., Paokka, P., Toivonen, H.: TANE: An Efficient Algorithm for Discovering Functional and Approximate Dependencies, http://citeseer.nj.nec.com/huhtala99tane.html

[18] Huhtala, Y., Karkkainen, J., Porkka, P., Toivonen, H.: Efficient Discovery of Functional and Approximate Dependencies Using Partitions. In: Proc. 14th Int. Conf. on Data Engineering, IEEE Computer Society Press, Los Alamitos (1998)

[19] Kramer, S., Pfahringer, B.: Efficient Search for Strong Partial Determinations (2001), http://citeseer.nj.nec.com/

[20] Kruse, R., Nanck, D., Borgelt, C.: Data Mining with Fuzzy Methods: Status and Perspectives (2001), http://citeseer.nj.nec.com/245408.html

[21] Liao, S.Y., Wang, H.Q., Liu, W.Y.: Functional Dependencies with Null Values, Fuzzy Values, and Crisp Values. IEEE Transactions on Fuzzy Systems 7(1), 97–103 (1999)

[22] Maimon, O., Kandel, A., Last, M.: Information-Theoretic Fuzzy Approach to Knowledge-Discovery in Databases. In: Roy, R., Furuhashi, T., Chawdhry, P.K. (eds.) Advances in Soft Computing - Engineering Design and Manufacturing, pp. 315–326. Springer, London (2001)

[23] Mitra, S., Pal, S. K., Mitra, P.: Data Mining in Soft Computing Framework: A Survey (2001), http://citeseer.nj.nec.com/mitra01data.html

[24] Savnik, I., Flach, P. A.: Discovery of Multi-valued Dependencies from Relations, report00135 (2000), http://citeseer.nj.nec.com/savnik00discovery.html

[25] The Insurance Company © Sentient Machine Research (2000), http://www.smr.nl

[26] Ullman, J.D., Widom, J.: A First Course in Database Systems. Prentice Hall, Inc., a Simon & Schuster Company (1997)

[27] Ullman, J.D.: Principles of Database and Knowledge-Based Systems. Computer Sciences Press Inc., Maryland (1988)

[28] Wang, S.L., Shen, J.W., Hong, T.P.: Incremental discovery of functional dependencies based on partitions. Intelligent Data Analysis (2002)

[29] Wei, Q., Chen, G.Q.: An Efficient Algorithm on Mining a Minimal Set of Functional Dependencies with Degrees of Satisfaction. In: De Baets, B., Kaynak, O., Bilgiç, T. (eds.) IFSA 2003. LNCS, vol. 2715, Springer, Heidelberg (2003)

[30] Wei, Q., Chen, G.Q.: Mining a Minimal Set of Functional Dependencies with Degrees of Satisfaction. In: International Conference on FIP 2003, Beijing, China, March 1-4 (2003)

[31] Wei, Q., Chen, G.Q.: Optimized Algorithm of Discovering Functional Dependencies with Degrees of Satisfaction. In: Ruan, D., D'hondt, P., Fantoni, P.F., De Cock, M., Nachtegael, M., Kerre, E.E. (eds.) Applied Artificial Intelligence Proceedings of the 7th International FLINS Conference, pp. 169–176. Word Scientific Press (2006)

[32] Wei, Q., Chen, G.Q., Kerre, E.E.: Mining Functional Dependencies with Degrees of Satisfaction in Databases. In: Proceedings of Joint Conference on Information Sciences, Durham, NC, USA (2002)

[33] Wei, Q., Chen, G.Q.: Efficient Discovery of Functional Dependencies with Degrees of Satisfaction. J. of Intelligent Systems 19, 1089–1110 (2004)

[34] Wijsen, J., Raymond T., Ng, Calders, T.: Discovering Roll-Up Dependencies (2001), http://citeseer.nj.nec.com/

[35] Wyss, C., Giannella, C., Robertson, E.: FastFDs: A heuristic-driven depth-first algorithm for mining functional dependencies from relation instances. Technical Report 551, CS Department, Indiana University (July 2001)

[36] Yang, Y. P., Singhal, M.: Fuzzy Functional Dependencies and Fuzzy Association Rules (2001), http://citeseer.nj.nec.com/

A Fuzzy Symbolic Representation for Intelligent Reservoir Well Logs Interpretation

Tina Yu[1] and Dave Wilkinson[2]

[1] Department of Computer Science
Memorial University of Newfoundland
St. John's, NL A1B 3X5 Canada
tinayu@cs.mun.ca
http://www.cs.mun.ca/~tinayu
[2] Chevron Energy Technology Company
San Ramon, CA 94583 USA
davidawilkinson@chevron.com

Abstract. Well log data are routinely used for stratigraphic interpretation of the earth's subsurface. This paper presents an automatic blocking scheme that transforms numerical well log data into a fuzzy symbolic representation. This representation maintains the character of the original log curves, which is essential for the stratigraphic interpretation, while making the interpretation task easier. Additionally, fuzzy symbols allow effective interpretation under uncertainty embedded in the data set and our knowledge of the earth's subsurface. We present the developed technique and test it on two sets of well logs collected from oil fields in offshore West Africa. The results give sensible well log blocking and resemble the original log curves reasonably well. Based on this fuzzy symbolic representation, an intelligent well logs interpretation system has been developed.

1 Introduction

In reservoir characterization, well log data are frequently used to interpret physical rock properties such as lithology, porosity, pore geometry, depositional facies and permeability [3, 6]. These properties are keys to the understanding of an oil reservoir and can help determining hydrocarbon reserves and reservoir producibility. Based on the information, decisions of where to complete a well, how to stimulate a field, and where to drill next, can be made to maximize profit and minimize risk.

Well log data, ranging from conventional logs, such as *spontaneous potential*, *gamma ray*, and *resistivity*, to more advanced logging technology, such as *Nuclear Magnetic Resonance* (NMR) logs, are sequence of curves indicating the properties of layers within the earth's subsurface. Figure 1 gives an example of *grammar ray*, *neutron* and *spontaneous potential* (SP) logs. The interpreted lithology is listed on the left-hand side.

Well log interpretation is a time-consuming process, since many different types of logs from many different wells need to be processed simultaneously. This paper

O. Castillo et al. (Eds.): Soft Computing for Hybrid Intel. Systems, SCI 154, pp. 417–426, 2008.
springerlink.com   © Springer-Verlag Berlin Heidelberg 2008

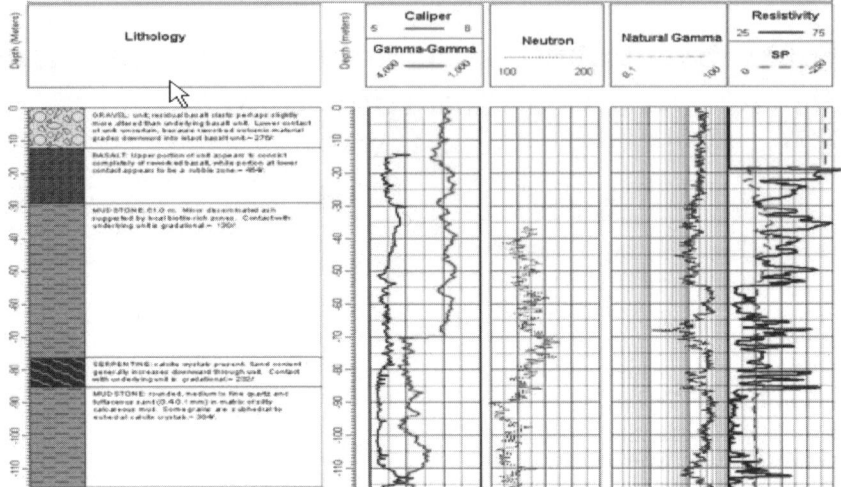

Fig. 1. An example of gamma ray, neutron and spontaneous potential logs. The interpreted lithology is listed on the left-hand side.

presents a technique to automatically transform well log data into fuzzy symbols which maintain the character of the original log curves. This simplified representation not only makes the interpretation task easier but also allows efficient interpretation under the uncertainty embedded in these data sets. The symbolic representation also has advantages over its numerical counter-part in that it is easier for computers to manipulate and process. Based on this representation, we have developed a computer system to perform automatic well-log interpretation. That work is reported in [7].

The paper is organized as follows. Section 2 presents the methodology to transform well log data into a fuzzy symbolic representation. In Section 3, we apply the method to two sets of well log data collected from an oil field in offshore West Africa and show the results. The intelligent well-log interpretation system is briefly described in Section 4. Finally Section 5 concludes the paper.

2   Methodology

The fuzzy symbolic representation is an approximation of well-logs that maintains the trend in the original data. The transformation process has four steps:

- Segmentation of the numerical well log data;
- Determine the optimal number of segmentation;
- Assign symbols to each segment;
- Symbols fuzzification.

These four steps are explained in the following subsections.

2.1 Well Log Segmentation

Well log segmentation involves partitioning log data into segments and using the mean value of the data points falling within the segment to represent the original data. In order to accurately represent the original data, each segment is allowed to have arbitrary length. In this way, areas where data points have low variation will be represented by a single segment while areas where data points have high variation will have many segments.

The segmentation process starts by having one data point in each segment. That is the number of segments is the same as the number of original data points. Step-by-step, neighboring segments (data points) are gradually combined to reduce the number of segments. This process stops when the number of segments reaches the predetermined number of segment.

At each step, the segments whose merging will lead to the least increase in *error* are combined. The *error* of each segment is defined as:

$$error_a = \sum_{i=1}^{n}(d_i - \mu_a)^2$$

where n is the number of data points in segment a, μ_a is the mean of segment a, d_i is the ith data point value in segment a.

This approach is similar to the Adaptive Piecewise Constant Approximation proposed by Keogh et. al. [4]. However, our method has an extra component that dynamically determines the number of segments (see Section 2.2). Another similar work using a different approach to determine the number of segments is reported in [1].

Figure 2 is an example of a well log with 189 data points, which are partitioned into 10 segments. The same data are partitioned into 20 segments in Figure 3. The average value of the data points within each segment is used to represent the original data.

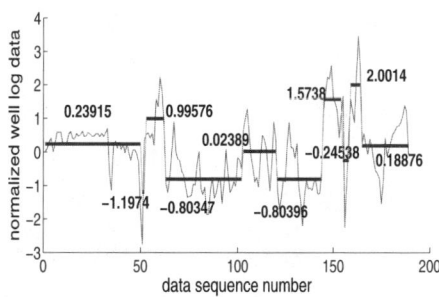

Fig. 2. 10 segments

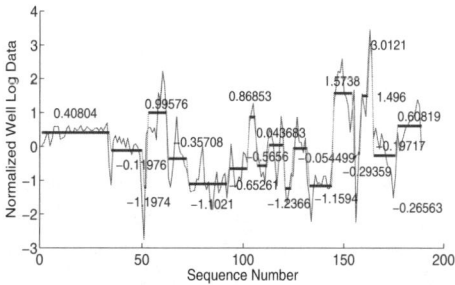

Fig. 3. 20 segments

2.2   Number of Segments

Although a larger number of segments capture the data trend better, it is also more difficult to interpret. Ideally, we want to use the smallest possible number of segments to capture the trend of the log data. Unfortunately, these two objectives are in conflict: the total *error* of all segments monotonically increases as the number of segments decreases (see Figure 4). We therefore devised a compromised solution where a penalty is paid for increasing the number of segments. The new *error* criterion is now defined as the previous total error *plus* the number of segments:

$$f = N + \sum_{i=1}^{N} error_i \qquad \text{where } N \text{ is the number of segment.}$$

During the segmentation process, the above f function is evaluated at each step. As long as this value f is decreasing, the system continues to merge segments. Once f starts increasing, it indicates that farther reducing the number of segments will sacrifice log character, hence the segmentation process terminates. For the log in Figure 2, 50 is the optimal number of segments for the 189 data points (see Figure 5).

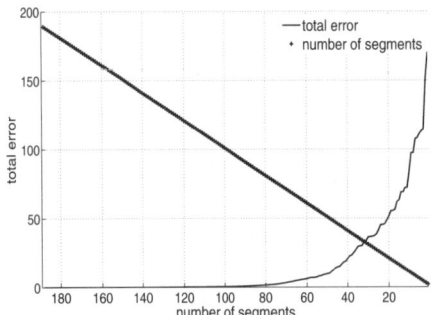

Fig. 4. Number of segments vs. total error

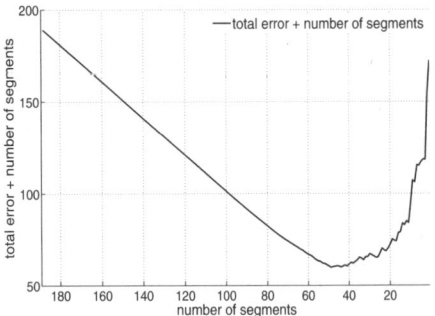

Fig. 5. A compromised solution

2.3   Symbols Assignment

Segmented well logs are represented as a set of numerical values, $\overline{WL} = \overline{s_1}, \overline{s_2}, \overline{s_3} \ldots$, where $\overline{s_i}$ is the mean value of the data within the ith segment. This numerical representation is farther simplified using symbols. Unlike numerical values, which are continues, symbols are discrete and bounded. This makes it easy for any subsequent computer interpretation scheme.

While converting the numerical values into symbols, it is desirable to produce symbols with equal-probability [2]. This is easily achieved since normalized sequence data have a Gaussian distribution [5]. We therefore applied z-transform to normalize the data and then determine the breakpoints that will produce n

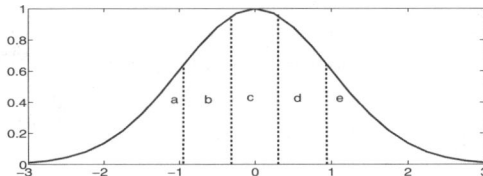

Fig. 6. Using 4 breakpoints to produce 5 symbols with equal probability

equal-sized areas under Gaussian curve. Figure 6 gives the four breakpoints -0.84, -0.25, 0.25 and 0.84 that produce 5 symbols, a, b, c, d, e, with equal probability. If only 3 symbols (a, b and c) are used, the breakpoints are -0.43 and 0.43.

Once the number of symbols, hence the breakpoints have been decided, we assign symbols to each segment of the well logs in the following manner: All segments have mean values that are below the smallest breakpoint are mapped to the symbol a; all segments have mean values that are greater than or equal to the smallest breakpoint and less than the second smallest breakpoint are mapped to the symbol b and so on. Figure 7 gives a well log that is transformed using 5 symbols.

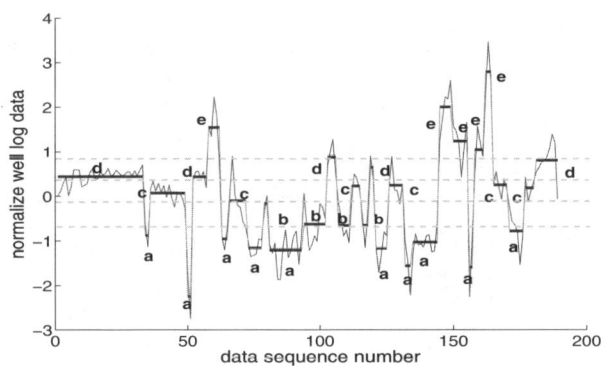

Fig. 7. A well log transformed using 5 symbols

2.4 Symbols Fuzzification

While some segments are clearly within the boundary of a particular symbol region, others may not have such clear cut. For example, in Figure 7, there are 3 segments lie on the borderline of a and b regions. A crisp symbol, either a or b, does not represent its true value. In contrast, fuzzy symbols use membership function to express the segment can be interpreted as symbol a and b with some possibility.

As an example, with crisp symbol approach, a segment with mean -0.9 is assigned with symbol a with 100% possibility (see Figure 8). Using fuzzy symbols designed by trapezoidal-shaped membership functions, the segment is assigned

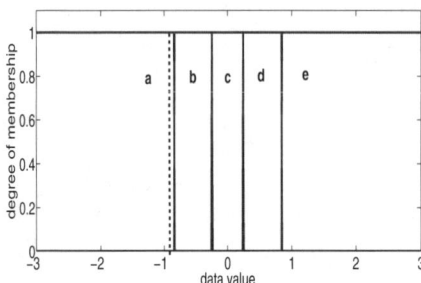 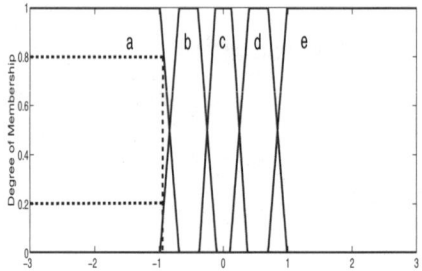

Fig. 8. The data value of -0.9 is transformed as a crisp symbol a

Fig. 9. The data value -0.9 is transformed as fuzzy symbol a (80%) and b (20%)

with symbol a with 80% possibility and symbol b with 20% possibility (see Figure 9). Fuzzy symbol representation is more expressive in this case.

In fuzzy logic, a membership function (MF) defines how each point in the input space is mapped into a membership value (or degree of membership) between 0 and 1. The input space consists of all possible input values. In our case, normalized well log data have open-ended boundaries with mean 0. Since 5 symbols are used to represent a well-log, we need to design 5 membership functions, one for each of the 5 symbols. Additionally, we used 3 symbols to represent core permeability. Three membership functions were also designed for these 3 symbols.

To design a trapezoidal-shaped membership function, 4 parameters are required: f_1 and f_2 are used to locate the "feet" of the trapezoid and s_1 and s_2 are used to locate the "shoulders" (see Figure 10). These four parameters are designed in the following way.

Let c_1 and c_2 be the breakpoints that define symbol n and $c_2 > c_1$:

$$d = \frac{c_2 - c_1}{4}$$
$$f_1 = c_1 - d; s_1 = c_1 + d; s_2 = c_2 - d; f_2 = c_2 + d$$

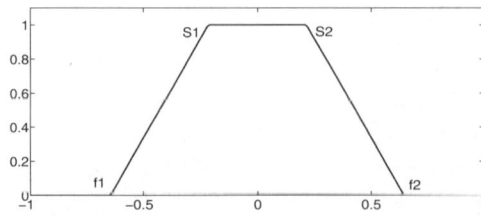

Fig. 10. 4 parameters, f1, f2, s1, s2, define a trapezoidal-shaped membership function

Table 1. Parameters used to design the trapezoidal-shaped membership function for each symbol

data	symbol	f_1	s_1	s_2	f_2
well-log	a	-3	-3	-0.9875	-0.6925
	b	-0.9875	-0.6925	-0.3975	-0.1025
	c	-0.375	-0.125	0.125	0.375
	d	0.1025	0.3975	0.6925	0.9875
	e	0.6925	0.9875	3	3
permeability	a	-3	-3	-0.645	-0.215
	b	-0.645	-0.215	0.215	0.645
	c	0.215	0.645	3	3

There are two exceptions: symbol a has $f_1 = c_1$ and symbol e has $f_2 = c_2$. Table 1 gives the four parameters used to design the membership functions for each symbol.

Once the 4 parameters are decided, the membership function f is defined as follows:

$$f(x, f_1, f_2, s_1, s_2) = \begin{cases} 0, & \text{if } x \leq f_1 \\ \frac{x-f_1}{s_1-f_1}, & \text{if } f_1 \leq x \leq s_1 \\ 1, & \text{if } s_1 \leq x \leq s_2 \\ \frac{f_2-x}{f_2-s_2}, & \text{if } s_2 \leq x \leq f_2 \\ 0, & \text{if } f_2 \leq x \end{cases}$$

Using the described fuzzy symbol scheme, 10 segments lie between two symbol regions in Figure 7 were mapped into fuzzy symbols shown in Figure 11.

In most cases, a reservoir well has multiple logs. To carry out the described transformation process, a reference log is first selected for segmentation. The result is then used to segment the other logs in the same well. After that, fuzzy symbols are assigned to each segmented data.

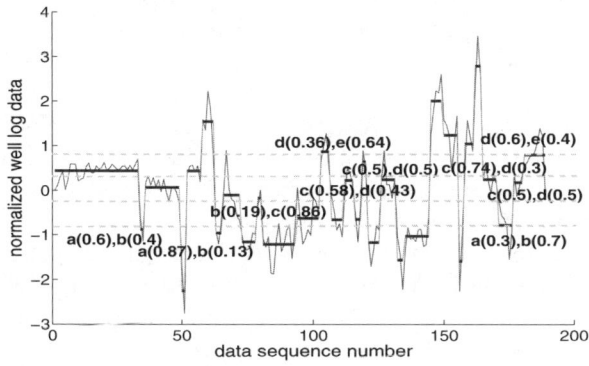

Fig. 11. A well log represented with fuzzy symbols

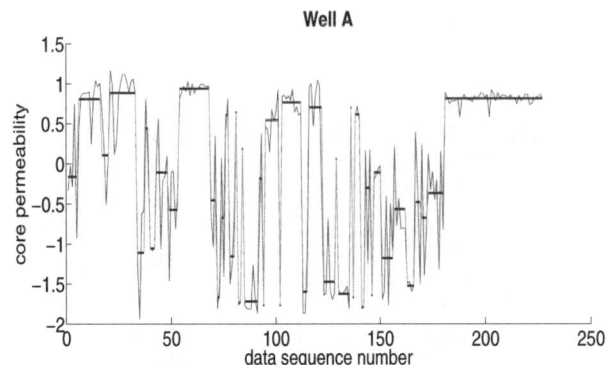

Fig. 12. The transformed core permeability (k)

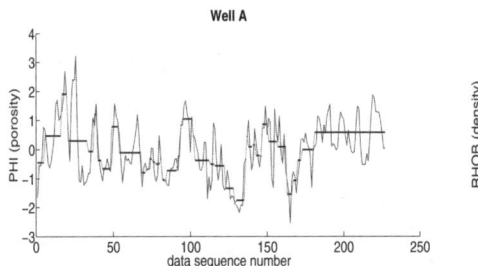

Fig. 13. Transformed PHI log   **Fig. 14.** Transformed RHOB log

3   Testing Results

We tested the developed transformation method on 2 sets of well log data collected from an offshore West Africa field. The first set is from Well A and contains 227 data points while the second set is from Well B and contains 113 data points. Each well has 3 different logs: PHI (porosity), $RhoB$ (density) and DT (sonic log). Additionally, V-shale (Volume of shale) information has been calculated previously [6]. We also have the corresponding core permeability data for these two wells.

In this case, permeability is the interpreted target. It is therefore chosen as the reference log to perform segmentation described in Section 2.1 and 2.2. Based on the segmentation results, the other 4 logs were segmented.

Figures 12, 13, 14, 15 and 16 give the segmented results for Well A. As shown, the results give sensible blocking and resemble the original log curves reasonably well. We do not show the transformed fuzzy symbols on these Figures as they take too much space. Also, due to space constraint, the results of Well B, which have a similar pattern, are not shown here.

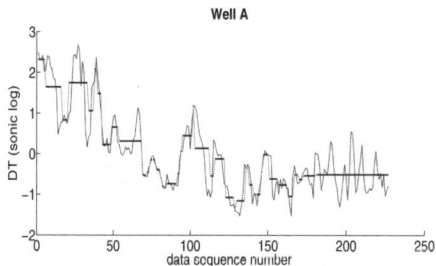

Fig. 15. Transformed DT log

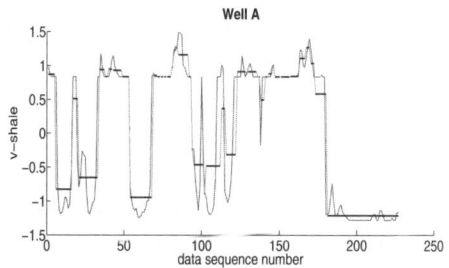

Fig. 16. Transformed V-shale log

4   Automatic Well Log Interpreter

Based on the fuzzy symbolic representation, we have developed an interpretation system that processes the fuzzy symbols and automatically interpret the permeability ranges that are associated with each well log segment. The system applies a co-evolutionary mechanism to evolve fuzzy rule sets. The fuzzy rule set is composed of two fuzzy rules, one classifies high-permeability log segments while the other identifies low-permeability segments. In this evolutionary system, two populations were maintained to evolve these two fuzzy rules simultaneously. The final fuzzy rule set were able to give sensible permeability interpretation for all well log segments. This work is reported in [7].

5   Conclusions

Well log interpretation is a routine, but time consuming, task in energy companies. With the increasing global energy demand, it is a natural trend to seek computerized well log interpretation techniques to provide results more efficiently. We have devised a method that maps the numerical well log data into a fuzzy symbol representation. This representation not only maintains the original well curve character but is more interpretable than its crisp numerical counter-part. The quality of this representation is verified using 2 sets of well logs data from offshore fields in South Africa. Based on the fuzzy symbolic representation, we have also implemented an intelligent well logs interpreter to interpret permeability with promising results [7]. We are currently applying this methodology to generate interpretation system for a wider ranges of reservoir properties, such as lithology and reservoir facies.

Acknowledgment

We thank Chevron for their permission to publish this work.

References

1. Abonyi, J., Feil, B., Nemeth, S., Arva, P.: Modified Gath-Geva Clustering for Fuzzy Segmentation of Multivariate Time-Series. Fuzzy Sets and Systems 149, 39–56 (2005)
2. Apostolico, A., Bock, M.E., Lonardi, S.: Monotony of Surprise and Large-Scale Quest for Unusual Words. In: Proceedings of the 6th International Conference on Research in Computational Molecular Biology, pp. 22–31 (2002)
3. Asquith, G., Krygowski, D.: Basic Well Log Analysis, 2nd edn, American Association of Petroleum Geologists (2004)
4. Keogh, E., Chakrabarti, K., Mehrotra, S., And Pazzani, M.: Locally Adaptive Dimensionality Reduction for Indexing Large Time Series Databases. In: Proceedings of ACM SIGMOD Conference on Management of Data, pp. 151–162 (2001)
5. Larsen, R.J., Marx, M.L.: An Introduction to Mathematical Statistics and Its Applications, 2nd edn. Prentice Hall, Englewood Cliffs (1986)
6. Yu, T., Wilkinson, D., Xie, D.: A Hybrid GP-Fuzzy Approach for Reservoir Characterization. In: Riolo, R.L., Worzel, B. (eds.) Genetic Programming Theory and Practice, pp. 271–290. Kluwer, Dordrecht (2003)
7. Yu, T.: A Co-Evolutionary Fuzzy System for Reservoir Well Logs Interpretation. In: Yu, T., Davis, D., Baydar, C., Roy, R. (eds.) Evolutionary Computation in Practice. Springer, Heidelberg (2007)

How to Solve a System of Linear Equations with Fuzzy Numbers

Rostislav Horčík[*]

Institute of Computer Science
Academy of Sciences of the Czech Republic
Pod Vodárenskou věží 2, 182 07 Prague 8, Czech Republic
horcik@cs.cas.cz

Abstract. The paper deals with a solution of a fuzzy interval system of linear equations, i.e. a system in which fuzzy intervals (numbers) appear instead of crisp numbers. We obtain general results and then use them for finding the united solution set in the case when all fuzzy interval occurring in the system have the trapezoidal shape.

1 Introduction

Fuzzy numbers and their arithmetic are investigated already for quite a long time. However, it seems that the fact that they are rather fuzzified versions of classical crisp intervals is not stressed very much. In this paper, we are going to reveal their interval nature by showing that many interesting results from classical interval analysis transfer also into the fuzzy case.

We will illustrate the problems when the fuzzy numbers are considered as numbers on the example of system of linear equations with fuzzy numbers discussed in [4]. In that paper, the authors considered a mechanical system which was described by a system of classical linear equations where some of the parameters were uncertain (given by fuzzy numbers). In order to solve this system, they used the common fuzzy arithmetic defined by means of Zadeh's extension principle. More precisely, let A, B be fuzzy numbers over reals and c be a crisp real number. The fuzzy numbers A and B represented unknown material parameters and c corresponded to a force acting on the system. The unknown fuzzy numbers X_1 and X_2 described the state of the mechanical system. In order to compute the state of the system, the authors solved the following system of linear equations:

$$\begin{pmatrix} A + B & -B \\ -B & B \end{pmatrix} \begin{pmatrix} X_1 \\ X_2 \end{pmatrix} = \begin{pmatrix} 0 \\ c \end{pmatrix}. \tag{1}$$

However, their method was incorrect. They solved the system symbolically by Gauss elimination, by means of the inverse matrix, and by means of SVD decomposition. Each method gives a different expression. Then they computed each

---

[*] The author was supported by the grant No. B100300502 of the Grant Agency of the Academy of Sciences of the Czech Republic.

of them using fuzzy arithmetic and obtained different results. Thus they concluded that there is an "artificial uncertainty" in fuzzy arithmetic and called for a different one.

In fact, they claimed that the reason why this is the case is that fuzzy arithmetic is not able to cope with dependency between the parameters. The same was mentioned also in [5] where the author somehow attempted to solve this problem. However, the problem with dependency is rather a secondary problem and formally can be solved by the constrained fuzzy arithmetic introduced by Klir in [6]. The main problem of this approach consists in usage of incorrect manipulations with the equations. Thus we focus here only on this problem.

There are two possible approaches how to define the solution set. The first one could be the following: find all possible fuzzy numbers X_1 and X_2 which satisfy the system (1). However, this is not a very interesting approach because of two reasons. First, the solution does not exist in almost all cases. Consider for example the second equation $-BX_1 + BX_2 = c$. It is clear that if $c \neq 0$ and B is not a crisp number, then the left-hand side either represents a fuzzy (not crisp) number or equals 0 if both X_1, X_2 are 0. Thus it cannot be equal to a crisp number c. Second, even if a solution exists then it need not represent what in fact the authors of [4] wanted to obtain. The more reasonable approach how to define the solution to the system (1) comes from classical interval analysis (see [8] for details). Let us firstly assume that A and B are classical crisp intervals. Our mechanical system is described by the following system of linear equations:

$$\begin{pmatrix} a+b & -b \\ -b & b \end{pmatrix} \mathbf{x} = \begin{pmatrix} 0 \\ c \end{pmatrix}, \tag{2}$$

where the entries a, b in the matrix are not known but we know that they belongs respectively to the intervals A and B, i.e., $a \in A$ and $b \in B$. Now it is natural to call a vector $\mathbf{x}$ *solution* if there exist $a \in A$ and $b \in B$ such that the system (2) is satisfied. In other words, we define the solution set as a set of all vectors $\mathbf{x} \in \mathbb{R}^2$ which solves the system (2) for some $a \in A$ and $b \in B$. Formally written, $\mathbf{x}$ is a solution if the following first-order formula is valid in classical logic:

$$(\exists a \in A)(\exists b \in B)\left(\begin{pmatrix} a+b & -b \\ -b & b \end{pmatrix} \mathbf{x} = \begin{pmatrix} 0 \\ c \end{pmatrix}\right).$$

Now, if A and B are fuzzy intervals (numbers), then the membership degrees of a in A and b in B may be in general less than one. Consequently, the solution set becomes a fuzzy subset of $\mathbb{R}^2$. However, the solution set in this case can be still defined by the same formula as before only interpreted in a suitable fuzzy predicate logic instead of the classical one:

$$\mathbf{x} \in \text{Solution set} \equiv_{df} (\exists a \in A)(\exists b \in B)\left(\begin{pmatrix} a+b & -b \\ -b & b \end{pmatrix} \mathbf{x} = \begin{pmatrix} 0 \\ c \end{pmatrix}\right)$$

Thus the solution set is a fuzzy set containing all states of the original mechanical system provided that a and b come from the fuzzy intervals A and B respectively.

In this paper we are going to discuss how to solve such systems of linear equations with uncertain parameters given by fuzzy intervals (numbers). Unlike the above-mentioned example we will assume that all parameters in the system are independent, i.e. each parameter (quantified variable) can appear only in one place of the formula describing the system of linear equations. The reason why we make this assumption is that it is likely a better approach to firstly develop a theory how to solve such systems where all parameters are independent and then try to generalize this theory to the case when a dependency may occur.

2 Preliminaries

2.1 Fuzzy Class Theory

Most of our results in this paper can be proved in any fuzzy logic that is at least as strong and expressive as MTLΔ[1]. Only at the end of the paper we restrict ourselves to a particular fuzzy logic. Let $\mathcal{F}$ be any fuzzy propositional logic extending MTLΔ. Thus $\mathcal{F}$ can be for instance Hájek's BL, Łukasiewicz logic, Gödel logic, or product logic (of course all of them extended by Baaz's delta operator Δ). For details on MTL see [2]. Details on fuzzy logics stronger than BL can be found in [3].

For dealing with fuzzy intervals we will use fuzzy class theory built over the logic $\mathcal{F}$. Originally this theory introduced in [1] was built over the logic ŁΠ. However, the definitions and basic results from [1] work in any logic extending MTLΔ. For convenience, we reproduce basic definitions of fuzzy class theory. Recall that from the point of view of formal logic, it can be characterized as Henkin-style higher-order fuzzy logic.

Definition 1 (Henkin-style second-order fuzzy logic). *Let $\mathcal{F}$ be a logic which extends MTLΔ. The Henkin-style second-order fuzzy logic over $\mathcal{F}$ is a theory in multi-sorted first-order logic $\mathcal{F}_1$ with sorts for atomic objects (lowercase variables) and classes (uppercase variables).*

Besides the logical predicate of identity, the only primitive predicate is the membership predicate $\in$ between objects and classes. The axioms for $\in$ are the following:

1. *The comprehension axioms $(\exists X)\Delta(\forall x)(x \in X \leftrightarrow \varphi)$, φ not containing X, which enable the (eliminable) introduction of comprehension terms $\{x \mid \varphi\}$ with the axioms $y \in \{x \mid \varphi(x)\} \leftrightarrow \varphi(y)$ (where φ may be allowed to contain other comprehension terms).*
2. *The extensionality axiom $(\forall x)\Delta(x \in X \leftrightarrow x \in Y) \rightarrow X = Y$.*

Convention 1. *The formulae $(\forall x)(x \in X \rightarrow \varphi)$ and $(\exists x)(x \in X \& \varphi)$ and the comprehension terms $\{x \mid x \in A \& \varphi\}$ are abbreviated $(\forall x \in X)\varphi$, $(\exists x \in X)\varphi$, and $\{x \in A \mid \varphi\}$ respectively. The formulae $\varphi \& \cdots \& \varphi$ (n times) are abbreviated φ^n. Let c be an atomic object. Then $\{c\}$ denotes the crisp class containing only c to degree 1, i.e. $\{c\} = \{x \mid x = c\}$.*

---

[1] Recall that MTLΔ is the logic of all left-continuous t-norms and their residua.

Definition 2 (Fuzzy class operations). *The following elementary fuzzy set operations can be defined:*

$$\emptyset =_{df} \{x \mid 0\} \qquad\qquad \textit{empty class}$$
$$V =_{df} \{x \mid 1\} \qquad\qquad \textit{universal class}$$
$$X \cap Y =_{df} \{x \mid x \in X \ \& \ x \in Y\} \qquad \textit{intersection}$$
$$X \sqcup Y =_{df} \{x \mid x \in X \vee x \in Y\} \qquad \textit{max-union}$$

Definition 3 (Fuzzy class relations). *Further we define the following elementary relations between fuzzy sets:*

$$\mathrm{Norm}(X) \equiv_{df} (\exists x)\Delta(x \in X) \qquad\qquad \textit{normality}$$
$$X \subseteq Y \equiv_{df} (\forall x)(x \in X \to x \in Y) \qquad \textit{inclusion}$$
$$X \parallel Y \equiv_{df} (\exists x)(x \in X \ \& \ x \in Y) \qquad \textit{compatibility}$$

2.2   Fuzzy (Interval) Arithmetic

Our intended universal class V of all objects is the set of real numbers $\mathbb{R}$ endowed with the usual structure of an ordered field. However, almost all our results hold over any ordered field. The field operations and the order between objects will be denoted in the usual way, i.e. $x + y$, $x - y$, $x \leq y$, etc.

Definition 4. *The following arithmetic operations and relations can be defined by Zadeh's extension principle for any fuzzy classes A, B, and a real number k:*

$$A + B =_{df} \{z \mid (\exists x \in A)(\exists y \in B)(z = x + y)\} \qquad \textit{addition}$$
$$A - B =_{df} \{z \mid (\exists x \in A)(\exists y \in B)(z = x - y)\} \qquad \textit{substraction}$$
$$kA -_{df} Ak -_{df} \{z \mid (\exists x \in A)(z = kx)\} \qquad \textit{scalar multiplication}$$
$$A \leq B \equiv_{df} (\exists x \in A)(\exists y \in B)(x \leq y) \qquad \textit{order}$$

Convention 2. *Tuples of the elements of the universe and fuzzy classes are denoted by* $\mathbf{x}, \mathbf{y}, \mathbf{z}, \ldots$ *and* $\mathbf{A}, \mathbf{B}, \mathbf{C}, \ldots$ *respectively. Matrices of elements are denoted by* $\mathbb{A}, \mathbb{B}, \mathbb{C}, \ldots$ *and matrices of fuzzy classes by boldface capital letters* $\mathbf{A}, \mathbf{B}, \mathbf{C}, \ldots$

Let $\mathbf{A} = (A_{ij})$ *be an* $m \times n$ *matrix of fuzzy classes. Then* $(\forall \mathbb{A} \in \mathbf{A})\varphi$ *stands for* $(\forall a_{11} \in A_{11}) \cdots (\forall a_{mn} \in A_{mn})\varphi$, *similarly for* $(\exists \mathbb{A} \in \mathbf{A})\varphi$. *We use the analogous conventions also for tuples of fuzzy classes.*

Finally, let us introduce basic definitions on tuples of fuzzy classes. Let $\mathbf{A} = \langle A_1, \ldots, A_n \rangle$ and $\mathbf{B} = \langle B_1, \ldots, B_n \rangle$ be tuples of fuzzy classes and $\mathbf{z} = \langle z_1, \ldots, z_n \rangle$ be a tuple of real numbers. Then we define

$$\mathbf{z} \in \mathbf{A} \equiv_{df} \&_{i=1}^{n} z_i \in A_i, \qquad \mathbf{A} \subseteq \mathbf{B} \equiv_{df} (\forall \mathbf{z})(\mathbf{z} \in \mathbf{A} \to \mathbf{z} \in \mathbf{B}),$$

where $(\forall \mathbf{z})\varphi \equiv_{df} (\forall z_1) \cdots (\forall z_n)\varphi$.

The addition and substraction of matrices or vectors of fuzzy classes is defined componentwise by means of Definition 4. The multiplication of an $m \times n$ matrix $\mathbf{A} - (A_{ij})$ with a vector $\mathbf{x} - \langle x_1, \ldots, x_n \rangle$ is defined as follows:

$$\mathbf{Ax} = \begin{pmatrix} A_{11}x_1 + \cdots + A_{1n}x_n \\ \vdots \\ A_{m1}x_1 + \cdots + A_{mn}x_n \end{pmatrix}.$$

3   Fuzzy Intervals

In this section we introduce the notion of a fuzzy interval. A fuzzy class A is said to be convex to the degree to which the formula $(x \in A \ \& \ y \in A \ \& \ x \le z \le y) \rightarrow z \in A$ holds. This formula is denoted by $\mathrm{Convex}(A)$.

Definition 5. *Let A be a fuzzy class. The degree of A being a fuzzy interval is given by the formula* $\mathrm{FInt}(A) \equiv_{\mathrm{df}} \mathrm{Norm}(A) \ \& \ \mathrm{Convex}(A)$.

Fuzzy classes which are fully fuzzy intervals can be characterized by means of its down-class and up-class. A down-class and an up-class generated by a class A are defined respectively by $A^{\downarrow} =_{\mathrm{df}} \{x \mid (\exists a \in A)(x \le a)\}$ and $A^{\uparrow} =_{\mathrm{df}} \{x \mid (\exists a \in A)(x \ge a)\}$.

Theorem 3. *Each fuzzy interval A to degree 1 is equal to the intersection of its down-class and up-class, i.e.,* $\Delta \mathrm{FInt}(A) \rightarrow A = A^{\downarrow} \cap A^{\uparrow}$.

4   Fuzzy Interval Linear System and Its Solution Set

In this section we formally define the solution set of a system of linear equations with uncertain parameters, i.e., the parameters which are known to belong to given fuzzy classes. Our intention is of course that these classes will be fuzzy intervals. However, some of our results hold generally for any fuzzy classes. Thus we define the solution set for arbitrary fuzzy classes.

Definition 6. *Let $\mathbf{A} = (A_{ij})$ be an $m \times n$ matrix of fuzzy classes (intervals) and $\mathbf{B} = \langle B_1, \dots, B_n \rangle$ be an n-tuple of fuzzy classes (intervals). Then the system $\mathbf{A}\mathbf{x} = \mathbf{B}$ is called fuzzy (interval) linear system.*

Thus a system of linear equations with fuzzy classes is called a fuzzy linear system whereas the system with fuzzy intervals is called a fuzzy interval linear system.

   The most common approach how to define a solution set $\varXi(\mathbf{A}, \mathbf{B})$ of a fuzzy linear system $\mathbf{A}\mathbf{x} = \mathbf{B}$ is the one which we described in the introduction, i.e.

$$\varXi(\mathbf{A}, \mathbf{B}) =_{\mathrm{df}} \{\mathbf{x} \mid (\exists \mathbb{A} \in \mathbf{A})(\exists \mathbf{b} \in \mathbf{B})(\mathbb{A}\mathbf{x} = \mathbf{b})\}.$$

Such solution set is usually called *united solution set*. However, it turns out that usage of the universal quantifiers in the definition is also meaningful. We will shortly present the main motivation for this coming from the very nice paper [8] on classical interval analysis. Consider a system which is to be controlled. This system is described by a system of linear equations $\mathbb{A}\mathbf{x} = \mathbf{b}$. Suppose that the entries of $\mathbb{A}$ corresponds to the inputs of the system and $\mathbf{b}$ to its outputs. Both inputs and outputs can be of two sorts. In the set of inputs we distinguish between

-   *perturbations:* the inputs on which we have no influence (e.g. noise, unknown material parameter, etc.), but we know that they belong to given fuzzy classes,

– *controls:* the inputs intended for a controller. We can set them arbitrarily but we are restricted by some constraints, i.e., they can be choosen only within given fuzzy classes.

In the set of outputs we distinguish between

– *stabilized:* the outputs which should be stabilized into given fuzzy classes (e.g. temperature of a heating),
– *controlled:* the outputs to which we must be able to attain any given values from prescribed fuzzy classes (e.g. it must be possible to put a robot's arm into any place in its operational space).

The vector $\mathbf{x}$ corresponds to a state of the system. Now we are interested in those states $\mathbf{x}$ for which for any perturbations and for any values of controlled outputs from the prescribed fuzzy classes, there exist suitable controls such that the stabilized outputs are within the given fuzzy classes and the controlled outputs attain the desired values. Such fuzzy class of vectors $\mathbf{x}$ will be for us the most general solution set of a fuzzy linear system.

Let $\mathbf{Ax} = \mathbf{B}$ be a fuzzy linear system. In order to define the formula describing the solution set, we split the matrix $\mathbf{A}$ and the tuple $\mathbf{B}$ into two disjoint parts according to the quantifiers. We define $\mathbf{A}^\forall = (A_{ij}^\forall)$, $\mathbf{A}^\exists = (A_{ij}^\exists)$, $\mathbf{B}^\forall = (B_i^\forall)$, and $\mathbf{B}^\exists = (B_i^\exists)$, where

$$
A_{ij}^\forall = \begin{cases} A_{ij} & \text{if } A_{ij} \text{ should be} \\ & \text{quantified by } \forall, \\ \{0\} & \text{otherwise,} \end{cases} \qquad
A_{ij}^\exists = \begin{cases} A_{ij} & \text{if } A_{ij} \text{ should be} \\ & \text{quantified by } \exists, \\ \{0\} & \text{otherwise,} \end{cases}
$$

$$
B_i^\forall = \begin{cases} B_i & \text{if } B_i \text{ should be} \\ & \text{quantified by } \forall, \\ \{0\} & \text{otherwise,} \end{cases} \qquad
B_i^\exists = \begin{cases} B_i & \text{if } B_i \text{ should be} \\ & \text{quantified by } \exists, \\ \{0\} & \text{otherwise.} \end{cases}
$$

Then we have $\mathbf{A} = \mathbf{A}^\forall + \mathbf{A}^\exists$, $\mathbf{B} = \mathbf{B}^\forall + \mathbf{B}^\exists$. Now we can write down the formal definition of the solution set.

Definition 7. *Let $(\mathbf{A}^\forall + \mathbf{A}^\exists)\mathbf{x} = \mathbf{B}^\forall + \mathbf{B}^\exists$ be a fuzzy linear system. Then its solution set is the following fuzzy class:*

$$
\Xi(\mathbf{A}^\forall, \mathbf{A}^\exists, \mathbf{B}^\forall, \mathbf{B}^\exists) =_{df} \{\mathbf{x} \mid ((\forall \mathbb{U} \in \mathbf{A}^\forall)(\forall \mathbf{u} \in \mathbf{B}^\forall)
$$
$$
(\exists \mathbb{E} \in \mathbf{A}^\exists)(\exists \mathbf{e} \in \mathbf{B}^\exists)((\mathbb{U} + \mathbb{E})\mathbf{x} = \mathbf{u} + \mathbf{e})\} .
$$

Finally, we generalize the fundamental theorem [8, Theorem 3.4] from classical logic to fuzzy logic. The theorem characterizes the solutions by means of the arithmetic defined on fuzzy classes in Subsection 2.2.

Theorem 4 (Fundamental theorem). *Let $(\mathbf{A}^\forall + \mathbf{A}^\exists)\mathbf{x} = \mathbf{B}^\forall + \mathbf{B}^\exists$ be a fuzzy linear system. Then a vector $\mathbf{x}$ belongs to the solution set $\Xi(\mathbf{A}^\forall, \mathbf{A}^\exists, \mathbf{B}^\forall, \mathbf{B}^\exists)$ to the same degree as the formula $\mathbf{A}^\forall \mathbf{x} - \mathbf{B}^\forall \subseteq \mathbf{B}^\exists - \mathbf{A}^\exists \mathbf{x}$ holds, i.e.,*

$$\mathbf{x} \in \Xi(\mathbf{A}^\vee, \mathbf{A}^\exists, \mathbf{B}^\vee, \mathbf{B}^\exists) \longleftrightarrow \mathbf{A}^\vee \mathbf{x} - \mathbf{B}^\vee \subseteq \mathbf{B}^\exists - \mathbf{A}^\exists \mathbf{x}.$$

The operations in $\mathbf{A}^\vee \mathbf{x} - \mathbf{B}^\vee$ and $\mathbf{B}^\exists - \mathbf{A}^\exists \mathbf{x}$ are the arithmetic operations defined on matrices and tuples of fuzzy classes in Subsection 2.2.

5   United Solution Set

The fundamental theorem serves as a good starting point for computing the solution set. Although it works for arbitrary fuzzy classes, we will restrict ourselves to fuzzy intervals in the rest of the paper. This restriction is necessary if we want to obtain results like in the classical interval analysis.

The second restriction we make in this section concerns the quantifiers in the definition of solution set. More precisely, we are going to describe the solution set for a fuzzy interval linear system in the case when all the quantifiers appearing in the system are existential. This restriction allows us to separate the particular equations in the computation of the solution set.

Assume that all quantifiers in Definition 7 of the solution set are existential. Then $\mathbf{A}^\vee$ is the matrix of crisp fuzzy classes $\{0\}$ and the same holds for $\mathbf{B}^\vee$, i.e. $\mathbf{A}^\vee + \mathbf{A}^\exists = \mathbf{A}^\exists$ and $\mathbf{B}^\vee + \mathbf{B}^\exists = \mathbf{B}^\exists$. Thus we will denote the solution set in this case by $\Xi(\mathbf{A}^\exists, \mathbf{B}^\exists)$ and call it the united solution set like in the classical interval analysis

Let $K = \{\uparrow, \downarrow\}^n$ be the set of all sequences of symbols $\uparrow, \downarrow$ whose length is n. The j-th component of $k \in K$ will be denoted by k_j. Further, we define $\varepsilon_{jk} = 1$ if $k_j = \uparrow$ and -1 otherwise. Let Q_k, $k \in K$, be the family of all orthants of $\mathbb{R}^n$, i.e., we have for each Q_k:

$$Q_k = \{\mathbf{x} \in \mathbb{R}^n \mid \varepsilon_{1k} x_1 \geq 0 \ \& \ \cdots \ \& \ \varepsilon_{nk} x_n \geq 0\},$$

where x_j stands for the j-th component of $\mathbf{x}$. Each Q_k is obviously crisp. Finally, we define $-k_j = \downarrow$ if $k_j = \uparrow$ and $\uparrow$ otherwise.

Theorem 5. *Let $\mathbf{A}^\exists \mathbf{x} = \mathbf{B}^\exists$ be a fuzzy interval linear system. Then*

$$\Xi(\mathbf{A}^\exists, \mathbf{B}^\exists) = \bigsqcup_{k \in K} \left(\bigcap_{i=1}^{m} (Q_k \cap S_{ik}^d \cap S_{ik}^u) \right),$$

where $S_{ik}^d = \{\mathbf{x} \mid (\sum_{j=1}^n A_{ij}^{k_j} x_j) \parallel (B_i^\exists)^\downarrow\}$, $S_{ik}^u = \{\mathbf{x} \mid (\sum_{j=1}^n A_{ij}^{-k_j} x_j) \parallel (B_i^\exists)^\uparrow\}$.

6   United Solution Set for Trapezoidal Fuzzy Intervals in Łukasiewicz Logic

In this section, we want to demonstrate how to use the results from the previous section for computing concrete solution sets. For this purpose we have to restrict ourselves to one concrete logic (we choose Łukasiewicz logic) and one special type of fuzzy intervals (namely those which are known in the literature under the name trapezoidal fuzzy numbers). We will work with the standard semantics of Łukasiewicz logic, i.e., with the standard MV-algebra on $[0, 1]$. Thus all the predicates can be viewed as $[0, 1]$-valued functions on reals, & is interpreted as Łukasiewicz t-norm, $\rightarrow$ as the corresponding residuum, $\wedge, \vee$ as min and max respectively.

We firstly introduce some notation for dealing with the chosen fuzzy intervals. Let $f : \mathbb{R} \rightarrow \mathbb{R}$ be a function. We define its truncation $\overline{f} = (f \vee 0) \wedge 1$. The trapezoidal fuzzy intervals form a certain subset of piecewise linear $[0, 1]$-valued functions. We firstly define their up-classes and down-classes. Let $d > 0$. Then

$$/a, d/(t) = \overline{\left(\frac{t - a}{d} + 1 \right)} = \begin{cases} 1 & \text{if } t \geq a, \\ \frac{t-a}{d} + 1 & \text{if } a - d \leq t \leq a, \\ 0 & \text{otherwise,} \end{cases}$$

$$\backslash a, d\backslash(t) = \overline{\left(\frac{a - t}{d} + 1 \right)} = \begin{cases} 1 & \text{if } t \leq a, \\ \frac{a-t}{d} + 1 & \text{if } a \leq t \leq a + d, \\ 0 & \text{otherwise.} \end{cases}$$

For $d = 0$ we define

$$/a, 0/(t) = \begin{cases} 1 & \text{if } t \geq a, \\ 0 & \text{otherwise,} \end{cases} \qquad \backslash a, 0\backslash(t) = \begin{cases} 1 & \text{if } t \leq a, \\ 0 & \text{otherwise.} \end{cases}$$

Note that we require in $/a, d/$ (resp. $\backslash a, d\backslash$) the coefficient d to be nonnegative.

Definition 8. *Let $a_1, a_2, d_1, d_2 \in \mathbb{R}$ such that $a_1 \leq a_2$ and $d_1, d_2 \geq 0$. Then the trapezoidal fuzzy interval $\langle /a_1, d_1/, \backslash a_2, d_2\backslash \rangle$ is the intersection $/a_1, d_1/ \cap \backslash a_2, d_2\backslash$.*

Before we state the final theorem describing the united solution set, we introduce a further notation. Let $\mathbf{T} = \langle T_1, \ldots, T_m \rangle$ be a tuple of fuzzy classes. Then $\bigcap \mathbf{T} = \bigcap_{i=1}^{m} T_i$. For a vector of reals $\mathbf{x} = \langle x_1, \ldots, x_n \rangle$, we define $|\mathbf{x}| = \langle |x_1|, \ldots, |x_n| \rangle$. Further, let $\mathbb{A} = (A_{ij})$ be an $m \times n$ matrix and $\mathbf{x} = \langle x_1, \ldots, x_n \rangle$ a vector of reals. Then

$$\mathbb{A} \odot \mathbf{x} = \begin{pmatrix} A_{11}x_1 \vee \cdots \vee A_{1n}x_n \\ \vdots \\ A_{m1}x_1 \vee \cdots \vee A_{mn}x_n \end{pmatrix}.$$

Thus $\mathbb{A} \odot \mathbf{x}$ corresponds in fact to the usual multiplication of a matrix and a vector where all sums are replaced by the operation $\vee$. Finally, define the following operation for $a, b \in \mathbb{R}$:

$$a \oslash b = \begin{cases} \overline{\frac{a}{b} + 1} & \text{if } b \neq 0, \\ 1 & \text{if } b = 0, \ a \geq 0, \\ 0 & \text{if } b = 0, \ a < 0, \end{cases}$$

Note that if $a, b \geq 0$ then $a \oslash b = 1$. We extend this definitions also for vectors of numbers component-wise, i.e. $\langle a_1, \ldots, a_n \rangle \oslash \langle b_1, \ldots, b_n \rangle = \langle a_1 \oslash b_1, \ldots, a_n \oslash b_n \rangle$. Let $\mathbb{A}$ be a matrix. The element in the i-th row and j-th column will be also denoted by $(\mathbb{A})_{ij}$.

Theorem 6. *Let* $\mathbf{A}^{\exists}\mathbf{x} = \mathbf{B}^{\exists}$ *be a fuzzy interval linear system such that* $\mathbf{A}^{\exists} = (A_{ij})$ *is an* $m \times n$ *matrix of trapezoidal fuzzy intervals* $A_{ij} = \langle /a_{ij}^{\uparrow}, c_{ij}^{\uparrow}/, \backslash a_{ij}^{\downarrow}, c_{ij}^{\downarrow} \backslash \rangle$ *and* $\mathbf{B}^{\exists} = \langle B_1, \ldots, B_n \rangle$ *is a tuple of trapezoidal fuzzy intervals* $B_i = \langle /b_i^{\uparrow}, e_i^{\uparrow}/, \backslash b_i^{\downarrow}, e_i^{\downarrow} \backslash \rangle$. *Then the solution set* $\Xi(\mathbf{A}^{\exists}, \mathbf{B}^{\exists})$ *can be described as follows:*

$$\Xi(\mathbf{A}^{\exists}, \mathbf{B}^{\exists}) = \bigsqcup_{k \in K} \left(Q_k \cap \bigcap \mathbf{T}_k \cap \bigcap \mathbf{R}_k \right),$$

where

$$\mathbf{T}_k(\mathbf{x}) = \left(\mathbf{b}_{\downarrow} - \mathbb{A}_k^{\uparrow}\mathbf{x} \right) \oslash \left(\mathbf{e}_{\downarrow} \vee \left(\mathbb{C}_k^{\uparrow} \odot |\mathbf{x}| \right) \right),$$

$$\mathbf{R}_k(\mathbf{x}) = \left(\mathbb{A}_k^{\downarrow}\mathbf{x} - \mathbf{b}_{\uparrow} \right) \oslash \left(\mathbf{e}_{\uparrow} \vee \left(\mathbb{C}_k^{\downarrow} \odot |\mathbf{x}| \right) \right),$$

$\mathbf{b}_{\downarrow} = \langle b_1^{\downarrow}, \ldots, b_n^{\downarrow} \rangle$, $\mathbf{b}_{\uparrow} = \langle b_1^{\uparrow}, \ldots, b_n^{\uparrow} \rangle$, $\mathbf{e}_{\downarrow} = \langle e_1^{\downarrow}, \ldots, e_n^{\downarrow} \rangle$, $\mathbf{e}_{\uparrow} = \langle e_1^{\uparrow}, \ldots, e_n^{\uparrow} \rangle$, *and* $\mathbb{A}_k^{\uparrow}$, $\mathbb{A}_k^{\downarrow}$, $\mathbb{C}_k^{\uparrow}$, $\mathbb{C}_k^{\downarrow}$ *are* $m \times n$ *matrices such that* $(\mathbb{A}_k^{\uparrow})_{ij} = a_{ij}^{k_j}$, $(\mathbb{A}_k^{\downarrow})_{ij} = a_{ij}^{-k_j}$, $(\mathbb{C}_k^{\uparrow})_{ij} = c_{ij}^{k_j}$, $(\mathbb{C}_k^{\downarrow})_{ij} = c_{ij}^{-k_j}$.

7   Example

In this section we are going to illustrate Theorem 6 on a concrete example. We will in fact consider a fuzzified version of a favorite example from classical interval analysis.

Example 1. Consider the following *classical* interval linear system:

$$\begin{pmatrix} [2, 4] & [-2, 1] \\ [-1, 2] & [2, 4] \end{pmatrix} \mathbf{x} = \begin{pmatrix} [-2, 2] \\ [-2, 2] \end{pmatrix}. \tag{3}$$

The united solution set of this system is depicted in Figure 1 (left).

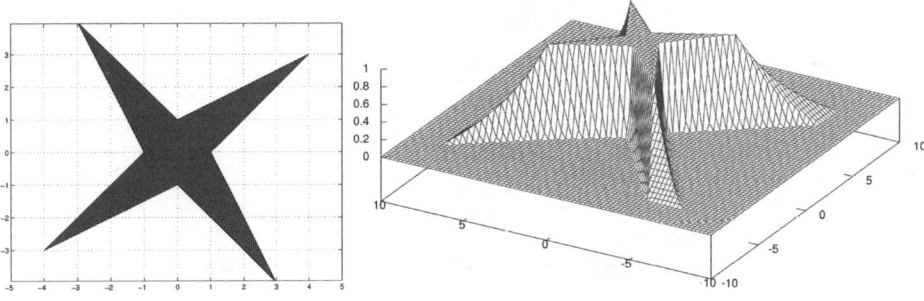

Fig. 1. The united solution set $\Xi(\mathbf{A}^{\exists}, \mathbf{B}^{\exists})$ of the systems from Examples 1, 2

Now we fuzzify the latter example and then describe its united solution set.

Example 2. Consider the following fuzzy interval linear system with trapezoidal fuzzy intervals:

$$\left(\begin{array}{cc} \langle/2,\frac{1}{2}/,\backslash 4,\frac{1}{2}\backslash\rangle & \langle/-2,\frac{1}{2}/,\backslash 1,\frac{1}{2}\backslash\rangle \\ \langle/-1,\frac{1}{2}/,\backslash 2,\frac{1}{2}\backslash\rangle & \langle/2,\frac{1}{2}/,\backslash 4,\frac{1}{2}\backslash\rangle \end{array} \right) \mathbf{x} = \left(\begin{array}{c} \langle/-2,\frac{1}{2}/,\backslash 2,\frac{1}{2}\backslash\rangle \\ \langle/-2,\frac{1}{2}/,\backslash 2,\frac{1}{2}\backslash\rangle \end{array} \right) \qquad (4)$$

In this case $n = 2$. Thus we have four orthants. Since the united solution set is in fact the union over all orthants, we describe the united solution set only in $Q_{\langle\downarrow,\uparrow\rangle}$. This means that we compute $\mathbf{T}_{\langle\downarrow,\uparrow\rangle}$ and $\mathbf{R}_{\langle\downarrow,\uparrow\rangle}$. We have

$$\mathbf{b}_{\downarrow} = \begin{pmatrix} 2 \\ 2 \end{pmatrix}, \quad \mathbf{b}_{\uparrow} = \begin{pmatrix} -2 \\ -2 \end{pmatrix}, \quad \mathbf{e}_{\downarrow} = \mathbf{e}_{\uparrow} = \begin{pmatrix} \frac{1}{2} \\ \frac{1}{2} \end{pmatrix}.$$

$$\mathbb{A}^{\uparrow}_{\langle\downarrow,\uparrow\rangle} = \begin{pmatrix} 4 & -2 \\ 2 & 2 \end{pmatrix}, \quad \mathbb{A}^{\downarrow}_{\langle\downarrow,\uparrow\rangle} = \begin{pmatrix} 2 & 1 \\ -1 & 4 \end{pmatrix}, \quad \mathbb{C}^{\uparrow}_{\langle\downarrow,\uparrow\rangle} = \mathbb{C}^{\downarrow}_{\langle\downarrow,\uparrow\rangle} = \begin{pmatrix} \frac{1}{2} & \frac{1}{2} \\ \frac{1}{2} & \frac{1}{2} \end{pmatrix}.$$

Then

$$\mathbf{T}_{\langle\downarrow,\uparrow\rangle}(\mathbf{x}) = \begin{pmatrix} 2 - (4x_1 - 2x_2) \\ 2 - (2x_1 + 2x_2) \end{pmatrix} \oslash \begin{pmatrix} \frac{1}{2} \vee \frac{1}{2}|x_1| \vee \frac{1}{2}|x_2| \\ \frac{1}{2} \vee \frac{1}{2}|x_1| \vee \frac{1}{2}|x_2| \end{pmatrix},$$

$$\mathbf{R}_{\langle\downarrow,\uparrow\rangle}(\mathbf{x}) = \begin{pmatrix} 2x_1 + x_2 + 2 \\ -x_1 + 4x_2 + 2 \end{pmatrix} \oslash \begin{pmatrix} \frac{1}{2} \vee \frac{1}{2}|x_1| \vee \frac{1}{2}|x_2| \\ \frac{1}{2} \vee \frac{1}{2}|x_1| \vee \frac{1}{2}|x_2| \end{pmatrix}.$$

The united solution set $\Xi(\mathbf{A}^{\exists}, \mathbf{B}^{\exists})$ is depicted in Figure 1 (right). Notice that the kernel of $\Xi(\mathbf{A}^{\exists}, \mathbf{B}^{\exists})$ is the same as the united solution set from Example 1.

References

1. Běhounek, L., Cintula, P.: Fuzzy Class Theory. Fuzzy Sets and Systems 154(1), 34–55 (2005)
2. Esteva, F., Godo, L.: Monoidal t-norm based logic: Towards a Logic for Left-continuous T-norms. Fuzzy Sets and Systems 124(3), 271–288 (2001)
3. Hájek, P.: Metamathematics of Fuzzy Logic in Trends in Logic, vol. 4. Kluwer, Dordercht (1998)
4. Hanss, M., Willner, K.: On Using Fuzzy Arithmetic to Solve Problems with Uncertain Model Parameters. In: Proc. of the Euromech 405 Colloquium, Valenciennes, France, November 17–19, pp. 85–92 (1999)
5. Hanss, M.: The Transformation Method for the Simulation and Analysis of Systems with Uncertain Parameters. Fuzzy Sets and Systems 130, 277–289 (2002)
6. Klir, G.J.: Fuzzy Arithmetic with Requisite Constraints. Fuzzy Sets and Systems 91, 165–175 (1997)
7. Mesiar, R.: Triangular-norm-based Addition of Fuzzy Intervals. Fuzzy Sets and Systems 91, 231–237 (1997)
8. Shary, S.P.: A New Technique in Systems Analysis under Interval Uncertainty and Ambiguity. Reliable Computing 8, 321–418 (2002)

Design and Implementation of a Hybrid Fuzzy Controller Using VHDL

Ismael Millán[1,a], Oscar Montiel[1], Roberto Sepúlveda[1], and Oscar Castillo[2]

[1] Centro de Investigación y Desarrollo de Tecnología Digital (CITEDI- IPN)
   Av. del Parque No.1310, Mesa de Otay, 22510, Tijuana, BC, México
   {o.montiel,r.sepulveda}@ieee.org
[a] M.S. Student of CITEDI-IPN
   millan@citedi.mx
[2] Division of Graduate Studies and Research,
   Calzada Tecnológico S/N, Tijuana, México
   ocastillo@hafsamx.org

Abstract. It is presented a novel hybrid controller that combines the benefits of classical controllers and fuzzy logic to improve the system response in tracking. The design was developed in VHDL for a posterior FPGA implementation. The code was simulated using Simulink and Xilinx System Generator (XSG) that allows to simulate the code of the final FPGA target. Several comparative experiments in soft-real time were conducted using a geared DC motor and the results are commented.

1   Introduction

The rapid development of digital technology and its decreasing cost in comparison with the analog counterpart has impacted the world of controllers by replacing analog solutions with digital proposals because they offer several advantages such as a considerable time reduction in the design stage, improvements on system reliability and performance, elimination of discrete tuning components, and the possibility of including various performance enhancements. Digital controllers can solve problems with enough high complexity to be tackled with analog technology [1, 2].

Nowadays, no matter how complicated the control of a plant might be, the majority of control loops in industrial control systems are using a Proportional-Integral-Derivative (PID) controller type or subtype. There are several ideas to implement a PID controller to overcome the disadvantages of linear PID controllers based on difference equations. Some of these ideas transform a linear PID controller into a PID-like structures of fuzzy controllers.

Digital controllers can be implemented in different hardware platforms, including personal/industrial computers, industrial boards based on discrete digital logic, Digital Signal Processor (DSP) or microcontroller systems, using dedicated hardware for specific applications like Application-Specific Integrated Circuit (ASIC) or in a Field Programmable Gate Array (FPGA) [3, 4].

O. Castillo et al. (Eds.): Soft Computing for Hybrid Intel. Systems, SCI 154, pp. 437–446, 2008.
springerlink.com

At present time, there are several digital techniques to implement a digital controller. Conventional PID controllers are still being a key component in industrial control because they are simple and provide useful solutions to many important industrial processes [5].

Although the extensive use of conventional PID controllers and the implementation of new techniques to improve them, they have significant limitations, because they work basically for linear processes. Some of the limitations are:

- They do not work effectively controlling complex processes that are nonlinear, time-variant, with major disturbances, uncertainties, and large time delays.
- They need to be tuned properly when the plant dynamics change. This is a frustrating and time-consuming experience if the nonlinearities of the plant becomes more accentuated. There are several algorithms to tune or auto-tune PID controllers, but in some situations, when the system becomes too complex, where the conventional PID controller would not work, no matter how it is tuned.
- A conventional PID cannot be used as the core of an smart control system.

The traditional structure of a PID controller has evolved, and the Fuzzy implementation of a PID structure (FPID) has demonstrated to be a successful idea for several reasons [6, 7], some of them are:

- FPID uses every day words to establish the Fuzzy Inference System (FIS). The linguistic variables *error*, *change of error*, and *integral of error* can be used to handle the same signals than the conventional PID controller uses.
- They can work as a linear or a non linear PID controller.
- They can be smart, working in combination with other soft computing techniques.
- The FPID has more parameters to adjust than conventional PID, but there are several effective methods to tune them [8].

Other ideas to implement digital controllers are focused in hybrid solutions in the sense of mixing conventional controllers with fuzzy, neural or neurofuzzy controllers [9]. This proposal presents a controller, that is based in an implementation of a PID controller in serial configuration, mixed with fuzzy logic to improve the system response tracking a signal.

This work is organized as follows: Section 2 presents the formal proposal of the controller which is basically an incremental PID with fuzzy adjustment of the proportional gain, Section 3 is devoted to explain the experiments and results, and finally in Section 4 are the conclusions.

2 Controller Design for FPGA Implementation

The standard representation of a PID controller is given in (1); where, three main components can be identify: They are the proportional, integral and derivative terms. The error signal $e(k)$ is defined as (2), it is the difference between the process output $y(k)$ and the set point $r(k)$. The derivative term was obtained using the backward difference method, and the integral term using the method of rectangular integration. In this equation, K_p, K_i, and K_d are parameters related to the gain of each term; T_s, T_i, and T_d are the sampling period, the integral time, and the derivative time, respectively.

$$u(n) = K_p e(n) + \frac{K_i}{T_i} \sum_{j=1}^{n} e(j)T_s + K_d T_d \frac{e(n) - e(n-1)}{T_s} \tag{1}$$

$$e(k) = y(k) - r(k) \tag{2}$$

The algorithm given in (1) is seldom used in practical applications since improvements in the overall performance can be obtained. In this work, a serial realization of this controller known as incremental controller was used. The first stage consists of a PD controller, hence (1) is transform to (3), the second stage is the integral part that is calculated with (4); in this way, the controller output can be rewritten as (5).

$$u_{pd}(n) = K_p e(n) + K_d T_d \frac{e(n) - e(n-1)}{T_s} \tag{3}$$

$$u_i(k) = \frac{K_i}{T_i} \sum_{j=1}^{n} e(j)T_s \tag{4}$$

$$u(k) = \frac{K_i}{T_i} \sum_{j=1}^{n} \left(K_p e(n) - K_d T_d \frac{e(n) - e(n-1)}{T_s} \right) T_s \tag{5}$$

In this proposal K_p is a fuzzy input-output mapping, equation (5) can be expressed in fuzzy terms where $F_{K_p}(r(k))$ is representing a FIS, hence the fuzzy output controller u_F is (6),

$$u_F(k) = \frac{K_i}{T_i} \sum_{j=1}^{n} \left(F_{K_p}(r(k))e(n) - K_d T_d \frac{e(n) - e(n-1)}{T_s} \right) T_s \tag{6}$$

Since the goal is to carry out the controller to an FPGA implementation, each stage of the controller was developed using VHDL codification [10]. Five entities (blocks) were identified and implemented: Proportional, Derivative, Incremental, Decimal Substraction, and the Sampling Period block.

2.1   Controller Simulink Model

Figure 1 shows the Simulink model developed to test the controller. The main block labeled PID is the serial implementation of the incremental controller described by (5). In Figure 1, there is a summing block that produces the error signal and it is connected to a display, we used this block to monitor the values of this signal for debugging, the same implementation is also into the PID block to achieve the controller action. The innovative part is in the K_p input of the PID block, where a fuzzy selection of this value is achieved according to the plant characteristic to improve the tracking of a reference signal. The plant is a geared DC motor model GM9236S025-R1 [11].

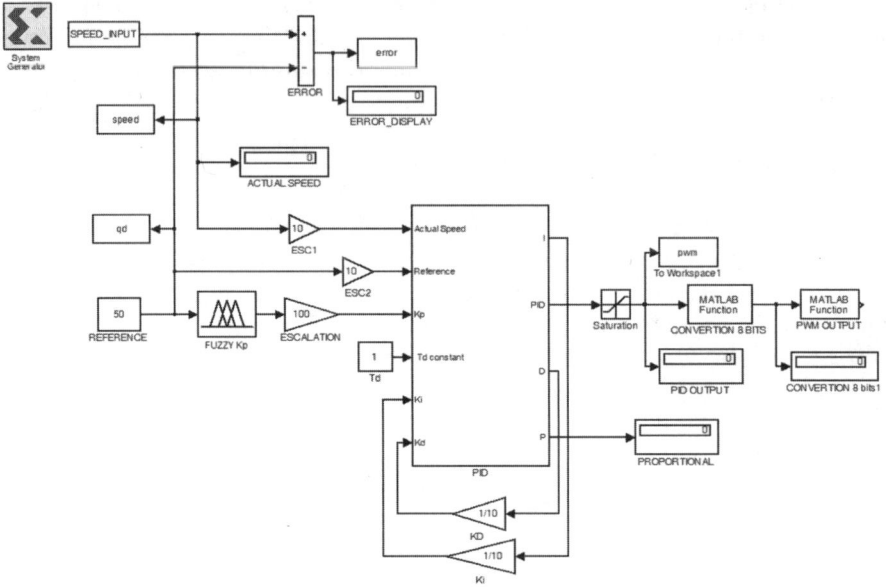

Fig. 1. Controller Simulink Model. The FIS is connected to the input K_p of the PID incremental controller.

2.2 Fuzzy Adjustment of K_p

The PID implementation based on equation (5) has the characteristic of having slow response tracking a signal. We are proposing to handle the K_p with a fuzzy inference system (FIS) to reduce the aforementioned problem. The universe

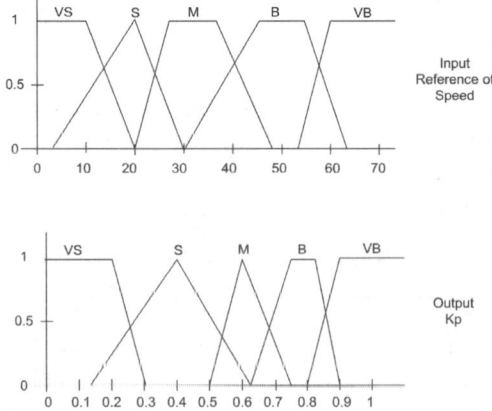

Fig. 2. Membership functions of Input and Output of the FIS to achieve the K_p adjustment

Table 1. Map of Fuzzy Rules

Reference $r(k)$	K_p
VS	VS
S	S
M	M
B	B
VB	VB

of discourse of the FIS is the domain [0,70] and it is related to the maximal revolutions per minute (RPM) of the motor. The fuzzy output is the variable K_p. Figure 2 shows the membership functions for the input and output of the FIS, and Table 1 shows the fuzzy rules.

3   Experimental Results

To achieve the experiments that we are going to explain next, we implemented a test platform with the next main components:

1. Matlab from Mathworks 7.1 (R14), service pack 3.
2. Xilinx ISE pack 8.2.03i
3. Xilinx System Generator v8.2
4. BASIC STAMP editor v2.4
5. Two PARALLAX Boards with BS2P24 microcontroller.
6. H-bridge module based on LMD18200 from National Semiconductor.
7. DC Servo Motor GM9236S025-R1 500 CPR.
8. Power Supplies.

The control goal of the experiments is to maintain the speed of the DC Motor and to eliminate the control error. In the first experiment, we used the classical controller to illustrate the tracking problem.

3.1   Experiment 1. Classical Controller

Figure 3 shows the system response when the target is to maintain a constant speed of 50 RPM, Figure 4 shows the response of the same controller tracking a signal, the changes of speed that we used are 50, 40,30,40,50 RPM. Figure 5 shows the tracking error when the classical controller was used in the aforementioned serial configuration. Figure 6 shows the effect of using different K_p gains values in a classical controller.

3.2   Experiment 2. Hybrid Fuzzy Controller

The control goals of this experiment are the same of Experiment 1, the difference is that the proposed hybrid fuzzy controller was used. Hence, Figure 7 shows the system response for tracking, Figure 8 presents the error when the controller

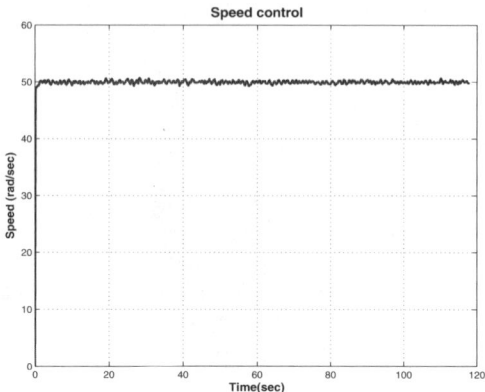

Fig. 3. Experiment 1. Constant speed of 50 RPM.

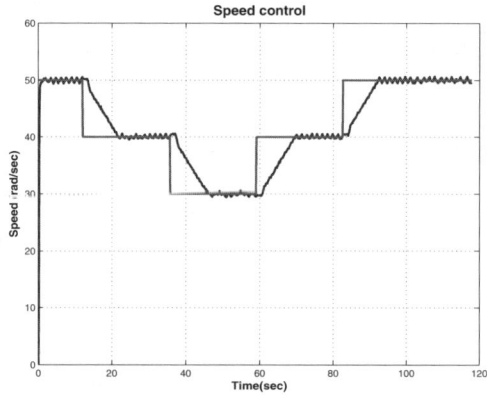

Fig. 4. Experiment 1. Speed control with changes of speed.

tracks the signal. It is implicit that the controller does not have any problem working as regulator. Figure 9 shows how the response change when the FIS is modified, and Figure 10 shows the control error with fuzzy adjustments of K_p.

3.3 Experiment 3. Comparisons

Figure 11 shows how the classical controller and the Hybrid fuzzy controller work tracking a signal. Figure 12 shows the corresponding tracking errors. Actually it is straightforward to appreciate which controller has the faster response tracking a signal.

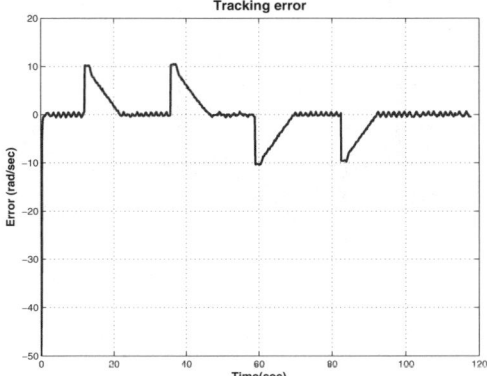

Fig. 5. Experiment 1. System response error when tracking a signal, using a classical controller.

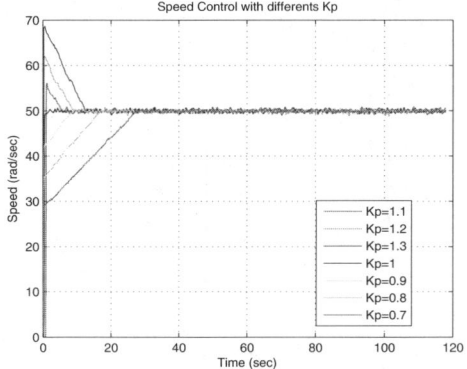

Fig. 6. Experiment 1. Effect of using different K_p gain values in a classical controller.

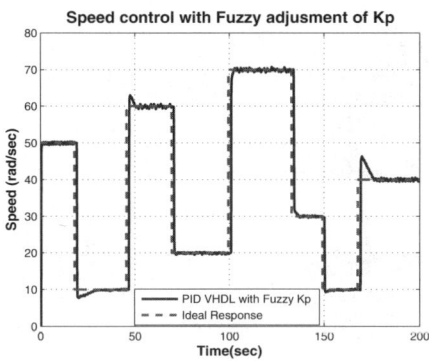

Fig. 7. Experiment 2. Control with changes of speed, fuzzy adjustment of Kp.

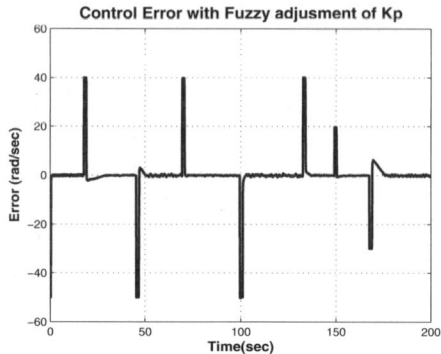

Fig. 8. Experiment 2. Control error.

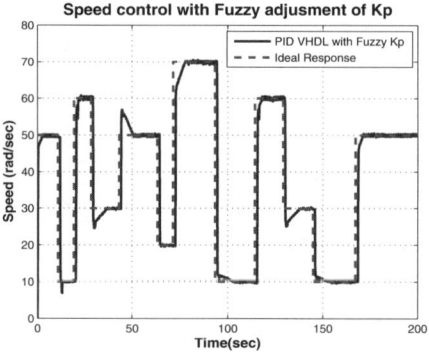

Fig. 9. Experiment 2. Control with changes of speed and fuzzy adjustment of Kp and different membership functions.

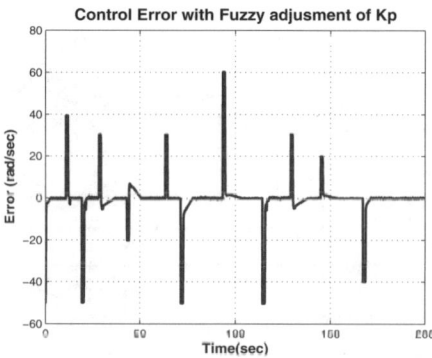

Fig. 10. Experiment 2. Control error.

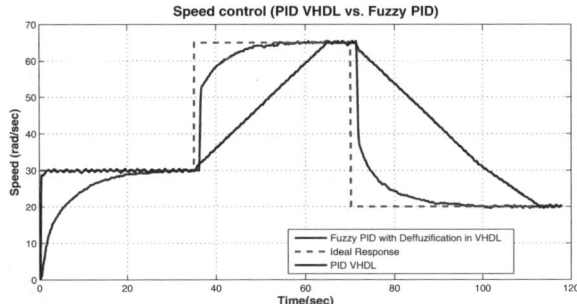

Fig. 11. Experiment 3. Tracking a speed signal (Classical controller in VHDL vs. Fuzzy PID in VHDL).

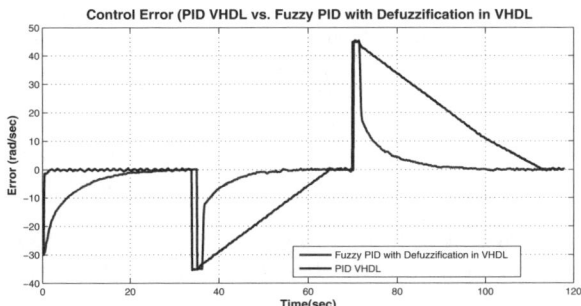

Fig. 12. Experiment 3. Tracking error of Figure 11.

4 Conclusions

This proposal showed that it is a good option to handle the proportional gain of a classical controller using a FIS to improve the system response when it is tracking a signal. We tested only one kind of classical PID controller to control the speed of a DC motor, and all the comparisons using the hybrid controller proposal were made against this controller. It was shown how the controller performance was increase when we hybridize the classical controller. All methods were tested using VHDL codification for a posterior FPGA implementation. The test platform worked well to achieve the experiments, the results can be improved by using a hard-real time platform. The use of Simulink and Xilinx System Generator is a good option for a fast prototyping since minors modification have to be done in order to have a functional controller embedded into an FPGA.

References

1. A Prodic and Dragan Maksimovic, Digital PWM Controller and Current Estimator for A Low-Power Switching Converter. In: 7th IEEE Workshop on Computers in Power Electronics, COMPEL 2000, Blacksburg, VA, July 16-18 (2000)

2. Wu, A.M., Xiao, J., Markovic, D., Sanders, S.R.: Digital PWM control: application in voltage regulation modules. In: 30th Annual IEEE on Power Electronics Specialists Conference, 1999. PESC 1999, vol. 1, pp. 77–83 (Auguest 1999)
3. Oldfield, J., Dorf, R.: Field-Programmable Gate Arrays. In: Reconfigurable Logic for Rapid Prototyping and Implementation of Digital Systems. John Wiley & Son, Chichester (1995)
4. Miguel, A., Rodríguez, V., Sánchez Pérez, J.M., Juan, A., Pulido, G.: Advances in FPGA Tools and Techniques. Microprocessors and Microsystems. Elsevier Science, Amsterdam (2005)
5. Quevedo, J., Escobet, T.: Digital Control 2000: Past, Present and Future of PID Control. In: Proceedings of the IFAC Workshop, Terrassa, Spain, 5-7 April (2000)
6. Jantzen, J.: Tuning of Fuzzy PID Controllers, Tech. report no. 98-H 871, Technical University of Dnmark, Department of Automation, (September 30, 1998)
7. Tang, K.S., Man, K.F., Chen, G., Kwong, S.: An optimal fuzzy PID controller. Industrial Electronics. IEEE Transactions 48, 757–765 (2001)
8. Montiel, O., Sepúlveda, R., Melin, P., Castillo, O., Porta, M.A., Meza, I.M.: Performance of a Simple Tuned Fuzzy Controller and a PID Controller on a DC motor. In: Proceedings of the 2007 IEEE Symposium on Foundations of Computacional Intelligence (FOCI 2007), Hawaii, U.S.A, pp. 531–537 (2007)
9. Cirstea, M.N., Dinu, A., Khor, J.G., McCormick, M.: Neural and Fuzzy Logic Control of Drives and Power Systems, Newnes, Great Britain (2002)
10. Chu, P.P.: FPGA Prototyping by VHDL Examples: Xilinx Spartan-3. Wiley-Interscience (2008)
11. Pittman, D.C.: Servo Motor, Data Sheet available,
    http://www.clickautomation.com/PDF/items/GM923GS025.pdf

Author Index

Printing: Krips bv, Meppel, The Netherlands
Binding: Stürtz, Würzburg, Germany